D1600074

Photoelectron and Auger Spectroscopy

MODERN ANALYTICAL CHEMISTRY

Series Editor: **David Hercules**
University of Georgia

ANALYTICAL ATOMIC SPECTROSCOPY
By William G. Schrenk • 1975

PHOTOELECTRON AND AUGER SPECTROSCOPY
By Thomas A. Carlson • 1975

In preparation

MODERN FLUORESCENCE SPECTROSCOPY
Edited by Earl L. Wehry

Photoelectron and Auger Spectroscopy

Thomas A. Carlson
Oak Ridge National Laboratory

PLENUM PRESS • NEW YORK AND LONDON

Library of Congress Cataloging in Publication Data

Carlson, Thomas A 1928-
Photoelectron and Auger spectroscopy.

(Modern analytical chemistry)
Includes bibliographical references and index.
1. Electron spectroscopy. I. Title.
QD96.E44C37 543'.085 75-28025
ISBN 0-306-33901-3

A Division of Plenum Publishing Corporation
227 West 17th Street, New York, N.Y. 10011

United Kingdom edition published by Plenum Press, London
A Division of Plenum Publishing Company, Ltd.
Davis House (4th Floor), 8 Scrubs Lane, Harlesden, London, NW10 6SE, England

Printed in the United States of America

Preface

In 1970 when I first seriously contemplated writing a book on electron spectroscopy, I recognized the impossibility of completely reaching my desired goals. First, the field was expanding (and still is) at such a rate that a definitive statement of the subject is not possible. The act of following the literature comprehensively and summarizing its essential content proved to be a divergent series. On the other hand, the field has increased to such a size that violent changes in its basic makeup no longer occur with the frequency that was present in its early days. Furthermore, the excitement of electron spectroscopy lies in its many-faceted interrelationships. In the era of specialization, electron spectroscopy is an open-ended subject continually bringing together new aspects of science. I wished to discuss not just one type of electron spectroscopy, but as many as would be possible. The book as it stands concentrates its attention on x-ray photoelectron spectroscopy, but also presents the basis of Auger electron spectroscopy and uv photoelectron spectroscopy, as well as mentioning many of the other branches of the field. A large, many-author volume might be an answer to some of these problems. However, though any one person possesses only a limited amount of expertise, I have always enjoyed books by a single author since what they lack in detailed knowledge they gain in a unified viewpoint. I hope the final product, though limited in its attainment of these goals, will still be of some merit.

The book was first completed at the beginning of 1974 and sent to the original publishers. Unfortunately, an explosion occurred in their office building and their schedule was severely upset, so that it was decided to switch publication to Plenum Press. However, since electron spectroscopy is such a volatile field, I felt a strong need to make the book as current as possible and made a thorough literature research in December 1974, carrying out a complete revision of the book that was finished in February 1975. I feel the book should now reflect the perspective of electron spectroscopy as it stands at the beginning of 1975.

This book is primarily for use of researchers and would-be researchers in the field of electron spectroscopy. It is hoped that it will also be of aid in the teaching and better understanding of science in general. The field of electron spectroscopy is unified by more than its method of measurement. It is unified by providing a basis for understanding the electronic structure of atoms and molecules. It brings together a broad scope of both chemical and physical phenomena, and demonstrates how science can move smoothly from the basic to the applied fields.

It is impossible to accomplish anything in life without a great deal of help and this book is no exception. I wish to thank my students for their help in surveying the literature, in particular Dr. Nils Fernelius, who while vice-president of RCI, conducted a literature search on which this book is based, and has since then kindly kept me updated. My thanks also go to Dr. George Schweitzer for making his literature research files on electron spectroscopy available to me. To Lee Pogue goes my admiration for the doubly Herculean task of reading my illegible scrawl and transcribing it somehow into the final typed form. Drs. R. G. Albridge, C. R. Brundle, J. C. Carver, M. O. Krause, and B. P. Pullen were kind enough to read the rough draft of the manuscript and made most helpful comments, but are not to be blamed for its final outcome. My wife, Effie, showed tolerance through the time of trials, and wrote her own book on musicology, which proved to be a competitive goad.

Perhaps the greatest debt of gratitude goes to my colleagues and friends whose research make up this book. One of the compensations for scientific work is the thrill of learning about new ideas from others, especially in such a rapidly growing field as electron spectroscopy. It makes one feel a part of a greater accomplishment. A complete list of those to whom I am indebted would be endless. Nevertheless, I feel I cannot leave unmentioned Prof. Kai Siegbahn, whose laboratory I visited many times while in Sweden in 1966–1967, for these visits formed an important point in my career.

Oak Ridge, Tennessee Thomas A. Carlson

Contents

Chapter 3
Fundamental Concepts 65

Chapter 4

Photoelectron Spectroscopy of the Outer Shells . 99

Chapter 5

Photoelectron Spectroscopy of the Inner Shells . 165

Chapter 6

Appendixes

Chapter 1

Introduction

1. HISTORY

During the last few years the growth of electron spectroscopy as a means of studying chemical systems has been so rapid that it might appear that a brand new physical tool for chemists has been invented. This impression, of course, is far from the truth.

Unlike the Mössbauer effect, for example, the field of ESCA (electron spectroscopy for chemical analysis) did not grow immediately after the discovery of a new physical phenomenon. The two most important phenomena on which ESCA is founded, photoionization and the Auger process, were already understood earlier in the century. Einstein,[1] in his classic paper of 1905, explained the photoelectric effect, while Pierre Auger[2] discovered in 1923 the effect named after him. Likewise, instruments capable of analyzing the kinetic energy of electrons had an early history. Even before the First World War, experiments on analyzing the energy of β rays had been made by use of magnetic fields.[3]

The need for nuclear physicists to study beta spectra and the discrete-energy electrons ejected during internal conversion gave a continuous impetus for workers to improve the techniques of electron spectroscopy. Indeed, much of the development and demonstration of the possible applications of electron spectroscopy to chemical analysis was carried out by those who came from the field of nuclear spectroscopy.[3,4]

The full potential for the use of electron spectroscopy in chemical problems, however, had to await the technology necessary for measuring with high resolution electrons of energies from a few tenths of an electron volt to a few kilovolts. During the 1960s this requirement was achieved. Perhaps the most important area in electron spectroscopy for chemical analysis has been the measurement of the discrete-energy electrons ejected by photons.

Pioneer efforts in this field were carried out in 1951 by Steinhardt and Serfass.[5,6] (The basic concept of x-ray photoelectron spectroscopy was actually conceived[7] as early as 1914.) Present-day high-resolution photoelectron spectroscopy, however, grew out of two parallel developments that took place principally in Uppsala, Sweden, and Imperial College, London, England. In Uppsala there was an interest in determining binding energies for the atomic shells by photoelectron spectroscopy, particularly for the less tightly bound orbitals, whose values from x-ray data were known only approximately. As a result of this program it was discovered that the binding energies of the core electrons were dependent on the chemical environment.[8] To study these chemical shifts, a spectrometer capable of measuring electrons of approximately 1 keV with a resolution better than 0.1% FWHM in energy is necessary. Fortunately, the beta spectrometers developed in nuclear physics could be easily adapted for the job.

Meanwhile, Turner and Al-Joboury in England[9] were looking for a method of directly studying the binding energies of the molecular orbitals. They also sought the solution to this problem in photoelectron spectroscopy, but rather than selecting soft x rays for their photon source, which is a requirement for the study of core electrons, they chose the He I resonance line as obtained in a gas discharge lamp. This radiation* had enough energy, 21.22 eV, to eject electrons from the outer orbitals (the ones which are usually the most important in determining the nature of the chemical bond) and at the same time had a natural linewidth of only a few millivolts. Spectrometer design was continually improved to take advantage of this small natural width, since the higher resolution permits measurements of the fine structure in the photoelectron spectrum arising from vibrational states. The electronic bands observed in photoelectron spectroscopy are made up of transitions to the various energy levels corresponding to the vibrational states in the singly charged molecular ion. The details of this vibrational structure have proven invaluable in the interpretation of the nature of molecular orbitals. The photoelectron spectroscopy of the outer shells has given researchers the opportunity to characterize the electronic structure in a relatively straightforward and comprehensive manner, such that molecular orbital theory has had its experimental basis increased to a far greater extent than was previously afforded by optical spectroscopy.

The field of Auger spectroscopy as applied to chemical problems has blossomed forth (like photoelectron spectroscopy) only in the last few years. Its first practical use in the study of surfaces took place in 1953[10] and the present-day interest began with the work of Tharp and Scheibner[11] and Weber and Peria.[12]

* It was W. C. Price who first suggested to Turner that the He I radiation might make a suitable source for studying the binding energies of the molecular orbitals.

When a vacancy is formed in an inner shell of an atom, it can be filled through means of x-ray emission or through a nonradiative transition (Auger process), in which a single electron is ejected with a unique energy, characteristic of the binding energies of the atomic orbitals involved in the process. Auger processes occur with high probability but are not detected as easily as x rays because of the high stopping power of matter for low-energy electrons. This disadvantage, however, has been changed to an advantage in the case of surface studies, since low-energy Auger electrons can escape only from the first few molecular layers. In addition, the fluorescence yield of *K*-shell vacancies for light elements is very small, and the most probable process for filling a *K* vacancy is a *K–LL* Auger transition. Thus the use of Auger spectra for locating and identifying elements (especially those of low *Z*) on surfaces has become an extremely important aspect of electron spectroscopy. In addition to surface studies, the investigation of high-resolution Auger spectra, particularly in the gas phase, has been most valuable in characterizing excitation processes and in studying the nature of doubly charged molecular ions.

Two other fields utilizing the discrete kinetic energy of electrons as their main source of data are electron impact spectroscopy and Penning ionization spectroscopy. In the former field, measurements of inelastic scattering processes have been most profitable in yielding information about the excited states of gaseous molecules, including transitions normally forbidden in optical spectroscopy. Penning ionization arises from the process $A^* + B \rightarrow A + B^+ + e^-$. The basic information gained concerns the energy required to remove an electron from B. Thus Penning ionization measurements parallel the data obtained from photoelectron spectroscopy. However, Penning ionization can lead to final excited states not reached by the photoelectric effect.

A comparison of the various forms of electron spectroscopy is shown schematically in Figure 1.1. This simplified, overall picture will hopefully be clarified in the more detailed discussions throughout the book.

2. SCOPE OF PRESENT BOOK AND REVIEW OF PAST BOOKS

Electron spectroscopy encompasses a variety of fields, and the number of recent papers is approaching a population explosion that is almost frightening. Thus, even a report of book length must circumscribe the areas to be covered, since a truly complete report seems outside the capacity of any one author or any one book. The criteria for what material should be covered in the present book have been empirically determined, for they essentially represent those areas related to research in which I have been actively engaged.

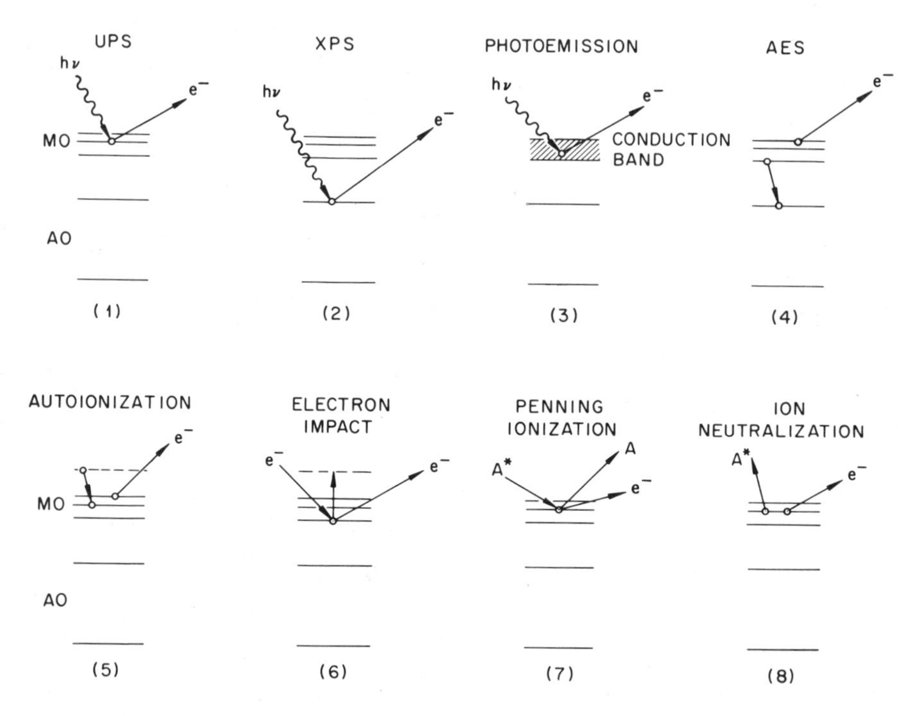

FIGURE 1.1. Schematic comparison of the various types of electron spectroscopies. MO and AO correspond, respectively, to molecular orbitals and atomic orbitals. (1) In UPS, uv radiation is used to eject photoelectrons from the valence shell or the molecular orbitals. (2) In XPS, x-rays can eject electrons from either the core shell (atomic orbitals) or the valence shell. (3) Photoelectron emission is used to measure the electron density of the conduction band. (4) Auger spectroscopy measures the ejection of an Auger electron as the result of a nonradiative transition to an inner shell vacancy. (5) Autoionization, like the Auger process, is a nonradiative readjustment, but not involving an initial core vacancy. (6) In electron impact a valence shell electron is promoted into an excited but bound state. (7) In Penning ionization excitation derived from an excited atom is used in the promotion of electron ejection. (8) Ion neutralization is a nonradiative readjustment to a vacancy in a foreign ion.

Fortunately, my interests have been rather eclectic, so that the criteria are not as restrictive as might at first appear and include the three most important areas of electron spectroscopy: photoelectron spectroscopy of the outer shell, photoelectron spectroscopy of the inner shell, and Auger spectroscopy. These three areas interact closely phenomenologically. Each deals essentially with the binding energies of atomic and molecular orbitals. Research in each of the areas can be carried out on the same spectrometer, with changes made only in the excitation source. The recognition of this possibility has induced the builders of some commercial electron spectrometers to include the potential for this versatility in their machines.

Before pursuing how these three areas will be covered in this book, let us review the books and long articles that are presently available: First, there are the two books[13, 14] on ESCA (electron spectroscopy for chemical analysis) by the Uppsala group. These are in essence progress reports of the research carried out at Uppsala under the direction of Kai Siegbahn. However, because of the comprehensiveness of these research efforts, the books offer a good review of the field of electron spectroscopy of the inner shells, though discussion of results on the application of Auger spectroscopy and spectroscopy of the outer shells is also made. They are certainly landmarks in the field of application of electron spectroscopy to chemical problems, and have done more than any other publication to awaken the scientific world to the potential of this field. However, because of their primary concern with presenting the data taken at Uppsala, the books do not offer a full account of other work done elsewhere, and with the continued increase of activity in ESCA throughout the world, a new, comprehensive appraisal is obviously needed.

What Siegbahn's books are to photoelectron spectroscopy of the inner shell, the book of Turner *et al.*[15] is to the photoelectron spectroscopy of the outer shell. The goal of Turner's book, like that of the Uppsala publications, is to present, in a comprehensive form, data taken at a single laboratory, this time, Oxford. In this work the He I resonance line was used exclusively as the photon source. Again the presentation is less general, and not pedagogical. In fact, the book was originally designed only to present data with a minimum of comment,* but was expanded to give more discussion; and there are many gems of general observation scattered throughout the book. Baker and Betteridge[16] have published a slim monograph on photoelectron spectroscopy. It covers material on both x-ray and uv photoelectron spectroscopy; however, its emphasis is on molecular orbital studies with the He resonance lines, and its most valuable chapter is on the interpretation of uv photoelectron spectra. Eland's[17] more recent book on photoelectron spectroscopy is excellent. It covers the literature of uv photoelectron spectroscopy up through 1972. Most important, it presents in a highly readable, cogent manner the basic principles of photoelectron spectroscopy of the valence shells of gases.

A sizable volume on electron spectroscopy has been written by Sevier.[4] The book is rather selective in its topics, although what it covers it does so with extensive documentation. It discusses very little regarding the use of electron spectroscopy for chemical systems. However, a number of topics will be at least of peripheral interest to the readers of this volume, such as instrumental methods, Auger processes in atomic systems, natural widths of atomic levels, atomic binding energies, and electron losses by scattering.

* C. R. Brundle, private communication.

Nevedov[18] has written a book (in Russian) on the use of x-ray photoelectron spectroscopy in the study of chemical problems. It contains his own rather extensive investigations as well as a comprehensive review of the literature, with numerous tables of core binding energies as a function of chemical environment.

Six extended lectures on various aspects of electron spectroscopy, which were given at a NATO summer institute held in 1972 at Ghent, have been published.[19] These are variable in quality and as a whole lack unity because of the nature of multiple authorship. However, the presentations have a strong and valuable pedagogical bent, and certain areas, particularly in theory, are given a full and careful coverage, which will not be found elsewhere. For an extended compilation of data on photoelectron spectroscopy data, Siegbahn *et al.*[10] have published three extensive tables, the first two on the valence shell and the third on core-shell electrons.

In addition to the above-mentioned books, a large number of review articles have been written. Particularly worthy of mention are those of Siegbahn,[21] Hercules and Carver,[22] Shirley,[23] Albridge,[24] Jolly,[25] and Gelius[26] covering primarily x-ray photoelectron spectroscopy. Also excellent are the reviews of Betteridge *et al.*,[27] Berry,[28] Turner,[29] Worley,[30] Brundle,[31] Heilbronner,[32,33] and Price,[34] which deal mainly with uv photoelectron spectroscopy. For discussion of Auger spectroscopy for surface analysis, see Chang[35,36] and Pennario.[37] Brundle's[38] articles on the use of electron spectroscopy for surfaces are also recommended.

Compilations of papers presented at the 1971 Asilomar conference,[39] the 1972 Faraday Society discussion,[40] and the Namur conference in 1974[41] offer a broad view of work in progress in the field of electron spectroscopy. Finally, the *Journal of Electron Spectroscopy* edited by C. R. Brundle and T. A. Carlson and published by Elsevier Press is devoted to all aspects of the field of electron spectroscopy—phenomenological, theoretical, and instrumental—and offers one of the best means for following the progress of this field. Informal discussion in electron spectroscopy may be found in the *CWRU ES Newsletter*, edited by G. D. Mateescu.*

The three areas, photoelectron spectroscopy of the outer shell, photoelectron spectroscopy of the inner shell, and Auger spectroscopy, will be covered in this book with varying degrees of completeness. I have attempted to survey extensively most of the research done on the photoelectron spectroscopy of the inner or core electrons. In the case of the outer shells, a nearly complete coverage of the various techniques and basic phenomena is given, accompanied by a substantial number of examples. Also, an outline of what has been done for the various basic chemical systems is sketched. However, no detailed evaluation of every molecular system has been made. Certainly,

* For further information write G. D. Mateescu, Case Western Reserve University.

the substantial work done in the evaluation of molecular orbitals by photoelectron spectroscopy should be critically examined and correlated with optical and other data. This correlation has been accomplished for specific systems in individual articles, and the book of Turner *et al.*[15] is a start in this direction. But a complete evaluation of all the uv photoelectron data would require a book of substantial volume unto itself. The potential of Auger spectroscopy as a surface tool is discussed in detail, but no attempt is made to comment on all the literature in this rapidly expanding field. It will be shown, however, how Auger spectra can be used to give basic information on molecular structure and excitation processes as well as elemental analysis. In general, I hope that most of the essential accomplishments and potentials of these three areas of electron spectroscopy are brought out. The literature has grown so quickly that only a fraction of it can be discussed in detail, but with the help of the references and appendices a comprehensive overview of the literature should be obtainable.

One chapter has been set aside to discuss the experimental aspects of electron spectroscopy. The main question we shall attempt to answer is: "What are my experimental needs for solving a given chemical problem?" An honest appraisal of the presently available commercial spectrometers will be made and a guess will be hazarded as to what one might expect in the future. A fair evaluation of what electron spectroscopy cannot do as well as what it can do will be a leading consideration here as well as throughout the book.

Though this book is undertaken primarily from an experimentalist viewpoint, the basic theoretical underpinnings are presented. Where more appropriate, certain theoretical considerations appear in the chapters dealing with one of the three special areas of electron spectroscopy. Chapter 3, however, collects the theoretical aspects which apply, in general, to the interpretation of data obtained from electron spectroscopy. Rather substantial application is made of the "sudden approximation" in which the probability of transition to a given final state can be obtained simply from the overlap integral between the initial and final states. This approach is used to clarify various problems encountered in photoelectron and Auger spectroscopy, such as those associated with Koopmans' rule, Franck–Condon factors, and the phenomenon of electron shakeoff.

The two main bodies of data gained from photoelectron spectroscopy as applied to chemical systems are (1) the energies of molecular orbitals and their identification and (2) the energy shift of the core electron of a given element as a function of its chemical surroundings. Copious examples of these data are presented in the main text. To supplement these, there is a compilation of chemical shifts and references to papers dealing with this question in Appendix 3.

This book is written primarily for those scientists actively engaged in the field of electron spectroscopy or those who plan to be. It is hoped that the book might also serve as a textbook for a short course in electron spectroscopy.

3. NAME-CALLING

In the development of any new field a new language or jargon is invented. Such vocabulary is useful in condensing and particularizing a new phenomenon, idea, or area of research. Unfortunately, a babel of confusion also results as new terms are met that are either undefined or ill defined. Language has an irresistable momentum of its own, and it is neither the desire nor within the power of this author to stem the tide. However, it seemed desirable to at least clarify the terms used in this book and indicate their present usage, and to present some of the more common ones in a convenient place for reference.

Thus, below we shall discuss a few general terms and at the end of the book give a general listing in Appendix 4, with definitions and references for a more detailed discussion where warranted. Perhaps the most used term coined for the material covered in this book is ESCA or electron spectroscopy for chemical analysis. Though it was first coined by Siegbahn and his coworkers at Uppsala(13) to cover all areas of electron spectroscopy, it has been used by many workers to refer to only work with soft x rays. I prefer to use ESCA in its more general sense. Varian has used IEE in connection with their electron spectrometer, which stands for induced electron emission. IEE in general conveys the same areas of research as ESCA, but the term has not found as much usage.

AES is commonly used for Auger electron spectroscopy, and PES (photoelectron spectroscopy) has been in use for some time but has been associated primarily with low-energy photons (uv and vacuum uv). To indicate the nature of the photon source, the expressions XPS (x-ray photoelectron spectroscopy) and UPS (uv photoelectron spectroscopy) are used. Generally I prefer to divide the research efforts phenomenologically rather than by the excitation sources employed and have used the initials PESOS for "photoelectron spectroscopy of the outer shell" to indicate those studies where the main interest is to obtain information on the binding energies and identity of the outer shell molecular orbitals, and have used the initials PESIS for "photoelectron spectroscopy of the inner shell," where the main interest lies in the energy shifts of the core electron as a function of chemical environment. One of the most amusing uses of initials is ESP (electron spectroscopy). The initials, however, have already been spoken for by others in another branch of science (extra sensory perception). Perhaps someone possessed the clairvoyance to foretell the coming of electron spectroscopy.

Other special terms referring to specific portions of our subject matter will be defined in the appropriate sections and, as mentioned, a collection of such terms is to be found in Appendix 4.

4. AREAS RELATED TO ELECTRON SPECTROSCOPY *NOT* TO BE DISCUSSED IN DETAIL

In this section we hope to fill out those areas closely related to the topics covered in this book, but for which a detailed description is outside its limited scope. First, we shall discuss those areas which are wholly within the domain of high-resolution electron spectroscopy, but are not covered in this book in detail. Second, we shall briefly sketch in a background to illustrate how electron spectrometry interacts with other disciplines endeavoring to study similar problems in molecular structure.

4.1. Electron-Impact Spectroscopy

The principle behind electron-impact spectroscopy (EIP) is that on impact an electron can lose a portion of its energy to inelastic scattering. This energy is equal to the difference between the ground state energy of the atom or molecule before collision and the excited state energy of that species after collision. If the initial electron beam has a well-defined energy, and the kinetic energy of inelastically scattered electrons is measured with a high degree of accuracy, one obtains precise information on the excited energy levels of the target gas molecule. The phenomenon was first noted by Frank and Hertz[42] in 1914. The time for high-resolution measurement was to come much later. In fact, the development of high-resolution electron-impact spectroscopy has paralleled the development of photoelectron spectroscopy. Resolution as good as 0.01 eV (FWHM) has been reported,[43] and a commercial instrument for the study of EIS has been developed.* Lassettre and his co-workers[44] have been particularly active in this field. A good survey of the subject can be found in the reviews of Berry[28] and of Trajamar *et al.*[45]

In obtaining data from EIS, one usually accepts electrons scattered through a given angle, determining their kinetic energy spectrum at this angle. Most data have been restricted to low-energy electrons scattered up to a few degrees. For the study of forbidden transitions and for electron-impact energies near the threshold of excitation, it has been desirable to work at larger scattering angles and to measure the energy spectrum as a function of

* McPherson Instrument Corp.

angle. Photoelectron spectroscopy and electron-impact spectroscopy are complementary to each other, since in the former case the final state is the ion, while in the latter it is the neutral species. Photoelectron spectroscopy is designed to study the binding energies of electrons in atomic and molecular orbitals, while EIS investigates the nature of excited states. In addition to the states that are optically allowed, EIS also reaches states that are optically forbidden, and it is the added information on these states that is the main value of EIS. Wight and Brion[46] have extended EIS studies on the valence shells to the excitation of core-shell electrons.

In addition to determining the energy of excited states, one may use data from electron-impact spectroscopy in several other ways: (1) The nature of the transition (i.e., electric dipole or quadrupole; forbidden or allowed) can be determined on the basis of the dependence of its probability on impact energy and angular distribution; (2) the oscillator strength can be evaluated from the intensity of the inelastically scattered electrons; (3) Franck–Condon factors may be determined from the vibrational structure; and (4) information on compound states made up of an electron and the target molecule or atom can be gathered.

We have restricted our discussion only to cases in which the impact process places an electron into a discrete state, because if a second electron is ejected into the continuum, both the energy of the scattered electron and that of the ejected electron must be measured in coincidence before the extent of excitation can be defined. Such measurements have indeed been accomplished by Ehrhardt and his co-workers.[47] Although in principle coincidence measurements in electron impact can yield data similar to PES, in practice this use of electron impact does not appear to be competitive. However, it does yield information on the excitation process and on states not excited by the photoelectric effect. Van der Wiel and Brion[48] have carried out experiments using a high-energy electron-impact source (3.5 keV) and made measurements in coincidence between low-energy electrons emerging at 90° and forward-scattered electrons. The incident energy of the electrons is high enough for the Born approximation to hold, and the forward scattering angle was chosen sufficiently small so that the momentum transfer is close to the optical limit, and the results should approach those seen with photoionization. One important advantage of the technique is that the energy loss of the forward-scattered electron can be varied at will and thus provides a study of "photoelectron" cross sections as a continuous function of energy.

Finally, some mention should be made of the practical use of electron-impact spectroscopy in gas analysis. The energy spectrum is of course characteristic of the molecule, and gases with concentrations as low as parts per million can in principle be detected by high-resolution equipment. In addition, Simpson[49] has constructed miniature spectrometers possessing excellent resolving power for use in space flight. A more complete discussion of the comparative values of EIS and ESCA for gas analysis will be made in later chapters.

4.2. Photoemission

The field called photoelectron emission is phenomenologically no different than photoelectron spectroscopy. In both cases one uses a monochromatic source of photons to eject photoelectrons, thereby learning about electronic structure through data on binding energies. However, the studies which group themselves under the title of photoemission have two important characteristics: (1) they deal primarily with the study of band structure, principally of metals, through the measurement of the electron density of states in the valence shells of solids, and (2) usually the photon source, rather than being a single characteristic line, is obtained from a continuous source in the uv region up to 11.6 eV, the cutoff point for a LiF window, although use of higher energy sources are quite common today. A source commonly employed is the Hinterregger-type lamp with H_2 gas and vacuum monochromator.[(50)] Studies are made as a function of photon energy. Only modest resolution is employed (~0.1 eV), but studies on the valence bands of solids do not require high resolution. Spicer and his co-workers[(51)] have done extensive work in this field, and Eastman[(50, 52)] has written several review articles.

It is obvious that data on XPS of the valence shells in solids closely overlap the interest of photoemission spectroscopy, and the subject will be brought up again in Chapter 4 on PESOS, where a brief outline of the nature of the field will be given. However, the large number of papers devoted to photoemission and the extensive theoretical efforts to interpret these data preclude any comprehensive or detailed treatment.

4.3. Penning Ionization Spectroscopy

Penning ionization is characterized by the following reaction:

$$A^* + B \rightarrow A + B^+ + e^-$$

Excited molecules or atoms (usually metastable rare gas atoms) are used to transfer in the gas phase a precisely known energy to species B, sufficient to promote ionization. From the measurement of the kinetic energy of ejected electrons, one can then obtain the ionization energy: $B \rightarrow B^+$. Thus, Penning ionization is closely related to photoelectron spectroscopy. For example, Brion *et al.*[(53)] have compared results on the transition probabilities to the vibrational states of the $X\,^1\Sigma^+$ state of NO^+ following Penning ionization and photoionization and found them to be in excellent agreement. For general application PES has thus far proved much more practical than Penning ionization, but the latter technique is useful in that it yields information not otherwise available. That is, the nature of excitation can lead to states inaccessible by photoelectron ejection and, in particular, the intermediate excited compound states $(AB)^+$ may decay to $AB^+ + e^-$, thus yielding information on the states of the molecular ion AB^+. Recent work in this field may be found in the papers of Čermák.[54]

4.4. Ion Neutralization Spectroscopy

If one bombards a surface with ions of low energy (for example, He^+, Ne^+, or Ar^+), an autoionization process can take place in which an electron from the surface drops into the potential well of the rare gas ion, while a second electron goes into the continuum. One may regard the process as analogous to an Auger process except that the initial hole is provided by another ion rather than an inner shell vacancy. It is analogous to Penning ionization in that excitation is transferred in a known amount, leading to electron ejection. The initial state of the target species is neutral (as in photoionization) but the final state is doubly charged (as with an Auger process). The main use has been in the study of surfaces. Hagstrum,[55] who has developed the method, has shown that the autoionization process takes place at or just above the surface.

5. FIELDS RELATED TO ELECTRON SPECTROSCOPY

Drawing a boundary about a field such as electron spectroscopy is useful in that it can help bring out its potentialities. But in attempting to solve a given problem or even to understand the data derived from a given technique, a full appreciation and understanding of all other related techniques is necessary. In fact, pursuing this notion to its logical conclusion, we arrive at the state of mind that to know anything we must know everything. However, since we are individuals, finite both in space and time, a compromise is made. Throughout the book, I hope at least to remind the reader of these connections, and in this section to collect some of the more obvious areas of interaction.

In the identification of energy levels both in atomic and molecular systems, optical spectroscopy is perhaps the most important ally that electron spectroscopy has. Historically, optical spectroscopy formed the principal basis for the understanding of the electronic structure of atoms and molecules in the valence shell. Photoelectron spectroscopy has substantially enlarged this basis, obtaining much data that were hitherto unavailable. There has also been much overlap of data, particularly in the case of simple molecules, and general agreement between the two approaches has helped strengthen our confidence in both fields.

Since the final states measured in electron spectroscopy are ions, mass spectrometry offers much data that are relevant, particularly the first appearance potential for the parent ion, which corresponds to the first ionization band in PESOS. Correlation between appearance potentials in mass spectra data and ionization bands above the first in PESOS occurs occasionally, but more often does *not*. Ions in mass spectroscopy are collected some 10^{-5} sec after the excitation event and represent the consequences of substantial energy readjustment. The appearance of fragment ions depends on the amount of energy deposited and on the nature of the energy reorganization and the subsequent fragmentation process. If energy could only be absorbed by the creation of vacancies in the molecular orbitals, there might be a break in the fragmentation pattern as one reaches the different ionization thresholds, but energy can be deposited through electronically excited states as well as photoelectron ejection. For a discussion comparing mass spectral data and PESOS, see Innorta *et al.*[56] In Auger spectroscopy the final states are usually the doubly charged ions rather than the singly charged species encountered in photoelectron ejection, but even here the limited data from optical and mass spectroscopy have been very useful.

With regard to the inner shell or core electrons, x-ray data have been invaluable in the determination of the atomic binding energies, and have been in the past the chief source for the ordering of the atomic structure of the inner atomic shells.[57] Photoelectron spectroscopy and Auger spectroscopy provide a means by which one can check and supplement these data, particularly for the less tightly bound orbitals, for which good x-ray data are less abundant. A great deal of other information from x-ray spectroscopy, such as photoelectron cross sections, level widths, and transition probabilities, has also been invaluable to the electron spectroscopist. More recently, data on shifts in x-ray energies as a function of chemical surroundings have been forthcoming and it has been most interesting to compare these shifts with those found in photoelectron spectroscopy.[58]

One of the main tasks of the photoelectron spectroscopy of the inner shells (PESIS) is the determination of electron density surrounding a given atom as deduced from the relative shifts in core binding energies. There exist other methods that also measure some aspect of electron density. For example, from neutron diffraction we can gain gross information about electron density distributions in solids.[59] From NMR and Mössbauer experiments we obtain information on changes in electron density near the nucleus or changes in the magnetic shielding of the nucleus, which in turn are reflections of changes in the wave functions of the bonding orbitals. In Chapter 5 on PESIS we discuss several cases where chemical shifts from photoelectron spectroscopy are correlated with those from other methods. Like the proverbial description of an elephant by blind men each examining the nature of the beast from a particular part, its trunk, tail, ears, etc., the use of only one method could lead to a gross error in our understanding. However, by combining our knowledge from various approaches we can subject the nature of the electronic structure of molecules to a rather profound analysis.

We have confined our attention in this section primarily to the overlap of different physical techniques with electron spectroscopy in regard to the study of chemical bonding. Electron spectroscopy is also a valuable analytical tool, yielding both elemental and molecular analyses. Its value is particularly apparent in gas analysis and in surface studies. There are many competitive alternatives, and it has been one of the tasks of this author to evaluate fairly the merits of electron spectroscopy as an analytical tool in comparison to other techniques, and to point out when electron spectroscopy has unique features and when it has particular failings. Again, no one approach is adequate in itself, but in combining various possible resources we may obtain success where failure occurs in isolation.

Chapter 2

Instrumentation and Experimental Procedures

It shall be the guiding purpose of this chapter to answer the question: What do I, as an experimentalist, need to get my job done? The answer to this question is indeed one of the overriding goals of the whole book, but in this chapter we shall concentrate our attention on the technical aspects of the problem. We shall first break down the basic operations of an electron spectrometer into their various components and discuss for each element the different ways in which the operation can be accomplished, and which of the ways is preferable in terms of results desired and cost, both in time and money. What can and cannot be done will be indicated, and at the conclusion I shall do some crystal-ball-gazing into what one might expect for the future in the way of new equipment and techniques.

Following the general introduction, there will be a review of some of the available commercial equipment relevant to electron spectroscopy, a sort of consumer's guide. To a large extent, this chapter will be devoted to the research scientist who may be attempting to set up or expand a new program in electron spectroscopy.

In essence, an electron spectrometer works in the following manner: An excitation beam (electrons, ions, x rays, or low-energy photons) irradiates a target sample; the ejected electrons enter the spectrometer and are separated according to their energies or momenta; they then strike a detector and are counted and the data are stored in a suitable manner. Figure 2.1 is a schematic drawing of an electron spectrometer divided into its essential components.

In a sense, then, an electron spectrometer has a beginning, a middle, and an end, dealing with the birth of an electron, its subsequent analysis, and its final detection. The breakdown of this chapter will be made along these three divisions. First, the source volume will be discussed. In the source volume

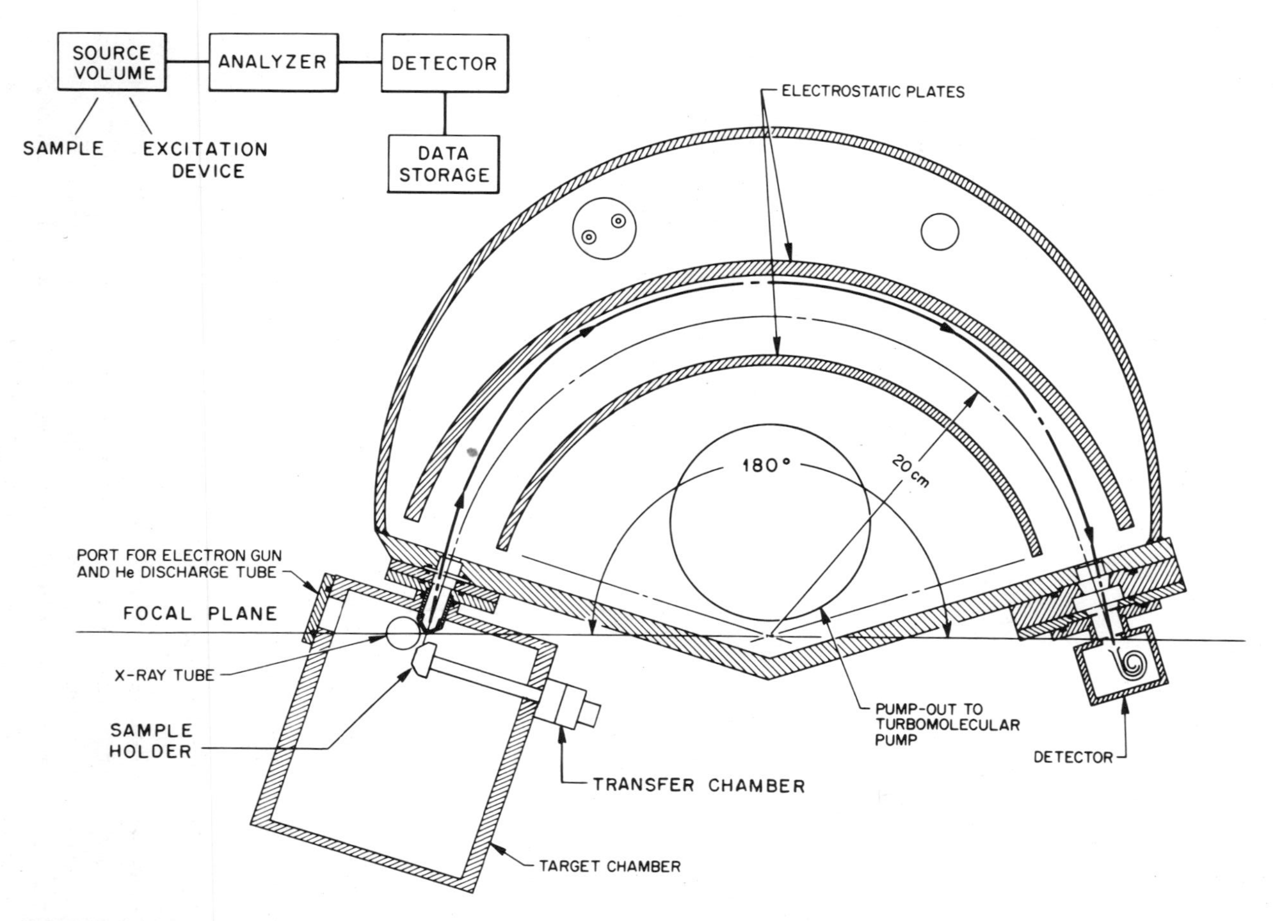

FIGURE 2.1. Diagram of an electron spectrometer in use at Oak Ridge National Laboratory. Insert shows schematic of

are the excitation devices for creating ionization and the target material, which undergoes ionization. After the electrons are ejected, they are separated according to their energies or momenta, either by retarding fields or, more likely, if high resolution is desired, by magnetic or electrostatic dispersion. The final stage is detection, in which the number of electrons at a given kinetic energy or momentum is determined and the information is stored for subsequent data analysis.

1. SOURCE VOLUME

The source volume will be defined as the chamber in which discrete-energy electrons are created. Two principal components make up the source volume: the excitation device and the target. The excitation device may be an electron gun, an x-ray tube, or a gas discharge lamp. The target may be a solid, a condensable vapor which is frozen on a cold finger, or a gas which is allowed to fill an appropriate volume that can be exposed to radiation, its pressure being controlled by differential pumping. Normally, the angle between the direction of the incoming radiation beam and outgoing ejected electron is fixed. However, it is desirable in some instances to study the intensity of the ejected electrons as a function of angle, and a discussion of chambers especially designed for this purpose will be made.

The ejected electrons can enter the analyzer through a slit. In some cases it may be advantageous, however, to either preaccelerate or decelerate the electrons before they reach the analyzer. In studies with low energy electrons, where resolution may be determined by factors other than the dispersion of the analyzer, preacceleration may enhance the intensity without affecting the overall resolution; while with electrons in the kilovolt range, deceleration may enable one to obtain a more favorable tradeoff between energy resolution and illumination. These remarks might seem contradictory, having a somewhat suspect promise of something for nothing, but hopefully they will be made clear in Section 1.1.4 of this chapter.

1.1. Excitation Devices

Three excitation devices that are possible for use with an electron spectrometer will be discussed: (1) electron gun, (2) x-ray tube, and (3) gas-discharge lamp. These three devices can be designed to be used interchangeably with the same electron analyzer, and thus provide one a great deal of versatility in carrying out a broad program of research. Because of geometric considerations, it is generally difficult (but not impossible) to mount all three excitation devices on the source volume at the same time. Alternatively, they can be designed to be interchanged quite easily. We shall also discuss the possible use of synchrotron radiation.

1.1.1. Electron Gun

An electron gun can be used for two basically different purposes. First, it can be used in the detailed study of inelastic electron scattering. For this use, the energy spread of the electron beam must not be much greater than 10 mV, which can be accomplished through the use of indirect heating of the electron source and the employment of an electron analyzer. As was explained in the introduction, this book will not deal with electron scattering phenomenon as a special technique. Second, the electron gun can be used as a means for creating inner shell vacancies, which are in turn filled by nonradiative processes in which a discrete-energy Auger electron is emitted. Our main concern in this regard is the Auger electrons.

The needs and design of the electron gun also depend on whether Auger spectroscopy is to be carried out on gases or solids. In the case of gases, I have used two designs, one of which is shown in Figure 2.2. This gun is a modification of a structure taken from a cathode ray tube. A tungsten filament is used as a replacement for the original electron source, which is not suitable at the higher pressures required for gas studies. The gun gives a well-focused beam, but rather low intensity. By enlarging the hole through which the electrons emerge to 2 mm, reasonable intensities were obtained. A second design was found to give even more satisfactory results. In this design the structure of the cathode ray tube is replaced by a BTI electron gun usually employed for high-temperature melting. It employs Pierce[1] geometry and gives a modest size focal spot of about 0.2 cm^2, but with excellent intensity. Both electron guns operate with the filament at high negative voltage. The electrons are drawn out through a small orifice toward the first structural element, which is at ground. The final kinetic energy is determined primarily by the voltage on which the filament is floating. The work function of the filament and the voltage drop across the filament make some small alterations to the total energy. The absolute kinetic energy of the electrons is generally not known with certainty, but changes in the voltage can be made with precision. With the use of heated filaments, the resistance drop across the filament and thermal tailing cause a slight broadening of the kinetic energy distribution ($\sim\frac{1}{2}$ V) for the extracted electron beam. This broadening, however, does not pose a problem when the beam is used for producing inner shell vacancies; and, in fact, the energy resolution is sufficiently sharp that the beam can be useful in the calibration of the spectrometer and in the evaluation of characteristic energy losses in gaseous samples.

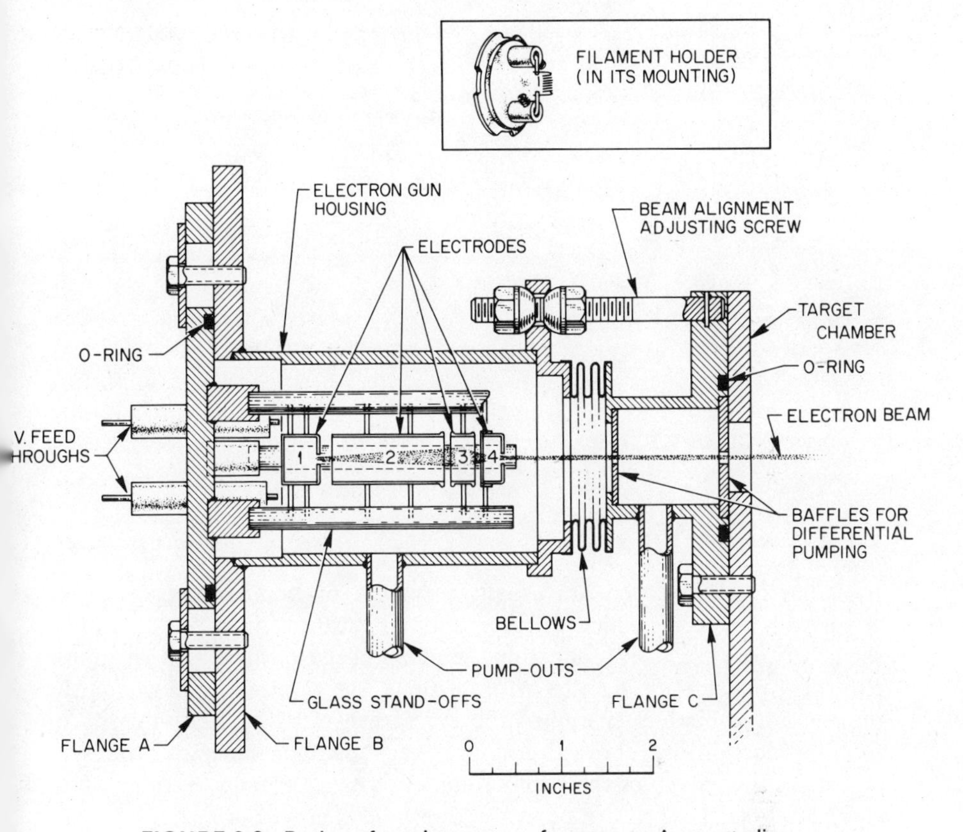

FIGURE 2.2. Design of an electron gun for gaseous Auger studies.

For work with gas targets, at least one and sometimes two stages of differential pumping are required for the electron gun, since low pressure is essential not only for preventing voltage breakdown, but also for helping to preserve the filament.

High spacial resolution for the electron beam is not required for gases, but Auger spectroscopy on solids can greatly benefit from a well-defined focal spot. One of the special advantages of Auger spectroscopy is its capability of examining a limited portion of the surface of a given solid target. Since for adequate counting rates solid targets require a less intense source of radiation than do gases, electron beam intensity can be sacrificed for a sharper focal spot. Normally, Auger spectroscopy on solids uses a beam with a diameter of less than 1 mm, and studies down to 1 μm resolution have been carried out. More details on electron guns may be found elsewhere.[2]

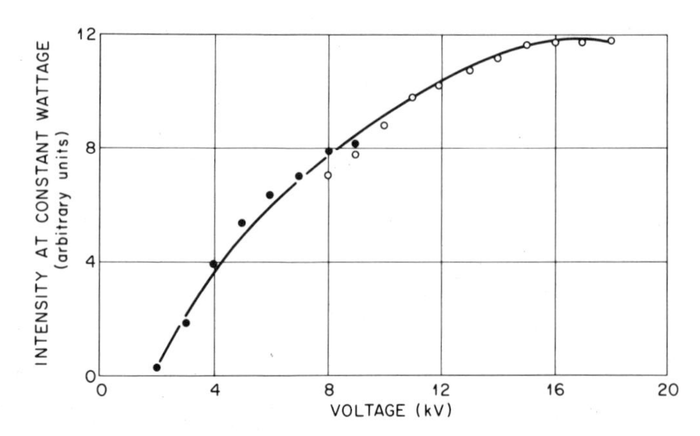

FIGURE 2.3. Plot of $K\alpha$ x-ray intensity vs. voltage on an Al anode. Wattage is held constant. [Unpublished data from J. C. Carver (○) and T. A. Carlson (●).]

1.1.2. X-Ray Tube

Characteristic x rays are usually produced by bombarding a suitable target with electrons, which will create initial vacancies in the desired inner shell. An x ray can then result from the filling of the vacancy with an atomic electron by means of a radiative transition. The threshold energy for the process is the binding energy of the inner shell electron. The intensity of x-ray production just above threshold is very low, but increases rapidly as one raises the energy of the bombarding electrons, normally leveling out at energies about three to five times the threshold energy. For example, if one wishes to produce Cu $K\alpha$ x rays (K binding energy $= 8979$ eV) more than 30 keV is desirable. For lighter elements (e.g., aluminum and magnesium) the leveling-out point is about 5–10 times the threshold energy. Figure 2.3 is a plot of x-ray intensity vs. tube voltage at constant wattage for an Al anode. Since the cost and inconvenience go up sharply with voltage, this factor must also be weighed before the optimum power supply is purchased. Some commercial x-ray tubes use a half-wave rectified voltage source, in which case the optimum voltage will be somewhat higher than when using a regulated dc power supply.

The two most important x-ray sources used in photoelectron spectroscopy are the characteristic $K\alpha$ x rays of Mg and Al. These yield the narrowest K x-ray lines that can be obtained practically (0.9 eV for Al and 0.8 eV for Mg). For higher Z (above 13) the natural width of the K shell increases so that by copper ($Z = 29$) it is 2.5 eV. It is difficult to fashion a suitable sodium anode*

* Although some success has been reported at McPherson Instrument Corp. (private communication).

($Z = 11$), neon ($Z = 10$) is a gas, and for lower Z elements the L shell is part of the valence shell, so that considerable band broadening occurs ($K\alpha$ x-rays occur as the result of a radiative transition in which a K vacancy is filled by an electron from the L shell). For example, carbon K x rays form a band with a width of approximately 4–8 eV, depending on the form of the material.[3]

In addition to K x rays, some work has been done with L and M x rays. These are less satisfactory than K radiation since their spectra are generally more complex. However, a few targets seem to have promise, namely the $M\zeta$ line of zirconium, which has an energy of 151.4 eV and a natural width of 0.8 eV, and the $M\zeta$ line of yttrium, which is 132.3 eV in energy with a natural width of only 0.47 eV.[4] Titanium and copper L lines have also been used, but their natural widths are in excess of 1 eV.[5]

A yttrium target is difficult to work with; it oxidizes easily, which spoils its resolution. The kinetic energies of the Y and Zr x rays are insufficient to study the core electrons of many important elements, e.g., carbon, nitrogen, oxygen, fluorine, sulfur, and chlorine. However, the small natural widths of these x rays makes them highly desirable for special studies.

Though harder x rays have larger natural widths, they can be invaluable for some special needs in ESCA. For example, they are useful for measurements of the binding energies of the innermost shells of the heavier elements. Also, the higher energy photoelectrons that are the consequence of shorter wavelength radiation have longer mean free paths in solids, and thus permit an examination of material that is farther below the surface than that probed by means of Al and Mg x rays.

Table 2.1 lists the energies and widths of a number of characteristic x rays that have been used in photoelectron spectroscopy. It is desirable to have an x-ray source with the capability for interchangeable targets, particularly if one wishes to study some property as a function of photon energy. Fellner-Feldegg *et al.*[6] have discussed various photon sources of characteristic energies as a function of their linewidths and intensities.

TABLE 2.1

X-Ray Sources Used in Photoelectron Spectroscopy

X rays	Energy, eV	Natural width, eV
Cu $K\alpha_1$	8048	2.5
Ti $K\alpha_1$	4511	1.4
Al $K\alpha_{12}$	1487	0.9
Mg $K\alpha_{12}$	1254	0.8
Na $K\alpha_{12}$	1041	0.7
Zr $M\zeta$	151.4	0.77
Y $M\zeta$	132.3	0.47

A simple x-ray tube design is shown in Figure 2.4. A desirable layout for a source volume is to have the position of the x-ray tube, the sample, and the entrance slit in close proximity (cf. Figure 2.1) in order to ensure the greatest solid angle between the x-ray anode and the sample, and the sample and the entrance slit. Any improvement in x-ray tube design should be balanced against loss in these solid angles. Alternatively, a crystal monochromator can assist in focusing the x rays at a distance, and electron optics can be used to transmit electrons from the sample target to the entrance slit without substantial losses. These devices will be discussed further when we examine monochromators and deceleration optics. It is desirable to have a window and a separate pumping system for the x-ray tube, and absolutely necessary if one is studying gases. A window prevents stray electrons from getting out of the x-ray tube and gases from getting in. It also filters out low energy bremsstrahlung. Foreign gases not only might cause voltage breakdown; they might attack the filament and contaminate the anode target. The window can be made of high purity foils of beryllium (0.5 mil) or aluminum (0.1 mil) if one is dealing with Al or Mg x rays. Although it is possible to make a window of beryllium leak-tight to atmospheric pressure yet still transmit the Al and Mg $K\alpha$ x rays, thinner windows are needed for softer x rays; and it is often good practice to use a bypass valve to equalize pressure in the x-ray source and source volume when the source volume is open to atmosphere. If lower energy x rays are desired, polystyrene, formvar, or carbon films can be used as windows with thicknesses as low as 10 $\mu g/cm^2$. If heat or radiation damage proves to be a problem, the carbon films are superior to the organic polymers. These films can be prepared on a metal screen for additional strength. They will easily hold pressures up to 1 Torr, but not atmospheric pressure.

For low energy x-ray sources it is imperative that the surface of the anode be kept clean. A contaminated surface will diminish the number of x rays extracted from the target (by orders of magnitude!); the contaminating material will also contribute undesired x rays, and there may be a broadening in the natural width of the x rays if the contaminating material actually changes the chemical composition of the surface of the target anode. Most of the contamination comes from the tungsten filament, in the form of both tungsten and absorbed impurities in the filament. I have found that when a new filament is put in, the operation of the tube is sometimes poor at first, but if the anode is recleaned, the tube will work well from that point on. The purging of the new filament seems to require full operation of the tube, including both heating and high voltage. The best way to avoid contamination from the filament is to use an electron gun that can be set back from the anode and even to bend the electron beam slightly so that the filament is not in direct line of sight to the anode.

The intensity of x rays produced depends on the voltage and current of

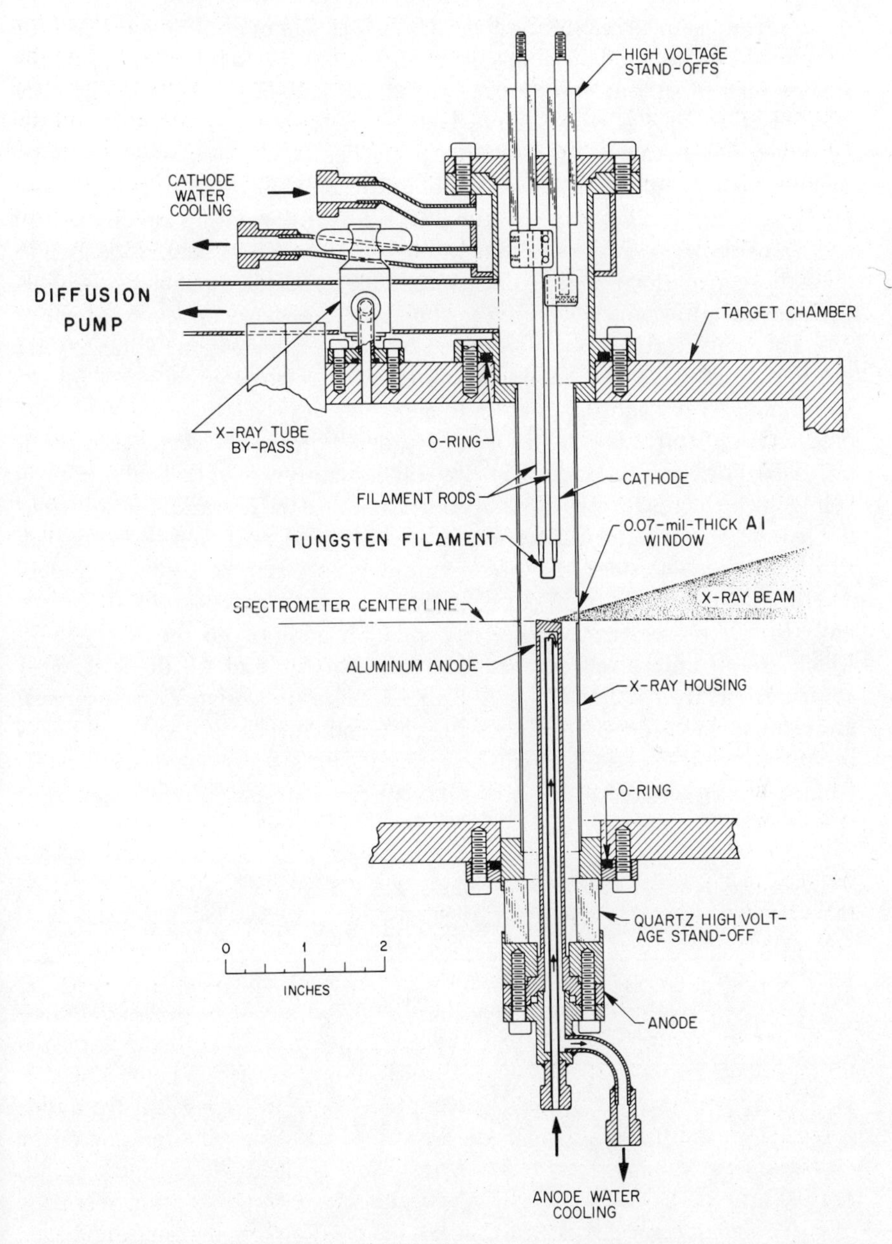

FIGURE 2.4. Design of a simple x-ray source. Cathode is usually kept at ground, though it may be operated at high voltage.

the electron beam striking the anode. Usually the amount of heat that can be dissipated by the anode (or power per unit area) limits the intensity of the x-ray source. Without special care an aluminum x-ray tube will normally operate at about 400 W while producing x rays from an area of 0.1 cm^2. Increased power can be obtained by special means of dissipating this heat, such as high pressure water cooling or a rotating anode. For example, the Uppsala group has developed[7] for ESCA a swiftly rotating anode ($\approx$5000 rpm) capable of power per unit area ratings of 1–2 orders of magnitude greater than the conventional x-ray tube. Such high-power x-ray sources are particularly advantageous when used in conjunction with a monochromator.

The characteristic x rays produced by bombarding Mg or Al targets are generally used without a monochromator. The relative intensity of bremsstrahlung or x-ray continuum to characteristic x rays is less important in the production of soft x rays than for hard x rays. For Mg and Al targets, about one-half of the x rays produced by electron bombardment at normal operating voltage, which pass through a Be or Al window, are the $K\alpha$ rays. The contribution to the photoelectron spectrum from continuous radiation is, in general, hardly noticeable, since the bremsstrahlung spectrum is distributed over several kilovolts while the K x rays are concentrated in a peak of about 1 V FWHM. Of more serious concern is that, in addition to the $K\alpha_{1,2}$ lines, which are an unresolved doublet in the case of Mg and Al, there are also contributions from other K x rays. Table 2.2 lists the designations, energies, and relative intensities of the K x rays for Mg and Al. The photoelectron peaks produced by these extra lines can be resolved, but are a problem when interferences occur between peaks from different elements and different subshells.

TABLE 2.2

Characteristic X Rays Produced from Al and Mg Targets[a]

	Mg		Al	
X ray	Energy, eV	Relative intensity	Energy, eV	Relative intensity
$K\alpha_1$	1253.7	67 } 100	1486.7	67 } 100
$K\alpha_2$	1253.4	33 }	1486.3	33 }
$K\alpha'$	1258.2	1.0	1492.3	1.0
$K\alpha_3$	1262.1	9.2	1496.3	7.8
$K\alpha_4$	1263.7	5.1	1498.2	3.3
$K\alpha_5$	1271.0	0.8	1506.5	0.42
$K\alpha_6$	1274.2	0.5	1510.1	0.28
$K\beta$	1302	2	1557	2

[a] M. O. Krause and J. G. Ferreira (to be published).

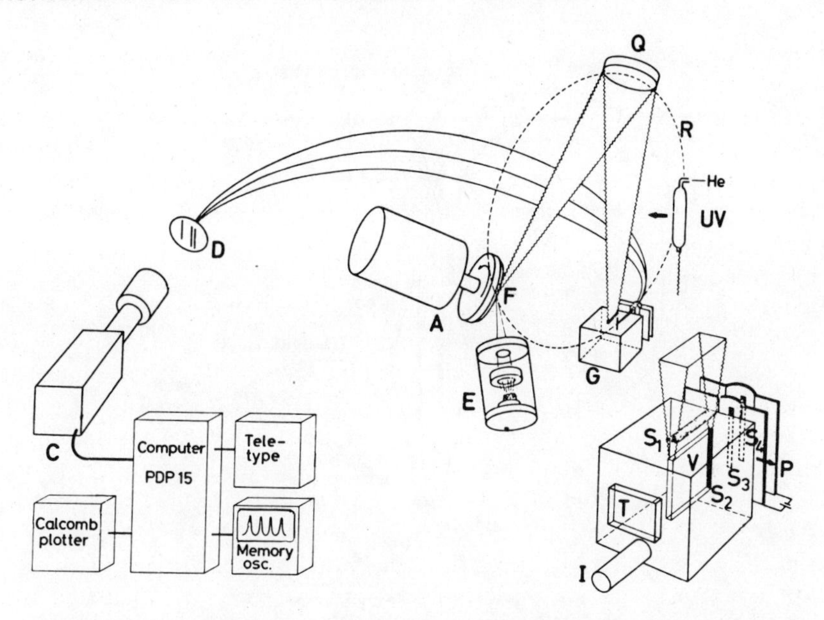

FIGURE 2.5. Schematic of electron spectrometer, which employs monochromatization and is designed for gases. E, electron gun; A, rotating anode; F, focal spot; Q, spherically bent quartz crystal; R, Rowland circle; UV, helium lamp; G, gas compartment; S_1–S_4, slits; V, effective irradiated gas volume; T, temperature raising device; I, gas inlet system; P, two-stage differential pumping system with an electron retardation step; D, multichannel plate detector; C, television camera. [Reproduced from U. Gelius, E. Basilier, S. Svensson, and K. Siegbahn, Univ. of Uppsala report UUIP-817, Figure 1.]

Also, one should be wary of x rays from contamination on the anode surface, as, for example, *M* x rays from tungsten.

A helpful review of practical problems with soft x-ray sources has been written by Henke and Tester.[8]

A monochromator can be used to ensure a single-line x-ray source. Even more important, a monochromator can be used to give a narrower x-ray line by taking only a portion of the $K\alpha$ band, which, as stated before, is nearly 1 V wide for Al and Mg. Finally, removing the bremsstrahlung and contributions from x-ray satellites is helpful in obtaining a very low background when studying the conduction bands of solids. By coupling a monochromator with Al radiation, an x-ray line as low as 0.2 eV FWHM has been obtained.[9, 10] There is considerable loss in source intensity using a monochromator, but improvements in x-ray sources, spectrometer design, and position-sensitive detector capabilities have helped make up for these losses. In addition, the monochromator serves as a focusing device for the x rays so they can be more effectively delivered to the sample target. Figure 2.5 shows the use of a quartz

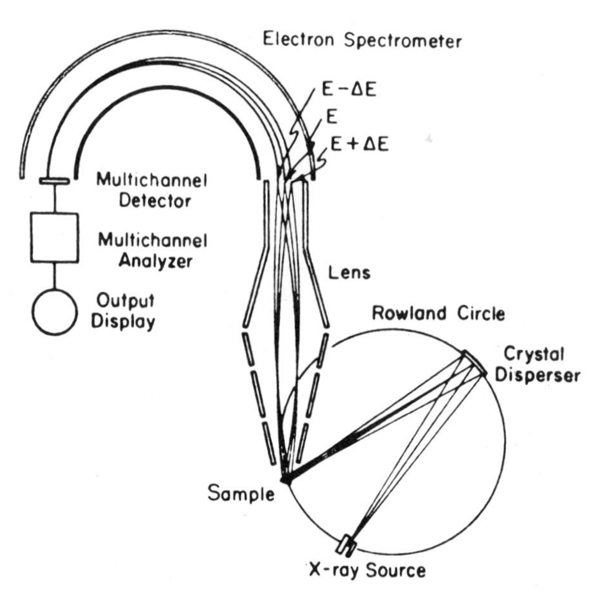

FIGURE 2.6. Schematic of monochromatization using dispersion compensation. [Reproduced from Siegbahn,[7] Figure 13.]

monochromator coupled with Al *Kα* radiation produced by a rotating anode. Slits admit only a portion of the Al radiation following the monochromatization.

A crystal monochromator will disperse an x-ray line of finite energy width along a Rowland circle. Instead of using a slit as indicated in the above paragraph to achieve a high-resolution x-ray source, a method known as dispersion compensation may also be employed.[9,10] (Cf. Figure 2.6.) The x rays are dispersed onto the target with photoelectrons coming out with slightly different energies, depending on their position along the target. This position and energy dispersion is compensated for in the spectrometer so that all photoelectrons will come to a common focus so long as they were ejected from atomic orbitals having the same binding energy. A quartz crystal with 100 plane is ideally suited for the Al *Kα* x ray. Such a device has been employed in the Hewlett-Packard electron spectrometer, and resolution down to 0.5 eV in the photoelectron spectrum of silicon has been achieved.[7]

1.1.3. Synchrotron Radiation

Synchrotron radiation provides a continuous source of photons from 10 eV to 10 keV. When a high energy electron is accelerated, energy is emitted in the form of electromagnetic radiation. This can occur even when the electron is not passing through matter. This energy loss, which is undesirable to high-energy physicists, has become an invaluable tool for those interested in atomic and molecular phenomena. Most applications of synchrotron radiation have been operated in a parasitical mode; that is, use of the radiation was made while other high-energy studies were in progress. However, machines designed principally for the use of synchrotron radiation have recently been built. Most important has been the development of storage rings, where a high-energy beam of electrons is injected into a ring and allowed to circulate for several hours with only small additions of power needed to maintain constant energy. The more powerful machines give out enough radiation to enable one to create with a proper monochromator a beam of monoenergetic photons that have the resolution and intensity equal to the best characteristic line sources. Most important, the desired wavelength can be selected from a continuous range.

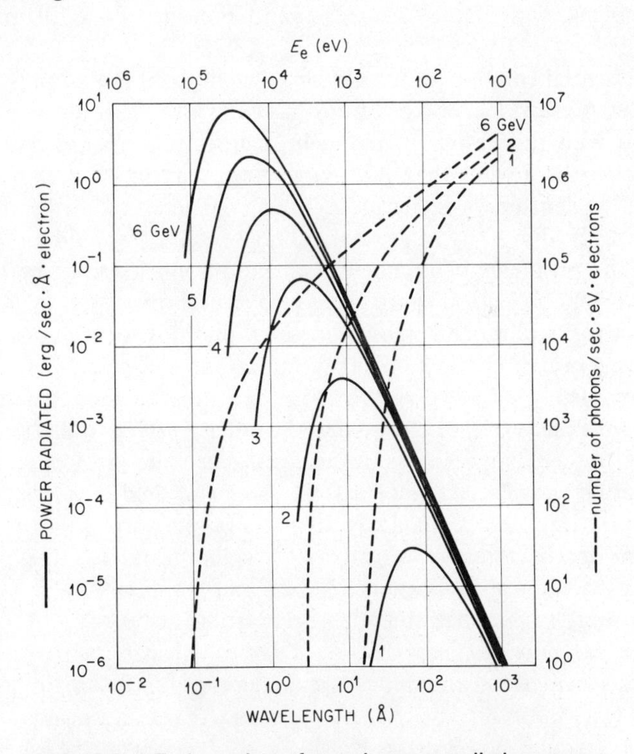

FIGURE 2.7. Intensity of synchrotron radiation as a function of the energy (wavelength) of the orbiting electron. Radius of orbit = 32 m. Solid lines give results in terms of power radiated as a function of wavelength. Dashed lines give results in terms of photons per sec as a function of energy in eV.

Figure 2.7 plots the intensity of synchrotron radiation in terms of power radiated as a function of the energy of an electron in orbit. Also plotted are the number of photons emitted per eV. It is obvious that the intensity is strongly related to the energy of the accelerated electron. For example, the energy radiated[11, 12] per revolution in a circular orbit is

$$\Delta E(\text{keV/rev}) = 88.5\,\frac{[E(\text{GeV})]^4}{R(\text{meters})} \tag{2.1}$$

If photons in the range of 1 keV or more are needed, the orbiting electrons should have energies greater than about 2 GeV.

If the energy of the desired radiation is above 100 eV, one ought to operate above 1 GeV. Machines performing at only 240 MeV can still yield useful beams for photons with wavelengths above 200 Å. For example, such a machine operating at 50 mA can deliver a beam of 30-eV photons having a 1 mm^2 spot size with 60 meV resolution and intensity of 4×10^{11} photons per second.[13]

An essential component for utilizing synchrotron radiation is the monochromator. A great deal of technology has gone into the solution of problems associated with the design of monochromators for synchrotron radiation. For wavelengths greater than 300 Å, gratings may be used at near normal incidence. At shorter wavelengths, a grazing incidence monochromator must be utilized. Below 20 Å, a crystal monochromator should be employed. Some of the problems of monochromatization which are of particular importance to photoelectron spectroscopy have been discussed by Thimm.[14]

Besides providing an intense source of photons, continuous in energy, synchrotron radiation has two other important characteristics. First, it is nearly completely elliptically polarized. Second, the light emerges nearly parallel to the plane of the orbit so that the intensity of the radiation decreases as the first rather than second power of distance from the source. Purcell[15] has suggested the placement of a fixed magnetic field whose strength and polarity vary in a regular manner along a straight portion of the electron's orbit. This modification is known as "wiggling" and has the purpose of producing an intense monoenergetic beam of photons and is particularly suited for short wave radiation in the region from 1 to 10 keV.

There are now a number of laboratories throughout the world using synchrotron radiation, and plans are underway for future improvements. Table 2.3 gives a list of some of these facilities. As yet only a small amount of work has been published (e.g., see Refs. 16 and 17) involving the use of synchrotron radiation in photoelectron spectroscopy, but it seems reasonable to expect a rapid change in this situation in the near future. At Stanford[18] x rays of 8 keV have been produced with a resolution of 0.2 eV (two parts in 10^5!) and a photoelectron spectrum taken of gold. At present the intensities are too

TABLE 2.3

Installations Providing Synchrotron Radiation

Installation	Type	Energy, GeV	Current, mA
NBS (Gaithersburg, Maryland)	Synchrotron	0.18	5
NBS (Gaithersburg, Maryland)[a]	Storage	0.24	50
Stoughton, Wisconsin	Storage	0.25	10
Stoughton, Wisconsin[a]	Storage	1.8	100
Hamburg, Germany	Synchrotron	7.5	16
Hamburg, Germany[a]	Storage	3	300
Tokyo, Japan	Synchrotron	1.3	50
Tokyo, Japan[a]	Storage	0.3	100
Frascati, Italy	Synchrotron	1.1	15
Daresbury, England	Synchrotron	5	20
Stanford	Storage	2.5	500
Orsay, France	Storage	0.5	100
Orsay, France[a]	Storage	1.8	500
Bonn, Germany[a]	Synchrotron	2.3	10

[a] Planned.

low for a practical source, but improvements in intensity are anticipated and such sources should open up new fields of electron spectroscopy. For further details on the theory and use of synchrotron radiation, see Refs. 11 and 12.

1.1.4. Vacuum-UV Sources

The most important sources of radiation for photoelectron spectroscopy of the outer shells are vacuum-uv lamps. The spectra created in these lamps contain sharp lines, formed as the result of deexcitation of excited free atoms or ions. These lines are intense, sufficiently energetic to eject electrons from the lesser bound orbitals of most molecules, and have a small natural width of only a few millivolts. A list of the more commonly used photon sources is given in Table 2.4. Frequently, one or two wavelengths are produced to the

TABLE 2.4

Vacuum-UV Sources Used in Photoelectron Spectroscopy[20]

Source	Energy, eV	λ, Å
He II	40.8	303.8
He I	21.22	584.3
Ne I	16.85	735.9
	16.67	743.7
Ar I	11.83	1048.2
	11.62	1066.7
H (Ly α)	10.20	1215.7

near exclusion of others and can be used without a monochromator. We shall discuss the different wavelengths separately. For a more detailed discussion of vacuum-uv sources, refer to Samson.[20]

1.1.4.1. He I (584 Å). The most widely used wavelength in the vacuum-uv region is the He I (584 Å, 21.22 eV) line. This photon is due to the $2p(^1P)$–$1s(^1S)$ transition following excitation of helium atoms into the resonance state. The natural width is only a few millivolts. When other conditions are favorable, this small width allows one to obtain photoelectron spectra in which not only the levels due to different electronic states are separated, but also the vibrational structure, and in some cases even rotational structure can be seen. Other factors, to be discussed later, limit the final resolution obtained in photoelectron spectroscopy, but lines as narrow as 4 mV have been observed.[21] The spectrum from a helium discharge lamp is remarkably free from other radiation, and thus experiments can be carried out without the use of a monochromator. The energies and approximate intensities of lines associated with the He I source are given in Table 2.5. Also listed are lines from common impurities of hydrogen and nitrogen.

Another important energy source is the He II radiation (that is, the principal resonance line from singly charged helium), which is at 304 Å (40.8 eV). This source will be discussed in detail in the next section. Generally, little confusion arises from these extra lines, but caution should be taken in interpretating low-intensity spectra.

Two basic methods have proven successful in producing the He I resonance line: use of (1) gas discharge lamp and (2) microwave discharge. No clear advantage is apparent for either method, although microwave discharge is reported to be more stable with regard to intensity output. Since the gas

TABLE 2.5

Radiation Associated with He I Lamp[20]

Line	Energy, eV	λ, Å	Approximate relative intensity
He I $2p \rightarrow 1s$	21.22	584.3	100
He I $3p \rightarrow 1s$	23.08	537.0	~2
He I $4p \rightarrow 1s$	23.74	522.2	~0.2
He II $2p \rightarrow 1s$	40.8	303.8	(strongly dependent on pressure)

Possible impurities

Element	Energy, eV	λ, Å
O	9.52	1302.2
H	10.20	1215.7
N	10.92	1135.0

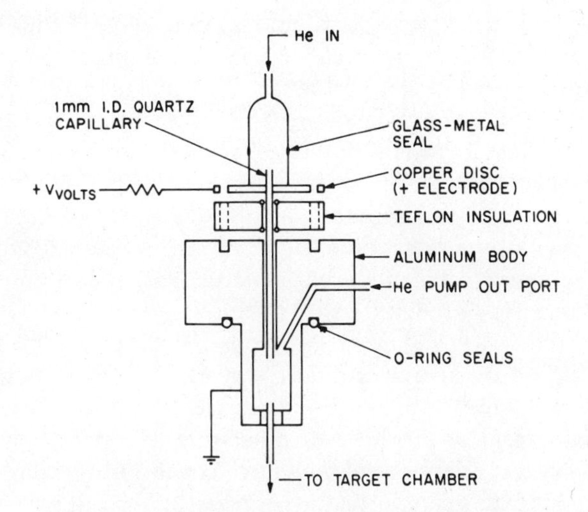

FIGURE 2.8. Design of a direct current discharge lamp for producing gaseous resonance lines in the vacuum untraviolet. [Reproduced from C. R. Brundle, *Appl. Spec.* **25**, 7 (1971).]

discharge lamp is the more commonly used source, it will be the one described here. Figure 2.8 shows a typical design. Helium is leaked through the tube at a controlled rate. The discharge takes place between the two electrodes in the capillary, which is made of quartz. The pressure at discharge is about 1 Torr. Differential pumping is applied to prevent excess He from getting into the source volume. The exit side of the tube is usually at ground, while the entrance is at either high positive or negative potential. If negative potential is used, the pressure of He should be sufficiently high to prevent an excess number of electrons from getting out of the tube. Cooling is necessary around the electrodes and desirable for the capillary tube. The starting voltage is usually about 50% higher than operating voltage, and a damping resistance is suggested for the initial power surge. Operating conditions depend on the geometry of the tube, but some typical figures are 3 kV starting voltage, 2 kV operating voltage with 100 mA of current. Reasonably good operations can be obtained over a wide variety of geometries and operating conditions. Samson[(22)] has pointed out, however, that optimum energy resolution occurs with a short capillary tube and suggests a length of about 3 cm.

1.1.4.2. He II (304 Å). Also produced in the He discharge lamp are ions of He whose excited states can give off radiation, the most important of which is the 304-Å He II $2p \rightarrow 1s$ transition. Under normal geometry and conditions for producing He I lines, only a small or negligible amount of He II radiation is formed. However, it is possible to operate a discharge lamp where the

He II 304-Å line is the dominant radiation. Poole *et al.*[23] and Burger and Main[24] have discussed how to preferentially obtain this radiation. This can be accomplished by enlarging the cathode volume and operating the discharge lamp at low pressure. For producing He II radiation it is, of course, necessary to concentrate helium ions in the discharge region. In addition, it is important to have very clean surfaces for the capillary; for this purpose boron nitride has been recommended rather than quartz.[25] The color of the discharge is a good indication of the radiation being produced. A yellowish white is seen when He I is primarily being produced, while a pinkish blue is observed under conditions where He II radiation is present. It is possible to use a window which passes the 304-Å radiation but stops the lower energy He I lines. Kinsinger *et al.*[26] have investigated the use of a 1500-Å Al window supported by 70 mesh nickel screen. A monochromator may also be employed.

The development of the 304-Å line has greatly enriched photoelectron spectroscopy since it offers an energy source capable of ionizing all but the most tightly bound molecular orbitals that make up the valence shell without much sacrifice in resolution. It also offers an opportunity to study the less tightly bound orbitals as a function of energy. Changes in intensity and angular distribution of the photoelectron spectrum associated with a given orbital as a function of energy can be of great help in determining the nature of that orbital.

1.1.4.3. Miscellaneous UV Line Sources. The 584-Å and 304-Å lines have been the most useful sources for photoelectron spectroscopy of the outer shell. Other sources are listed in Table 2.4. The neon and argon resonance lines can be produced simply by running the respective gases into the same gas discharge lamp as was used to produce the He I lines (cf. Figure 2.8). These lines are mainly employed to study the effect of photoionization as a function of energy. They have been useful in revealing the presence of autoionization states.

In the search for higher energy sources, researchers (see Ref. 20, p. 163) have used spark discharges to form multiply charged ions with fair intensity. However, because of the complexity of the spectra arising from this method, a monochromator is generally required. Duoplasmatrons have produced a variety of useful uv radiation.[27] The presence of strong magnetic fields, however, is incompatible with the need for field-free operation required for electron analysis.

1.2. Target Sample

In principle, any solid or gas sample can be studied by electron spectroscopy. However, liquids require very special techniques. In order to analyze the kinetic energy of electrons ejected from a sample target, the electrons must travel from the target to the detector without collision, and this requires operating the spectrometer at low pressures. In the case of solids, the source volume as well as the analyzer is kept at as low a pressure as possible (assuming sublimation does not occur). Gas samples can be studied by maintaining the source volume at moderate pressures (1–1000 μm) with the help of differential pumping. Generally, liquids will simply evaporate before measurements can be made. However, liquids with low vapor pressures (a practical limit is about 1 mm) can be studied if differential pumping is adequate.

Liquids in an asperated form might be studied, but even if possible, one would be examining the surface of a boiling liquid rather than the liquid under normal conditions. Siegbahn[7] has reported the study of liquid beams. A substance which is a liquid at normal temperatures can, of course, be investigated in the frozen state.

Even though electron spectroscopy cannot be used to study everything, there is a rather wide world of chemistry that can be investigated. We shall divide this section on target samples into three sections: gases, solids, and condensed vapor and liquids, since each of these different types of target has its own particular requirements.

1.2.1. Gases

One of the principal problems in studying gases is that a high pressure is desirable in the region of ionization in order to obtain reasonable signal strength. At the same time, the pressure should be sufficiently low so that the electron does not suffer a collision, particularly an inelastic one, between the time it is ejected from the atom or molecule until it strikes the detector. This is accomplished to a large extent by the differential pumping that is afforded by the entrance slit. The gas to be studied is leaked into the source volume. For a typical slit size of 1 cm $\times$ 0.2 mm a differential pressure of 10^3 can be maintained if the spectrometer is being pumped at the speed of 100 liter/sec. Thus, a pressure of 10^{-5} Torr can be maintained in the analyzer while the pressure in the source volume is 10^{-2} Torr. The ionizing beam is positioned so that it is close to the slit and runs parallel to the slit in such a way that the volume of the ionization cone that is being analyzed is at a maximum. The beam should be close to the slit not only to take advantage of a larger solid angle, but also so that the electron will travel in the high-pressure region as small a distance as possible. Under these conditions, good intensities can normally be obtained with a minimum of undesirable collisions.

Additional differential pumping can be applied between the slit and its baffles. Studies by Siegbahn *et al.*[28] have been carried out with gas pressures as high as 1 Torr. However, their spectra at this pressure contained considerable contributions from collision peaks that made analysis of low intensity satellite lines difficult. If secondary peaks due to collisions are to be avoided, the pressure in the source volume should be no greater than about 1 μm. Two procedures may be used to determine which part of the spectrum may be due to collision processes. First, an electron gun can be used to study the inelastic scattering of monoenergetic electrons. It is difficult to reproduce the identical path that occurred in the photoelectron experiment. However, the spectrum of elastically scattered electrons from a beam directed toward the entrance slit should give a good representation of the collision peaks if it passes through a gas pressure that is maintained under the same conditions as the experiment. A characteristic spectrum is produced (cf. Figure 2.9) for a given pressure of the sample gas and electron energy. Second, one can study the electron spectrum as a function of pressure. Collision peaks should vary as the square of the pressure (but not always, if, for example, the pressure is too high!), while the spectrum arising from electrons that have not undergone collisions should have only a first-power dependence on pressure.

The gas to be studied is allowed into the source volume from a manifold by means of a variable leak. The leak should be steady and easily controllable. A good choice for such a valve is a Granville-Phillips series 203 variable leak valve. The gas sample ought to be in sufficient amount so that there is little change in pressure during the run. This is usually not difficult to achieve, since in a typical 24-hr run, only about 10 ml of gas at STP is consumed, although this amount may vary considerably.

The kinetic energy of an electron emerging from the source volume depends on the work function of the chamber. This work function in turn

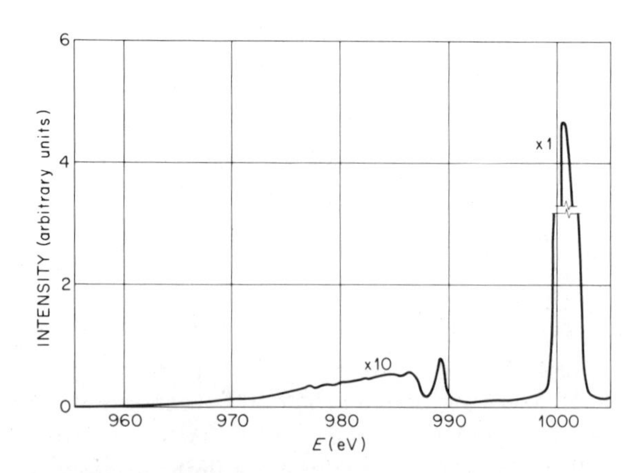

FIGURE 2.9. Characteristic energy loss spectrum. Electron source with kinetic energy of about 1000 eV passing through argon gas at pressure of about 10^{-2} Torr. [D. P. Spears, unpublished data.]

depends on the amount of ionizing radiation striking the walls and the pressure and nature of the gas in the chamber. In addition, the cone of ionization will have a net positive charge, since the ions formed in the photoionization process move more slowly than the electrons. Thus, electrons emerging from the ionization cone may be slightly retarded. Under these conditions an absolute determination of kinetic energy is difficult. However, by taking a photoelectron spectrum or Auger spectrum of an admixture of the sample gas with a rare gas or some other suitable molecule where binding energies are known from other measurements, one can obtain a reliable calibration.

If shifts or fluctuations due to changes in the work function or plasma, as described above, occur during a run, there is also a deterioration of the resolution. It is particularly important to keep the pressure constant during a run. A change of a factor of two in pressure can cause a shift of 0.1 eV. Even if conditions can be maintained very stable, microfluctuations occur. If the ionization source becomes too intense, line broadening also occurs. One very important factor is to have all the surfaces of the inside chamber walls and slits at a constant work function with good conductivity. In particular, one would like to avoid a nonconducting oxide layer. This can be done by gold plating, or, even better, by applying a coating of carbon, for example, Aquadag (a colloidal suspension of graphite), over all the surfaces and rubbing the surface smooth with a cloth.

1.2.2. Solids

A solid target, in general, is mounted in a fixed position as close to a photon source and entrance slit as possible, or close to an optical system so that the ejected electrons are transmitted into the analyzer with high efficiency. In photoelectron spectroscopy of solids the choice of the angle between the direction of the incoming photon and direction of the outgoing photoelectron is generally not critical. Although the distribution of photoelectrons ejected from a pure *s* wave is peaked in the direction perpendicular to the incoming beam according to $\sin^2 \theta$, the angular distribution of electrons ejected from molecules is usually more isotropic. In addition, extensive inelastic scattering in the solid destroys the original sense of direction. Some investigations have been carried out on solids as a function of angle, primarily to learn about channeling (cf. Chapter 5, Section 5.3). When the interest of the study is the surface monolayer, a shallow angle between the exciting source and target is desirable. At such angles a higher percentage of the ejected electrons arise from the surface. In Auger surface spectroscopy, a common angle for the impact electron is 10°.

One of the prime considerations in studies of solids is surface cleanliness, since the kinetic energy of the electrons generally dealt with in ESCA allows these electrons to escape only from or close to the surface (e.g., 1-keV electrons have a mean range of 20 Å in aluminum). For the study of perfectly clean surfaces, special care must be taken. For example, Brundle *et al.*[29] have discussed an ultrahigh vacuum electron spectrometer for surface studies. At pressures of 10^{-9} Torr, a surface layer is covered in about 1 hr, if one assumes a sticking factor of unity. Thus, any study which focuses its attention on the surface layer will require a vacuum of about 10^{-9} Torr or better. To obtain such a vacuum requires that all seals should be metal. The system ought to be built of material that can be repeatedly baked (usually stainless steel, but one should take care to remove the possible residual paramagnetism). Efficient cold traps are necessary, especially in the source volume. Recent developments have produced very efficient cryogenic refrigeration systems, which are easy to install and maintain, with temperatures down to 4°K. Such cold traps act as a pumping system by themselves.

The main pumping system should be clean. Until recently, most purists preferred ion pumps. However, with the development of better pumping oils that will not easily decompose if accidentally exposed to the atmosphere at high temperature, and traps and baffles that essentially eliminate back-streaming, diffusion pumps are again serious contenders for producing ultrahigh vacuums. Ion pumps when used should be operated only under high vacuum conditions in order to reduce the chance of overloading. Ion pumps sometimes have the tendency to accidentally give back adsorbed gases. Other disadvantages are that ion pumps produce magnetic fields and are impractical if gas studies are to be carried out with the spectrometer. Turbomolecular pumps can maintain a pressure of 10^{-9} Torr, but require traps for better results. They can pump gases at high pressure without injury to the pump. Their disadvantages are that they can be noisy and are relatively expensive. For initial rough pumping, a cryostatic sorption pump is recommended.

Differential pumping can help ease the problem of ultrahigh vacuum considerably. It would be permissible, for example, to have the pressure in the main portion of the analyzer at 10^{-7} Torr if the source volume could be maintained by differential pumping to 10^{-9}–10^{-10} Torr. Since the mean free path is very long at low pressures, a gas molecule emerging from the analyzer through the entrance slit will strike the target if the target and the slit are in a direct line of sight. This effect can be reduced by moving the target back from the slit. If the electron emerging from the target can be transferred to the slit by an electrostatic lens, the reduction in solid angle need not be accompanied by a substantial loss in the number of electrons that can be analyzed.

It is insufficient simply to maintain a clean vacuum system; one must also clean the surface of the target *in situ*. Usually, an ion gun is used for this purpose, or the sample is sputtered by raising it to a high voltage (~1 keV) under an argon pressure of 10^{-3}–10^{-4} Torr, or the sample is heated to a high temperature, or mechanical scraping is employed. The sample may also be bathed in a reducing atmosphere if cleaning a metal free from oxides is the central concern.

The procedures used in ultrahigh vacuum techniques put severe limitations on high-resolution spectroscopy in terms of sample handling and prevention of undesirable magnetic fields. In many cases when one is interested in bulk properties and where the sample is not susceptible to attack by small partial pressures of residual gases such as oxygen or water vapor, reasonable vacuum in the range of 10^{-6}–10^{-8} Torr is tolerated. Under these conditions, only a thin film of carbonaceous material, usually hydrocarbons from the *O*-rings and vacuum grease, appears on the surface. Unless one is interested in the surface *per se*, this film does not usually present any major difficulties. The film reduces the signal only slightly, and the carbon line can be used for energy calibration. (See discussion below.)

A valuable accessory for sample handling is a double-valved transfer compartment. With this device one can mount the samples in an inert atmosphere, transfer them from a dry box to the spectrometer without exposure to air, pump out the inert gas, and introduce the sample into the source volume without breaking vacuum. In this way, the sample remains essentially uncontaminated (of course, it is still not possible to maintain the standards set for working in ultrahigh vacuum), and there is no reason to let the source volume up to air. If samples are introduced in this fashion, data taking can begin in a few minutes.

It may be most convenient to introduce several samples at one time. New targets can be studied simply by raising and lowering or turning the sample holder from the outside. Such a device not only saves time between introducing each sample, but one can quickly recheck a standard sample without introducing any change in the conditions of measurement.

Several different methods have been used to attach the samples. Easiest has been the use of double-stick Scotch tape. If powder is used, care must be taken not to allow the adhesive material to cover the exposed surface of the sample. The sample can also be pressed into a wire gauze. Concern here should be for any changes wrought on the sample by the pressure and possible contamination from the press. It should be ascertained beforehand whether any exposed portions of the sample holder would contribute Auger or photoelectrons that would interfere with the desired spectrum. Powder samples can also be pressed into soft metal surfaces roughened by etching. Samples of odd shapes and reasonable bulk can be analyzed directly without pulverizing. Friedman *et al.*[30] have reported on the employment of a special sample holder for odd-shaped objects. The study of odd-shaped samples is not feasible for all spectrometer designs, some of which may require the target to be located in a well-defined plane.

As the result of photon or electron bombardment, a nonconducting material can charge up to an undetermined potential. It is, therefore, necessary to calibrate each new material under the specific conditions in which it is studied. This is done by measuring the electron spectrum of a known substance that is mixed with the sample. Any shift in the potential due to charging up of the sample will be reflected by a shift in the kinetic energy of all the electrons being ejected from the solid so long as there is electrical contact between the different substances. A common photoelectron peak used for calibration is the carbon $1s$ line from the hydrocarbon contaminating the surface. The use of this peak poses some difficulties because the exact nature of the hydrocarbon is unknown and because carbon from other sources of contamination originally in the solid (filter paper, CO_2, etc.) may interfere. A flash of gold onto the surface has been used by some as a standard. The gold may also be mixed into the sample in a powdered form. Hnatowich *et al.*(31) have demonstrated the efficacy of gold and that it follows the potential changes in the solid. Difficulties with gold as a standard have been commented upon by several authors.(32) The carbon in Scotch tape has been employed by some.(33) Swartz(34) has advocated the use of MoO_3 as a standard. Graphite and KCl have also been used. Ogilvie and Wolberg(35) recommend Al_2O_3 for use with catalysts. Confidence in the final analysis is best gained by the verification of more than one type of calibration. Alternatively, one may wish to measure the binding energies relative to the Fermi level, which is the onset to the photoelectron spectrum. This is more easily discerned with conductors than for insulators, and a highly resolved monochromatic source of photons is desired. Finally, one can use the presence of secondary electrons to determine the zero energy point or spectrometer settings at which electrons leave the sample with zero kinetic energy. Binding energies fixed to this point are then referred to the vacuum potential. For further discussion, see Evans.(36)

Perhaps the most serious drawback to photoelectron spectroscopy of solids is line broadening. This broadening results from three main sources: (1) width of the photon source, (2) width of the subshell from which the electron is photoejected, and (3) broadening due to charging. Let us discuss in turn each of these factors.

1. The width of uv lines is negligible compared to other problems involved with solids. The width of the Al and Mg $K\alpha$ x rays is, however, not negligible. The widths are in part due to the lifetime of the K shell vacancy and in part due to spin–orbit splitting between the $2p_{1/2}$, $2p_{3/2}$ levels. The net full-width at half-maximum is about 0.9 eV for Al $K\alpha_{1,2}$ and 0.8 eV for Mg $K\alpha_{1,2}$. As discussed in Section 1.1.3 this source of broadening can be overcome by use of a monochromator.

2. Line broadening occurs as the result of the finite lifetime of the vacancy

created in the atom of the material under study. The lifetime decreases with an increase in binding energy. Also, if Coster–Kronig transitions (cf. Chapter 6 and Appendix 4 for definition) can fill the vacancy, the lifetime is considerably shortened. Thus, photoelectrons ejected from the orbital with the highest angular momentum for a given principal quantum number are likely to have the smallest natural widths since Coster–Kronig transitions cannot occur with these orbitals. In solids the valence band is generally broad, and phonon broadening can also occur. In spite of all these sources of line broadening, one usually expects to find at least one level among the core shell electrons that has a natural width of only a few tenths of an electron volt. In practice, these expectations are not fully realized. For metals, the broadening in the $2p_{3/2}$ level in silicon has been measured to as low as 0.5 eV.[10] For gases, the *K* shell of carbon in CO has been determined to be less than 0.15 eV. However, for nonconducting solids photoelectron peaks have rarely been found narrower than about 1 eV, largely because of charging.

3. For nonconducting solids one of the most serious sources of line broadening is due to the uneven charging of the surface as the result of photoionization. It has been suggested by Siegbahn[9] that some of the charging is reduced by the radiation damage itself, creating current-carrying centers between the solid and conducting sample backing. This aspect has been further discussed by Citrin and Thomas.[37] However, this process of neutralization is far from being completely efficient, and fluctuation in the charging process and localization of the charge probably occur. Grimvall[38] has examined a theoretical model for charging. Flashing the surface with gold does not seem to help, nor does the mixing of conductive graphite with the sample give a narrower peak. It has been suggested[39] that maintaining the sample at a constant potential by spraying the solid with a controlled flood of electrons will help with the problem of charging. Another solution is to limit studies to very thin films. For example, Clark *et al.*[40] have studied organic molecules condensed from the vapor phase on a cold finger with gold backing. So long as the film was still thin enough to see the gold lines, the sample did not appear to charge up. Under this condition, the spectrum for 1*s* carbon in benzene was 1.15 eV, indicating a linewidth from the carbon of no more than 0.6–0.7 eV. When the sample thickened, the photoelectron spectrum shifted due to charging, and the width of the photoelectron peaks broadened.

1.2.3. Condensed Vapors, Liquids, and Targets at Other Than Room Temperature

It is most useful to be able to heat or cool the sample holder during a run. In this way, ESCA can be used to study a chemical change as a function of temperature. Temperature control has also been used extensively to examine materials that cannot be practically measured at room temperature.

Fadley and Shirley,[41] for example, have studied the surfaces of metals under a slight pressure of H_2 (10^{-3}–10^{-2} Torr) with the sample at elevated temperatures as provided by a specially designed oven. At 700–900°C the hydrogen reacts with any oxygen adsorbed chemically or physically to the metal, and thus they were assured of clean surfaces for their study of the density of states near the Fermi level.

Liquids with fairly high vapor pressures have been studied by allowing the vapor to enter the source volume and condensing the vapor onto a cold finger, specifically a liquid nitrogen trap containing a surface that could be exposed to the x-ray beam in the position of a normal sample target. The results were most satisfactory, since a continually renewed surface was deposited, which ensured a minimum of contamination. In addition, as pointed out in the previous section, thin films can be easily studied by this method. Under special conditions even liquids can be examined. Siegbahn *et al.*[42] have shown that this can be accomplished by use of a liquid beam which enters the sample compartment under high vacuum via a nozzle.

1.3. Chamber for Angular Distribution Studies

Most of the studies carried out with electron spectroscopy are done at a fixed angle. Studies as a function of angle were mentioned briefly in the section on solids. Angular studies with gases are, however, even more profitable. As will be brought out in Chapter 4, the study of the intensity of photoelectrons as a function of angle between the direction of the incident beam and ejected photoelectron is of great help in assigning the nature of the molecular orbital. If one wishes to compare total intensities integrated over all angles of the ejected photoelectrons, he is advised to make the measurements at the "magic" angle of 54.7°, whereupon the intensities from randomly oriented gas molecules are independent of the angular parameter β (cf. Chapter 3, Section 1). Various experimental arrangements for angular studies using a gas discharge lamp have been reported by Berkowitz and Ehrhardt,[43] Samson,[44] Vroom *et al.*,[45] Carlson and Jonas,[46] Mason *et al.*,[47] and Morgenstern *et al.*[48] The latter three experimental groups have coupled their angular chambers with high-resolution dispersion spectrometers. Krause[49] has studied angular distribution using a soft x-ray source, and Mitchell and Codling[16] have employed synchrotron radiation.

An example of an angular chamber is shown in Figure 2.10. The chamber is separated into two parts by an *O*-ring so that the lower half containing the discharge lamp can be moved continuously from 20 to 140° with an angular

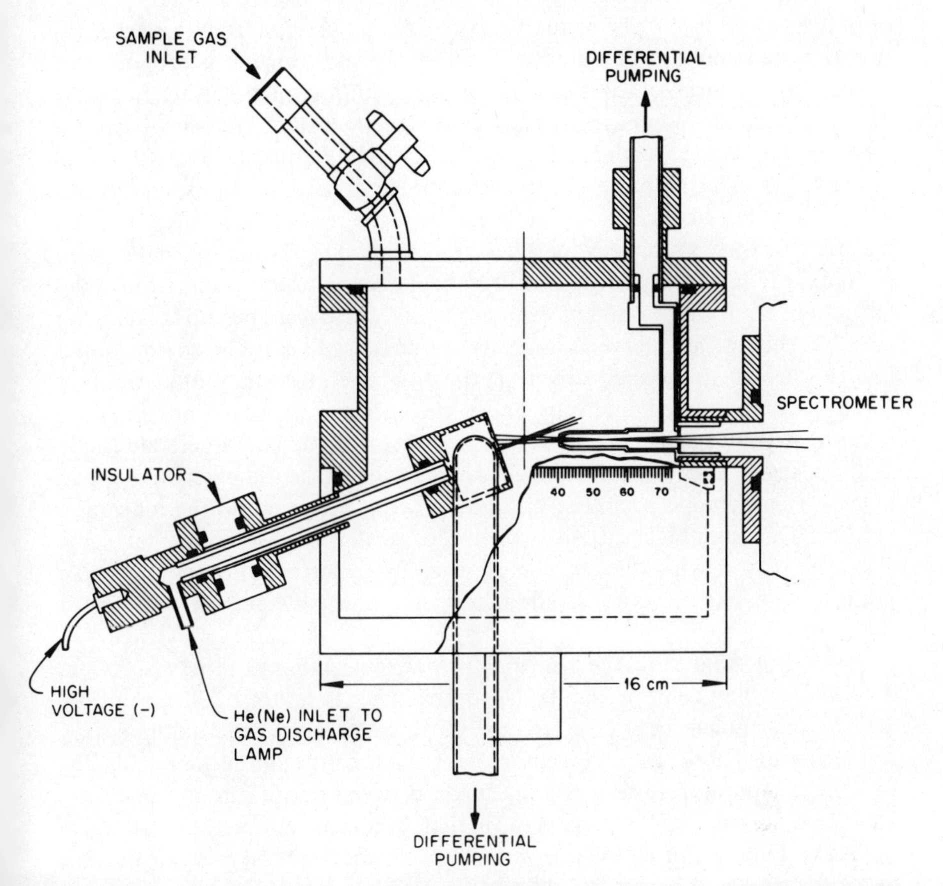

FIGURE 2.10. Chamber used for angular distribution measurements. [Reproduced from Carlson and Jonas,[42] Figure 1.]

resolution of 3° or less. Because of the restriction imposed by obtaining good angular definition, the intensities obtainable in angular measurements are substantially lower than for a fixed angle, but good signals may be obtained if the energy resolution requirements are relaxed slightly. Ames *et al.*[50] have modified their spectrometer so as to allow for a two-angle measurement.

1.4. Preacceleration and Deceleration

Under certain circumstances it is advantageous to accelerate or decelerate the electrons before they enter the analyzer. Preacceleration has been used on gas samples when the photoelectrons being analyzed have kinetic energies of only a few volts.[51] The primary advantage that accrues from this acceleration is more intensity: The window width of the spectrometer is proportional to the kinetic energy of the electrons, and, in addition, the focusing character of the preacceleration fields can enhance the intensity. The signal from a 1-eV photoelectron was increased 80-fold by giving an additional 10 V preacceleration. The absolute energy resolution afforded by the spectrometer will worsen, but at these low kinetic energies, the resolution is often dictated by other factors. If the spectrometer is set for 0.1% resolution, data taken at 10 eV will have a FWHM of only 10 mV, a value which is quite tolerable. One might argue that to gain more intensity one could just open up the defining slits, but the slits also serve as a means of differential pumping, so one would like to keep them stopped down. Thus, preacceleration may offer the best solution for increased intensity under these circumstances. Care must be taken that the electrons are born in a field-free volume and accelerated only as they leave the chamber. This will be achieved if the diameter of the grid holes is small compared to the distance between the grid and the ionization volume. One should be attentive in the case of preacceleration that some electrons may reach sufficient energy whereby inelastic collisions may take place, giving rise to extra satellite lines not normally present with He I radiation.

When the analyzer is the principal factor in determining the net resolution, deceleration may be desirable. If the spectrometer is set for 0.5% resolution, a 1-kV electron can be decelerated to 100 V, which gives an absolute energy resolution of 0.5 eV, or a percentage resolution of the initial kinetic energy of 0.05%. A compromise is usually made between resolution and intensity. Helmer and Weichert[52] have shown that by deceleration, the product of these two factors can be made a minimum. As mentioned before, if the slits also help serve as a device for aiding differential pumping, as is the case when handling gas samples, opening up the slits to achieve greater transmission of the electrons may not be a desirable option.

Deceleration can be applied by placing the sample holder at negative voltage. Although this method works perfectly satisfactory for conducting solids, nonconducting materials sometimes show considerable broadening in their peaks due to an uneven potential on the surface. A much preferred procedure is to maintain the entire volume around the sample holder as well as the sample holder itself at constant potential, so that the electrons are not decelerated until after they have emerged some distance from the target. Deceleration should ideally take place before the electrons reach the focal plane of the entrance slit. Electrostatic lenses are commonly used to both decelerate the emerging electrons and transfer them, without substantial losses, to the entrance plane. Weiss[53] has discussed the properties of a three-element cylindrical electrostatic lens for this use.

2. ANALYZER

The second basic portion of the electron spectrometer is the analyzer. Before we consider the various types of analyzers, we should consider an important auxiliary piece of equipment, a device for canceling the magnetic fields that surround the analyzer both from the earth and other sources.

2.1. Cancellation of Magnetic Fields

It is desirable to have magnetic fields canceled in the regions of the source volume and detector, but the time an electron spends in the analyzer between the entrance and exit slits is particularly critical. For example, a magnetic field of 1.0 mG will bend a 1-keV electron 0.2 mm during a flight path of 50 cm. The dispersion for a spherical electrostatic analyzer with a 20-cm radius is about 0.2 mm for a 1-keV electron measured with 0.04% energy resolution. Thus, one would like to keep the magnetic field to at least less than a milli-gauss. (The earth's field is approximately 500 mG.) Actually, if the extraneous field is constant and uniform, the requirements for cancelation are not so severe; but if the field is asymmetric or changing, this requirement is necessary.

There are two basic methods for canceling the earth's and other external magnetic fields: (1) Helmholtz coils and (2) ferromagnetic shielding.

2.1.1. Helmholtz Coils

Two circular coils of wire wound in opposite directions and placed directly over each other in a parallel position will produce a large region of equal field strength when equal currents are placed through the wires. Square coils when placed at a distance 0.57 times the side will behave much like a pair of circular coils whose diameter equals the side of the square, and are easier to construct. To cancel the fields in all three directions, three sets of coils are used. For some analyzers, such as those utilizing spherical sector plates, the most critical pair is the horizontal set, since it cancels the vertical field, which controls the motion of the electrons along the energy dispersion plane. It is advisable to monitor the field with a magnetometer, and by means of a feedback mechanism to automatically change current for these coils, so as to keep the field constant. This can be easily done, at least with an electrostatic spectrometer, with the field being held to ±0.02 mG.

One advantage of a set of Helmholtz coils is the accessibility to the whole spectrometer. One can easily step inside these coils to work, and new additions to the spectrometer can be made without much concern for dimensions. One of the chief disadvantages is the inability of the coils to shield against rapidly oscillating magnetic fields such as may be set up by a strong source of alternating current. In addition, although some asymmetry in the magnetic fields can be corrected by passing an unequal current through one member of a pair of coils or by tilting, strong asymmetry is hard to cancel.

2.1.2. Magnetic Shielding

The volume to be shielded from magnetic fields may also be surrounded by a highly paramagnetic substance. The availability at reasonable cost of large quantities of ferromagnetic alloys (e.g., conetic or μ metal) whose ratio of permeability is several orders of magnitude that of iron, has made this method very popular. A cylinder that has a diameter twice the dimensions of the spectrometer and a length three times can be employed. The cylinder is degaussed after construction to eliminate any spurious spots of magnetism. In commercial models, often the housing is closely fitted to the shielding. To achieve greatest efficacy, the shield is wrapped in two or more layers with magnetically insulating material between the layers. The advantages of ferromagnetic shielding for field canceling are: (1) It is usually less expensive than the Helmholtz coils, and (2) it is quite successful for shielding fields set up by alternating currents and arising from changes in nonuniform magnetic fields around the room in which the spectrometer is placed. The disadvantages are that: (1) It is sometimes more cumbersome to work with and less amenable to changes in spectrometer design, and (2) it is often difficult to completely degauss the shielding, and localized areas of magnetism can accrue with time due to accidental blows to the shielding or by contact with strong magnetic fields.

2.2. Types of Analyzers

There are two basic types of analyzers commonly used: (1) retarding potential and (2) dispersion. Dispersion instruments are generally used for high resolution work, and thus will be given more attention here, although some recent developments in the use of retardation warrant a reappraisal of this method. With the dispersion method, both magnetic and electrostatic fields have been employed. However, because of cost, electrostatic instruments have been much more popular. Each of the different basic types of analyzers that are being used in electron spectroscopy for chemical analysis will be described below. Details of design criteria for low energy electron spectrometers can be found elsewhere.[54–56]

In comparing the various designs of the different analyzers, the proper question to ask is not, "What is the ultimate resolution?", since the ultimate resolution in terms of Δ energy at full-width half-maximum is usually determined by factors other than the analyzer. Rather, the question should be,

"What is the intensity of ejected electrons measured at a desired resolution?" This factor depends on several considerations, namely the solid angle of the ejected electrons accepted by the spectrometer, the target area (or volume, in the case of gases), whether focusing and deceleration need to be employed between the target and the analyzer, and if there exists a focal plane suitable for a position-sensitive detector. In addition to intensity and resolution, convenience needs to be considered: Can the analyzer be used with both solids and gases, and what are the requirements for target positioning—in which energy range does the analyzer work best in, and can it be employed for a variety of jobs? Usually the three variables of resolution, intensity, and convenience can be traded off against one another, but each analyzer will have its own optimum set of values.

2.2.1. Retarding Grid

The principle of the retarding grid analyzer is simple, and much of the initial work done in the field of ESCA was carried out with such a system; even today in situations where other considerations, such as surface cleanliness, are more important than high resolution, use is made of the retarding grid. Analysis of the kinetic energy of an electron is achieved by the retarding potential method by requiring the electrons to pass through a retarding potential before they are collected. For example, if a beam of ionizing radiation is being created at the center of a sphere, the electron will feel a retarding potential between two spherical grids. If the net negative potential is greater than the kinetic energy of the electron, the electron will not emerge; but if it is less, the electron will emerge and be counted. Such a spectrometer is highly efficient and simple to construct, but the ultimate resolution achieved is considerably less than for a dispersion instrument. Figure 2.11 shows plots of the type of spectra normally obtained by retarding grid and dispersion methods. By differentiating the data obtained from a retarding grid, a more sensitive analysis can be achieved, but as one goes to lower kinetic energy, the spectrum becomes harder to analyze because of the ever-increasing background. The background is a greater problem with a retarding grid spectrometer because in a single-pass retarding grid analyzer the background is summed over all energies down to the retarding potential, while a dispersion instrument passes only electrons of energies that fall within a finite window width ΔE.

A device commonly used in surface Auger spectroscopy is shown in Figure 2.12. It consists of a retarding grid system. The innermost grid establishes the potential field between the analyzer and the sample. The second and third grids are used as retarding grids, the use of two improving the resolution. A fourth grid, held close to ground, is sometimes employed to shield the

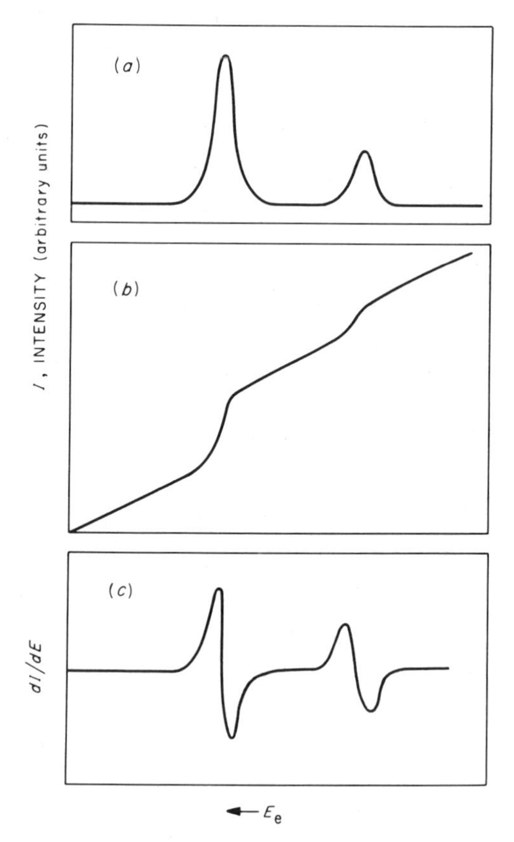

FIGURE 2.11. Comparison of idealized plots of photoelectron spectra taken with dispersion and retarding grid methods. E_e is photoelectron energy. (a) Dispersion spectrometer, (b) retarding grid spectrometer, (c) first-order differential intensity from retarding grid spectrum.

collector against the ac perturbation applied to the analyzing grids. To improve the sensitivity for filtering out the small signal offered by the Auger lines above secondary electron background, an electronic detection method is used, by which the analyzing energy is modulated and the output is detected synchronously. This allows for an electronic differentiation and avoidance of noise by (1) reducing the effective bandwidth and (2) using a modulation frequency which is in a quiet region of the noise system. In this scheme a perturbing voltage $\Delta E = k \sin \omega t$ is superimposed on the analyzer energy so that the collected current $I(E)$ is energy-modulated. The analyzer can then be tuned to ω or 2ω. All other frequencies will produce a time-varying signal with

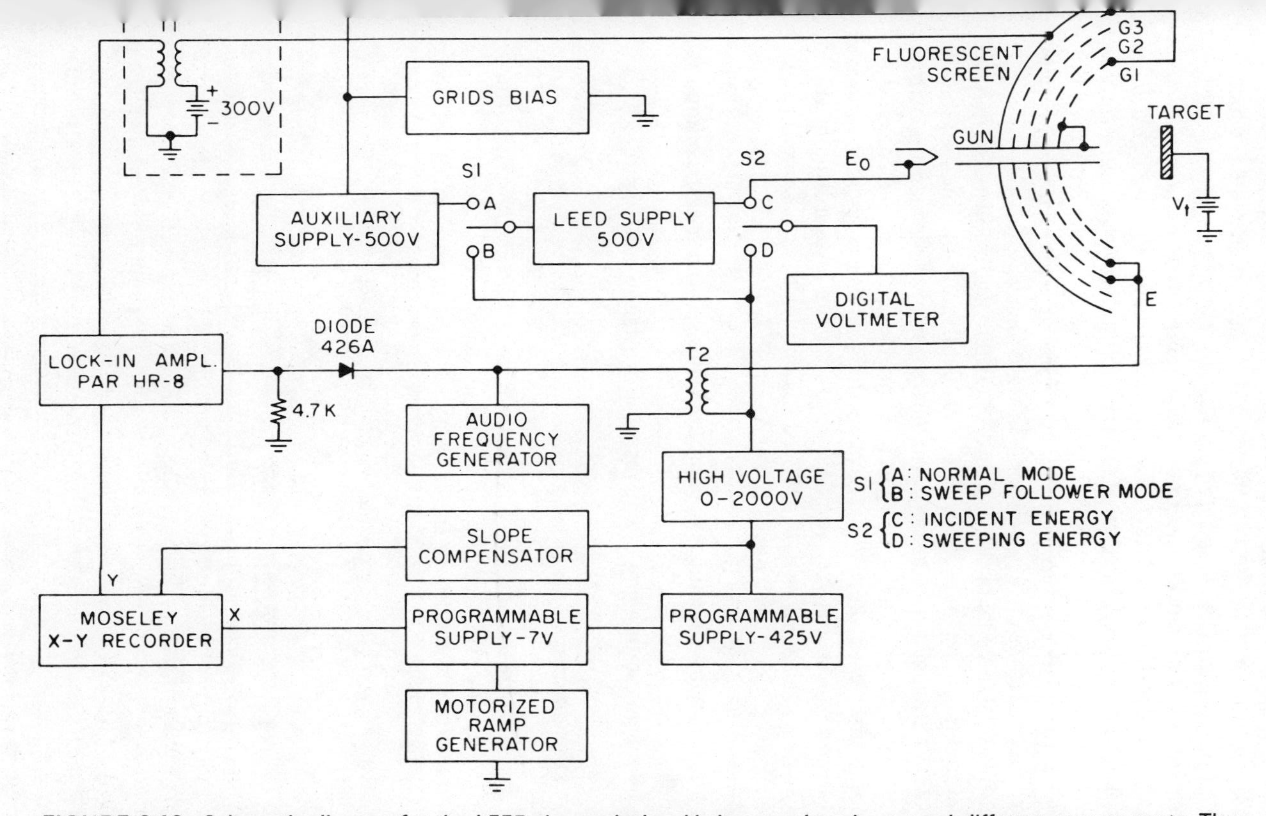

FIGURE 2.12. Schematic diagram for the LEED-Auger device. Various workers have used different arrangements. The sweep follower mode produces spectra with all loss peaks eliminated. A variable V_t is useful for making sure that an observed peak originates from the target. The slope compensator enables subtraction of a background slope from the data. The diode arrangement provides for tuning the HR8 to ω or 2ω. [Reproduced from C. C. Chang, *Surf. Sci.* **25**, 53 (1971), Figure 6.]

zero average dc values and will be "filtered out." The frequency ω corresponds to $N(E)$, while tuning to 2ω will give the first derivative $N'(E)$, which is usually employed. Synchronous detection has also been employed for dispersion instruments (e.g., a cylindrical mirror) yielding $N'(E)$ by tuning to ω.

Several designs for retarding grid spectrometry have been used. A cylindrical system has been used, but is limited in resolution because not all electrons are ejected at right angles to the retarding field. Better results have been obtained with spherical grids, or with plane retarding grids that employ slits to ensure that electrons are accepted only at right angles.

Recent developments in the design of retarding grid spectrometers have demonstrated that they may also be used for high-resolution studies. Golden and Zecca[57] designed a retarding grid instrument for resonance electron scattering experiments. At low energies it is capable of 20–50 meV resolution with good transmission. Lee[58] has developed a double-energy filter spectrometer. Electrons to be analyzed are first sent through a low-pass filter, where only those with energy $<E_e$ will be allowed to continue their passage through the spectrometer. Only electrons with energies $>E_e$ will go through the high-pass filter and be detected. Before analysis the electrons are retarded. The spectrum is swept by changing the retarding voltage, and high resolution is achieved by having the analyzer set to pass only electrons of a low, fixed voltage. A schematic drawing of the instrument is shown in Figure 2.13.

2.2.2. Dispersion

2.2.2.1. Magnetic Field. Electrons can be dispersed or separated according to energy by having them move in either a magnetic or electrostatic field. A dispersion analyzer should have the capacity for focusing at a point (or line)

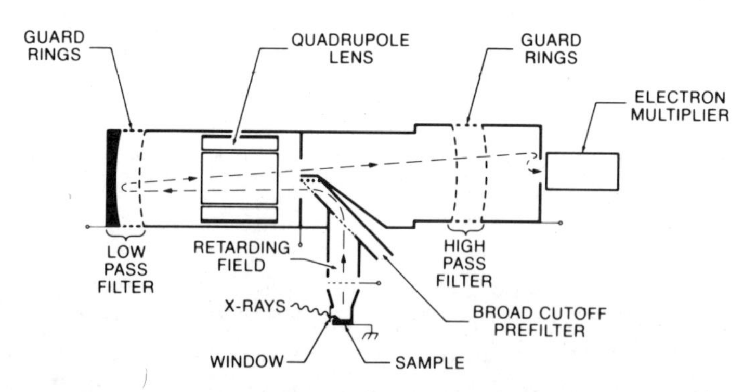

FIGURE 2.13. Schematic drawing of a double-pass nondispersion electron spectrometer (Du Pont 650 electron spectrometer). [Reproduced from Riggs and Fedchenko,[58] Figure 6.]

a beam of electrons of the same kinetic energy. For high kinetic energies (>5000 eV) magnetic fields are superior to electrostatic fields for use in electron analysis because of the practical problem in supplying the necessary field, and because the optics needed for handling relativistic electrons is better understood for magnetic than for electrostatic fields. Thus, most of the technology used in beta spectroscopy has evolved around the magnetic analyzer, and early applications to chemical problems involved the use of these instruments. Magnetic and electrostatic devices have comparable resolution and general performance characteristics. The principal difference is cost. The cost of a magnetic analyzer of equal quality to an electrostatic analyzer is about an order of magnitude greater. Other minor disadvantages are: (1) less operating room for the source volume, (2) greater difficulties in monitoring the stray magnetic fields, and (3) care needed in eliminating hysteresis in the magnetic field.

The type of magnetic spectrometer that has been used successfully most often for ESCA has been the double-focusing, iron-free spectrometer. In this spectrometer, the position of focus is 254° from the entrance slit. The magnetic analyzer is a momentum analyzer, in contrast to an electrostatic analyzer in which the path of the electron is proportional to the kinetic energy. The path of an electron in a magnetic field having cylindrical symmetry is given by

$$B(\rho_0) = mv/e\rho_0 \tag{2.2}$$

where B is the magnetic field, m and v are the mass and velocity of the electron, and ρ_0 is the radius of the circular path in which the electron travels. In comparing performance of the different spectrometers, the reader is reminded that the resolution for a magnetic instrument is given in terms of momentum, and for low energy $\Delta p/p \approx \frac{1}{2}\Delta E/E$.

Further description of magnetic spectrometers may be found in books by Sevier[59] and Siegbahn.[9,60] An excellent review of principles as well as a description of a rather elegant design for a magnetic spectrometer is given by Fadley *et al.*[61] For this spectrometer the authors claim an ultimate energy resolution of 0.01%. The source volume is accessible, and the coils are built so they can be easily separated from the vacuum system when it is required that the spectrometer be baked for outgassing. The spectrometer also has a large focal plane, which maintains excellent resolution and permits the use of position-sensitive detectors.

Betteridge *et al.*[62] have used a set of Helmholtz coils to obtain a single-focusing magnetic analyzer. Its use has been restricted to the low-energy range suitable for He I and He II radiation. Its advantages are that it is inexpensive to build (in contrast to the magnetic spectrometers discussed above), convenient for handling low-volatity materials, and may be used in conjunction with a gas–liquid chromatograph.

2.2.2.2. Electrostatic Field. Most of the present electron spectrometers now being constructed for chemical analysis are dispersion instruments with electrostatic fields. Cylindrical plates have been used extensively for PESOS. The plates are relatively easy to machine and, in principle, should give an energy resolution better than 0.05%, generally more than sufficient for use with the He resonance lines. In practice, the analyzer has been generally required to operate only at about 0.2% resolution, which for a 10-eV electron is only 20 mV. The intensities obtained with the use of gas discharge lamps are quite satisfactory for most purposes.

Spherical plates are superior to cylindrical plates since they can obtain the same resolution but with greater transmission because they focus electrons in two directions. Spherical sector plates have been quite popular (cf. Figure 2.14). They can be made by milling with a high degree of precision,

FIGURE 2.14. Spherical sector plates of double-focusing electron spectrometer. Cf. Figure 2.1.

ensuring high ultimate resolution (at least 0.02%). These plates are not expensive to prepare, but do require an excellent machine shop.

Full spheres and hemispheres have also been employed. This, of course, greatly increases the solid angle available for study. It is generally harder to design a spectrometer utilizing such plates; thus, the resolution is not as good as that which can be obtained with spherical sectors. However, by decelerating the electrons before they reach the analyzer (cf. Section 1.4), one can achieve the desired energy resolution ΔE without undue loss in intensity. One important advantage that the spherical sector plates have over plates with 2π geometry is that they afford a focal plane that can easily be adapted to a position-sensitive detector (cf. Section 3.2). An additional advantage is that with sector plates, sample loading is generally much more flexible.

The path of an electron in a spherical analyzer is governed by

$$V = \frac{E}{e}\left(\frac{R_2}{R_1} - \frac{R_1}{R_2}\right) \tag{2.3}$$

where E for nonrelativistic velocities is the kinetic energy of the electron, and V is the potential difference between the two spheres of radii R_1 and R_2. Although the resolution of the photoelectron peaks is essentially unaffected by the relativistic effect for electrons of energy below 2 keV, the change in plate setting is not quite linear with kinetic energy as suggested by equation (2.3). If photoelectrons of different kinetic energies E_1, E_2 are to be compared at different plate settings V_1, V_2, the following expression can be used, where k is an instrumental constant:

$$E_2 = E_1 + k(V_2 C_2 - V_1 C_1) \tag{2.4}$$

where

$$C(E) = 2\frac{c^2}{v^2}\left\{\frac{1}{[1-(v^2/c^2]^{1/2}} - 1\right\}$$

Although the relativistic correction C is not negligible in the region of interest for ESCA (at 1000 eV it is 0.29%) it is nearly linear with kinetic energy. In practice, a spectrometer constant k' can be obtained empirically which includes the relativistic effect and equation (2.4) becomes

$$E_2 = E_1 + k'(V_2 - V_1) \tag{2.5}$$

Electrons are focused with spherical plates 180° from the source. In contrast, a cylindrical analyzer brings the electrons to focus at 127°. The resolution for both spherical and cylindrical analyzers is proportional to the slit width and radius of the flight path. That is,

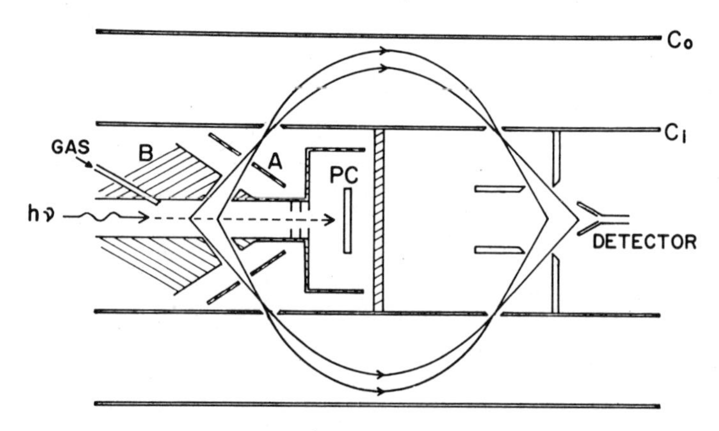

FIGURE 2.15. Schematic of cylindrical mirror analyzer. [Reproduced from Gardner and Samson,[63] Figure 1.]

$$\Delta E/E \propto (\Delta l)/r \tag{2.6}$$

Thus, a 40-cm spectrometer with a 0.4-mm slit will have the same resolution as a 20-cm spectrometer with 0.2-mm slit.

Another important configuration used for electrostatic plates is the cylindrical mirror (Figure 2.15). These plates are easy to machine and, by proper choice of angle, can in principle give the best theoretical limit to resolution.* The solid angle from a point source through which electrons can be analyzed is generally larger than that for the spherical sector, but the area (or volume if a gas is studied) that can be analyzed for ejected electrons may be smaller. One limitation to the design is the absence of a simple focal plane suitable for position-sensitive detectors.

An electrostatic analyzer with a rather odd trajectory has been designed by Allen *et al.*[64] (Figure 2.16). Its chief advantage is simplicity in construction; it has the shape of a pill box.

2.2.2.3. Other Analyzer Designs. In addition to retarding grid and magnetic and electrostatic dispersion instruments, electron energy analysis may, in principle, be carried out effectively for ESCA applications by other methods.

For example, there is the crossed magnetic and electrostatic field design known as the Wien filter.[65]

* See Refs. 55 and 63 for detailed discussion of the optics and design of cylindrical analyzers.

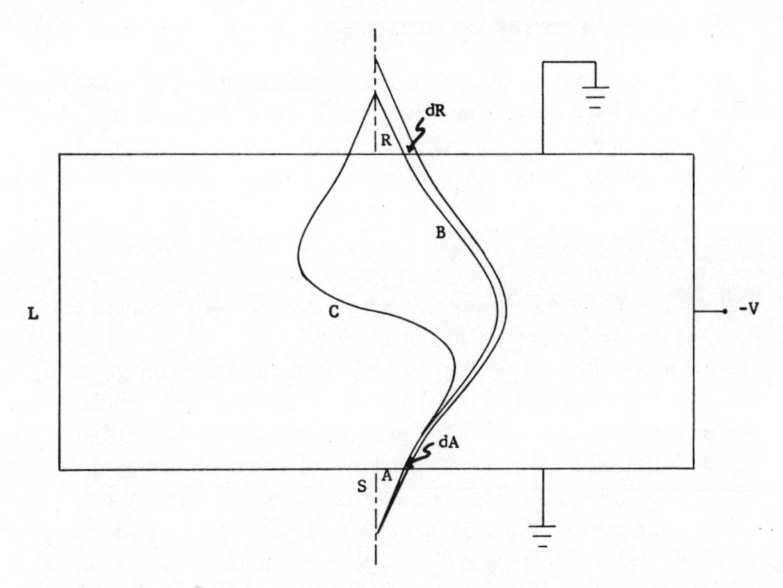

FIGURE 2.16. Trajectory for proposed "Bessel Box." [Reproduced from Allen *et al.*,[64] Figure 1.]

A time-of-flight design has proven to be very useful for the study of low velocity electrons in the range from 0 to 1 eV and has been applied to photoelectron spectroscopy.[66]

Spectrometers have also been designed especially for studies at high temperature[67] and for coincidence studies between ions and electrons formed in the gas phase.[68] A study of droplets by use of a Millikan condenser method has been described by Ballard and Griffiths[69] for use with uv radiation

3. DETECTOR SYSTEMS AND DATA ANALYSIS

As a consequence of having passed through the analyzer, electrons of a given kinetic energy are spatially separated from electrons of different energies. These electrons must now be detected and recorded, together with information which determines their energy, such as the electrostatic or magnetic fields applied to the analyzer. This information must be stored, and some method devised for scanning over a spectrum of electron energies. Finally, one may wish to process the data for plotting, curve fitting, subtracting background, etc. We shall now discuss, in order, each of these topics.

3.1. Single-Channel Detector

It is often convenient to use an electrometer with some retarding grid analyzers, since electrons are being collected over a large surface area. The sensitivity for such a device is much poorer than that of an electron multiplier. The latter essentially counts every electron as an independent event, with a dark current of about 0.1 count/sec, whereas to observe a signal above background with an electrometer, one usually needs the order of 10^4 electrons. Proportional counters have been used with electron spectrometers, but since they require a window, low-energy electrons must be post-accelerated several kilovolts just before they enter the detector.

More commonly used for low-energy electrons are electron multipliers. The multistage devices combine 8–15 stages with activated surfaces such as Be–Cu or Ag–Mg alloys and were used extensively in the earlier days of ESCA, but these multipliers are especially subject to damage and must be kept under vacuum at all times. The new multipliers are very hardy devices, such as the channeltron made by Bendix or the channel electron multipliers produced by Mullard, which are glasses impregnated with metal, such as lead or vanadium. A potential field set up in this material allows for a substantial electron cascade. These glassy multipliers can be exposed to atmospheric pressure for long periods of time without damage. They count electrons with high efficiency even at essentially zero kinetic energy, and the background is 0.1 count/sec or lower. Those equipped with a funnel have an effective counting area of about 0.5 cm^2. Some have been specially equipped with slits 1.5 cm long. Their only drawback is that a high count rate (>20,000 counts/sec) will cause a saturation effect, and they must be allowed to rest. A third type of multiplier is that put out by W. Johnson, Inc., which consists of a gridded wafer. The Johnson multiplier should not be allowed to stand at atmospheric pressure, since it may lose its efficiency, but it can be easily reactivated. Its main advantage is a large effective counting area.

3.2. Position-Sensitive Detector

The most exciting development in detectors is the position-sensitive detector. An electron spectrometer can have a rather substantial focal plane over which electrons, having passed through the analyzer, are distributed according to their kinetic energy. Usually, only a small portion of the spectrum is sampled; that is, those electrons passing through a single exit slit (whose size is matched to the entrance slit) will be counted by one electron multiplier. The focal plane covers a range of about 5% of the mean energy being analyzed. Thus, if one were operating at a resolution of 0.05%, only one part in 100 of the available data is being recorded. For example, the 20-cm spherical sector electron spectrometer at Oak Ridge normally uses a 0.2-mm slit. It was found that the resolution did not deteriorate over a 2-cm focal plane, which represented a 5% spread in energy. In the case of a cylindrical mirror or full spherical spectrometer, the image for a given energy is in the form of an annulus with radial dispersion of the kinetic energy of the electrons. Superposition of high and low energy electrons will occur unless half the emerging annular beam is blocked. In any case, such an image would be difficult to analyze with a position-sensitive detector. Thus, at the present time, designs such as the spherical and cylindrical sector spectrometers which offer line spectra are the most suitable analyzers for use with position-sensitive detectors.

In principle, one could place a film at the focal plane and take a picture of this spectrum, but the film sensitivity is no match for an electron multiplier. Bendix has perfected a multichannel plate (now distributed by Galileo) which is a matrix of electron multipliers similar, in principle, to the solid-state channeltrons described earlier. Each cell has a radius as small as 0.015 mm and a gain of 10^4 with low dark current. Two plates can be combined, one cut at an angle to prevent ion back-current, to give a gain of 10^7 electrons. Several other companies (e.g., Varian) are also producing similar devices.

The signal from the position-sensitive detector may be read in several ways: (1) The electrons are collected into anode strips. The pulses are amplified and recorded from each strip individually. Such a method has been successfully tried out both at Berkeley and at Oak Ridge. Even with the microcircuitry available, such an arrangement is both expensive and cumbersome. (2) The electron charge from each microchannel can also be recorded by a vidicon scanner, either directly or by first impinging on a fluorescent screen. Hewlett-Packard has built such a device for their spectrometer. (3) The electrons can be directed toward a resistor plate, and the ratio of the current read at the two ends of the plate can be converted into position information. Moak *et al.*[(70)] have demonstrated the use of this device for electron spectroscopy. The resistor strip method gives excellent resolution and low background and is simple to construct. Its limitation is, however, in the total counting rate that can be handled by present-day electronics, which is about 20,000 counts/sec.

Workers at Uppsala[(6)] have developed a small parallel-plate multiplier that can be used individually and stacked together as a position-sensitive detector. The chief advantage of this device is that the individual multipliers can be replaced if damaged. The chief disadvantage is the time consumed in fabricating the large number of individual multipliers, and the lost counting area between each unit.

For a position-sensitive detector, then, data are stored according to both the voltage on the plates and the position on the focal plane that yields a unique kinetic energy identification for the electron being analyzed. Thus, a whole energy spectrum is recorded simultaneously.

Another approach to the position-sensitive detector is through the Hadamard transform, which has been used successfully for infrared spectroscopy[71] and is now proposed for electron spectroscopy.[72] In this method a single counter is used with a large counting area. A mask is employed which allows, on the average, 50% transmission. The mask has a designed pattern of transmission and opacity and the pattern is changed in fixed intervals, the total number of counts being recorded at each setting, until the pattern reaches a full cycle. In a sense, then, we have a series of simultaneous equations representing the spectrum displayed along the focal plane. Solving these equations on a digital computer will yield the desired spectrum. However, if the signal (including background inherent to the measurement, such as inelastically scattered electrons from the target) is much larger than the multiplier background, as is usually the case with electron spectrometry, there is no significant improvement in the statistical deviations of the data for equal counting times, as has been shown from random number calculations.[73] Thomas[74] has also pointed out the difficulty in using the Hadamard transform for electron spectroscopy.

3.3. Scanning the Spectrum

For a given setting of the analyzer, electrons of one kinetic energy are counted by the detector (unless, of course, we are dealing with a position-sensitive detector). To scan over the spectrum, the electrostatic or magnetic field is altered, and the counting rate is determined as a function of the field, which in turn is proportional to the kinetic energy (or momentum in the case of a magnetic field) of the electron. One of the more inexpensive ways to carry out this scan is to couple the output of the multiplier to a strip chart recorder. The principal drawback of this method is that the data cannot be stored for processing, and it is generally harder to analyze the resultant spectrum for real and spurious peaks. A better but more expensive method is to scan the spectrum with the aid of a computer, which advances the voltage in discrete steps and stores the data, the total number of counts being recorded at each incremental step in a given channel. The data once stored can be processed by subtracting background, by assigning errors according to the $\sqrt{n}$, where n is the number of counts, and by deconvolution of overlapping peaks.

Another alternative method is to sweep over the spectrum with a saw-tooth voltage in coincidence with the channel advance of a multichannel analyzer. An entire spectrum is swept every 100 msec. Stability can be controlled to one part in 10,000. Such rapid scanning ensures that essentially no drift occurs during the scan. By setting the sweep on top of a bias voltage and adjusting the size of the sweep, any energy region can be scanned. Still, a computer is more versatile in being able to choose different step sizes and to refer back to standard peaks.

Finally, if a position-sensitive detector is employed, the enormous data storage problem is best facilitated by a small computer. Krause and Tarrant,[75] however, have designed a system to which the sawtooth sweep mode described above has been interfaced to a position-sensitive detector by the use of delay-time circuitry.

3.4. Data Analysis

Most electron spectrometers use on-line computers for operating their data taking and collection. It is also desirable to correct the data for background, distortion of collection efficiency as a function of electron energy, and contributions from satellite x rays if monochromatization is not employed. Finally, it is most valuable to deconvolute the spectral data. The principal problems that beset the deconvolution are knowledge of the line shapes and background, which are often quite difficult to ascertain with precision. Nevertheless, deconvolution is a necessary element of data analysis, particularly in x-ray photoelectron spectroscopy. The Dupont curve resolver, which uses analog circuitry, has been used effectively. Its value is the ease in evaluating a variety of parameters; its weakness lies in the lack of an objective criterion for the goodness of the analysis. Mathematical methods have been discussed[76] for the deconvolution and smoothing of data with particular concern for data encountered in x-ray photoelectron spectroscopy.

4. NEW DEVELOPMENTS

One cannot really predict the future, but rather one extrapolates the present and says more of the same, but better (or worse). The author has no greater facility for prognostication than anyone else, so a look into the crystal ball for new technology will be no more than an attempt at reasonable extrapolation.

Eventually, the ultimate resolution (1 or 2 mV) of the He resonance lines will be achieved, so that rotational structure as well as vibrational and electronic structure will be an important topic for study. With higher resolution, vibrational analysis can be made of the more complex molecules, helping to clarify the nature of their many orbitals. Besides the He I and II line sources, other sources with small natural widths, but still higher energies, will be developed, so that all molecular orbitals can eventually be studied with high resolution.

The variety of available soft x-ray sources will also increase, some with natural widths of only a few tenths of an electron volt. Monochromators will be coupled to higher energy x-ray energy sources (3–10 keV) so that studies can be made at lower surface depths without undue loss in resolution. Synchrotron radiation should come into its own, and studies will be possible over a large, continuous range of photon energy (10–10,000 eV) with good resolution. Studies of electrons ejected at a few volts above threshold may provide photoelectron spectroscopy with substantial enhancement in sensitivity.

The removal of line broadening in nonconducting materials will be achieved in many cases by better sample preparation and through the study of thin films. Uses of monochromatized x-ray sources in the gas phase will enable one to measure the photoelectron spectra of core electrons with a resolution of 0.2 eV, allowing the full usefulness of measuring subtle chemical shifts to be realized at last.

Work with clean surfaces at 10^{-10} Torr pressure without sacrifice of high resolution will come to be more common, and methods for working with microgram samples or smaller will be devised. Photoelectron spectra will routinely be accumulated as a function of photon energy and angular distribution. Detailed studies will be made on molecular beams of oriented molecules using polarized radiation. Methods will be worked out for determining the location beneath the surface of elements in an inhomogeneous solid.

Most of these developments require greater intensities. This can be accomplished through more efficient spectrometer design, position-sensitive detectors, and more intense excitation sources. It is not unrealistic to expect an increase of 10^3 over what is now accepted as a good performance.

At the same time, more convenient and less expensive electron spectrometers will be made for practical applications in the field for environmental monitoring and industrial uses.

The above comments were made with the optimism valve full open, and it is doubtful that they will all come to pass, but they are not at all unrealistic. Much has already occurred since I first wrote these paragraphs. It is also not unrealistic to hope that some completely unforeseen breakthroughs in electron spectroscopy will also occur.

5. REVIEW OF COMMERCIAL INSTRUMENTS

In this section the various commercial instruments available for electron spectroscopy will be reviewed. Such information becomes very quickly outdated, but it seemed valuable to at least view the situation at some point in time. (My last revision of this section took place in January 1975.)

It is rarely possible to establish truly objective criteria. To give the resolution is meaningless without some statement as to intensity. Intensity, in turn, is a function of the sample, sample holder, and the intensity of the excitation source as well as the transmission of the spectrometer. An instrument using a position-sensitive detector may yield a spectrum more quickly than a single-channel device at ten times the counting rate. In addition to considerations of intensity and resolution, one must consider adaptability. Some spectrometers are designed to work primarily with soft x rays, some for work with gas discharge lamps. Some spectrometers are better suited for gases, some for solids. The ease in loading samples varies greatly, as well as the suitability for surface cleanliness. Then, of course, there are the nebulous questions of reliability and trouble-free operation.

There is obviously no point system that will work; rather a general description is given with critical comments for each machine. In most cases, I have been able to see at least some version of the instrument, and to talk with researchers actually using the machine. I have given the manufacturers a chance to correct or rebut my criticisms. However, it has been impractical to make comparative tests or acquire all the information necessary for a truly fair comparison. Prices shown are only very approximate and given to yield a proper perspective. Also, the operational characteristics are hearsay and are again offered to give a qualitative picture.

5.1. AEI*

One of the most widely sold of the x-ray electron spectrometers is built by AEI. It combines good counting rates, good resolution, and relatively easy target mounting. The spectrometer is hemispherical in design, and electrons are delivered to the analyzer by a well-designed optical system. AEI offers a large number of optional features, including an ion gun for heavy samples and a cryostatic cold finger. Its cost runs about $80,000–120,000. The manufacturer hopes to employ a position-sensitive detector and demountable monochromator, but these hopes have yet to materialize commercially. The latest version of the spectrometer has a source volume pressure of about 10^{-9}–10^{-10} Torr. A uv lamp is offered as an optional accessory for solids, and gases can be studied with x-ray photoelectron spectroscopy. High-resolution uv photoelectron spectroscopy has not been demonstrated for gases but it has been used effectively for solids.

5.2. Du Pont†

Du Pont offers the highest counting rates for the price in x-ray photoelectron spectroscopy. Through use of their nondispersion double-pass

* AEI, 500 Executive Boulevard, Elmsford, New York 10523.
† Du Pont Company, Instrument Products Division, Wilmington, Delaware.

design (cf. Section 2.2.2.3) an instrument of very high sensitivity and modest resolution has been built. The company claims a counting rate of 300 kc/sec on Au $4f$ at a resolution of 1.2 eV (FWHM). Most remarkable is the price, which is only about $40,000.

5.3. Hewlett-Packard *

Perhaps the most ambitious design of the present electron spectrometers is that of Hewlett-Packard. It provides both a position-sensitive detector and a monochromator using dispersion compensation. Thus, a photon beam (extracted from the Al $K\alpha$ line) can be produced with a width of only 0.3 eV. This, combined with the absence of bremsstrahlung and satellite lines, makes the Hewlett-Packard machine very useful for studying the valence band structure of metals by x rays. Earlier, the instrument had a very serious drawback in its performance with nonconducting materials, due to the lack of secondary electrons to stabilize charging. With the addition of an electron flood gun, this problem has been remedied. The sample mounting is restricted since the sample must conform to the Rowland circle. The sample chamber is well constructed, and the pressure kept at better than 10^{-8} Torr. The cost of the spectrometer runs from $100,000 (without computer) up to $150,000 with full accessories. The servicing is reported to be very good.

5.4. McPherson†

The spectrometer of McPherson is based on the early 35-cm double-focusing spherical sector instrument of Siegbahn. Unlike the other instruments used for x-ray photoelectron spectroscopy, McPherson's spectrometer does not use a retarding potential to obtain the needed high resolution, normally operates at 0.05% FWHM in energy, and can easily operate down to 0.02%. This design is the most versatile of all the commercial instruments in terms of sample mounting. Any reasonable size or shape of object can be studied. Up to eight powder samples can be mounted and automatically run in sequence. The counting rates are modest, 10–20 kc/sec on the most intense line of gold at a spectrometer resolution of 1 eV, but it has a large focal plane ideally suited for a position-sensitive detector. However, the company does not as yet provide such a detector, By use of an effectively designed cold trap maintained at pressure in the source, volume can be maintained in the 10^{-10} Torr region. The target area is, however, directly exposed through the slit to the spectrometer vacuum. In practice, then, the target probably sees an effective pressure of the order of 10^{-8}–10^{-9} Torr. The McPherson instrument

* Scientific Instruments, 1601 California Avenue, Palo Alto, California 94304.

† GCA/McPherson Instrument Corp., 530 Main St., Acton, Massachusetts 01720.

has also demonstrated its use in the study of gases both with the uv lamp and with an x-ray source. The spectra that I have seen are quite satisfactory. The price with computer is from $100,000 to $125,000.

5.5. Perkin-Elmer*

Perkin-Elmer has concentrated its efforts on the uv photoelectron spectrometer. It is a cylindrical sector analyzer with 127° focusing, a design first built by Turner. For scientists interested solely or primarily in UPS, this is the recommended instrument. Electron spectroscopy using a He resonance lamp can, in principle, be carried out on the same analyzer that uses x rays (this is indeed what I have done in my research), but an instrument designed for XPS is so much more expensive that it does not seem a wise choice if one wishes to do only UPS. In addition, Perkin-Elmer has designed uv lamps for both the He I and He II radiations, and the spectrometer is optimized to work with gas samples. In addition, the recent apparatus has been designed to handle relatively nonvolatile materials. The instrument sells for about $40,000. Such a device normally operates with a calibrated flowchart-type recorder. A desired improvement might be to digitalize the output data by means of a computer, but the added expense would be considerable.

5.6. Physical Electronics†

Physical Electronics builds an electron spectrometer especially for surface Auger spectroscopy, and boasts the services of a number of pioneers in the field, e.g., Weber and Palmberg. The spectrometer is a cylindrical analyzer. Although the instrument is a dispersion analyzer, the spectra are differentiated electronically. There is a high degree of sensitivity, and the principal peaks in the spectrum can be generated in $\frac{1}{10}$ sec. The resolution is only 0.5%, but the vacuum is excellent, 10^{-10} Torr, and the sample can be cleaned in place with an ion gun. A higher resolution version using a double-pass cylindrical mirror analyzer gives 0.1%. The sample may be oriented perpendicularly to the electron beam or at a grazing angle. The sample can be maneuvered *in situ*; and with help of an external microscopy, a small and well-defined portion of the sample can be studied by Auger spectroscopy. A recent version combines scanning electron microscopy and Auger spectroscopy with a resolution of better than 5 μm.

In general, it is a meticulously built machine, and sells from about $50,000 to $80,000. The company also markets an x-ray photoelectron

* Perkin-Elmer, Beaconsfield, Bucks, England.
† Physical Electronics Industries, Inc., 6509 Flying Cloud Drive, Eden Prairie, Minnesota.

spectrometer. Its performance is satisfactory, though not outstanding. Its chief advantages are that it can be incorporated with the present Auger spectrometer and it has excellent vacuum capabilities.

5.7. McCrone-RCI *

A 20-cm, double-focusing, spherical-sector spectrometer, similar to that built by the author, is being readied for marketing by MCR. It employs cryogenic pumping and a multielement anode for easy change of x-ray source. Most important, however, it will include a position-sensitive detector which increases its data-taking capacity nearly 100-fold, thus making it one of the most sensitive instruments on the market. With computer it will be priced in the range of $100,000–$150,000.

5.8. Vacuum Generators, Inc.†

VG is continually upgrading its spectrometer. The most recent model (ESCA-3) is one of the most versatile instruments. The vacuum in the source volume is better than 2×10^{-10} Torr, and VG has thus achieved truly high vacuum conditions with x-ray photoelectron spectroscopy. The counting rate is claimed to be 160,000 counts/sec on gold at 1.3 eV resolution, operating the x-ray source at 480 W. A uv lamp, an x-ray source with twin anodes, and an electron gun can all be fitted into the source volume at the same time. Gases as well as solids have been studied, and the uv lamp has successfully produced satisfactory He II and Ne II radiation. The source volume is equipped with an argon ion gun, and with an evaporating source for surface studies. The price is about $100,000.

5.9. Varian‡

Varian was the first of the commercial firms to build an x-ray photoelectron spectrometer, and under the direction of J. C. Helmer, added much to the state of the art in electron spectroscopy. It utilized a full spherical analyzer. The spectrometer had good sensitivity with modest resolution, e.g., 60,000 counts/sec on the Au $4f_{7/2}$ with a spectrometer resolution of about 1.4 eV. The sample mounting was awkward, and the source volume was maintained at only modest vacuum, 10^{-7} Torr. Many of the "bugs" in the first model were removed; however, Varian has discontinued further development and sale of their photoelectron spectrometer.

* MCR Corp., 2820 S. Michigan Avenue, Chicago, Illinois.
† Vacuum Generators, Inc., Charlwood Road, East Grinstead, Sussex, England.
‡ Varian Analytical Instrument Devices, Palo Alto, California 94302.

Varian also pioneered in supplying an Auger spectrometer to fit their LEED instrument at a modest extra cost. It uses a spherical grid retarding design. As an extension to the effective use of low-energy electron diffraction, this is most commendable, but as an Auger spectrometer *per se*, one is advised to look elsewhere. More recently Varian has produced a high-resolution Auger spectrometer.

5.10. Others

Other firms have indicated an interest in making an electron spectrometer available commercially, e.g., Hitochi and JASC in Japan and Leybold-Heraeus in Germany. In general, the market is very competitive and improvements and changes are being made continually. The specific comments made above are best used as a guide for the types of problems the consumer should be concerned with, rather than any fixed evaluation.

Chapter 3
Fundamental Concepts

In this chapter we shall discuss portions of the theoretical basis upon which much of the activity of electron spectroscopy is founded. Specific theoretical problems will be taken up in their proper places in Chapters 4–6. Here we shall confine our discussion to subjects that span all or most of our attention in this book. First, the basic nature of the photoelectric effect will be presented. (A discussion of the Auger process is contained in Chapter 6.) Second, we shall turn our attention to the nature of binding energy. Third, a discussion will be made of distribution of final states from the viewpoint of the "sudden approximation." Fourth, we shall include in this chapter a brief section on self-consistent-field atomic wave functions and their use in electron spectroscopy. Finally, a qualitative description will be given of molecular orbital theory as it is applied in electron spectroscopy as well as definitions of some of the more common symbols used in characterizing atomic and molecular orbitals.

1. PHOTOELECTRIC EFFECT

The photoelectric effect is a dipole interaction in which all the energy of the photon is expended in the ejection of a bound electron. The beauty of studying photoelectron ejection is that a known quantum of excitation is put into the system, and from conservation laws we need only to measure the energy of the ejected photoelectron to obtain an exact energy definition of the final state. Photons can also undergo inelastic scattering processes (Compton effect), but for $h\nu < {\sim}5$ keV the cross section for this process is negligible compared to the photoelectric cross section, and for the purpose of this book the Compton effect can be neglected.

Since one of the principal aims in ESCA is the determination of the binding energy for each of the molecular and atomic orbitals, we are concerned with the partial cross sections; that is, the cross section for each subshell, rather than the total cross section. Hall[1] has published a review of atomic photoelectron cross sections that were calculated through the use of screened hydrogenic wave functions and the Born approximation. A comprehensive tabulation of cross sections using this method was made by Bearden.[2] In recent years more sophisticated calculations have become available. Manson and Cooper[3] have, for example, adopted an approach using Hartree–Fock–Slater wave functions. Kennedy and Manson[4] have extended this treatment to include exchange in the continuum wave function.

For a single-electron model using polarized incident light the intensity of ejection will be

$$\frac{d\sigma(\varepsilon)}{d\Omega} = \frac{\sigma_T(\varepsilon)}{4\pi} [1 + \beta P_2(\cos \Phi)] \tag{3.1}$$

where $\sigma_T(\varepsilon)$ is the total cross section for ionization of an electron with photoelectron energy ε and Φ is the angle between the direction of polarization and direction of the outgoing photoelectron. If the radiation is unpolarized, as is normally the situation for the photon sources employed in ESCA, then the expression becomes

$$\frac{d\sigma(\varepsilon)}{d\Omega} = \frac{\sigma_T(\varepsilon)}{4\pi} [1 - \tfrac{1}{2} \beta P_2(\cos \theta)] \tag{3.2}$$

where θ now is the angle between the incident photon beam and the direction of the ejected photoelectron. If one considers a system of randomly oriented molecules as would occur in the gas phase, Tulley *et al.*[5] and Cooper and Zare[6] have shown that equation (3.2) is also applicable to molecular systems. The parameter β depends on the nature of the given molecular orbital. For a spherically symmetric distribution of charge, such as we have for an atomic s orbital, the value of β is +2, the photoelectron coming out with a preferred direction at right angles to the photon beam. For photons at low energy, which is usually the case in ESCA, we can normally neglect the effect that the initial momentum of the photon has on the angular distribution. For orbitals of higher angular momentum the outgoing photoelectron can be described by a continuum wave function with angular momentum $l + 1$ or $l - 1$. The interference of these waves (or if you wish, the availability of more than one outgoing channel) leads to a lower value of β, to zero (isotropic distribution), or even to a negative value, which indicates a preferred forward or backward direction of the photoejected electrons. A plot of β as a function of energy and subshell is given in Figure 3.1 for Kr. For molecules the problem is more complex in that l is not a good quantum number for molecules. The β value, however, provides an important clue as to the nature of the molecular orbital, and its application to this problem will be discussed more thoroughly in Chapter 4.

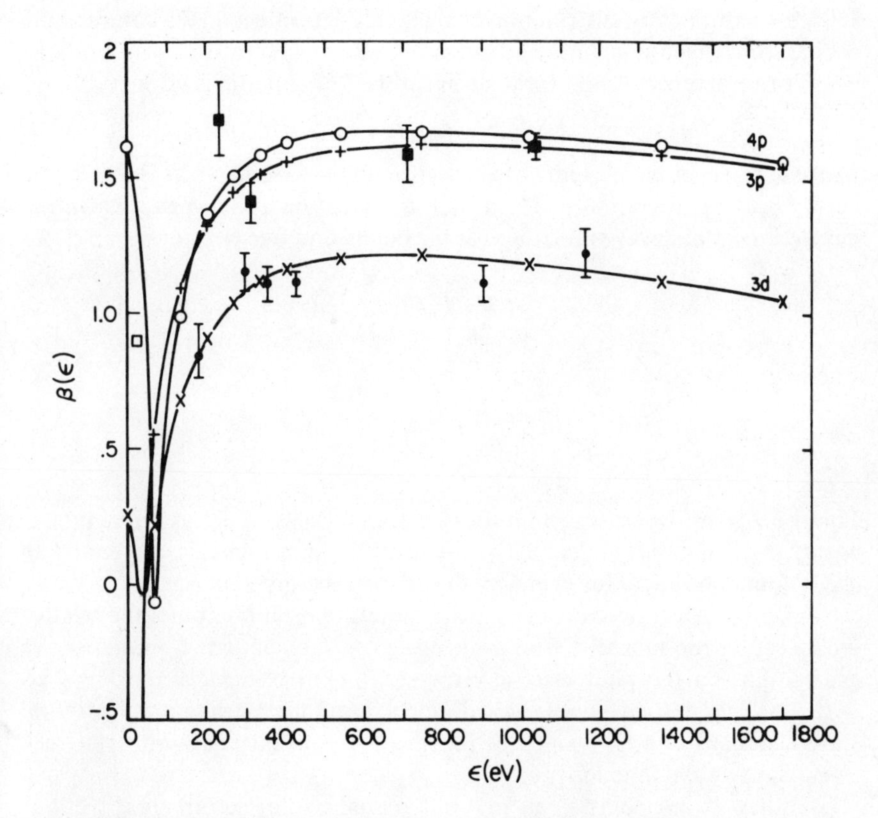

FIGURE 3.1. Plot of β for Kr as a function of photoelectron energy and subshell. [Reproduced from J. W. Cooper and S. T. Manson, *Phys. Rev.* **177**, 157 (1969), Figure 4.] β for *s* shells should be $+2$ and independent of energy.

Incidentally, the relative intensities for photoelectrons ejected from orbitals with different β values when measured at the angle $\theta = 54.7°$ are equal to the relative intensities integrated over all the angles, so long as the momentum imparted to the recoiling ion is negligible.

If the molecule can be oriented in space (as, for example, in a polarized beam or when molecules are adsorbed on a well-defined surface) and polarized radiation is used, then the intensity of ejected photoelectrons as a function of angle of ejection may be strongly dependent on the orientation of the molecule relative to the direction of the incoming beam of photons. For example, Purvis[7] has shown that when the angle θ between the photon beam and the direction of the ejected photoelectrons is kept constant (e.g., 90°), and the axis of the molecule (e.g., N_2) is rotated in the plane made by the incoming photon beam and outgoing photoelectron beam, then the intensity of the observed photoelectron will go through several maxima and minima between 0 and

180°, the nature of the distribution being highly dependent on the nature of the orbital from which the photoelectron is ejected.

The total cross section for a given subshell of an atom is given by

$$\sigma_{nl}(\varepsilon) = \tfrac{4}{3}\pi\alpha a_0^2(2l+1)(\varepsilon - \varepsilon_{nl})[lR_{\varepsilon,l-1}^2 + (l+1)R_{\varepsilon,l+1}^2] \tag{3.3}$$

where ε_{nl} is the binding energy of an electron in the nl shell, and ε is the kinetic energy of the photoelectron. The $R_{\varepsilon,l\pm1}$ are radial matrix elements which depend on the radial wave functions for the bound and free electrons:

$$R_{\varepsilon,l\pm1} = \int_0^\infty P_{nl}(r)rP_{\varepsilon,l\pm1}(r)\,dr \tag{3.4}$$

where

$$\int_0^\infty P_{nl}^2(r)\,dr = 1$$

Figure 3.2 gives the cross section for various subshells of Kr as a function of energy. Note the shift in the relative intensity. At higher energies the 3*s* and 3*p* shells dominate, but at lower energy the 3*d* shell is more important. In general, the lower the photoelectron energy, the more important becomes the relative cross section for subshells with a high angular momentum. Comprehensive calculations on the photoelectric cross section for elements from $Z=2$ to 54 at low energies have been carried out by McGuire[(8)] using nonrelativistic wave functions and by Scofield[(9)] for $Z=2$–103 using relativistic Hartree–Fock–Slater wave functions.

Fadley[(10)] has pointed out that differential photoelectron cross sections calculated by using wave functions that are unrelaxed in the final state really calculate the total cross section for both one-electron and multielectron processes. The calculated cross sections are thus too high by the shakeoff plus shakeup probabilities, which amount to about 20%. If one wishes to compare photoelectron cross section with the observed main photoelectron peak, a wave function corresponding to the relaxed single vacancy ion must be used for the final state.

2. BINDING ENERGY

The principal use of photoelectron spectroscopy is to determine the binding energies of the atomic and molecular orbitals by means of the following relationship:

$$E_B = h\nu - E_e - [-w + r_e] \tag{3.5}$$

where E_B is the binding energy, $h\nu$ the photon energy, and E_e the photoelectron

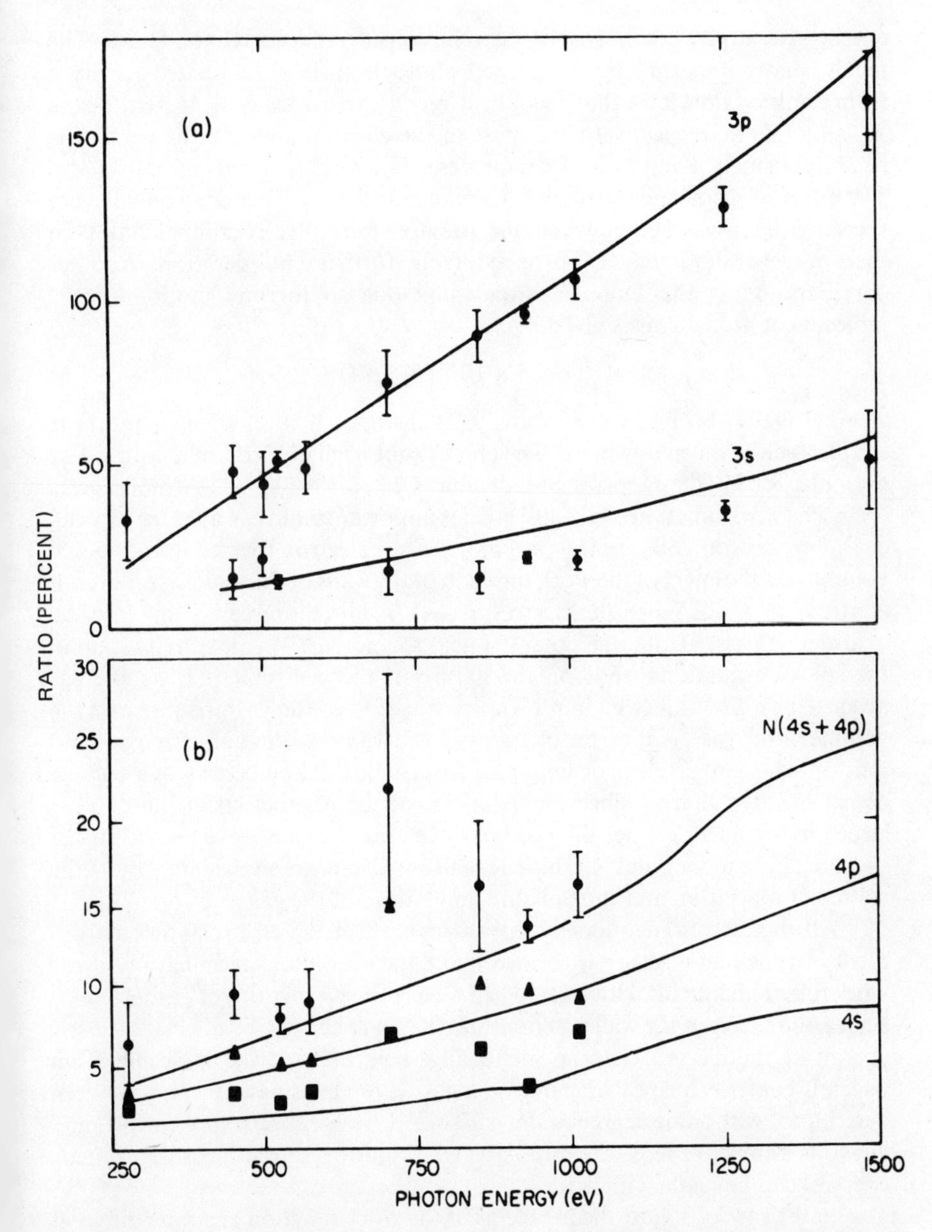

FIGURE 3.2. Plot of the relative cross section for various subshells of Kr as a function of photoelectron energy. [Reproduced from J. W. Cooper and S. T. Manson, *Phys. Rev.* **177**, 157 (1969), Figure 3.] Ratios are given relative to the 3*d* cross section (at 90°). (a) 3*p* and 3*s*. (b) 4*p* and 4*s* and their sums. Theoretical ratios are plotted as solid lines, experimental data are given as points with uncertainties shown as error bars.

energy. The expression in brackets can be considered a correction factor. The recoil energy imparted by the ejected photoelectron is designated r_e. Since from conservation laws the fraction of recoil carried away is M_e/M_i, where M_e and M_i are, respectively, the mass of the electron and ion, the value r_e is negligibly small. Even in an extreme case, H_2, M_e/M_i is only about 0.03%. The effect of recoil energy enters into photoelectron spectroscopy only in very special situations, such as when one attempts to resolve rotational states, so that in general r_e may be dropped from further consideration. Another consideration is the Doppler broadening due to thermal motion of the molecule or atom. This is given by

$$\Delta E_e = 7.5 \times 10^{-4} (E_e T/M)^{1/2} \tag{3.6}$$

where T is the absolute temperature, M is the mass in atomic units, and E_e is the photoelectron energy in eV. The effect is not negligible for light atoms. For example, for H_2 the Doppler broadening is 22 meV when 584-Å radiation is used. The work function w is of still greater importance and can alter the binding energy by several volts. In the case of a gas, an electron formed in the source volume sees the effect of the work function of the walls of the vessel as an overall electrostatic shell. When the electron passes from the source volume into the analyzer, it passes from one potential field to another. The potential set up in the source volume depends on the nature of the wall material, on the gas absorbed on the wall (which in turn is a function of the gas pressure and the volume of the gas), and on the charging of the wall by secondary electrons and ions. In addition, an electron emerging from an ionization beam sees a volume of net positive charge, since the velocities of the photoelectrons and Auger electrons are much greater than those of the ions and thus leave the vicinity of the ionization more quickly. This effect also will depend on the intensity of the source of ionization and the pressure and nature of the gas.

All the factors mentioned above which affect the electrostatic environment in which an electron is formed make the absolute determination of the work function difficult. However, when we examine a mixture of gaseous molecules, the effects of the wall potential and the space charge will affect the kinetic energy of the ejected electron identically, regardless of the molecule from which it comes. Thus, by introducing a gas whose binding energies have been determined with some degree of dependability, whose spectrum is simple, and which is unreactive (e.g., the rare gases), calibration of the measured kinetic energies can be made with certainty.

In the case of solids, the problems of the work function are quite different than those encountered in the study of gases. In the case of conductors the problem is fairly straightforward, since if the analyzer and specimen are in good electrical contact, the Fermi levels for the two will be the same, and binding energies for different materials will be referred to this Fermi level. An electron being ejected from the metal will have to overcome the work function of that metal, but when it enters the spectrometer chamber it adjusts to the work function of the spectrometer. If the binding energy E_B is referred to the Fermi level, then the correction w in equation (3.5) will be the work function of the material out of which the spectrometer is constructed. This value, w_{sp}, will be a spectrometer constant, and barring any changes on the surface of the spectrometer chamber, will be the same regardless of the sample being examined. (Cf. Figure 3.3.)

For nonconducting materials the problem is more complex. First, the Fermi level in a nonconducting material is not as clearly defined. Second, the electrical connection between the sample and the metal holder is not completely satisfactory. Charge carriers created during x-ray irradiation such as used in photoelectron spectroscopy of the inner shell (PESIS) help alleviate this problem, but the factors affecting the efficiency of this process are not clearly understood. The same material mounted at different times can give widely varying results. It is usually necessary to calibrate the observed electron spectra by having a known material mixed intimately with the substance to be studied so that the electrons removed from the known substance will see the same macropotential as that from the unknown. Even so, the Fermi level for a nonconducting material is not well defined theoretically.

After applying corrections due to the work function, one may now turn to the meaning of the binding energy. The binding energy for a given molecular or atomic orbital is the difference in total energy between the initial and final states of the system in which one electron has been removed from the given orbital. In solids we generally refer the initial and final states to their Fermi levels; in gases we refer to the vacuum potential. For free molecules, we need to consider the vibrational and rotational states as well as the electronic states. The separation of rotational states in electron spectroscopy (at the present state of the art) has been achieved for only a few special cases and may generally be disregarded. The contributions of the vibrational levels are not inconsequential and are, in fact, a very important part of the analysis in photoelectron spectroscopy of the outer shell (PESOS). Usually, the molecule under investigation is in the ground state for both the electronic and vibrational levels. The adiabatic ionization potential or orbital binding energy for a free molecule is defined as the difference in total energy between the ground state of the neutral molecule and the ground state of the final ion in which an electron has been removed from a given orbital. When there is a band of vibrational peaks corresponding to the various vibrational levels in the molecular ion, the lowest binding energy peak, or onset of the band if the vibrational structure is not resolved, corresponds to the adiabatic binding energy. The peak of the electronic band is often called the vertical ionization potential. When photoionization occurs in the inner shells, a nonbonding electron is removed so that the internuclear

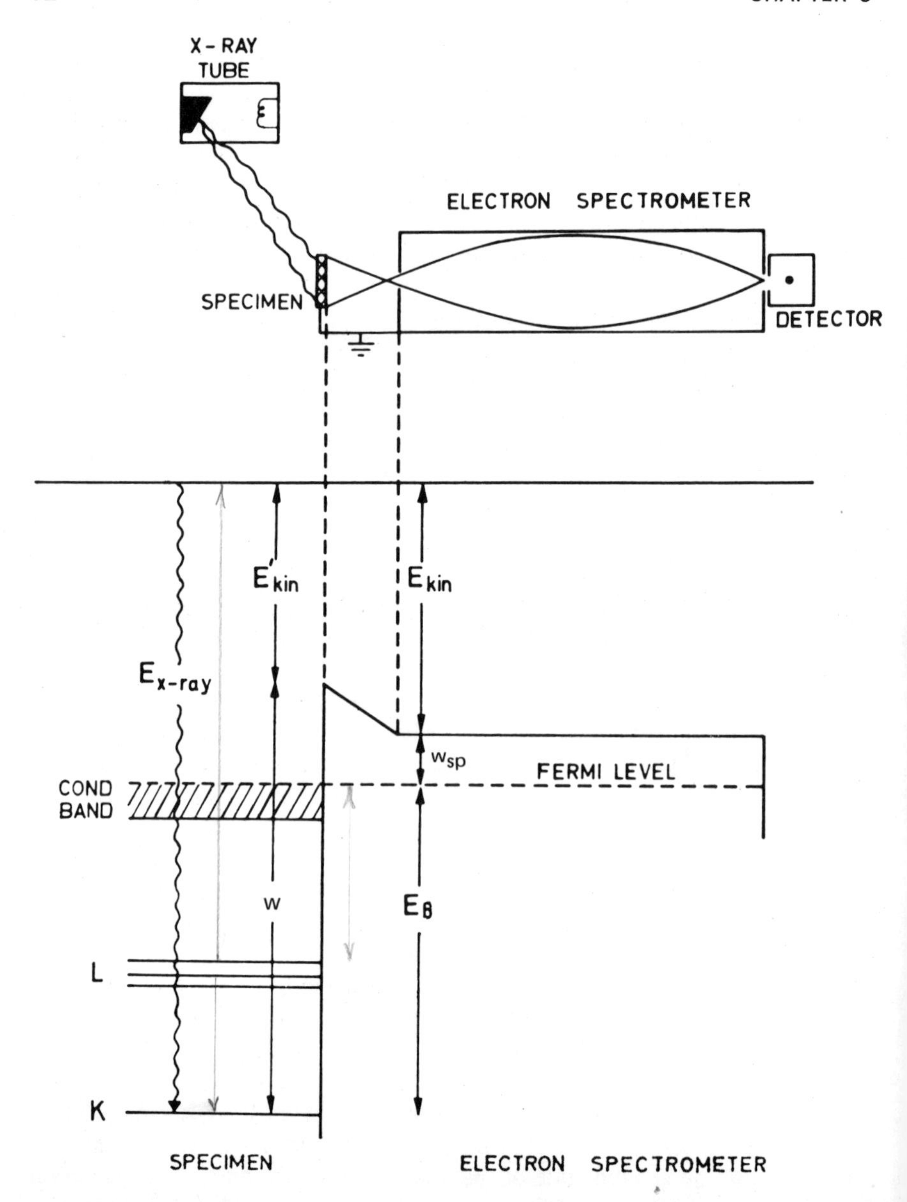

FIGURE 3.3. Schematic of binding energy as measured in a conducting solid in electrical contact with the spectrometer. $E_{\mathrm{x\,ray}}$ is energy of the x ray filling the *K* vacancy. E'_{kin} is kinetic energy of the photoelectron ejected from the sample. E_{kin} is the kinetic energy of the electron measured in the spectrometer. *w* is the work function of the specimen and w_{sp} is the work function of spectrometer. E_B is the binding energy of the specimen referenced to the Fermi level. [Reproduced from of K. Siegbahn *et al.*, "ESCA." *Nova Acta Regiae Soc. Sci. Upsaliensis, Ser. IV*, **20**, 1967, Figure II.5.]

distance is not disturbed, and nearly all of the transitions go to the ground state vibrational level.

The total energies for the initial and final states T_i and T_f can in principle be calculated using a Hartree–Fock method. Though this may be accomplished without too much difficulty for atomic systems, the problems involved for a molecular system are too large for any routine computation. Instead, application is made of the eigenvalues for the solution of the neutral molecule. From Koopmans' theorem[11] the negative of the eigenvalue for a given orbital is equal to its binding energy or energy needed to extract the orbital electron to infinity so long as there is no readjustment of the other electrons in the atom or molecule. The other electrons remain in frozen orbitals. The difference between the eigenvalue and the adiabatic binding energy is known as the atomic relaxation energy and has been obtained using atomic Hartree–Fock wave functions (cf. Table 3.1), i.e.,

$$E_R = -\varepsilon_{nlj} - [T_f - T_i] \tag{3.7}$$

There is no physical state corresponding to Koopmans' frozen orbital, but as will be pointed out in the following section, Koopmans' binding energy does correspond to the energy averaged over the possible final electronic states, including the ground and various excited states, according to the sudden approximation. The eigenvalue in principle should thus be slightly larger than the adiabatic binding energy. This fact has led some to associate the vertical ionization potential with the eigenvalue, but there is no physical basis for this association except that both energies are slightly higher than the adiabatic binding energy. The vertical ionization potential is higher than the adiabatic ionization potential because of the distribution of excited vibrational states; the eigenvalues are higher than the adiabatic binding energies because of the distribution of the excited electronic states as viewed in the sudden approximation.

Just how important is the error due to Koopmans' theorem? Table 3.2 gives binding energies calculated for argon using Hartree–Fock wave functions. Columns three and five give the binding energies corresponding to the differences in total energies between neutral argon and singly charged argon with a vacancy in the appropriate orbital; columns two and four give the binding energies from the eigenvalues; and column six gives the experimental binding energies. The percentage difference between the two methods is small for the inner shells, but the absolute value is large compared to the chemical shifts which are measured in PESIS. Fortunately, the shifts in the binding energies of the inner shells due to changes in the valence shell (cf. Chapter 5, Table 5.3) are essentially the same for the two methods of calculating the binding energy. Also note that the agreement between experiment and the difference in total energy is not always substantially better than when using the eigenvalue. For

TABLE 3.1
Relaxation Energy for Atoms[a] (eV)

Element	Z	$1s$	$2s$	$2p$	$3s$	$3p$	$3d$	$4s$	$4p$	$4d$	$4f$	$5s$	$5p$	$5d$	$6s$
He	2	1.5													
Li	3	3.8	0.0												
Be	4	7.0	0.7												
B	5	10.6	1.6	0.7											
C	6	13.7	2.4	1.6											
N	7	16.6	3.0	2.4											
O	8	19.3	3.6	3.2											
F	9	22.1	4.1	3.9											
Ne	10	24.8	4.8	4.7											
Na	11	23.3	4.1	4.7	0.3										
Mg	12	24.6	5.2	6.0											
Al	13	26.1	6.1	7.1	1.0	0.2									
Si	14	27.1	7.0	8.0											
P	15	28.3	7.8	8.8											
S	16	29.5	8.5	9.6	1.4	0.9									
Cl	17	30.7	9.3	10.4											
Ar	18	31.8	9.9	11.1	1.8	1.4									
K	19	31.2	9.1	10.5											
Ca	20	32.0	9.6	11.1											
Sc	21	33.8	11.5	12.9											
Ti	22	35.4	13.0	14.4	3.9	3.4	2.0	0.3							
V	23	37.0	14.5	16.0											
Cr	24	38.6	15.9	17.4											
Mn	25	40.1	17.2	18.8	—	—	3.6	0.4							
Fe	26	41.6	18.5	20.2	5.7	5.3									
Co	27	43.2	19.8	21.6	—	—	4.1	0.0							
Ni	28	44.7	21.1	22.9	6.7	6.3									
Cu	29	48.2	23.7	25.7	7.7	7.2	5.3	0.3							
Kr	36	53.6	26.1	28.9	8.7	8.9	9.0	1.7	1.3						
I	53	66.9	34.8	38.2	14.7	15.2	16.3	5.6	5.3	4.6	—	1.3	0.7		
Eu	63	76.5	43.5	46.6	24.8	25.3	25.3	12.2	11.4	9.5	6.2	3.2	2.8	—	0.2
Hg	80	94.7	58.5	60.7	35.4	36.5	49.3	19.6	17.1	14.4	12.9	5.7	4.8	2.6	1.9
U	92	103.9	63.7												

[a] Results of elements He–Cu taken from U. Gelius and K. Siegbahn, *Faraday Disc. Chem. Soc.* **54**, 257 (1972); results of Kr–U derived from relativistic HFS calculations of A. Rosen and I. Lindgren, *Phys. Rev.* **176**, 114 (1968). Results of different j subshells are averaged. Calculations are based on equation (3.7).

the deeper shells the relativistic effects must be considered, and for the outer shells electron correlation is very important. Sometimes too much concern is given the error involved in estimating the binding energy by Koopmans' rule. In perspective, it is only one of a number of uncertainties in obtaining good agreement between theory and experiment, and not necessarily the most important. The concept of single-electron orbitals with specific binding energies is in itself an ideality. Basically, one has only the difference in total energies between different states. But the concept of a molecule built up of energy levels corresponding to the binding energies of its molecular orbital is a highly productive one which should not be abandoned because it contains some half-truths. For the many-body problem, which is at the heart of understanding atoms and molecules, what indeed is the complete truth? Lundqvist and

TABLE 3.2

Comparison of Binding Energies (eV) Obtained from Eigenvalues and Total Energies for Argon[a]

Shell	$-\varepsilon$(NR)	ΔSCF(NR)	$-\varepsilon$(R)	ΔSCF(R)	EXP
$1s$	3227.4	3195.2	3240	3209	3205.9
$2s$	335.3	324.8	336	327	326.3
$2p_{1/2}$	260.4	248.9	261	250	250.6
$2p_{3/2}$			259	248	248.5
$3s$	34.8	33.2	34.9	33.3	29.3
$3p_{1/2}$	16.1	14.8	16.1	14.8	15.94
$3p_{3/2}$			15.9	14.6	15.76

[a] ε(NR), eigenvalues from nonrelativistic Hartree–Fock wave functions [P. S. Bagus, *Phys. Rev.* **139**, A619 (1965)].
ΔSCF(NR), difference in total energies for atoms and single vacancy ion using nonrelativistic solution [P. S. Bagus, *op. cit*].
ε(R), eigenvalues from relativistic HFS (optimized) wave function [A. Rosén and I. Lindgren, *Phys. Rev.* **176**, 114 (1968); K. Siegbahn *et al.*, "ESCA", *Nova Acta Regias Soc. Sci. Upsaliensis, Ser. IV* **20** (1967)].
ΔSCF(R), difference in total energies for atoms and single vacancy ion using relativistic solution [A. Rosén and I. Lindgren, *op. cit.*].
EXP, experimental values [see K. Siegbahn *et al.*, *ESCA Applied to Free Molecules* (North-Holland, Amsterdam, 1969)].

Wendin[12] have reviewed the application of many-body theory to x-ray photoelectron spectroscopy.

The importance of estimating the extra atomic relaxation energy for the core shells of molecules has attracted increasing attention in the last few years.[13, 14] The problem will be brought up again in Chapters 5 and 6. Hedin and Johansson[15] found that the relaxation energy could be obtained to a good approximation as

$$E_R(i) = \tfrac{1}{2}\langle i | V^* - V | i \rangle \tag{3.8}$$

where E_R is the relaxation energy for orbital i, and V^* and V are the Fock potentials for the relaxed hole state and ground state. It will be shown in Chapter 5 how this relationship can be easily calculated with very approximate potentials. Basch,[13] using perturbation theory, has derived an expression similar to equation (3.8), which can also be used as a starting point for evaluating the relaxation energy.

3. FINAL STATES AND THE SUDDEN APPROXIMATION

As the consequence of photoionization (or an Auger process) the resultant ion will find itself in one of a number of final possible electronic states.

The energy of the photoelectron (or Auger electron) is governed by the difference in total energies of the initial and final states and will be equal to E_e given in equation (3.5) if we let $E_B = T_f - T_i$, where T_i and T_f are the total energies of the initial and final states. The width of the photoelectron line is related to the lifetime of the final state according to the Heisenberg uncertainty principle,

$$\Delta E = \hbar/\Delta t \tag{3.9}$$

where Δt is the mean lifetime of the final state. When photoionization occurs in the valence shell, Δt is generally quite large, $>10^{-8}$ sec, and the photoelectron peak is sharp, $\Delta E < 10^{-2}$ eV. Final states in which vacancies are formed in the inner shells or core electrons have much greater natural widths. Most of the inner shell vacancies with which we shall be concerned in this book will be filled by nonradiative transitions, and the half-lives for states containing these inner shell holes will be dependent essentially on Auger transition rates. (Radiative transitions become competitive with Auger transitions only when the transition energy is greater than about 10 keV.) When a vacancy cannot be filled by a Coster–Kronig transition (i.e., an Auger transition which involves an electron from the same principal shell as the vacancy, e.g., L_{I}–$L_{\mathrm{III}}M_{\mathrm{III}}$) the natural width of the state containing that vacancy is the order of a few tenths of an eV. Because of the greater overlap between subshells of the same principal quantum number, Coster–Kronig transition rates are an order of magnitude greater, and natural widths for shells containing vacancies where such transitions can occur are usually several electron volts wide. Usually, sharp photoelectron lines in which there is an absence of Coster–Kronig transitions occur only when vacancies are found in the highest filled subshell of a given principal quantum number or when Coster–Kronig transitions are not energetically possible.

Normally, photoelectron ejection will give rise to a single final electronic state corresponding to the removal of an electron from a given atomic or molecular orbital. For each atomic or molecular orbital the photoelectric process yields a single characteristic final electronic state. The four principal exceptions to this generalization that result in multiple final states are: (1) Coupling of the vacancy produced by photoionization with a partially filled orbital, (2) Jahn–Teller splitting, (3) electron shakeoff, and (4) multiple excitation due to configuration interaction. We shall now give a short discussion of each of these phenomena. In addition, we shall include in our discussion spin–orbit splitting.

3.1. Spin–Orbit Splitting

Spin–orbit splitting arises from a coupling of the spin and orbital angular momentum. The true nature of this process is best understood by Dirac's theory, in which the principles of relativity and wave mechanics are combined. The energy level splitting arises naturally out of this theory. Spin–orbit separation is thus already present in the initial state and does not arise necessarily as the result of photoionization. However, I have chosen to discuss this subject in the section on final states because it can be profitably considered as a perturbation of a degenerate single-energy nonrelativistic orbital, and partly because a thorough understanding of the final state is necessary to the understanding of spin–orbit splitting, particularly with other phenomena occurring, such as Jahn–Teller splitting and multiplet splitting.

In atoms, spin–orbit splitting has been well characterized by relativistic self-consistent wave functions, at least for the inner shells. For a closed shell of $l > 0$ two subshells arise with j quantum numbers $l + \frac{1}{2}$ and $l - \frac{1}{2}$. No spin–orbit splitting, of course, occurs with s subshells. The occupancy and thus the relative intensities of the photoelectron lines are $2l$ and $2l + 2$. The splitting increases with Z roughly as a function of Z^5. For a given Z the splitting is percentage-wise the same magnitude for each of the subshells with $l > 0$. Refer to Table 3.3 for comparison of calculated and experimental results. In the valence shell, calculation of the spin–orbit splitting is complicated by consideration of electron correlation, and additional coupling if the valence shell is only partly occupied.

Problems involved with spin–orbit splitting in molecular orbitals will be deferred to Chapter 4.

3.2. Multiplet Splitting

Unfilled shells containing unpaired electrons and a net spin $\neq 0$ are a common occurrence for free atoms, but when molecular bonds are formed,

TABLE 3.3

Comparison of Experimental[a] and Theoretical[b] Splitting (eV) in the $l + \frac{1}{2}$ and $l - \frac{1}{2}$ Subshells

	Ar		Kr		Xe	
Shell	$\Delta\varepsilon$(calc)	$\Delta\varepsilon$(exp)	$\Delta\varepsilon$(calc)	$\Delta\varepsilon$(exp)	$\Delta\varepsilon$(calc)	$\Delta\varepsilon$(exp)
$2p$	2.23	2.03	55.1	52.5	329.6	319.9
$3p$	0.17	0.18	8.2	7.8	63.2	61.5
$3d$	—	—	1.40	1.21	13.5	12.6
$4p$	—	—	0.72	0.65	12.8	—
$4d$	—	—	—	—	2.13	1.97
$5p$	—	—	—	—	1.41	1.30

[a] Cf. K. Siegbahn *et al.*, *ESCA Applied to Free Molecules* (North-Holland, Amsterdam, 1969).
[b] Cf. Lu *et al.*[41]

these spins are usually coupled. There do exist, however, a large number of cases where molecules are formed with unpaired electrons. For gaseous molecules we have, for example, NO and O_2.

In solids there are the transition metals with their unfilled *d* orbitals, and the rare earths and actinides with their unfilled *f* orbitals. When an additional vacancy is made by photoionization, there can be a coupling between the unpaired electrons left behind following photoelectron ejection and the unpaired electrons in the originally incompletely filled shell. For example, in Fe_2O_3 there are five unpaired electrons in the 3*d* shell. Following photoionization in the 3*s* shell there are two possible final states. This is shown schematically in Chapter 5, Figure 5.24. The energy difference between states arising from the coupling of singly unpaired *s* electrons and the 3*d* orbital is given by the Slater exchange integral,

$$G^2(3s, 3d) = C^2(0, 0, 2, 0) \int_0^\infty \int_0^\infty R_{3s}(r_1) R_{3d}(r_2) R_{3s}(r_2) R_{3d}(r_1) \frac{r_<^2}{r_>^3} \, dr_1 \, dr_3 \qquad (3.10)$$

where R_{3s} and R_{3d} are the radial wave functions for the respective 3*s* and 3*d* orbitals, and $C(0, 0, 2, 0)$ is a Clebsch–Gordan coefficient. The relative intensity of the multiplet lines is given by the multiplicity, namely,

$$I \propto (2L + 1)(2S + 1) \qquad (3.11)$$

When a vacancy is formed in an inner shell orbital whose angular momentum is greater than zero, the number of multiplet states greatly increases. Table 3.4 lists the number of possible multiplet states following photoionization in a

TABLE 3.4

Number of Possible Final States Arising from Multiplet Splitting in the Photoionization of Transition Metals

Final state	Without configuration interaction	With configuration interaction
$s^1 d^n$	2	2
$p^5 d^1$	6	6
$p^5 d^2$	6	12
$p^5 d^3$	6	19
$p^5 d^4$	6	18
$p^5 d^5$	2	4
$p^5 d^6$	6	18
$p^5 d^7$	6	19
$p^5 d^8$	6	12
$p^5 d^9$	6	6

p orbital for atoms containing different numbers of electrons in an unfilled d shell (the situation is similar for photoionization in the d and f core shells). The number of possible states listed is based on selection rules for photoelectron emission, i.e., $\Delta L = 0, \pm 1$ are for an L–S coupling scheme. If configuration mixing is allowed in the d shell, extra states are possible. When the two vacancies that couple have the same principal quantum number, then considerable correlation can arise. It was found[16] that when additional configuration states, which involve two-electron excitation states, were included, substantial improvement was obtained between experiment and theory. The subject of multiplet splitting, particularly with regard to molecular bonding, will be taken up in detail in Chapter 5, section 5.1.1.

3.3. Jahn–Teller Splitting

In molecules that possess a high degree of symmetry, e.g., methane, the ejection of a photoelectron will destroy that symmetry, and the state of the positive ion will undergo what is termed Jahn–Teller splitting. A variation of this phenomenon leading to the breakdown in degeneracy for a linear molecule due to photoionization is Renner–Teller splitting. These subjects will be taken up in more detail in Chapter 4.

3.4. Electron Shakeoff and Shakeup

Two other phenomena leading to a multiplicity of final electronic states are called electron shakeoff and electron shakeup or sometimes monopole excitation and monopole ionization. As the result of a sudden change in the central potential of an atom, an electron in a given orbital may go into an excited (electron shakeup) or continuum (electron shakeoff) state. This change in potential may occur as the result of a change in the nuclear charge, such as in beta decay; or it may result from a change in shielding, such as is felt by the outer shell electron when a vacancy is suddenly produced in one of the inner shells by photoelectron ejection. That is, the effective charge seen by the outer shell electron is

$$Z_{\text{eff}} = Z - \sigma \tag{3.12}$$

where Z is the nuclear charge and σ is the electron screening. The theory of electron shakeoff was first developed by Migdal[17] and Feinberg[18] for beta decay and was based on the sudden approximation. The same approach was later applied to electron shakeoff in Auger processes by Wolfsberg and Perlman.[19] It was first applied to photoionization by Carlson *et al.*[20,21] and to electron impact by Åberg.[22]

The sudden approximation is based on the assumption that the initial and final states are well defined, and that the change in the Hamiltonian is instantaneous. If these conditions are fulfilled, the probability for an electron in a given single-electron orbital (n, l, j) going from an initial to final state is

$$(P_{i\to f})_{n,l,j} = |\psi_f^* \psi_i \, d\tau|^2 \tag{3.13}$$

where ψ_i is the single-electron wave function of a given orbital of the ground state of the atom or molecule before the event that will give rise to a change in the central potential, and ψ_f is the single-electron wave function of one of a series of possible orbitals in which the electron may be found as the result of this sudden change. In the case of photoionization the wave function ψ_f represents states in which a vacancy has been formed in one of the orbitals. As can be seen from equation (3.13), the matrix element appears to be missing. Actually, it is unity. This is a monopole transition. Only the principal quantum number changes. The spin and angular momentum cannot change. (In photoionization the total change is $\Delta L = \pm 1$ because photoionization is a dipole interaction, but the extra monopole excitation due to the sudden loss of shielding electrons does not alter the angular momentum further.) In molecules, parity and symmetry are conserved.

In addition to monopole excitation one can conceive of a direct collision process as the ejected photoelectron (or β particle) leaves the atom. Though this process can occur, it is usually deemed negligible compared to monopole excitation.[17,23] Since the direct collision mechanism is dependent on the velocity of the ejected electron, while in the sudden approximation monopole excitation is independent of E_e, the independence of extra ionization on the photoelectron energy is taken as evidence for the relative unimportance of direct collision.[24]

Calculations of the probability and energy of the states encountered in monopole excitation and electron shakeoff have been carried out for atoms using Hartree–Fock wave functions. It is easiest to obtain good wave functions for the ground states of the initial atom and final ion. Thus, the total probability for electron shakeoff from a given orbital $P(n,l,j)$ is usually derived from the following expression:

$$P(nlj) = 1 - [|\psi_f^*(nlj)\psi_i(nlj)\, d\tau|^2]^N - P_F \tag{3.14}$$

where $\psi_f(nlj)$ and $\psi_i(nlj)$ are the single-electron wave functions for the orbital nlj in the initial atom and final ion in which a change in the inner shell electron configuration has taken place. That is, we have calculated the probability that electron shakeoff *will not* take place and subtracted from unity to find out the chance that it *will* take place. P_F is a correction for transitions to filled states, not allowed by Pauli's exclusion principle. Table 3.5 lists the electron shakeoff probabilities for producing vacancies in the various shells of Kr. Note that the

TABLE 3.5

Probability[a] for Electron Shakeoff (%) for the Various Subshells in Krypton n_s as the Result of a Sudden Vacancy in Orbital n_0

n_0 \ n_s	$1s$	$2s$	$2p_{1/2}$	$2p_{3/2}$	$3s$	$3p_{1/2}$	$3p_{3/2}$	$3d_{3/2}$	$3d_{5/2}$	$4s$	$4p_{1/2}$	$4p_{3/2}$	$\sum P$
$1s$	0.002	0.060	0.089	0.18	0.22	0.38	0.75	1.43	2.13	1.80	4.39	9.06	20.5
$2s$	0.000	0.002	0.007	0.014	0.89	0.15	0.31	1.36	2.04	1.50	3.78	7.97	17.2
$2p_{1/2}$	0.000	0.008	0.006	0.026	0.099	0.18	0.36	1.42	2.14	1.53	3.86	8.13	17.8
$2p_{3/2}$	0.000	0.008	0.011	0.018	0.097	0.17	0.36	1.41	2.12	1.52	3.84	8.10	17.7
$3s$	0.000	0.000	0.000	0.000	0.006	0.020	0.045	0.32	0.50	1.22	3.26	6.97	12.3
$3p_{1/2}$	0.000	0.000	0.000	0.000	0.012	0.010	0.048	0.33	0.51	1.21	3.26	6.96	12.3
$3p_{3/2}$	0.000	0.000	0.000	0.000	0.011	0.020	0.034	0.31	0.48	1.20	3.24	6.94	12.2
$3d_{3/2}$	0.000	0.000	0.000	0.000	0.012	0.023	0.050	0.27	0.55	1.19	3.22	6.88	12.2
$3d_{5/2}$	0.000	0.000	0.000	0.000	0.012	0.022	0.049	0.35	0.45	1.18	3.21	6.87	12.1
$4s$	0.000	0.000	0.000	0.000	0.000	0.000	0.000	0.003	0.004	0.14	1.33	3.02	4.5
$4p_{1/2}$	0.000	0.000	0.000	0.000	0.000	0.000	0.000	0.001	0.002	0.20	0.50	2.31	3.0
$4p_{3/2}$	0.000	0.000	0.000	0.000	0.000	0.000	0.000	0.001	0.001	0.18	0.96	1.66	2.8

[a] Calculations from Carlson and Nestor.[25] Calculations include shakeoff plus shakeup, and utilize single-electron relativistic Hartree–Fock–Slater wave functions. $P_{nlj} = 1 - [|\int \psi^*_{nlj}(A_0)\psi_{nlj}(A)\,dr|^2]^N - P_F$ where $\psi_{nlj}(A)$ represents the orbital nlj in the neutral atom and $\psi_{nlj}(A_0)$ represents the orbital nlj in the ion A_0 whereby a single vacancy has been created in a given subshell of atom A. N is the number of electrons in orbital nlj. The quantity P_F is a correction which arises from the condition that electron shakeup transitions to occupied levels are not physically allowed.

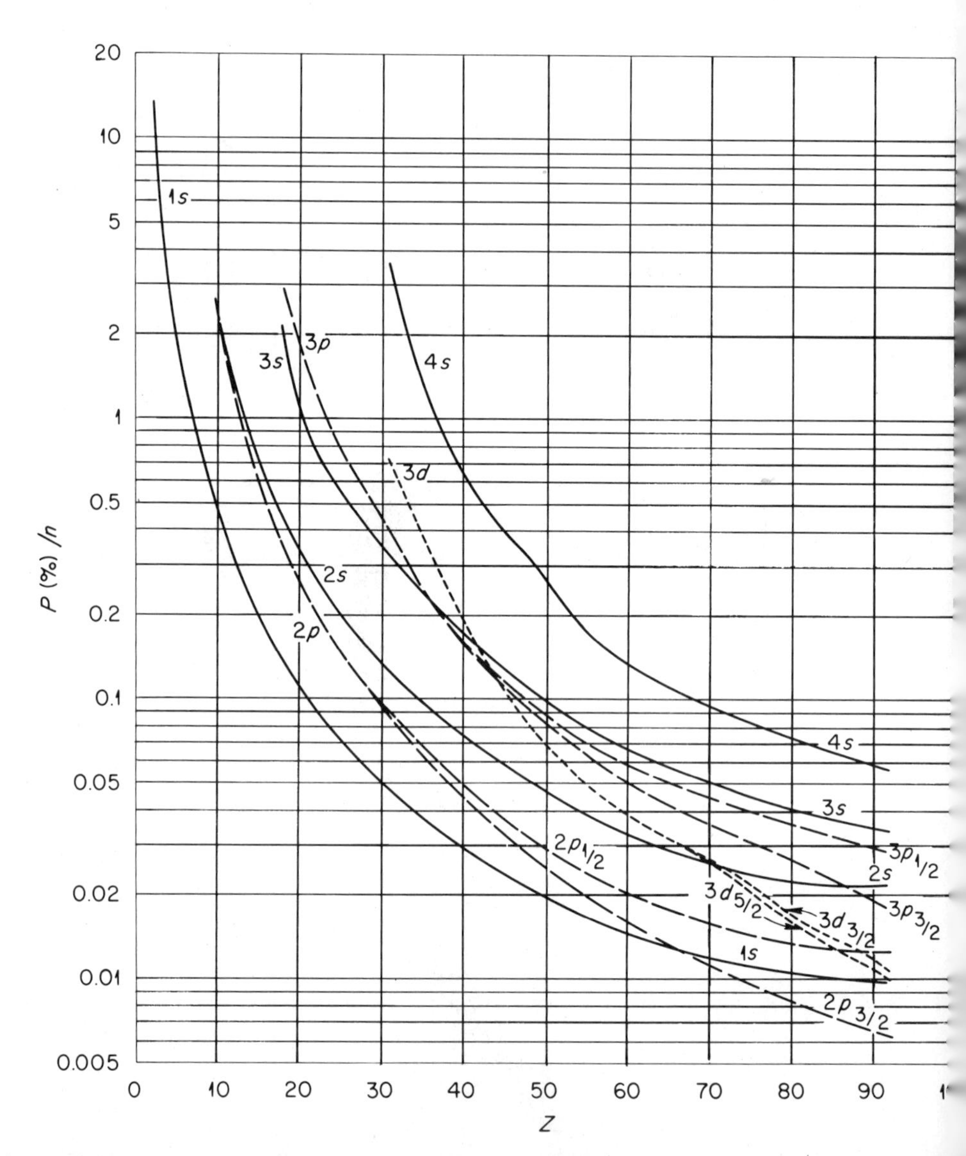

FIGURE 3.4. Calculated percentage probability for electron shakeoff per shell divided by numb of electrons in given shell, plotted as a function of Z for subshells 1s to 4s. [Reproduced fro Carlson *et al.*,[21] Figure 2.]

probability for electron shakeoff for shells below the initial vacancy is negligible (there is essentially no change in effective charge for these shells) and that the probability increases as one goes to the outer or valence shell. Note also that the probability for shakeoff in the valence shell is essentially independent of which core shell is vacated. (It rises slightly, the lower the principal quantum number.) This follows from the fact that the change in shielding is nearly the same. Comprehensive calculations of shakeoff probability for atoms will be found in the literature.[25]

Figure 3.4 shows calculated shakeoff probabilities following β decay for the different subshells as a function of Z. The probability for a given shell decreases rapidly as a function of Z. However, for a given atom the probability increases with increasing principal quantum number. As a consequence, regardless of which element one deals with, the orbital most likely to be excited is in the valence shell; and the total shakeoff probability shows no overall trend with Z. (Cf. Table 3.6.)

The question of "how sudden is sudden?" has been examined experimentally, and in photoionization[24] the probability for electron shakeoff reaches an asymptotic value when the energy available for secondary electron ejection is several times above that required for the additional ionization. (The energy available for electron shakeoff in photoionization is that amount above the binding energy of the photoejected electron that normally goes into the kinetic energy of the photoelectron. If electron shakeoff occurs, the kinetic energy of the photoelectron is correspondingly diminished.) Near the threshold for photoionization the probability for electron shakeoff approaches zero.

If inner shell ionization is created by charged particles rather than by x

TABLE 3.6

Total Electron Shakeoff Probability as a Function of Z[21]

Z	Element undergoing β^- decay	Total shakeoff, %
2	He	26.9
10	Ne	22.5
18	Ar	25.3
21	Sc	37.7
26	Fe	31.0
31	Ga	29.7
36	Kr	21.8
46	Pd	21.8
54	Xe	21.3
61	Pm	27.1
70	Yb	24.2
79	Au	19.7
83	Bi	22.6
92	U	26.7

rays, the extent of electron shakeoff and shakeup will be the same if velocities of the impact particles and the ejected electron are high enough to satisfy the sudden approximation. That is, if a vacancy is created fast enough, it makes no difference how it is formed. This statement has been proven valid both in the case of electron impact[26] and proton impact.[27]

When photoionization occurs in the valence shell, extra ionization and excitation take place[28] that can qualitatively be explained in terms of monopole transitions, but the probability is much higher than that calculated using single-electron wave functions, and wave functions explicitly including electron correlation must be used.[29]

3.5. Configuration Interaction

In a true many-body problem, individual, separable wave functions for single-electron orbitals do not really exist. To improve on a wave function made up of single-electron orbitals it is common practice to consider an admixture of excited, unoccupied orbitals as configuration mixing. When proper weighting is given to these added orbitals, a solution can often be obtained that gives a greater total negative energy, indicating a more realistic wave function. If an ion that results following photoionization can be described by a wave function that includes excited configurations, then it is possible that the ground state may interact with these excited states, and the final state may end up as one of these excited states. In a sense, then, electron shakeup is a special case of excitation by means of configuration interaction. In addition, other excited states may occur, such as two-electron excitation. Such a situation has already been discussed under multiplet splitting. Many theoreticians, such as Basch,[13] would prefer to view the whole problem of final states in terms of configuration mixing.* Specific examples of configuration interaction will be brought up in the case of photoelectron spectroscopy in Chapters 4 and 5.

3.6. Koopmans' Theorem and the Sudden Approximation

An interesting relationship exists between Koopmans' theorem and the sudden approximation. Manne and Åberg[30] have shown that the adiabatic ionization potential I_0 and the binding energy as calculated from Koopmans' theorem I_K, are related by the following expression:

$$I_K = I_0 + \sum_{i=1}^{\infty} |\langle\psi_i|\psi_0\rangle|^2 (I_i - I_0) \tag{3.15}$$

* Recently, Martin and Shirley (to be published) showed that in calculating electron shakeup probabilities, configuration interaction is important in the initial as well as in the final state.

where ψ_0 are single-electron Hartree–Fock wave functions for the initial ground state of the atom or molecule, and ψ_i are single-electron wave functions for the various final states in which an ion can find itself following electron removal from an orbital corresponding to the ionization potential I_0. The value $(I_i - I_0)$ corresponds to the energy difference between the ground state of the ion in which one vacancy has been produced in a given orbital and the various excited states of this ion. In brief, Koopmans' binding energy is related to the adiabatic binding energy by the energy-weighted average of the shakeoff probabilities. This has been tested by comparing Koopmans' binding energy for the K shell of neon (i.e., the eigenvalue for the $1s$ shell as obtained from a Hartree–Fock solution), which equals 892 eV, and the experimental binding energy (870 eV) plus the experimentally obtained average energy from the electron shakeoff spectrum (16 eV) which gives 886 eV, which is in reasonable agreement with I_K. Relaxation energies have been estimated[25] with the help of shakeoff calculations and shown to agree reasonably well with values obtained from Hartree–Fock calculations.

Application of the sudden approximation using single-electron wave functions to the problem of Koopmans' theorem is not strictly applicable to the valence shell, because of the importance of electron correlation. (The same problem was noted above with regard to electron shakeoff in the photoionization of the valence shell.)

3.7. Vibrational and Rotational Final States

When one is dealing with a molecule, the rotational and vibrational levels will be superimposed on the final electronic state. Except in rare cases, e.g., H_2, rotational bands are not resolved in electron spectroscopy. However, peaks due to vibrational structure are clearly seen when photoelectron studies are carried out with photon sources such as the He I (584 Å) resonance line, whose natural width is only a few millivolts. In most instances, the neutral molecule is in the ground vibrational state, so that the vibrational structure observed in photoelectron spectra arises from transitions to the various vibrational states of the resultant ion. The relative intensities of the various vibration peaks are primarily determined by the Franck–Condon factors. The Franck–Condon factors are based on the Born–Oppenheimer approximation that the motion of the orbital electron is fast compared to motion of the nuclei, and thus the Hamiltonian governing nuclear motion (the vibrational energy) can be separated from the Hamiltonian governing the electronic motion. (Itikawa[31] has recently examined the Franck–Condon factors without recourse to the Born–Oppenheimer approximation.) In calculating the photoelectron cross section, the electronic contribution ($G_{ee'}$) can be separated from the vibrational contribution, i.e.,

$$P_{vv'} = [G_{ee'}]^2 \, [\psi_{v'}\psi_v \, dr]^2 \tag{3.16}$$

where $\psi_{v'}$ and ψ_v are the wave functions for the final and initial vibrational states. The electronic contribution various slightly with the photoelectron energy, so that the observed intensity ratios of the vibrational bands in photoelectron spectra are slightly distorted from the Franck–Condon factors as calculated from $(\int \psi_{v'}\psi_v \, dr)^2$. The overlap integral essentially depends on the

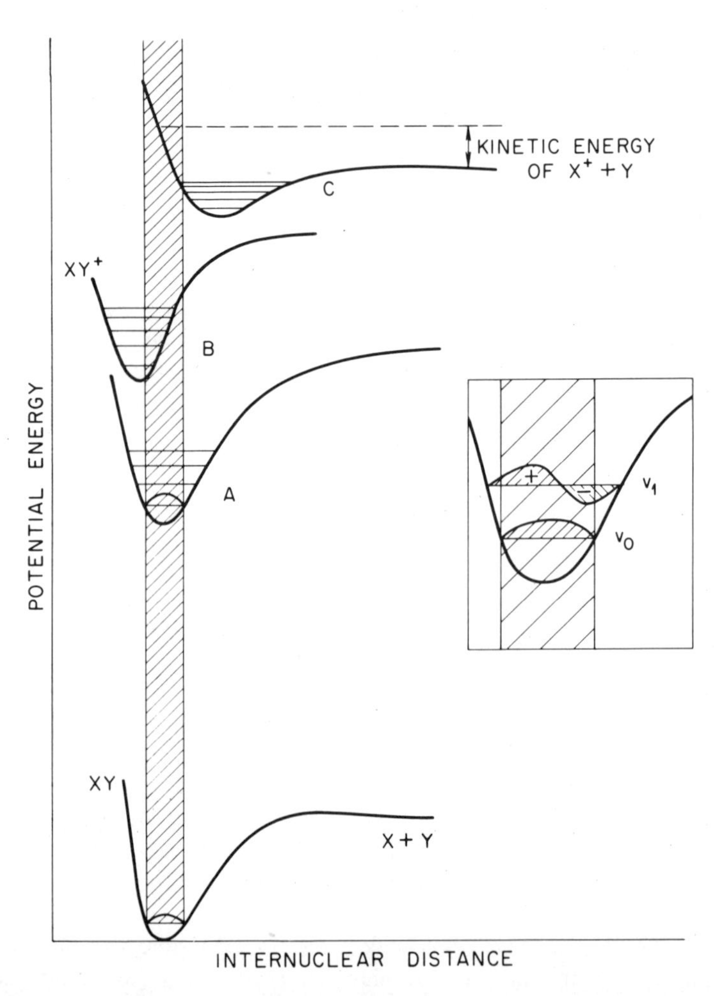

FIGURE 3.5. Schematic of Franck–Condon transitions in diatomic molecule XY. (Insert shows effect of overlapping wave function of vibrational states for A). A, No change in internuclear distance; transitions mainly to v_0. B, Shift in internuclear distance, most intense transition to excited vibrational state. C, Transition to dissociation state (vibrational states are in the continuum).

change in the potential well of the new electronic state, particularly with regard to the internuclear distance, and can be appreciated best by considering Figure 3.5. The parallel lines show the range of possible overlap. If the internuclear distance is unchanged in the ion, the overlap with v_0 will be large, the Franck–Condon factor being essentially unity. For the higher vibrational states the negative portion of the wave function will essentially cancel the positive portion, and the transition probability will be very small. (However, it will not be zero even if there is no change in the internuclear distance, so long as the potential curve does not represent a pure harmonic oscillator.) If the internuclear distance changes as in state B of Figure 3.5, the overlap will actually be higher for an excited vibrational level. In some cases where there is a substantial change in the internuclear distance, it may be very difficult to observe transitions to the ground state. Finally, transitions can take place as in the case of state C to dissociative states where there is a continuum of vibrational states and no structure can be seen.

When vacancies are produced in the core electrons, transitions occur primarily to the ground state vibrational level, since the core electrons are in nonbonding orbitals. However, it is now appreciated[32,33] that much of the line broadening found with PESIS is due to vibrational excitation (with free molecules) or to phonon excitation (lattice vibrations) when dealing with solids.

4. ATOMIC WAVE FUNCTIONS

Results from atomic wave function calculations serve as a basis for much of our discussion during this book on the interpretation of data from electron spectroscopy, and it seemed propitious to offer a few comments on the nature of atomic self-consistent wave functions and their use.*

Hartree first formulated the solution of a many-body atom by means of a self-consistent method. In this method the total wave function was assumed to be separable. Each electron moves in its own orbital, feeling the effects of the other electrons as an average screening of the stationary nuclear charge. Each electron thus experiences a central potential with an effective charge Z_{eff}, where

$$Z_{eff} = Z - \sigma \tag{3.17}$$

Z is the nuclear charge and σ is the screening. The radial part of the wave function under these conditions is separable from the angular portion, the latter being a fixed coefficient. For a spin-restricted Hartree–Fock solution, electrons with the same principal and angular momentum quantum numbers are considered identical. A numerical solution for the radial wave function is obtained

* See Ref. 34 for a review.

by assuming a starting solution, e.g., hydrogenic wave function, and adjusting the wave function of a given orbital according to the screening of the other electrons as calculated from their electron density distribution. This readjustment is done for each orbital reiteratively until the total energy of the atom reaches a self-consistent minimum.

A form of the Hartree–Fock equation which is solved is as follows [see Ref. 35, p. 8, equation (17.17)]:

$$\underset{\text{(A)}}{-\nabla_1^2 u_i(1)} - \underset{\text{(B)}}{\frac{2Z}{r_1} u_i(1)} + \underset{\text{(C)}}{\left[\sum(j) \int u_j^*(2)(2\sqrt{r_{12}})\, u_j(2)\, dv_2\right] u_i(1)}$$

$$\underset{\text{(D)}}{-\left[\frac{\sum(j)\, \delta(m_{si}\, m_{sj}) \int u_i^*(1) u_j^*(2)(2/r_{12}) u_j(1) u_i(2)\, dv_2}{u_i^*(1) u_i(1)}\right]} u_i(1) = \varepsilon_i u_i(1) \qquad (3.18)$$

where u_i is a solution of a Schrödinger equation with a Hamiltonian operator that is the sum of (A) the kinetic energy, (B) the potential energy in the field of the nucleus, (C) the potential energy in the field of N electrons distributed in the orbitals occupied in the determinational wave function, and (D) a correction term involving exchange integrals; ε_i is the eigenvalue for orbital i. The exchange term (D) may be treated in three different ways: (1) It is omitted (called a Hartree solution); (2) it is solved correctly (Hartree–Fock); (3) it is approximated by a fixed potential (Hartree–Fock–Slater). If a given shell is partially filled (filled s, p, d, and f shells should have respectively 2, 6, 10, and 14 electrons), more than one state may exist, depending on the coupling. The partially filled shell can be treated as filled, giving rise to an average value; or a coupling scheme can be assumed and solutions for each possible term value obtained separately. In the latter case, for example, L–S coupling can be employed by using F and G integrals in the Hamiltonian with their appropriate coefficients as given by Slater (Ref. 35, Appendix 21).

Hartree–Fock solutions omit electron correlation (except for the gross effect of screening). This can in part by accounted for by realizing that the concept of single-electron wave functions is an assumption, and that an admixture of other wave functions of different configuration may give a true account of the many-body problem. A given orbital can thus be taken as a sum of different configuration states and the mixing coefficients varied in a self-consistent manner in the Hamiltonian.

Thus far, we have implicitly discussed only a nonrelativistic solution. Electrons near the atomic nucleus travel at speeds close to the speed of light and thus must be treated relativistically. This problem grows rapidly as a function of Z (approximately as Z^4) and even for the lighter elements it is not necessarily negligible. Table 3.7 shows results of comparative eigenvalues

TABLE 3.7

Ratio of Eigenvalue Energies Obtained from Relativistic and Nonrelativistic Hartree–Fock–Slater Wave Functions[a]

Shell	$Z=9$ F	$Z=29$ Cu	$Z=50$ Sn	$Z=70$ Yb	$Z=80$ Hg	Z = 92 U	$Z=103$ Lw
$1s$	1.0011	1.0118	1.0368	1.0780	1.1074	1.1535	1.2119
$2s$	1.0023	1.0216	1.0603	1.1255	1.1712	1.2421	1.3341
$2p_{1/2}$	1.0020	1.0208	1.0588	1.1232	1.1682	1.2380	1.3284
$2p_{3/2}$	0.9977	0.9983	0.9991	1.0041	1.0068	1.0106	1.0152
$3s$	—	1.0327	1.0632	1.1353	1.1799	1.2441	1.3327
$3p_{1/2}$	—	1.0370	1.0627	1.1351	1.1792	1.2425	1.3305
$3p_{3/2}$	—	1.0024	1.0016	1.0163	1.0215	1.0282	1.0386
$3d_{3/2}$	—	0.9985	0.9953	1.0109	1.0155	1.0215	1.0315
$3d_{5/2}$	—	0.9681	0.9774	0.9790	0.9755	0.9717	0.9702
$4s$	—	1.0272	1.0734	1.1711	1.2088	1.2567	1.3538
$4p_{1/2}$	—	—	1.0760	1.1799	1.2151	1.2590	1.3570
$4p_{3/2}$	—	—	0.9935	1.0230	1.0231	1.0254	1.0416
$4d_{3/2}$	—	—	0.9638	1.0166	1.0136	1.0150	1.0326
$4d_{5/2}$	—	—	0.9307	0.9670	0.9599	0.9583	0.9634
$4f_{5/2}$	—	—	—	0.6901	0.8875	0.9227	0.9363
$4f_{7/2}$	—	—	—	0.5990	0.8527	0.8963	0.9055
$5s$	—	—	1.0536	1.1794	1.2559	1.2755	1.4361
$5p_{1/2}$	—	—	1.031	1.1862	1.2825	1.2830	1.4591
$5p_{3/2}$	—	—	—	0.9754	1.0016	1.0045	1.0411
$5d_{3/2}$	—	—	—	—	0.9233	0.9742	1.0232
$5d_{5/2}$	—	—	—	—	0.8042	0.9005	0.9203
$5f_{5/2}$	—	—	—	—	—	0.5242	0.6320
$5f_{7/2}$	—	—	—	—	—	—	0.5389
$6s$	—	—	—	1.0878	1.2425	1.2910	1.5258
$6p_{1/2}$	—	—	—	—	—	1.2960	1.5678
$6p_{3/2}$	—	—	—	—	—	0.9304	0.9464
$6d_{3/2}$	—	—	—	—	—	0.743	0.659
$7s$	—	—	—	—	—	1.181	1.308

[a] Taken from T. A. Carlson and B. P. Pullen, Oak Ridge National Laboratory report, ORNL-4393 (1969).

using relativistic and nonrelativistic solutions. Relativistic corrections based on hydrogenic wave functions are usually inadequate. Electrons in orbitals with higher principal quantum number do not move with relativistic velocities, but they are profoundly affected by the behavior of electrons in the deeper shells, and only a SCF calculation can bring out this effect. It has proven a better approach to substitute for the Schrödinger wave function the Dirac equation (e.g., see Ref. 36) for a single particle in a spherically symmetric potential field:

$$[-c\alpha \cdot p - \beta mc^2 + V(r)]\psi = E\psi \tag{3.19}$$

where α and β are Dirac 4×4 matrices. The wave function ψ is expressed in a four-component form. Whereas in the nonrelativistic limit we had a single wave function for a close shell orbital n, l, the relativistic solution yields two orbitals $n, l, j = l + \frac{1}{2}; n, l, j = l - \frac{1}{2}$. Thus, the spin–orbit splitting arises naturally out of the solution of the Dirac equation. Each relativistic wave function has a major and a minor component $F(r)$ and $G(r)$ such that for a normalized radial wave function

$$\int (F^2(r) + G^2(r)) = 1 \tag{3.20}$$

In the nonrelativistic limit the nonrelativistic wave function equals the major component of the relativistic solution.

In recent years it has been possible to make comprehensive SCF calculations on atoms and ions. Nonrelativistic Hartree–Fock–Slater wave functions have been tabulated by Herman and Skillman.[37] Nonrelativistic Hartree–Fock parameters including eigenvalues, total energies, and radial expectation values have been listed by Mann,[38] Froese,[39] and Clementi.[40] Relativistic Hartree–Fock–Slater results from $Z = 2$ to 126 are given by Lu *et al.*[41] Relativistic Hartree–Fock solutions have been compiled by Mann *et al.*[42] Codes are available for both nonrelativistic[43a],* and relativistic[44] atomic wave functions.

5. MOLECULAR ORBITAL THEORY

There will be no attempt in this book to develop the basic concepts of molecular orbital theory. However, since the prime concern of PESOS is to provide an experimental basis for the testing of theoretical calculations, and description of results obtained in PESIS and Auger spectroscopy is often in terms of molecular orbitals, it seems prudent to give a listing with a descriptive evaluation of some of the more fashionable theoretical models in current use. Also in this section an attempt will be made to give a glossary of some of the more common symbols used in characterizing atomic and molecular orbitals.

5.1. Theoretical Models

Most calculations used to correlate the experimental energies derived from PESOS with theory are based on the molecular orbital approach. In simplest terms a molecular wave function is derived from a linear combination of atomic orbitals. The eigenvalues that result from the solution of the molecular wave

* For a general review of Hartree–Fock wave function codes see Ref. 43*b*.

function can be compared with the experimentally determined binding energies by evoking Koopmans' rule. As discussed in Section 3.6, Chapter 3, eigenvalues should give slightly higher values than the adiabatic ionization potentials. A more fundamental comparison between experiment and theory is to use the difference in total energies calculated for the initial and final states, but this entails a much more difficult calculation, particularly if one is involved with the excited states of the ion. The molecular orbital methods most commonly employed in the understanding of photoelectron spectroscopy data group themselves into two main categories: *ab initio* calculations and semiempirical calculations.

5.1.1. *Ab Initio* Calculations

In principal, *ab initio* calculations are derived from basic physical concepts without reference to any empirical data. For molecules the *ab initio* method is generally based on the same fundamental approach as that used by Hartree and Fock for the calculation of atomic orbitals, which has been described in Section 4. As with atoms, the *ab initio* approach for molecules calculates the energy for all electrons, including core electrons. Rather than a one-center problem that is encountered with atoms, the solution of a molecular wave function is a many-center problem, depending on the number of atoms in the molecule. This makes for a highly time-consuming problem, but improvements in high-speed computers have extended the range of calculations from the simple diatomics and triatomics to larger, more complex molecules, e.g., pyridine.

The quality of different *ab initio* calculations can vary tremendously depending on the completeness of the basis sets. Some contraction of the basis set is generally necessary for all but the simplest molecules, and in such a situation the solution is not a true Hartree–Fock solution. On the other hand, improvements on the Hartree–Fock solution are often employed. The Hartree–Fock method does not treat electron correlation explicitly (except for shielding), but this can be taken account of in part by inclusion of configuration mixing.

A theoretical method which has gained some popularity is the multiple-scattering method, which makes use of a muffin-tin potential.[45] A further modification[46] is the X_α approximation in which the exchange terms of the Hartree–Fock total energy are expressed as exchange potentials that are localized or made proportional to the cube root of the total electron density. This method is much easier than the full Hartree–Fock approach, yet gives results that are superior to semiempirical calculations.

The eigenvalues obtained from molecular Hartree–Fock calculations are usually a couple of volts higher than experimental ionization potentials, and the discrepancy increases with binding energy. This behavior is generally ascribed to the failure of Koopmans' theorem. In the few cases where it has been possible to calculate the total energies for the neutral molecule and singly charged ion, the theoretical results were found to be slightly lower than the experimental binding energies. Here the discrepancy was explained in terms of the failure to describe electron correlation correctly.

As an example of the degree of uncertainty in MO calculations, Figure 3.6 shows a comparison of theoretically obtained eigenvalues for C_3O_2 together with experimental ionization potentials. No theoretical method gives a

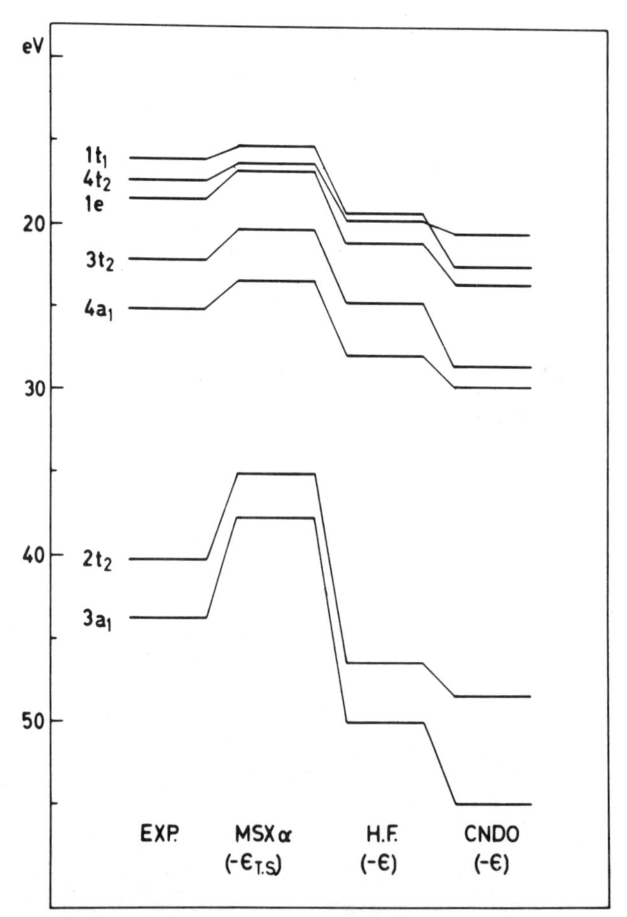

FIGURE 3.6. Comparison of eigenvalues (and ionization potentials) for C_3O_2. EXP, experiment; MSX_α, multiple scattering with exchange correction; H.F., Hartree-Fock; CNDO, complete neglect of differential overlap. [Reproduced from Connolly *et al.*,[46] Figure 10.]

close fit either to experiment or to each other, although the Hartree–Fock and multiple scattering in the X_α approximation agree as to the correct ordering.

Cederbaum and others[47] have suggested an alternative to Koopmans' theorem by means of Green's functions in an expansion using perturbation theory. With this method they are able to include the effects of both relaxation energy and correlation. The calculations are difficult to perform and have thus far been restricted to simple molecules, but the results are very encouraging. The authors believe they can account for about 80% of the difference between the eigenvalues of a Hartree–Fock solution and the true ionization potential. Cederbaum and Domcke[48] have also extended their method to the calculation of transition probabilities to excited vibrational states and their vibrational frequency by using one-body coupling constants.

Purvis and Öhrn[49] have made use of the electron propagator to obtain information about atomic and molecular electronic structure. The electron propagator is expressed as a ground state average of the N-electron system of a time-ordered product of electron field operators. From calculations of the electron propagator, the authors are able to extract binding energies and intensities of ejected photoelectrons. Of particular interest is the natural way in which the satellite structure, which arises, for example, from electron shakeup, is extracted.

5.1.2. Semiempirical Calculations

As described above, *ab initio* calculations are quite involved and it is most important to have a more readily calculable method. This is generally achieved by (1) treating electrons other than valence electrons as a fixed core potential, (2) treating the valence electrons as Slater orbitals, and (3) using parameters in the formulation that are fitted empirically.

5.1.2.1. Extended Hückel Method. The extended Hückel method was proposed by Hoffmann[50] and was based on the earlier HMO (Hückel molecular orbital) method. The simple Hückel method was used for predicting π-orbital energies for conjugated hydrocarbons. Hoffmann extended the method by including overlap. The diagonal matrix elements H_{ii} are set equal to valence state ionization potentials and the off-diagonal elements are estimated by the Mulliken approximation

$$H_{ij} = 0.5K(H_{ii} + H_{jj})\, S_{ij} \tag{3.21}$$

where K is an empirical constant. The extended Hückel method offers a quick and simple solution. Brogli and Heilbronner[51] have discussed the Hückel method with regard to its recent application to photoelectron spectroscopy of organic molecules.

Cox *et al.*[52] have used the Hückel method with success by employing an extended Slater-type orbital basis when calculating the required overlap integrals. In addition, they claim that any LCAO-MO method using a single Slater-type valence–orbital basis will be unreliable in its predictions concerning nonbonding orbitals.

5.1.2.2. Semiempirical SCF-MO Method. The most widely used methods for calculating orbital energies for comparison with photoelectron spectra have acronyms that sound like characters from a Gilbert and Sullivan operetta: CNDO, INDO, MINDO, and SPINDO.

Pople, Santry, and Segal[53] developed CNDO (complete neglect of differential orbitals). In this method all integrals involving differential overlap are neglected. An improved version, CNDO/2, employs empirical atomic matrix elements and neglects electron repulsion integrals. A refinement of the CNDO method is INDO (intermediate neglect of differential overlap). In the latter form all one-center exchange integrals are retained. This allows INDO to make a distinction between multiplet states. In general, CNDO/2 and INDO calculations give orbital energies that are the order of 5 V higher than experimental ionization potentials. However, they usually give the correct ordering in the energy levels of a given molecule or between groups of related molecules. Friedman[54] has reviewed some of the underlying assumptions present in these semiempirical methods.

Dewar *et al.*[55] have developed still another related approach, a modified INDO, called MINDO. In the latter method (1) the Slater–Condon parameters used to describe the one-center-electron-interaction integrals are all evaluated empirically from atomic spectral data, (2) the two-center integrals are treated empirically, and (3) the two-center resonance integrals H_{ij} are determined empirically from a small number of reference molecules. The orbital energies from MINDO have, in general, given better agreement with experimental binding energies than those from CNDO and INDO calculations.

Since the semiempirical methods rely on a firm empirical basis, only molecules containing elements from the first row, i.e., B, C, N, O, F, can be calculated with any degree of dependability. Empirical parameters for estimates of heavier elements are being developed, however. The second row elements have been evaluated and values have been obtained up to $Z = 35$.[56]

Lindholm *et al.*[57] have discussed the problem of the semiempirical method for organic molecules and proposed a procedure that is called SPINDO (spectroscopic-potential-adjusted INDO).

5.2. Basis Set Extension and MO Mixing

Most LCAO (Linear Combination of Atomic Orbitals) methods use a limited basis set for obtaining a molecular orbital solution. One apparent means for improving the calculation is to include more atomic orbitals in the basis set. In one instance of basis set extension, molecular orbitals from LCAO are altered by adding atomic *d* character to those elements whose principal quantum number is greater than 2, for example, silicon and sulfur. Bull *et al.*,[58] Bassett and Lloyd,[59] and Hillier and Saunders,[60] for example, have incorporated sulfur or silicon 3*d* Slater-type orbitals into their calculations to obtain better agreement between theory and experiment. Green *et al.*[61] and Brundle *et al.*[62] have suggested that the 3*d* participation may be overexaggerated.

Mixing of molecular orbitals is also offered as a qualitative explanation of the behavior of ionization potentials in a series of homologous compounds. For example, note the σ–π mixing found in organometallic compounds.[63]

5.3. Atomic and Molecular Orbital Nomenclature

Nomenclature, like language, continuously proliferates, and not always in the most logical manner. This section gathers a glossary of some of the more commonly used symbols and their definition for individual orbitals and energy states of atoms and molecules.

5.3.1. Atoms

The individual atomic orbitals are designated by two numbers and a lower case letter, for example, $3d_{3/2}$. The first number stands for the principal quantum number, in this example $n = 3$. The lower case letter gives the angular momentum quantum number: $s\,(l = 0), p\,(l = 1), d\,(l = 2), f(l = 3)$. The fraction given in the subscript is one of the two possible j values for a given closed shell, namely $l + \frac{1}{2}$ and $l - \frac{1}{2}$. Atomic orbitals are also often given names derived from x-ray spectroscopy. Using our example, we might also assign the symbol M_{IV}. The capital letter stands for the principal quantum number: $K\,(n = 1)$, $L\,(n = 2)$, $M\,(n = 3)$, and so forth. The subscript Roman numeral gives the angular momentum quantum number and j value, I ($s_{1/2}$), II ($p_{1/2}$), III ($p_{3/2}$), IV ($d_{3/2}$), V ($d_{5/2}$), VI ($f_{5/2}$), VIII ($f_{7/2}$).

The symbol given to describe the total state of the atom (or term value) is made up of two numbers and a capital letter. For example, ${}^2D_{5/2}$. The capital letter stands for the total angular momentum L: $S\,(L = 0), P\,(L = 1)$, $D\,(L = 2)$, $F\,(L = 3)$, $G\,(L = 4)$. The left-hand superscript number designates the multiplicity $(2S + 1)$ and the subscript gives the total J value $(|L + S|)$.

5.3.2. Molecules

As with atoms, both individual molecular orbitals and the overall state of the molecule are given distinctive symbols, lower case letters usually being reserved for the orbitals, while capital letters are characteristically used for the overall molecular state.

The diatomic and linear polyatomic molecules have their own special symbols since the projection of the angular momentum λ along the internuclear axis is a good quantum number. The orbital is σ if $\lambda = 0$ and π if $\lambda = 1$. The orbital is designated u (*ungerade*) if there is a change in sign of the wave function with respect to the axis perpendicular to the internuclear axis, and g (*gerade*) if there is no change in sign. An asterisk is added to indicate an antibonding orbital. If the molecular orbital is primarily derived from a given atomic orbital, this atomic orbital is sometimes indicated. Orbitals are given numbers in the order in which they occur, starting from the most tightly bound orbital. (Some authors start with the $1s$ shell, others with the most tightly bound valence orbital.) For example, the ground state configuration for O_2 is

$$(1\sigma_g\,1s)^2\,(1\sigma_u\,1s)^2\,(2\sigma_g\,2s)^2\,(2\sigma_u\,2s)^2\,(3\sigma_g\,2p)^2\,(1\pi_u\,2p)^4\,(1\pi_g\,2p)^2$$

The number above the parentheses tells the number of electrons in each orbital. The $\pi_g\,2p$ orbital may also be written $1\pi_g^*$ since it is the first time a π_g orbital appears and the orbital is antibonding.

The principal symbol standing for the state of a molecule (molecular ion) is a capital Greek letter corresponding to $\Lambda = \Sigma_u \lambda_u$, where for $\Lambda = 0, 1, 2$ the letter used is, respectively, Σ, Π, Δ. A superscript number to the left of the letter indicates the multiplicity. A plus or minus superscript to the right indicates whether the total wave function is symmetric (+) or antisymmetric (–) under σ_v operation (that is, reflection in a plane of symmetry containing the principal axis). A subscript g or u is given corresponding to *gerade* or *ungerade* symmetry. (A direct product of any number of g functions or even number of u functions is always g). Before the main symbol is often placed the capital letter X, A, B, C, etc., indicating the ground, first excited, second excited, third excited, etc., states. For example, the ground state of N_2^+ is $\tilde{X}\,^2\Sigma_g^+$. (The tilde above the X indicates that this is a singly charged ion.) The single unpaired spin gives a multiplicity of $2S + 1 = 2$, while the unfilled shell is *gerade*. In the case of spin–orbit splitting, the J value will be indicated. For example, in HCl the first ionization band contains the doublet $^2\Pi_{3/2}$ and $^2\Pi_{1/2}$.

Polyatomic molecules have even a wider variety of symbolic designations, most of which derive from character tables for symmetry groups (cf. Cotton[64]) and are called Mulliken symbols. Again lower case letters stand for individual orbitals and capital letters for the total state of the molecule. We shall use capital letters in the following discussion. Substitute lower case letters when describing individual orbitals. Nondegenerate states are designated A or B. Doubly degenerate states are called E, and triply degenerate states are denoted. T. One-dimensional representations of the orbital that are symmetric with respect to $2\pi/n$ rotation about the principal C_n axis are A and those that are antisymmetric are B. Subscripts 1 or 2 indicate states symmetric or antisymmetric to a vertical plane of symmetry. Subscript j occurs when the wave function is symmetric to the z axis. Primes and double primes indicate those states that are, respectively, symmetric and antisymmetric with respect to σ_n or reflection in the plane of symmetry perpendicular to the principal axis. The symbols g and u are attached to indicate symmetry or antisymmetry with respect to inversion if the molecule (or orbital) has a center of inversion.

It is also useful, particularly in organic compounds, to label the orbitals of a polyatomic molecule σ or π, a σ orbital or bond being defined as one in which there is no nodal surface that contains the bond axis, while a π bond has one nodal surface or plane containing the bond axis. A subscript is appended to indicate the order in which a π orbital is found. For example, the highest filled orbital in the ground state of benzene, that is, the orbital that has smallest binding energy, is designated as $1e_{1g},\pi_3$. The designation indicates it is doubly degenerate, is the first orbital with symmetry properties designated by e_{1g}, and is the third π orbital in order of decreasing binding energy. For another example, the ground state for cyclopropane is ${}^1A_1{}'$ indicating a singlet state; it is nondegenerate and symmetric with respect to rotation about the principal axis, and with regard to both vertical and horizontal planes of symmetry.

Chapter 4

Photoelectron Spectroscopy of the Outer Shells

1. INTRODUCTION

In this chapter we shall be concerned with the valence shell or outer shell of the molecule. To give a precise, unambiguous definition of the outer shell is a difficult task. In atoms, the outer shell contains those electrons that make up the highest occupied principal shell. In molecules, the outer shell is made up of those orbitals that are derived from atomic outer shell orbitals. Electrons in the outer shell are characterized by large radii and low binding energies. The outer *d* orbitals of transition metals and the outer *d* and *f* orbitals of the rare earths and actinides are examples of the difficulty of unambiguously characterizing the outer shell, since electrons in these orbitals have low binding energies and are often involved in chemical bonding; yet these orbitals are not part of the outermost principal shell and have radii that are considerably smaller than electrons in this shell. Nevertheless, dividing electron orbitals into inner or outer shells is, for the most part, a useful concept. It is the outer shell which includes those electrons that are involved in chemical bonding, and our interest here lies in the intrinsic nature of the molecular orbitals. These orbitals, unlike core levels, are not characteristic solely of a particular atom but of the overall molecule. Essentially, we desire the complete energy level scheme of the molecule. In principle, one ought to examine all the electrons in a molecule, including even the inner shell or core electrons. The inner shell electrons, however, behave primarily like electrons in a free atom, feeling the chemical environment as a perturbation. Thus, the behavior of the inner shell electrons is more profitably discussed from an entirely different viewpoint, and will be treated in a separate chapter (Chapter 5).

The main aim of PESOS is the provision of a basis upon which to test the theories of electronic structure of free molecules. This basis is an energy level scheme, where each level corresponds to the binding energy of a molecular orbital in the neutral molecule. In determining such a level scheme, one needs not only the energy but also the correct identification of the orbital. The binding energies of the outer shells usually do not exceed 50 eV, and most of the valence shell molecular orbitals have binding energies less than 25 eV. Thus, photon sources such as the resonance lines of helium, He I (21.22 eV) and He II (40.8 eV), are the most frequently used for PESOS, since they have small natural widths while still possessing enough energy for the investigation of most molecular orbitals of interest, and can be produced with high intensity. X-ray photon sources have also been employed, but their natural widths (order of 1 eV) severely limit their resolution. X rays, however, have proved useful for giving information on the relative cross sections at high energy, which can be useful in identifying orbitals. With the use of a monochromator, resolution down to 0.2 eV can be achieved, and x rays may play a more important role in the future.

Experiments of photoelectron spectroscopy of the outer shell in solids have also been carried out. However, with solids, unlike free gaseous molecules, there often do not exist discrete levels of small natural widths for the outermost molecular orbitals, but rather bands. The shape and nature of these bands can, however, be studied by photoelectron spectroscopy. It has been particularly valuable to carry out these studies as a function of photoelectron energy so as to elicit information as to the character of the angular momentum in the band. The field of photoemission is rather extensive and, as explained in Chapter 1, no attempt will be made to cover the subject thoroughly in this book. However, a brief discussion of the field seems in order, particularly in the use of higher energy photon sources commonly used in photoelectron spectroscopy.

The photoelectron spectra of molecular orbitals, however, can be clearly seen in the solid state when the valence electrons are not part of the solid state band structure but have their own molecular integrity. Two areas of current interest are the photoelectron spectra of the outer shells for molecules adsorbed on surfaces and for large anions in crystalline material. These topics will also be discussed in Section 6 of this chapter.

2. ENERGY LEVEL SCHEME

2.1. Binding Energy

Ideally, when photoelectron spectra are taken with a monochromatic photon source, one and only one photoelectron peak will occur corresponding to each of the molecular orbitals, since the binding energy is related to the photoelectron energy as expressed in the simple relationship of equation (3.5).

The binding energy of a given molecular orbital is the difference between the total energies of the initial and final states, the initial state being the neutral molecule, and the final state being the molecular ion in which an electron from the given orbital has been removed. If the molecular ion is in the ground state of the single hole configuration, the binding energy is identical to the adiabatic binding energy. Rather than using calculations of the total energies of the initial and final states in order to compare with experimental binding energies, one usually employs eigenvalues calculated for the molecular orbitals of the neutral molecule. The use of eigenvalues implies Koopmans' approximation, which asserts that the binding energy of a given molecular orbital equals the eigenvalue (with reverse sign) of that orbital if we consider all other electrons in this molecule to remain frozen in their initial orbitals. In reality, the other electrons relax as the result of the vacancy created by photoionization, and the eigenvalue is larger than adiabatic binding energy by the relaxation energy.

Two complications, however, can arise. First, there sometimes exist more than one final electronic state. This problem will be discussed in detail in the next section. Second, transitions will go to a series of vibrational states governed by Franck–Condon factors (see Section 3.2). The photoelectron energy is thus given as

$$E_e = h\nu - [(\varepsilon_f + v_f + r_f) - (\varepsilon_i + v_i + r_i)] \tag{4.1}$$

where ε, v, and r are the electronic, vibrational, and rotational energies for the initial neutral molecule and final ionic state. The rotational bands in PESOS are generally not resolved and will be neglected in most of our discussions.

The adiabatic ionization potential or adiabatic binding energy is for the transition that goes to the ground state vibrational level $v_0 \to v_0'$. (The phrase "adiabatic binding energy" is normally redundant, since the definition of binding energy usually includes the concept of adiabaticity.) When the vibrational structure can be easily resolved, and transitions to the ground state have a reasonably high probability, the adiabatic ionization potential can be determined with some accuracy, since it corresponds to the lowest kinetic energy peak in the spectrum. Often the vibrational structure is not well resolved, due to the presence of substantial contributions from more than one degree of vibrational freedom or from the fact that the final electronic state is a dissociative one in which the vibrational states are in the continuum. In this case we assign the adiabatic ionization potential to the onset of the smooth band. If there is a large difference in the internuclear spacing between the initial and final electronic states, the Franck–Condon factor for the transition to the ground vibrational state may be zero or so close to zero that it cannot be observed experimentally. Under such conditions the onset of an electronic band seen in a photoelectron spectrum may be at a slightly higher energy than the true adiabatic ionization potential.

Occasions also arise in which a photoelectron peak may fall at an energy slightly below the adiabatic ionization potential. This occurs when the initial neutral molecule may occasionally be in an excited vibrational state. The probability for this is small, generally less than 1 %. Its probability I is given by

$$I = I_0 \exp\,[-(\varepsilon_{v_1} - \varepsilon_{v_0})/kT] \tag{4.2}$$

where k is Boltzmann's constant and T is the temperature. Because the filling of the higher vibrational state is temperature dependent, the appearance in photoelectron spectra of a vibrational band due to an initially excited vibrational state is called a "hot band." The vibrational band due to a $v_1 \to v_0'$ transition will lie $\Delta E = \varepsilon_{v_1} - \varepsilon_{v_0}$ below the adiabatic ionization potential, where ε_{v_1} and ε_{v_0} are, respectively, the energies of the v_0 and v_1 vibrational states of the neutral molecule.

An even greater problem in locating the onset of the electronic band occurs when the electronic bands overlap with one another. One way to untangle such a situation is to observe the photoelectron spectrum as a function of the angle between the direction of the photon beam and outgoing electron. Each ionization band will have a characteristic angular distribution, given in terms of the parameter β [cf. equation (3.2)]. By determining β over the whole photoelectron spectrum one can untangle the different electron bands. For example, Figure 4.1 shows a portion of the photoelectron spectrum of benzene together with the angular analysis.

Because of the difficulty in ascertaining at times the adiabatic ionization potential, it is also useful to determine the vertical ionization potential, which is the peak of the electronic band. The vertical ionization potential thus represents the transition to the vibrational state that has the greatest probability of being reached and is dependent on the values of the Franck–Condon factors. The vertical ionization is more appropriate to a theoretical treatment which neglects the change in internuclear distances as the result of photoionization, although the peak of the ionization band is equivalent neither to the adiabatic binding energy nor to the eigenvalues for the neutral species via Koopmans' theorem (cf. Chapter 3, Section 3.6).

2.2 Final States

One of the basic experimental premises in photoelectron spectroscopy is that there exists one and only one photoelectron peak (or band) corresponding

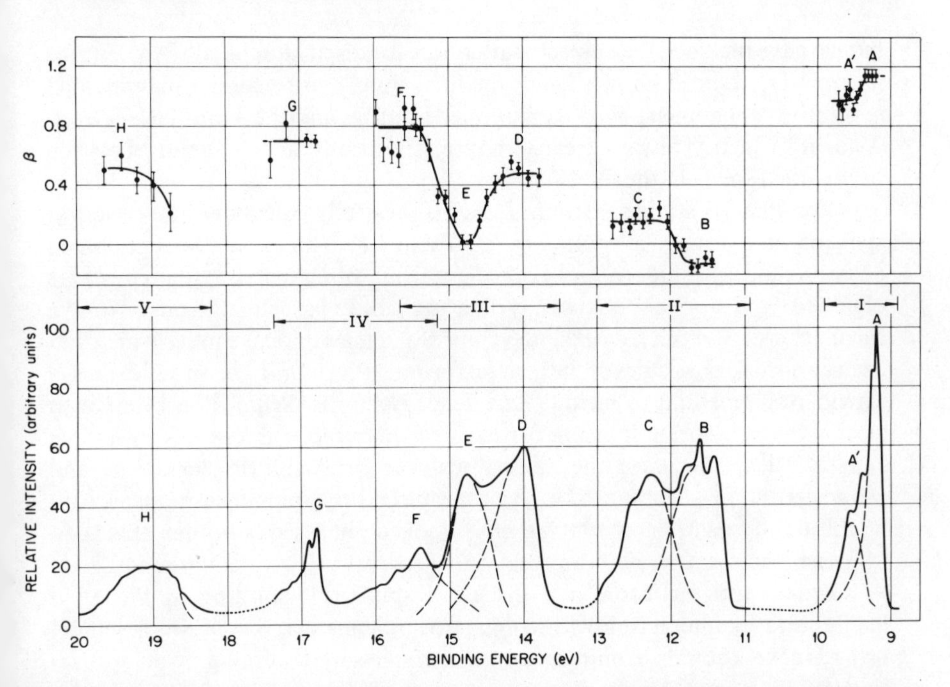

FIGURE 4.1. Photoelectron spectrum of benzene using $h\nu = 584$ Å, deconvoluted through help of angular distribution analysis. The angular parameter β is given as a function of binding energy in the upper portion of the figure. The lower portion is a spectrum taken at $\theta = 90°$. Dotted lines show deconvolution. [Reproduced from Carlson and Anderson,[82] Figure 1.]

to a given molecular orbital. In spite of many important exceptions, which will be discussed in this section, the generalization is remarkably valid. For nearly every molecule studied a band corresponding to each of the orbitals has been found so long as the photon used had the required energy to eject an electron. The intensities between the bands vary, and occasionally some bands are missed, but the intensities rarely are so low that a careful investigation will not reveal them. In analyzing a photoelectron spectrum, one should attempt to account for each molecular orbital by assigning a portion of the spectrum to that orbital (or assume the corresponding photoelectron peak lies outside the observed energy range). In most instances only one final electronic state will occur as the result of photoejection of an electron from a given molecular orbital. Let us now examine exceptions to this rule.

2.2.1. Spin–Orbit Splitting

As explained in Chapter 3, spin–orbit splitting strictly speaking arises from relativistic theory and is not solely a characteristic of the final state, but we have included the subject matter here for discussion because to describe it properly, we need knowledge of the final state, and because it may interact with other phenomena such as multiplet splitting and the Jahn–Teller effect. Also, it is nearly always treated nonrelativistically as a coupling between partly filled shells in the final state.

For PESOS studies, Grimm[1] has successfully calculated the observed splitting for a variety of molecules (cf. Table 4.1). The calculations are based on a method of Ishiguro and Kobori,[2] using observed coupling constants obtained from atomic spectroscopy data and molecular orbitals from CNDO/2 calculations. Berkosky *et al.*[3] developed a semiempirical model which has been applied to molecules of the general formula PX_3, while Lee and Rabalais[4] extended the method to include *nd* valence electrons. Wittel[5] has examined closely the spin–orbit splitting for each of the molecular orbitals in iodine. Jungen[6] also discussed the anomalous spin–orbit splitting in iodine and proposed the $^2\Pi_u$ states of I_2^+ to be perturbed by spin–orbit-induced configuration interaction. Wittel *et al.*[7] looked into second-order effects in spin–orbit interaction with application to several mercury compounds.

Brogli and Heilbronner[8] discussed spin–orbit splitting in the alkyl halides and its interaction with conjugated systems based on a simple model using Hückel orbitals. Conjugation reduces spin–orbit splitting by delocalization and by mixing the localized orbitals of the halide with semilocalized orbitals from the hydrocarbon.

TABLE 4.1

Comparison of Theoretical and Experimental Values of Spin–Orbit Splitting Following Photoionization[1]

		Spin–orbit splitting, cm^{-1}	
Ion	Final state	Experiment	Theory
CO^+	$A\ ^2\Pi$	118	125
HF^+	$X\ ^2\Pi$	240	314
F_2^+	$X\ ^2\Pi$	337	298
ClF^+	$X\ ^2\Pi$	637	620
Cl_2^+	$X\ ^2\Pi$	645	626
Br_2^+	$X\ ^2\Pi$	2904	2508
IBr^+	$X\ ^2\Pi$	4678	4859
ICl^+	$X\ ^2\Pi$	4678	4700
CO_2^+	$X\ ^2\Pi$	160	155
$C_4H_2^+$	$X\ \Pi_g$	33.3	32.1
CS_2^+	$X\ ^2\Pi$	440	432
CH_3Cl^+		627	585
CH_3Br^+		2560	2291
CH_3I^+		5050	4627

2.2.2. Multiplet Splitting due to Spin Coupling

If a molecular orbital is only partially filled, then a vacancy created by photoionization in a second orbital will lead to the presence of two orbitals with uncoupled spins. These two orbitals will couple, giving rise to more than one final state. Examples of such molecules are O_2,[9] NO[10] and some of the transitional metal complexes.[11] For instance, O_2 has the configuration $KK(\sigma_g\, 2s)^2(\sigma_u\, 2s)^2(\sigma_g\, 2p)^2(\pi_u\, 2p)^4(\pi_g\, 2p)^2\, {}^3\Sigma_g^-$. The $(\pi_g\, 2p)^2$ shell is only half-filled and contains two unpaired electrons. Nine final states are possible from the removal of an electron from a valence shell. One state arises from the removal of an electron from the $(\pi_g\, 2p)$ orbital, while two states each arise from the removal of an electron from each of the other four valence orbitals, depending on whether the spin of ejected electron is parallel or antiparallel to the spin of the electron in the $(\pi_g\, 2p)$ orbital. For example, the second and third photoelectron bands with thresholds at 16.1 and 17.0 eV both arise from the $(\pi_u\, 2p)$ orbital.

Rules for band intensities in the photoelectron spectra of open shell molecules have been discussed by Cox *et al.*[12] and Rabalais *et al.*[11] They are as follows.[12]

1. If a closed shell is ionized, all states arising from the coupling of the single hole will occur, giving rise to photopeaks whose intensities are proportional to their spin–orbital degeneracy.

2. If the open shell is ionized, the probability for producing different final ionic states will reflect the squares of the fractional parentage coefficients rather than the spin–orbital degeneracy.

3. The total intensities of all photoelectron peaks arising from ionization in a given subshell will be proportional to the one-electron cross section of that subshell times the occupancy of the subshell.

2.2.3. Jahn–Teller Effect

If a molecule has complete symmetry, e.g., CH_4, then a removal of an electron from one of its degenerate orbitals will break down the electronic symmetry and consequently its degeneracy, and the molecular ion may be found in more than one final electronic state. This is known as the Jahn–Teller effect, and the energy separation between the final states as the Jahn–Teller splitting. Figure 4.2 plots the bands for the *t* orbital in CH_4, SiH_4, and GeH_4. A double peak is readily seen. Also, evidence is present for a third contribution seen as a shoulder on the low-energy portion of the structure. The extent of overall Jahn–Teller broadening decreases with a decrease in the force constant: $CH_4 > SiH_4 > GeH_4$. Dixon[13] and Grimm and Godoy[14] have analyzed the case of methane. The triple degeneracy is removed by distortion to a D_{2d}

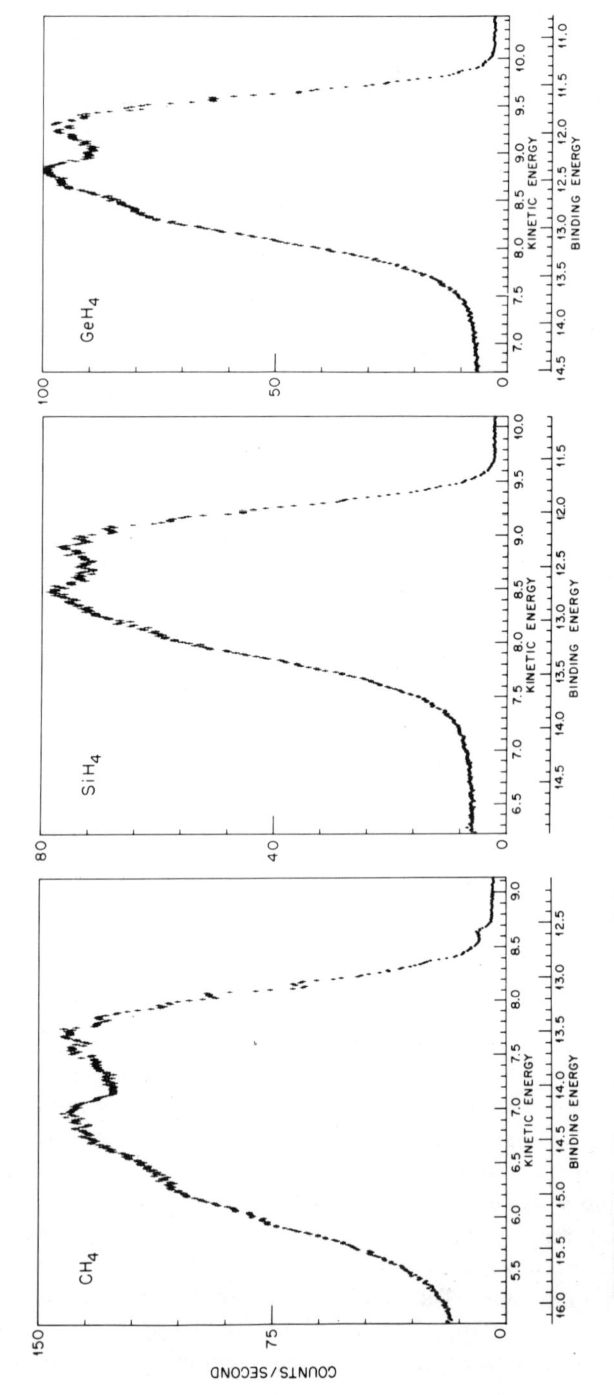

FIGURE 4.2. Photoelectron spectra of the *t* orbital in CH_4, SiH_4, and Ge_4H. [Reproduced from Pullen *et al.*,[136] Figure 6.]

conformation, with two low-lying states: $(1e)^4(1b_2)^1\,{}^2D_2$ and $(1e)^3(1b_2)^2\,{}^2E$. Approximate calculations by Dixon on the vibrational envelopes to be expected for methane predict three broad, equally spaced maxima separated by about 0.7 eV, in reasonable agreement with experiment. Potts and Price[15] have argued for D_{2d} conformation for silane, germaine, and stannane as well as methane from analysis of the vibrational structure and ionization potentials.

The Jahn–Teller effect is small when heavier atoms are substituted for hydrogen, as, for example, in CF_4, but is again noted when the central atom becomes very heavy, as in Pb $(CH_3)_4$.[16,17]

For the heavier elements, interaction between spin–orbit splitting and Jahn–Teller splitting becomes important. This interaction in CBr_4 is discussed later in this chapter. Although linear molecules do not undergo Jahn–Teller distortion, vibrational levels in which one or more quanta of degenerate bending vibrations are excited can be split into several components by interaction of electronic and vibrational motion. This is known as the Renner–Teller effect and its consequences have been measured for water.

For a more detailed description of both the Jahn–Teller and Renner–Teller effects in photoelectron spectroscopy, see Eland.[18]

2.2.4. Electron Shakeoff and Shakeup

As the result of a vacancy found in a given orbital due to photoionization, electrons in the same and other orbitals see a change in the effective nuclear charge due to an alteration in the electron screening. This change in effective charge can give rise to an excited state in which an electron may undergo a transition into either the continuum (electron shakeoff) or a discrete state (electron shakeup). The transitions normally obey the rules of monopole excitation. This process has been described elsewhere in some detail (Chapter 3, Section 3.4). In situations such as are covered in this chapter where photoionization occurs in the outermost or valence shell, simultaneous electron excitation or ionization of a second electron is strongly influenced by correlation effects; and the probability appears to be considerably higher than that calculated from single-electron wave functions. The presence of electron shakeoff/up will occur in the photoelectron spectra as lower energy satellite peaks or bands displaced from the main peaks by the exact amount needed for excitation. Monopole ionization or transitions to the continuum appear as a continuum spectrum and will probably disappear into the background, but monopole excitation or transitions to discrete states may be seen as well-resolved discrete lines in the spectrum.

The probability for monopole ionization and excitation usually decreases when the energy of the ejected electron is less than about three times that necessary for double ionization. However, due to electron correlation there can be a sudden rise just before threshold in the case of photoionization in the valence shell.[19] The probability for monopole excitation is the order of 10% of the total photoionization cross section as indicated from studies on rare gases using x rays.[20, 21] Some peaks have been observed in the photoelectron spectrum of the valence shells of molecules using x rays, which appear to be due to monopole excitation or electron shakeup.[22] However, little evidence has been obtained for such processes using the He I resonance line (21.22 eV), although Cradock and Duncan[23] believe that weak bands in the spectra of CS_2 and CSe_2 are due to electron shakeup. The general absence of shakeup spectra probably results because 21.22 eV is insufficient to simultaneously cause both photoionization and excitation. The use of He II (40.8 eV) radiation, however, ought to show the presence for structure due to shakeup and there is in fact recent evidence for such behavior.[24]

2.2.5. Configuration Interaction

As a consequence of photoionization in a given orbital the molecule may go into a final state which is still allowed by selection rules for photoionization, but in which one or more electrons in other orbitals are rearranged. The general terminology for such behavior is configuration interaction. Electron shakeup is special case of configuration interaction in which only one electron is excited. Weak transitions to doubly orthogonal states may also occur. These have been discussed by Lorquet and Cadet[25] for N_2O and H_2S. A weak, diffuse band sometimes reported as doubtful or even a poorly defined increase in the background of the high-energy region of the spectrum may be evidence for such states. Such may also be evidence for monopole excitation.

2.2.6. Resonance Absorption

In addition to giving rise to photoionization, a monoenergetic photon might also be absorbed into a resonant state without ionization. To effect the transition with any degree of probability, the energy of the photon must match the energy difference between the ground and excited states to within the natural widths of the photon and resonant state, which are the order of a few millivolts. Near the ionization limit for a given orbital there are often a large number of closely spaced Rydberg states, so that if the energy of the absorbed photon is at or just slightly below the binding energy of one of the molecular orbitals (or even slightly above, if the excess energy can be taken up by an excited vibrational state) resonance absorption is likely to occur.

The resonant state will most likely decay by autoionization to a final electronic state that could also be reached by direct photoionization. The energy carried away by the autoionized electron is the difference between the

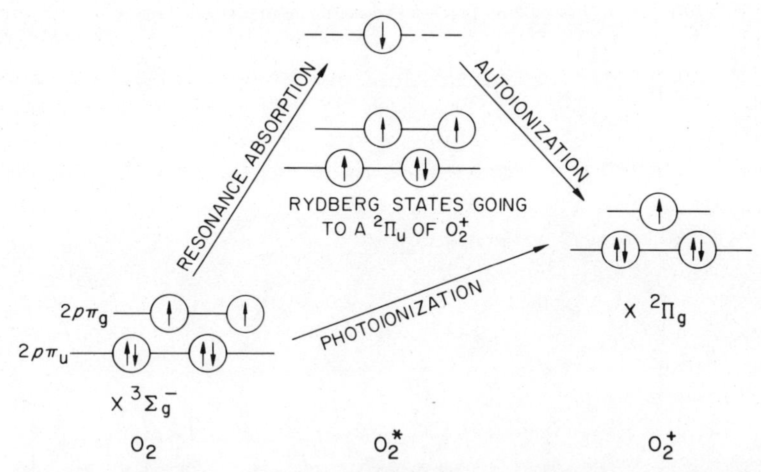

FIGURE 4.3. Diagram representing autoionization in O_2. [Reproduced from Carlson *et al.*[83] Figure 7.]

initial and final electronic states and in this sense is indistinguishable from a direct photoionization process. For example, see Figure 4.3 for the case of resonance absorption of the 16.85-eV Ne I radiation in oxygen. Thus, in this section we are dealing less with the production of a new final electronic state than with another method of reaching the same final state. But although the electronic states may be the same, the distribution of vibrational states or Franck–Condon factors is entirely different. The energy spacings of the vibrational states will reflect the nature of the final electronic state, but the relative intensities of the vibrational bands will have been determined by the overlap of a vibrational level in the resonance state with those in the final state. (Note that in resonance absorption of a monochromatic photon the energy of the transition must match the total energy for both the electronic and vibrational states, and thus the transition to a resonance level will be to only one given vibrational state.) In addition, the angular distribution for the ejected electrons relative to the direction of the photon beam will depend on entirely different considerations in accordance with which of the two modes, photoionization or autoionization, the electron was produced. Figure 4.4 shows the vibrational structure found for the first ionization band in O_2 using the Ne I 16.85 (16.67) eV resonance lines. Note that there are obviously two vibrational envelopes with different angular distributions. The band at higher kinetic energy is due to photoionization, while the other arises from autoionization.

Resonance absorption is not commonly found with He I and II radiation, but is seen quite often with the lower energy photons of Ne I (16.87, 16.67 eV) and Ar I (11.62 eV). See, for example, the work of Collin and others.[26–31]

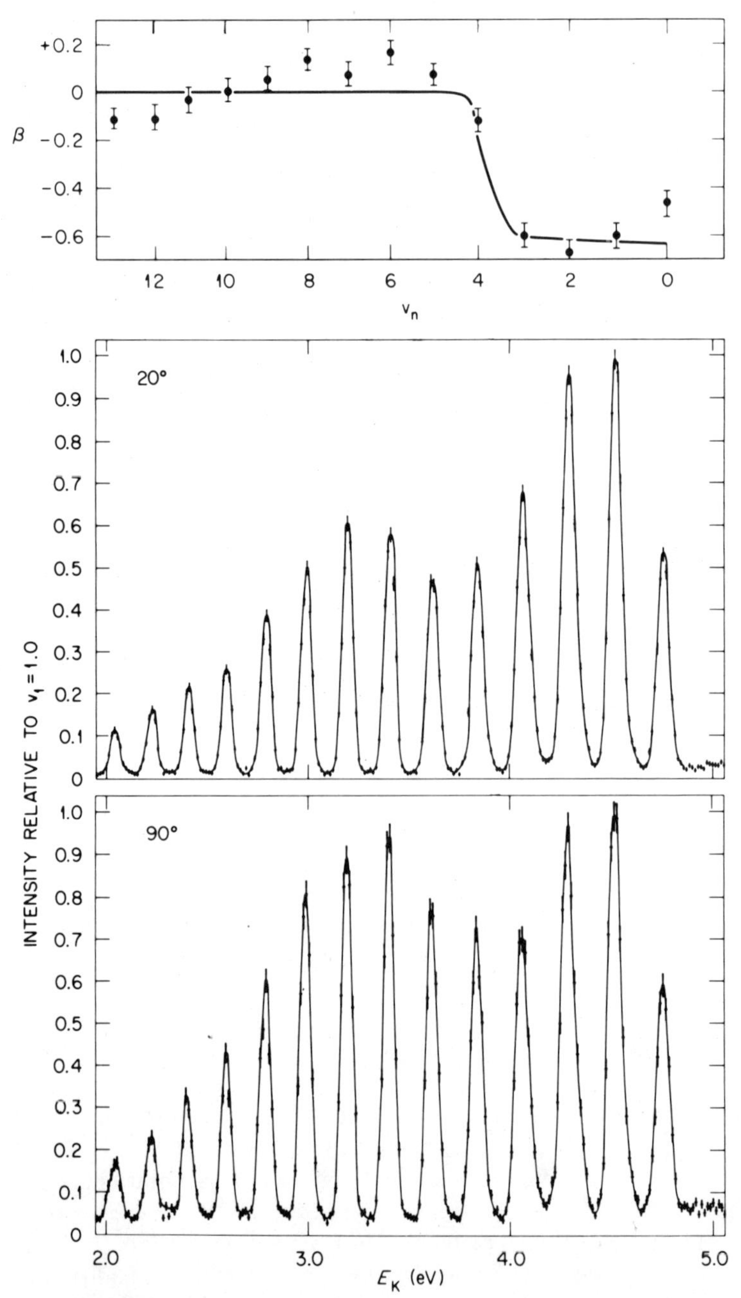

FIGURE 4.4. Photoelectron spectrum of first ionization band of O_2 taken at $\theta = 90°$ and $20°$ using $h\nu = 16.85$ eV, together with the angular distribution parameter β as a function of photoelectron energy E_K. [Reproduced from Carlson,[81] Figure 2.]

What one really requires is a continuous source of radiation without loss in high resolution, so one might run over the resonances. Such a source would be possible with synchrotron radiation. In the meantime Gardner and Samson[32] have carried out experiments on N_2O with a series of discrete-energy photon sources between 133 Å (threshold) and 972 Å.

2.2.7. Collision Peaks

Before reaching the spectrometer, electrons may undergo inelastic collision with other molecules. If these collisions result in discrete losses, lines will appear in the electron spectra that may be confused with the normal photoelectron spectrum. This problem is not too acute when the He I (584 Å) line is used, since most of the photoelectrons ejected do not possess sufficient energy for impact excitation, but it occasionally does occur. In particular, spectra appearing at low kinetic energy may be due to collision losses. The problem is more serious when higher energy sources are employed. The contribution from collision losses may be detected and corrected for by (1) study of the spectra as a function of pressure and (2) determination of the nature of the collision losses by separate electron impact experiments (cf. Chapter 2, Section 1.2.1). Streets *et al.*[33] have used the photoelectron emitted from argon as the result of He I irradiation (5.46 and 5.28 eV) to study collision losses at high pressure. This "natural" source of high-resolution electron beams yields information that compares favorably with results obtained by conventional electron impact techniques.

3. IDENTIFICATION OF THE ORBITAL

The principal task for PESOS is the determination of the nature of molecular orbitals. It is the characterization of the photoelectron spectrum in terms of the molecular orbitals out of which photoelectron ejection has occurred, namely, the identification of which ionization band corresponds to which orbital. We shall now consider how this identification can be made with the help of photoelectron data. We shall consider, in order, the ionization potentials, the vibrational structure, the photoelectric cross section, and the angular distribution of photoelectrons.

3.1. Ionization Potentials

In this section we shall examine the adiabatic and vertical ionization potentials themselves for a clue to the nature of the molecular orbital. The most important clue comes from comparing the experimentally determined binding energies with the ordering obtained from theory. The naming of the orbitals comes from symmetry considerations; the binding energies arise out of solutions to the wave functions for the molecule. The use of theory would appear to beg the question since the purpose of photoelectron spectroscopy is to provide an experimental basis for the testing of molecular orbital calculations. However, the approach to establishing the correct electronic structure for a molecule does not proceed by two independent paths, theory and experiment, which are then compared at the end of the road, but rather consists of a continuous interaction between the two until the most reasonable solution is achieved. Theory will usually narrow down the choice of the orbital corresponding to a given ionization, and experiment will be required to establish which one of the few choices is correct.

The general procedure for analyzing the spectrum is to first list all the possible molecular orbitals derived from the valence electron. Their number should equal the number of valence electrons divided by two if the spin-up and spin-down states are not to be distinguished. Sometimes two or three pairs of orbitals will be degenerate. Next, a list of theoretical binding energies is made up corresponding to the orbitals, usually from eigenvalues derived from solutions of the wave functions. Finally, one prepares a list of ionization potentials from the photoelectron spectra. A matching of the experimental and theoretical

TABLE 4.2

Analysis[a] of the Photoelectron Spectrum of Benzene

Number of electrons in subshell	Orbital	Type	Experiment, vertical ionization energy, eV	Theory[b] $-\varepsilon$, eV
4	$1e_{1g}$	$p\pi$	9.3	10.1
4	$3e_{2g}$	$p\sigma$	11.4	14.3
2	$1a_{2u}$	$p\pi$	12.1	14.6
4	$3e_{1u}$	$p\sigma$	13.8	16.9
2	$1b_{2u}$	$p\sigma$	14.7	17.8
2	$2b_{1u}$	$s\sigma$	15.4	18.0
2	$3a_{1g}$	$p\sigma$	16.9	20.1
4	$2e_{2g}$	$s\sigma$	19.2	23.0
4	$2e_{1u}$	$s\sigma$	22.5	28.2
2	$2a_{1g}$	$s\sigma$	25.9	31.8
Total 30				

[a] Based on Åsbrink *et al.*[141] and B. Jonsson and E. Lindholm, *Ark. Fys.* **39**, 65 (1968).
[b] Theory based on eigenvalues obtained from Roothaan-type SCF calculations [J. M. Schulman and J. W. Moskowitz, *J. Chem. Phys.* **47**, 3491 (1967)].

binding energies is rarely a simple matter. Besides the error implicit in Koopmans' approximation, the calculations may contain many other uncertainties which lead not only to incorrect absolute binding energies but even errors in the correct ordering of the orbitals. Experimentally, overlapping of the bands may make the assignment of the ionization potentials uncertain. A typical example of such listings is given for benzene in Table 4.2. In spite of a great amount of work done on benzene, the order of some of the orbitals, even for this molecule, is still in question. To ascertain the correct analysis of the photoelectric peaks and consequently the correct experimental binding energy for each of the molecular orbitals, correlation with other evidence is brought into account, such as will be discussed below. For further discussion of techniques used in the analysis of photoelectron spectra the reader is directed to Eland[(18)] and Betteridge and Thompson.[(34)]

3.1.1. Characteristic Ionization Bands

A distinctive feature of an otherwise complex spectrum that can be easily identified is that due to the lone pair. Since the lone-pair orbitals are highly localized and nonbonding, they exhibit very little vibrational structure, but give a sharp band structure. Besides being easily identified, the behavior of the lone pair (its change in energy relative to the other bands) and its broadening due to a loss in lone-pair character can serve as a valuable guide to the nature of the electronic structure of the molecule as a whole. Sweigart and Daintith[(35)] have discussed splitting of the lone-pair orbitals in a series of six-membered sulfur heterocyclic compounds and point out that the splitting can arise from direct overlap or indirect coupling through CH_2, CC, and CS σ network. In Chapter 4 of their book,[(36)] Baker and Betteridge give a short review of the behavior of the photoelectron peaks associated with lone-pair orbitals and how they may interact with π orbitals. They also demonstrate linear relationships between the ionization potential for lone-pair orbitals and electronegativity.

Spin–orbit splitting will also yield a tell-tale pattern. The extent of spin–orbit splitting depends on the initial splitting in the atomic orbital which makes up the final molecular orbital. The splitting in the atomic orbital depends strongly on the nuclear charge of the atom. Table 4.3 lists the energy splitting of the $p_{1/2}$–$p_{3/2}$ levels in atoms for the valence shell as a function of Z. More detailed information on the nature of spin–orbit splitting can be found in Chapter 3 and in Section 2.2.1 of this chapter. When molecular orbitals involve the p electrons of a heavier atom, splitting in the photoelectron spectra can be quite apparent. Spin–orbit splitting will be associated with those orbitals that are highly localized about the heavier atoms. These orbitals in turn will usually be lone-pair orbitals with a sharp band structure.

TABLE 4.3

Energy Separation between the $p_{1/2}$–$p_{3/2}$ Levels in the Valence Shell of Atoms[a]

Atom	Z	$\Delta\varepsilon$, eV
N	7	0.02
O	8	0.04
F	9	0.08
Ne	10	0.12
P	15	0.06
S	16	0.10
Cl	17	0.14
Ar	18	0.18
As	33	0.22
Se	34	0.43
Br	35	0.57
Kr	36	0.73
Sb	51	0.71
Te	52	0.92
I	53	1.15
Xe	54	1.41
Bi	83	2.16
Po	84	2.77
At	85	3.40
Rn	86	4.07

[a] As obtained from eigenvalues based on relativistic HFS wave functions using average configuration for open shells [C. C. Lu, T. A. Carlson, F. B. Malik, T. C. Tucker, and C. W. Nestor, Jr., *Atomic Data* **3**, 1 (1971)].

Thus, from Figure 4.5, we can assign the sharp line doublet found in methyl bromide to the $^2E_{1/2}$, $^2E_{3/2}$ states, since they arise from orbitals that have been built up from the halogen outermost p shell, and one can clearly see[(37)] the spin–orbit splitting increasing as one goes from chlorine to iodine. Brogli and Heilbronner[(8)] have discussed the extent of spin–orbit splitting in terms of conjugation. Another splitting in the electronic band that can be useful for identification is the Jahn–Teller splitting, which arises from the breakdown in degeneracy of a highly symmetric molecule due to photoionization (cf. Section 2.2.3). For example, in the case of methyl fluoride, Jahn–Teller splitting is observed with the first ionization band, indicating that the orbital associated with this band is the $e(CH_3)$ orbital.[(38, 39)] Brogli *et al.*[(40)] have studied the Jahn–Teller splitting in allene as a function of alkyl substitution,

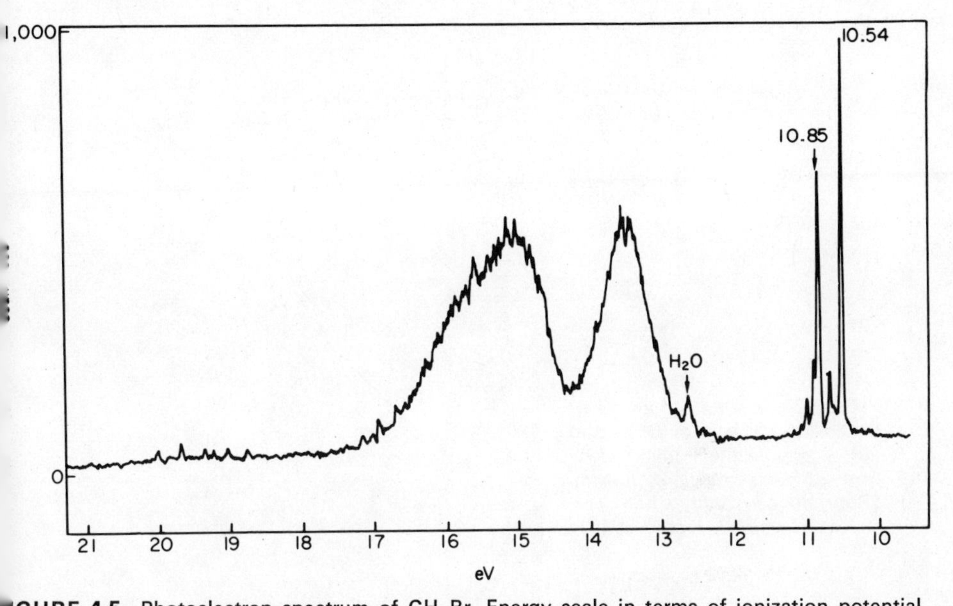

FIGURE 4.5. Photoelectron spectrum of CH_3Br. Energy scale in terms of ionization potential. [Reproduced from Turner *et al.*,[37] Figure 8.5.]

and Mines and Thomas[41] have evaluated the Jahn–Teller distortion in the photoelectron spectrum of SO_3.

3.1.2. Effects of Substituents

One source for clarification concerning which ionization potential is associated with which orbital is to follow the pattern of binding energies through a series of homologous compounds. If the experimental data appear to follow the same pattern as predicted by theory, they give confidence to the assignment. Several examples of this will be found later in a survey of the literature. One of the most direct approaches of this technique is to examine the photoelectron spectrum as a function of substituent. One effect that can be followed is the change in ionization potential of a highly localized orbital with the electronegativity of the substituent. Correlations have been obtained between the observed ionization potentials and the electronegativity of the substituent,[42,43] or the partial charge derived from electronegativity.[34] See, for example, Figure 4.6. Betteridge and Thompson[34] have also successfully made correlations with Taft σ^* values for organic compounds. Yoshikawa *et al.*[44] have related the ionization potentials to the basicities and *s* character of the nitrogen

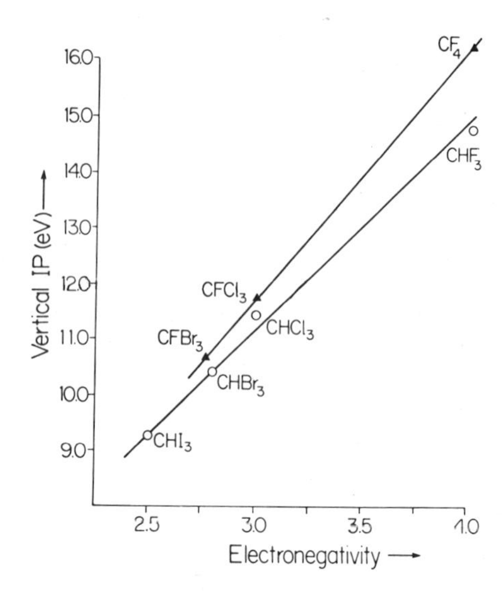

FIGURE 4.6. Plot of vertical IP of the $1a_2$ orbitals of CFX_3 and CHX_3 against the Pauling electronegativity of the halogen atom. [Reproduced from Chau and McDowell,[43] Figure 4a.]

lone-pair electrons. The inductive effect of carbonyls has also been recently studied.[45] In addition to short-range inductive effects that can be directly related to the electronegativity of the neighbor atoms, there can be long-range inductive effects and mesomeric interaction due to the substituents. The interaction of these three factors has been discussed by Streets and Williams[46] for the dihaloanthracenes. Finally, there are the concepts of "through space" and "through bond" interaction between localized or semilocalized orbitals as formulated by Hoffmann.[47] These interactions have been clarified through a systematic study of the photoelectron spectra of heterocyclic compounds.[48]

3.1.3. Sum Rule

Kimura and co-workers[49] have made extensive use of a sum rule that arises from the fact that the sum of all the diagonal elements in a secular matrix will be equal to that of the resultant eigenvalues. Thus, they assume

$$\sum(-I_i) = \sum \varepsilon_i{}^0$$

where I is taken to be the vertical ionization potential for molecular orbital i, and the $\varepsilon_i{}^0$ are obtained empirically from fitting data of simple molecules. Table 4.4 gives a listing of $\varepsilon_i{}^0$. The partial sum for each symmetry should be independent of orbital interaction as well as the total sum over all orbitals. The correlation of results by this sum rule is rather impressive, although there is some

TABLE 4.4

Energy Parameters ε_i^0 Used in Calculation of Sum Rule[a]

Orbital	$-\varepsilon_i^0$,eV	Orbital	$-\varepsilon_i^0$,eV
πCC	10.51	σN–N	14.90
πCO	14.50, 15.19	nN	10.88, 10.80
πCN	14.42	πNH_1	15.77, 16.40
σCN	14.42	σOH	15.57, 16.60
πCH_3	14.30	σO–O	16.10
πCH_2	14.30	nO	12.61
σCF	17.14	nO[b]	13.76
σCC	11.75	nF	16.05
πC≡C	11.40	nCl	13.05, 12.78
σCCl	14.45, 14.42	nBr	11.89, 11.84
σCBr	13.52, 13.49	nI	10.72
σCI	12.50		

[a] Taken from Kimura *et al.*[(49)] In some cases more than one value is listed when derived from different set of compounds.
[b] Nonbonding orbital in OH.

arbitrariness in choosing a proper set of ε_i^0. The sum rule is found to hold best for localized orbitals.

3.1.4. The Perfluoro Effect

The perfluoro effect, though limited in its scope, is a powerful method for distinguishing the nature of molecular orbitals for a selective group of molecules. In essence the perfluoro effect is the observation of systematic changes that occur between molecules in which all the hydrogens have been replaced by fluorine.[(50)]

Specifically, the substitution of fluorine for hydrogen in a planar molecule has a much larger stabilizing effect on the σ MO's than on the π MO's. For example, see Figure 4.7 comparing ethylene and perfluoro ethylene. The effect is most apparent for nonaromatic hydrocarbons, in which fluorination shifts the π molecular orbitals 0–0.5 eV, while the photoelectron peaks due to σ orbitals change by 2–3 eV. The effect is not as distinct when atoms other than carbon are in the molecule. Fluorination of the methyl derivatives of linear molecules causes an increase in the ionization potentials for all photoelectron peaks regardless of the nature of the molecular orbital. In fact, this is the behavior in general for nonlinear molecules. Perfluorination makes all nonbonding electrons much less so regardless of whether they are σ or π. The rules governing perfluoro shifts of π and σ molecular orbitals apply equally in nonaromatic and aromatic compounds, although the shift in the π orbitals may

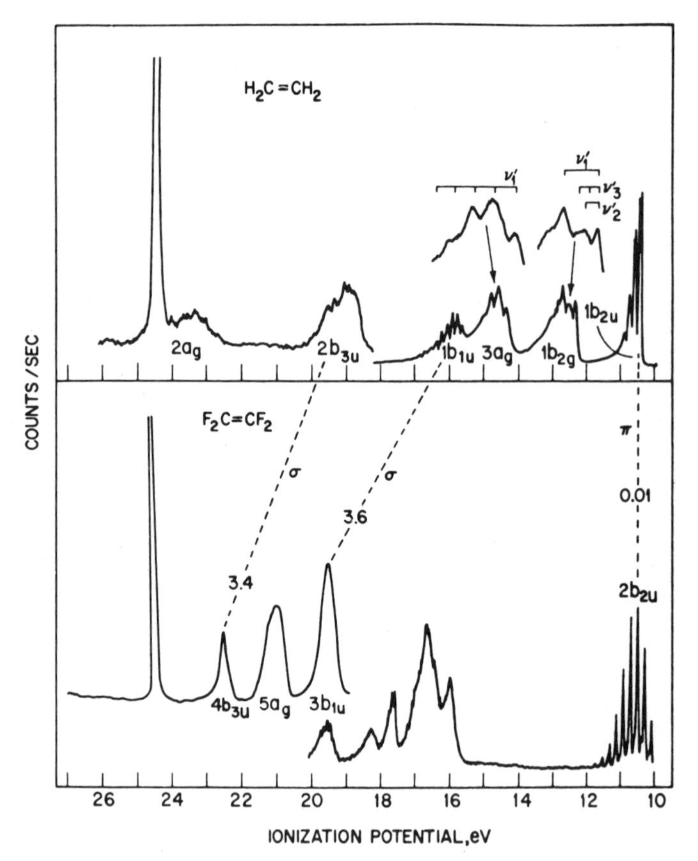

FIGURE 4.7. Comparison of perfluoro effect in the photoelectron spectra of ethylene and perfluoro ethylene. [Reproduced from C. R. Brundle, M. B. Robin, N. A. Kuebler, and H. Basch, *J. Am. Chem. Soc.* **94**, 1451 (1972), Figure 1.]

be slightly greater for the aromatics. Comparison of the photoelectron spectra of an aromatic hydrocarbon and its perfluoro derivative can be used to demonstrate whether the compounds are planar or not.

3.1.5. Dependence on Steric Effects

Through Figure 4.8 Maier and Turner[51] were able to establish for a series of biphenyls a simple empirical relationship involving the steric property of the molecules, namely the dihedral angle θ between the planes of the two rings and the energy separation of the first and fourth ionization potentials (π_6 and π_3). These types of studies have been extended to phenyl ethylene, aniline, and other compounds.[52–55] Baker *et al.*[56] have measured the splitting between the $p\pi$ and $p\pi^*$ orbitals that are formed from the outermost atomic p orbitals of disulfides and used these values to calculate the dihedral angle, while Nelsen and Buschek[57] have discussed the relationship between the photoelectron peak representing a lone pair and the angle between the atoms on which the lone-pair orbital resides.

Cowley *et al.*[58] have suggested that for molecules such as A_2X_4 the lone-pair splitting should be much greater for *trans* isomers than for *gauche*. They have applied this criterion to a series of dimethyl phosphines and arsines. They feel that PESOS offers great promise for the rapid assay of rotational isomerisms in acyclic species containing adjacent electron pairs.

Finally, Boshi *et al.*[59] have pointed out that in polycyclic aromatic hydrocarbons a vibrational progression of 13.50 cm^{-1} is seen in the first ioniza-

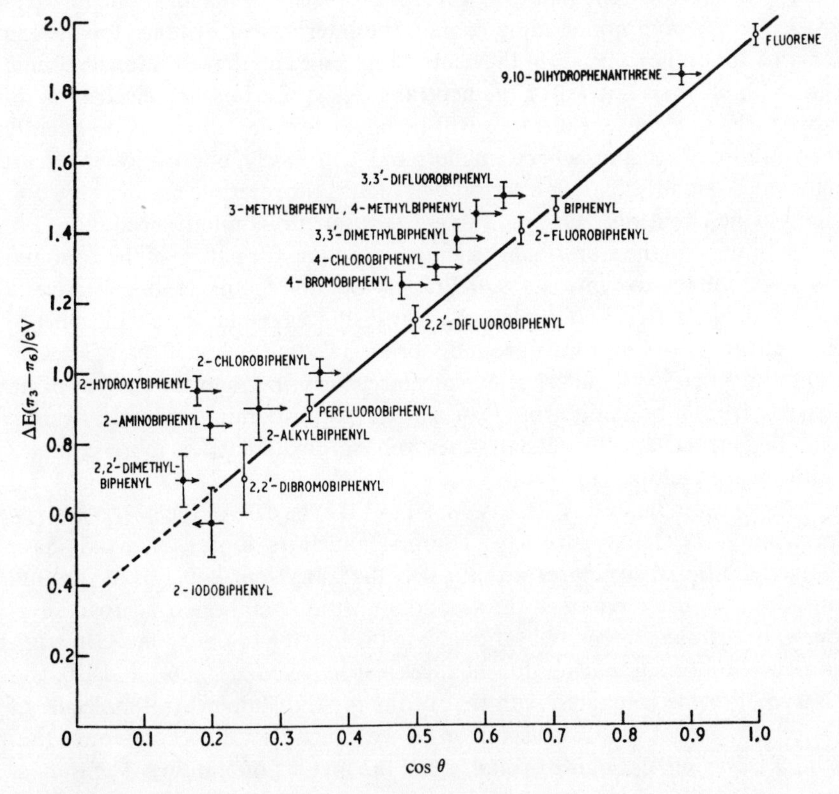

FIGURE 4.8. Plot of separation in ionization potentials corresponding to the π_3 and π_6 orbitals as a function of the dihedral angles for biphenyl. [Reproduced from Maier and Turner,[51] Figure 2.]

tion band of planar molecules, which becomes blurred in sterically overcrowded systems.

3.2. Identification of Orbitals by Vibrational Structure

One of the principal advantages of high-resolution photoelectron spectroscopy has been the ability to separate vibrational bands, since the vibrational structure offers a powerful method for the identification of the orbital associated with a given ionization band.

The first source of identification comes from the spacing of the vibrational peaks. This spacing is governed by the energy separation of the vibrational levels of the final state. The final state is the molecular ion corresponding to the configuration formed following a single-electron ejection from a given orbital. For a diatomic molecule the following generalization can be made: If one removes an electron from a nonbonding orbital, the internuclear distance for the ion remains about the same as for the neutral molecule and the vibrational spacing is also about the same as for the neutral species; if a bonding electron is removed, the internuclear distance will be larger for the ion and consequently the vibrational spacings will be smaller; and, conversely, but to a lesser extent, removal of an antibonding electron will reduce the internuclear distance and increase the vibrational spacing. For a diagrammatic summary, see Figure 3.5.

Not only do the vibrational spacings depend on the nature of the bonding, but so do the relative intensities or Franck–Condon factors. If one removes a nonbonding electron and the internuclear distances remain constant, most of the transitions will go to the ground vibrational state; while if there is a substantial change in the internuclear distance due either to the removal of an electron from a bonding orbital (or a strongly antibonding orbital), overlap with higher excited vibrational states will be greater and a more complex vibrational envelope will occur.

As an example of this discussion, let us take the case of the photoelectron spectrum of N_2 (cf. Figure 4.9). The first and third ionization bands have simple vibrational structures with most of the transitions going to the ground vibrational state. In contrast, the second ionization band offers a broad envelope with the most intense vibrational band belonging to the v_2' level. In addition, the vibrational spacings for the three respective bands are 2100, 1810, and 2340 cm^{-1}, to be compared with that of the ground state neutral molecule of 2345 cm^{-1}. From the above discussion we see clearly that the second ionization band arises from a bonding orbital, while the first is nonbonding. The third is antibonding; its spectrum is not much more complex than that of the nonbonding orbital since the change in the potential is rather small. Edwards[60] has proposed that the change in potential energy between an atom pair for a

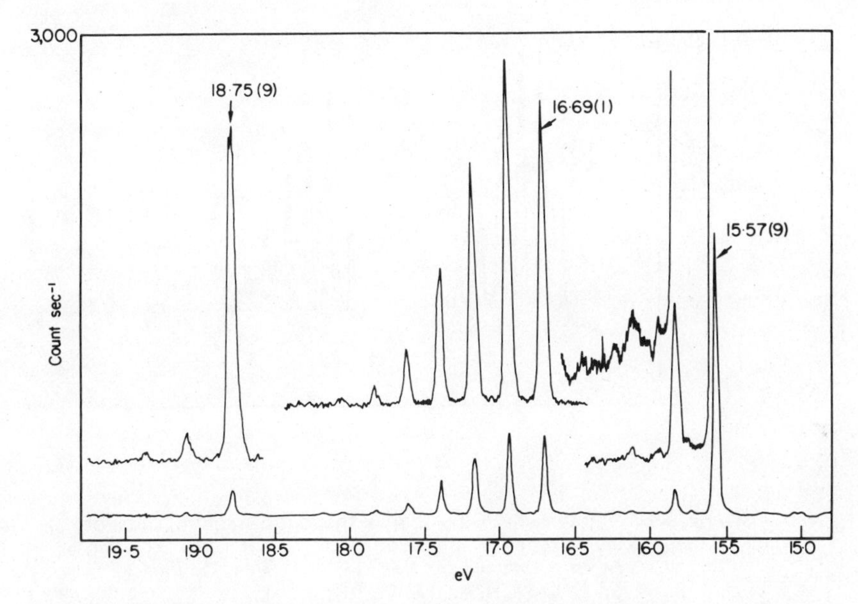

FIGURE 4.9. Photoelectron spectrum of N_2. Energy scale in terms of ionization potential. [Reproduced from Turner *et al.*,[37] Figure 3.7.]

particular molecular orbital be calculated using the CNDO/2 method. From this calculation the change in vibrational frequencies can be properly evaluated.

The vibrational spectra for polyatomic molecules are much more complex. Diatomic molecules have only one vibrational mode, while the number of vibrational degrees of freedom is $3N - 5$ for a linear molecule and $3N - 6$ for a nonlinear molecule, where N is the number of atoms making up the molecule. However, fortunately, only one or two of the vibrational degrees of freedom are usually activated, and its identification gives a good indication of the orbitals involved in the photoelectron spectrum. For example, the fourth ionization band of BF_3 can be seen in Figure 4.10. Although there exists six degrees of freedom, only one vibrational mode has been excited. The spacing is 725 cm^{-1}, which corresponds to the ν_1 888-cm^{-1} spacing in the neutral molecule. The vibrational mode is involved in symmetric stretching, and the orbital which is most likely to be involved with the band is the $a_2''(\pi)$.

The vibrational spacings in the final state seen in photoelectron spectra rarely change more than 50% from the corresponding vibrational mode for the the ground state neutral molecule, and the largest changes are observed when the bond involves a heavy and a light atom, as in the case of carbon and hydrogen. The change in polyatomic molecules is nearly always toward a wider spacing in the molecular ion.

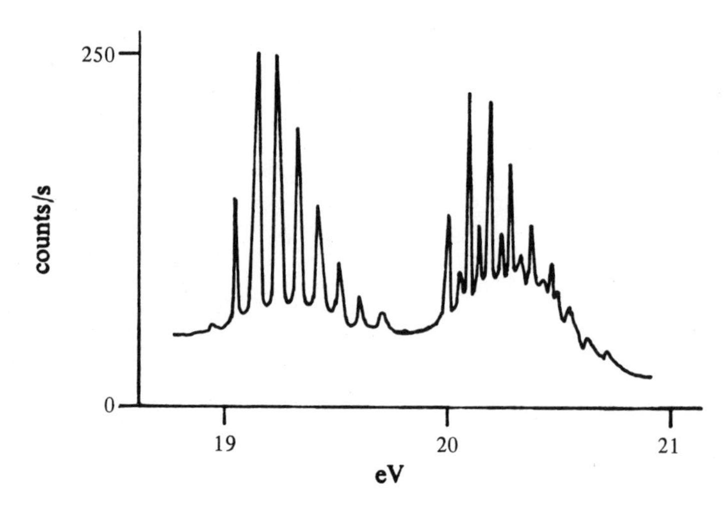

FIGURE 4.10. Photoelectron spectrum of the fourth and fifth bands of BF_3. Energy scale in terms of ionization potential. [Reproduced from King *et al., Faraday Disc. Chem. Soc.* **54**, 70 (1972), Figure 4.]

Even the absence or presence of vibrational structure can provide helpful information in the analysis of orbital structure. For example,[61] in the spectrum of CF_4 the first and second ionization bands show no vibrational structure while the third band shows two vibrational sequences. From mass spectroscopy data we know that the ground state of CF_4^+ is unstable toward dissociation. Thus, it is not surprising to find most of the bands without vibrational structure since the final states are probably dissociative ones with no discrete vibrational structure. However, by removal of an electron from a nonbonding orbital one may be able to go to an electronic state that does not immediately dissociate. The most likely orbital in CF_4 to have this characteristic (and still not be too strongly bound) is the *e* orbital leading to the ^2E final state. Following this logic further, the fourth band of SiF_4 was also designated as the *e* orbital, since it was found that of the first four ionization bands seen with SiF_4, only the fourth had vibrational structure, which, like the case of CF_4, was composed of two series of vibrational lines. (The assignment of the fourth band of SiF_4 has been questioned, however, on grounds of molecular orbital calculations.[62])

In addition to utilizing the vibrational spectra to determine the nature of the molecular orbital, the data can be used to evaluate molecular geometry of the ground state of the ion, such as shown in the work of Hollas and Sutherley[63] on $C_2H_2^+$ and XCN^+. Berkowitz[64] has used the vibrational data from photoelectron studies to determine the potential curves of an ion as in the case of the X $^2\Pi$ and $^2\Sigma^+$ states of HF^+.

3.3. Identification of Molecular Orbitals from Intensities of Ionization Bands

Although the intensities for the various ionization bands will vary from band to band, a fairly safe generalization is that these intensities are usually the same order of magnitude. One ought not generally to ascribe the apparent absence of a band to the assumption that its intensity was simply too low to be seen experimentally, and a photoelectron band that is more than an order of magnitude lower than other bands formed in the spectra should be viewed with suspicion as possibly due to impurities or other spurious causes. Like most generalizations, exceptions occur, but it forms a sensible point of departure at the start of the analysis.

One obvious factor determining the intensity of a photoionization band is the number of electrons in a degenerate state. This is, $t > e > a, b$. Assignments based on statistical weights are tenuous, since the photoelectron cross section depends on other considerations, namely the energy of the outgoing photoelectron and the symmetry and angular momentum of the molecular orbital. In comparing orbital intensities, it is generally more prudent to only trust comparison of orbitals whose photoelectron energies are close and whose molecular orbital character is similar. In general, comparison of data taken

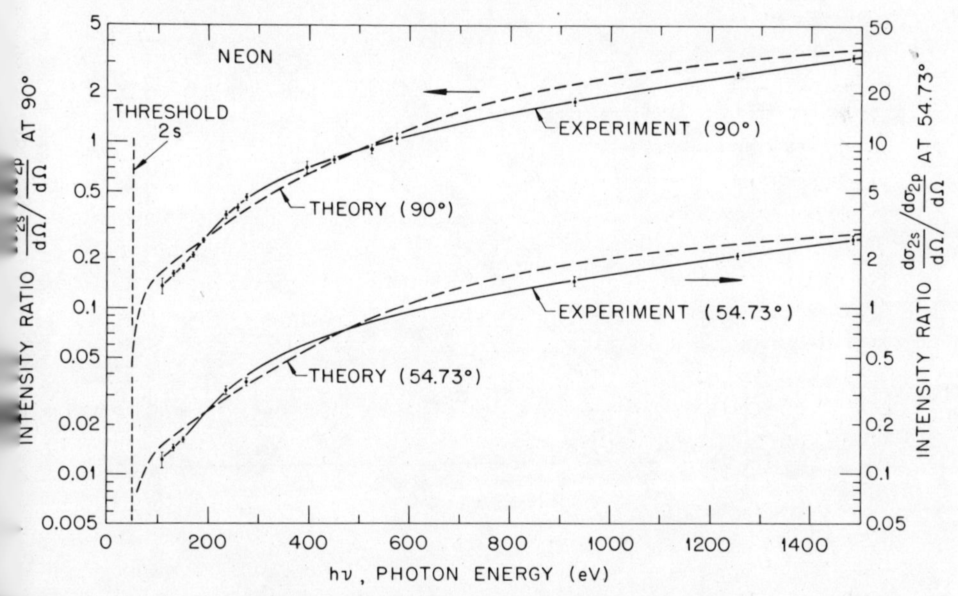

FIGURE 4.11. Ratio of $2s$ to $2p$ cross sections for neon at $\theta = 90$ and 54.7°. Theoretical values given as dashed lines, experiment by solid lines. [Reproduced from F. Wuilleumier and M. O. Krause, *Phys. Rev. A* **10**, 242 (1974) Figure 4.]

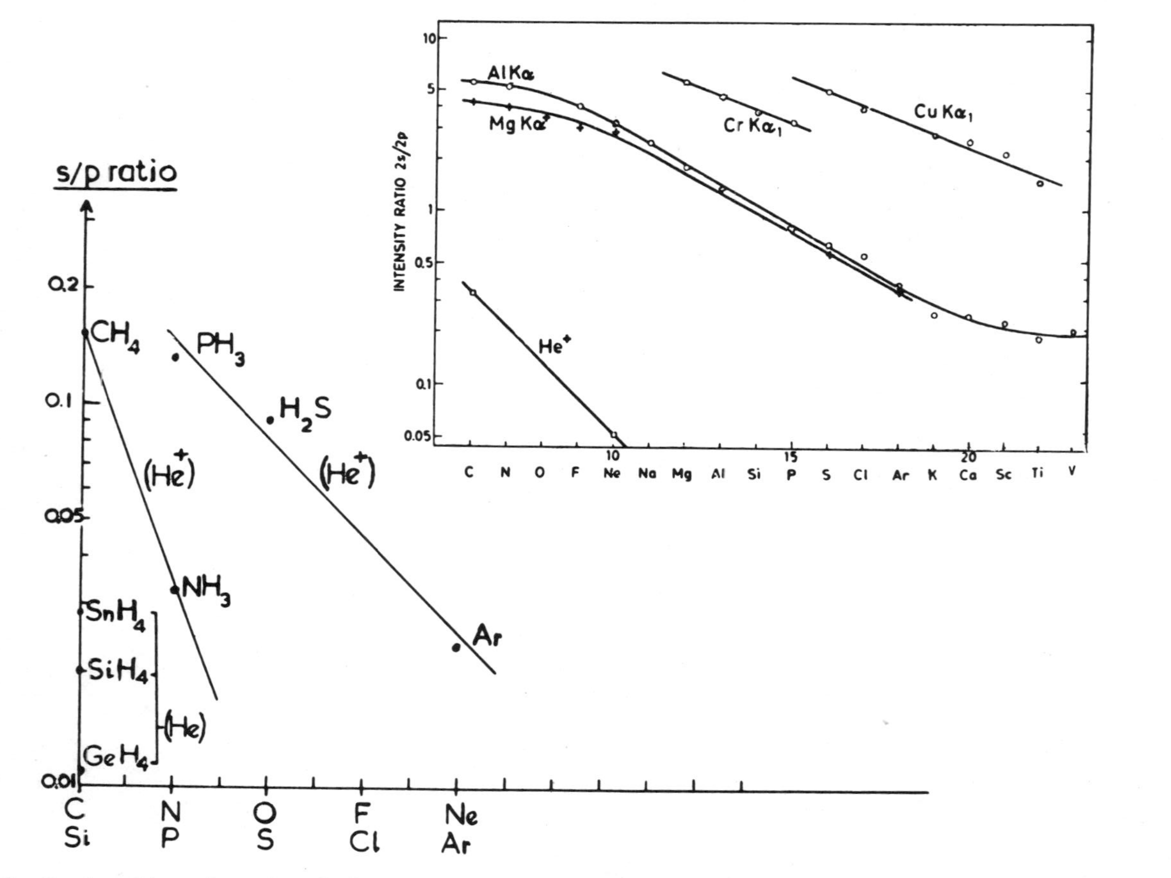

FIGURE 4.12. Graphs of intensity ratios of *s*/*p* type bands plotted against atomic number for different ionizing photon energies. [Reproduced from Price *et al.*,[65] Figure 3.]

with the He II resonance line is more reliable since the percentage change in energy between the different bands is smaller.

Figure 4.11 shows the differential cross sections for the $2s$ and $2p$ orbitals of neon as a function of photoelectron energy. At higher energies the relative cross section for the s orbital is greater, but the reverse is true at lower energies.

Price *et al.*[65] plotted (Figure 4.12) experimental ratios of s/p type bonds for a number of atoms and molecules. Angular momentum is not a good quantum number for molecular orbitals, but an MO may be construed as being made up of varying amounts of atomic s or p character.

Gelius[66] pointed out that at higher photoelectron energies, of the order of 1 keV, such as produced when using Al and Mg $K\alpha$ lines, the total photoelectron cross section for a given orbital is a linear combination of atomic s and p contributions. Based on a theoretical analysis by Lohr,[67] the relative intensity for a photoelectron band corresponding to a given molecular orbital j is

$$I_j^{\mathrm{MO}} \propto \left[2+\frac{\beta}{2}\right] \sum_{\mathrm{A}} \left[\sum_{\lambda} C_{\mathrm{A}\lambda_j} \left(\frac{\sigma_{\mathrm{A}\lambda}}{\sigma_{\mathrm{A}_0\lambda_0}}\right)^{\frac{1}{2}}\right]^2 \tag{4.3}$$

where $\sigma_{\mathrm{A}\lambda}/\sigma_{\mathrm{A}_0\lambda_0}$ is the relative atomic cross section for a given atom A and given angular momentum λ. A list of these values is given in Figure 4.13

$\sigma_{A\lambda}/\sigma_{A_0\lambda_0}$ ($A_0\lambda_0$ \ $A\lambda$)	C 2s	C 2p	N 2s	N 2p	O 2s	O 2p	F 2s	F 2p	Ne 2s	Ne 2p	S 3s	S 3p	A 3s	A 3p
C 2s	1		1.2		1.4		2.0				0.47			
C 2p	13	1												
N 2s			1											
N 2p			11	1										
O 2s					1									
O 2p					8.8	1								
F 2s							1							
F 2p							10	1						
Ne 2s									1					
Ne 2p									8.7	1				
S 3s											1			
S 3p											1.1	1		
A 3s													1	
A 3p													1.4	1

FIGURE 4.13. Estimated relative photoionization cross sections $\sigma_{\mathrm{A}\lambda}/\sigma_{\mathrm{A}_0\lambda_0}$ for atomic valence electrons. The table may be completed for all the atoms, except Ne and Ar, by combining the $\sigma_{\mathrm{A}s}\ \sigma_{\mathrm{C}s}$ ratios in the first row with the $\sigma_{\mathrm{A}s}/\sigma_{\mathrm{A}p}$ ratios in the diagonal. [Reproduced from Gelius,[66] Table I.]

[σ(H 1s) $\approx$ 0]. The $C_{A\gamma_j}$ are the relative atomic densities (Mulliken densities) derived from the MO-LCAO model. That is,

$$\phi_j = \sum_{A,\lambda} C_{A\lambda_j} \phi_{A\lambda} \tag{4.4}$$

The expression [2 + (β/2)] is a correction for photoelectron spectral data taken at 90° to the photon beam, where β is the angular parameter. For an example of the method, see Figure 4.14 on SF_6. The solid line is the calculated curve based on equation (4.3). The method is extremely valuable for discernment between orbitals having various amounts of s and p character. Perry and Jolly[68] have made use of the method to analyze a group of silicon compounds. Huang and Ellison[69] have used a model based on the plane-wave approximation to obtain differential cross sections for molecular orbitals with soft x rays. Comparison with Gelius' semiempirical approach is not perfect but gives semiquantitative agreement.

Price *et al.*[65] discussed the problem of photoelectron cross sections for molecular orbitals in some detail, evaluating qualitatively the effects of the

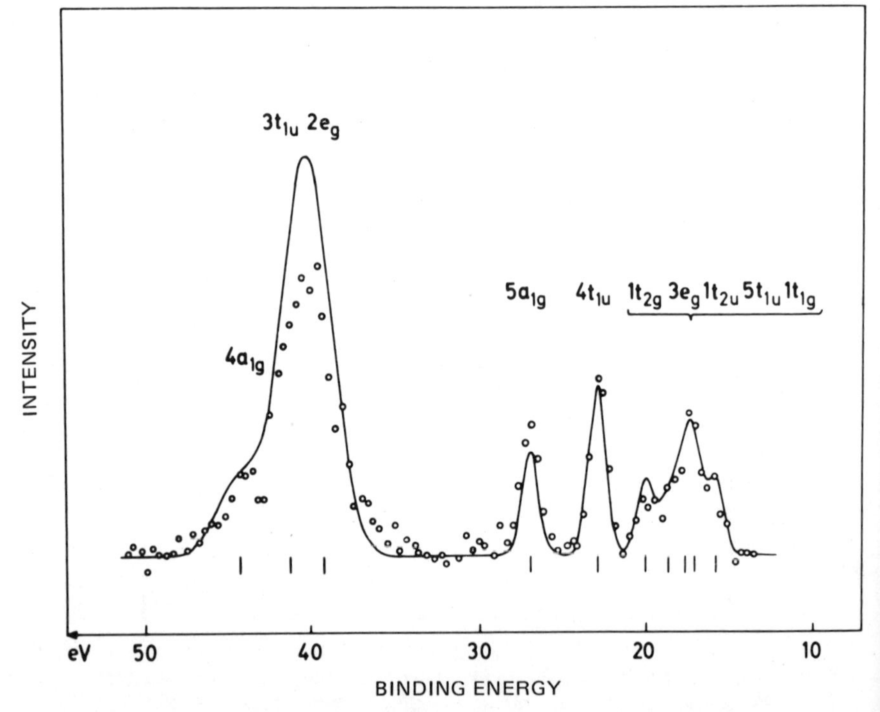

FIGURE 4.14. X-ray photoelectron spectrum of SF_6 using Mg $K\alpha$ x rays. Solid line is theoretical plot using equation (4.3). [Reproduced from Gelius,[66] Figure 11.]

size of the orbital, the number of nodal surfaces, and the wavelength of the photoelectron on the transition moment of a photoionizing process. Price *et al.* also pointed out that when molecular orbitals are formed from in-phase and out-of-phase combinations of atomic orbitals, it is to be expected that the transition moment for photionization will be affected by the size and symmetry of such orbitals, and they demonstrated that in the paraffins the ionization bands grow progressively stronger with the increase in the number of nodal surfaces in the orbitals concerned.

A rigorous calculation of the photoelectric cross section for H_2 has been carried out by Kelly,[70] but for more complex molecules a simpler approach is required. Lohr[67] has utilized a plane wave for the continuum wave function in his calculations and this approach has been exploited by Schweig and Thiel.[71] Using their notation, the photoelectron cross section of orbital n is

$$\sigma_n = Ap_n\left[\sum_a Q_a + \sum_{a<b} Q_{ab}\right]/h\nu \tag{4.5}$$

where A is a constant, p_n is the momentum of the ejected photoelectron, and $h\nu$ is the photon energy. Q_a and Q_{ab} are the one- and two-center terms composed from individual basis atomic orbitals and their combination. The agreement between calculated cross sections using the plane wave approximation and the relative areas of the photoelectron peaks obtained with 584-Å and 304-Å radiation is at best qualitative. Substantial improvements occur when the continuum wave function is orthogonalized to the molecular orbitals.[72–75] The plane wave approach for cross section calculations is very uncertain, particularly with sources of low photon energies such as the He I and He II resonance lines. Nevertheless, it represents a start toward understanding the more complex molecules. The following conclusions were drawn for the first row elements[73]: (1) The cross sections rise sharply near the ionization threshold and then decrease monotonically for higher photon energies, the peak maximum being at about $h\nu \sim 2\text{IP}$, where IP is the ionization threshold; (2) there seems to be no simple rule for cross sections in the uv region, (3) in the soft x-ray region the s/p ratios do, indeed, solely determine the relative cross sections for the different orbitals, if one refers to the percentage of orthogonalized $2s$ character.

It is particularly instructive to study the relative intensities of the photoelectron peak as a function of photon excitation, and comparison of data taken with the He I and He II resonance lines has been most helpful.[76,65] Murell[77] sought to explain changes in band shapes occurring with He I (584 Å) and He II (304 Å) radiation. Such alterations in the band shape should occur when there are large geometry changes upon ionization. The effects of variation of the photoelectron cross section over the vibrational envelope have been calculated by Lee and Rabalais.[78]

3.4. Identification of Molecular Orbitals by Angular Distribution

A powerful method for determining which molecular orbital is associated with a given ionization band is through the measurement of the intensity of the photoelectron as a function of the angle θ between the direction of the incoming photon and outgoing photoelectron. The angular distribution is given in terms of the parameter β. From equation (3.2) one obtains the relationship between the intensity of a photoelectron peak I, and the angle θ,

$$I \propto 1 + \tfrac{1}{2}\beta\left(\tfrac{3}{2}\sin^2\theta - 1\right) \quad (4.6)$$

The parameter β is a function of the photoelectron energy and the nature of the molecular orbital. Some 40 different molecules have been examined with regard to angular dependence of their photoelectron spectra.[79–91]

It has been found, in general, that photoelectron bands corresponding to different final states produced by removing an electron from a given orbital will have similar values of the angular parameter β. For example, the shape of the first ionization band in methane shows no angular dependence even though the Jahn–Teller splitting is clearly resolved. Peaks attributed to spin–orbit splitting have essentially identical β values, differences arising only from slight differences in the photoelectron energies. Photoelectron bands that correspond to states arising from the spin coupling of unfilled shells, such as in the case of O_2 and NO, also exhibit β values that more closely relate to the initial orbital than to the nature of the final state.[83]

Specific calculations of β for molecules have been made by Berry and others[92] on H_2 and N_2 using a pseudopotential model. The results for H_2 are in good agreement with experiment, but only fair agreement was obtained in the case of nitrogen. Lohr[67] has made calculations on acetylene using plane waves for the continuum wave function. The calculations indicate that at high photoelectron energies, β for σ orbitals approaches +2, and for π orbitals, −1. At lower energies, which includes data taken with He I resonance sources, the theory is at complete variance with experiment. Recent improvements by Iwata and Nagakura,[93] using continuum wave functions based on a hydrogen-like potential and orthogonalized to the bound orbitals promise better success at lower energies. Rabalais et al.[94] have also carried out a number of calculations on β using orthogonalized plane waves. Richie[95] has suggested an approach to solving β which would be suitable to extension to polyatomic systems, but is applicable only to fast photoelectrons. At lower energies the β values are theoretically harder to calculate, but the β values are also more sensitive to the details of the molecular orbital, and the data can be taken with much better resolution. It would in any case be desirable to make measurements of β as a function of photoelectron energy, since it would certainly enhance the uniqueness in the behavior of a given orbital. The β value is expected[94,96] to go through more than one maximum and minimum as one measures the photoelectron energy from threshold to about 200 eV. A complete energy study of β ought to be rather unique for each molecular orbital. If molecules could be orientated and polarized radiation used, a more complex angular distribution would result, containing much more information about the nature of the molecular orbital. Kensinger and Taylor[97] have already made measurements on benzene using polarized radiation from a synchrotron source.

Attempts are being made to calculate β by a semiempirical approach, but as yet success has not been forthcoming. Some other simple, yet limited correlations have been devised. The β has been correlated[82–84] for symmetric molecules with a description of the molecular orbital in terms of "angular momentum character." Carlson *et al.*[85,86] have pointed out that molecular orbitals of similar symmetry which contain significant atomic contributions from atoms in the same group of the periodic table have higher β values when the Z is higher. White and Carlson[89] have correlated the β values obtained for the chloromethanes by dividing the contributions into different groups. That is,

$$\beta(E) = P_{ji}\beta_i(E) \tag{4.7}$$

where $\beta_i(E)$ is associated with a given group, in this case either 1, C–H bonding; 2, C–Cl bonding; or 3, Cl lone pair; and P_{ji} is the fraction of group character as derived from Mulliken density calculations [cf. equation (4.4)] associated with orbital j. Such procedures as above are too highly parameterized to yield a deep understanding of β. However, they give hope that eventually a useful semiempirical approach to β will be forthcoming.

Examination of photoelectron spectra as a function of angle can be of aid simply in unscrambling complicated spectra. For example, when there is an overlap of vibrational structure of different electronic bands, angular measurements can be of help in detecting the presence of these bands and even in estimating their shapes so long as (1) the electronic bands have β values that are sufficiently different so as to distinguish them, (2) the bands do not identically overlap one another, and (3) the β value for a given band does not vary over the vibrational structure. As an example, see Figure 4.1, particularly noting the fourth, fifth, and sixth ionization bands of benzene (D, E, and F in the figure) and the analysis of the spectrum in terms of β.

In general, the relative intensities of the vibrational bands making up a given photoelectron band are independent of the angle θ. The angular parameter β is dependent only on the nature of the electronic structure. From the Born–Oppenheimer approximation it follows that wave functions due to nuclear motion can be factored from those due to electrons, and thus the Franck–Condon factors or the relative intensities for transitions to the various vibrational states are given simply in terms of the overlap integrals of the wave functions for the vibrational states and are completely independent of θ. [Strictly speaking, according to Herzberg,[99] in the Born–Oppenheimer approximation one neglects only the finer interaction of electronic motion and vibration (type *b* vibronic interaction). One should also consider the dependence of the electron wave functions on the nuclear distances (type *a* vibronic interaction).]

The relative intensities of the vibrational bands can be strongly affected by the presence of a second type of excitation. That is, instead of direct photoionization in which an electron is removed from a given orbital into the continuum, $A \to A^+$ resonance absorption can occur in which an electron, whose binding energy is greater than that of the photon, is promoted into a virtual state that subsequently decays by autoionization to the same final electronic state as was formed by photoionization, $A \to A^* \to A^+$. As far as the energy difference between the electronic states is concerned, the two processes are identical. However, the Franck–Condon factors and the angular distribution will be different, since different orbitals are involved. (Cf. Section 2.2.6 and Figure 4.3 for further details on this subject.)

In several cases,[81, 84, 86] e.g., N_2 and CO, the relative intensities of the vibrational lines were found to be strongly dependent on θ in spite of the fact that there appears to be no convincing evidence[100] for resonance absorption. (However, Potts[101] has results on N_2 suggesting that resonance absorption for N_2 may be indeed be taking place with the 584-Å line.) In the series CO_2, COS, CS_2 there seems to be a pattern associated with the vibrational mode. In principle, a breakdown in the Born–Oppenheimer approximation or the presence of type (a) vibronic interaction could lead to a dependence of vibrational structure on θ. However, no quantitative explanation for this behavior has as yet been advanced.

Niehaus and Ruf[88] studied the angular distribution in hydrogen vibrational bands using Ne I resonance lines. Under these conditions they were able to resolve the rotational bands, and they found that the different rotational bands gave different β values. Sichel[102] and Dill[103] have discussed this point theoretically in the case of hydrogen. There may also be a connection between the rotational behavior and the unexplained variation in β for the vibrational bands mentioned above.

4. COMPARISON OF PESOS WITH OTHER EXPERIMENTAL DATA

Data from any one physical method should not stand alone. The value of any physical method is not fully realized until it is united with data obtained from other related techniques. This proposition was discussed in general in Chapter 1. An excellent example of the correlation of the various techniques of PES and optical and mass spectroscopy can be found in the series of articles by the group at the Royal Institute of Technology in Stockholm.[104] Here, I should like to discuss two fields of particular importance to photoelectron spectroscopy: optical spectroscopy and mass spectroscopy.

4.1. Optical Spectroscopy

Photoelectron spectroscopy is an outgrowth of optical spectroscopy in the sense that it attempts to extend the areas of knowledge where optical spectroscopy can advance only slowly and with great effort; that is, a complete set of binding energies for all the molecular orbitals. In many cases (but nothing like the prodigious amount of data compiled by PES) it has been possible to obtain the binding energy from optical data through the identification of Rydberg series. In addition, it has been possible to obtain vibrational spacings in various electronic states of the molecular ion. A major source for optical data are the volumes by Herzberg.[99,105,106] Also available for comparison with photoelectron spectra are the Franck–Condon factors determined from fluorescence data. The data are not strictly comparable, however, since the intensities of the vibrational bands seen in photoelectron spectra are dependent not only on the Franck–Condon factors but also on the photoelectron cross sections, which are energy dependent.

Haselbach and Schmelzer[107] have compared ionization potentials obtained from a series of molecules with excitation energies from the same series of initial molecular orbitals. They find, in principle, that such a comparison can yield qualitative information about the shape of orbitals without recourse to theoretical calculations.

Optical measurements concurrent with photoelectron spectroscopy data are being attempted. Analysis of the fluorescence spectrum of excited ionic species, especially if done in coincidence with analysis of the photoelectron energy, would lead to a deeper understanding of the final state following photoionization. Some initial experiments on fluorescence spectra following photoionization by He I irradiation have been carried out on a few simple molecules.[108]

A complementary technique to photoelectron spectroscopy that holds strong promise is the study of soft x-ray fluorescence. For example, a hole in the core shell of an atom can be filled by x radiation from the valence shell. The final state of such a process will have a vacancy in the valence shell and consequently will be identical with certain final states formed by direct photoionization. Transition probabilities to the final state by x-ray fluorescence, and consequently the x-ray spectrum, will depend on the location of the core vacancy, the overlap with the valence orbitals, and the dipole selection rules for x radiation. Thus, the x-ray spectrum will help clarify the nature of the valence orbital in which the final hole is created.

The limitation in the past for making greater use of x-ray fluorescence has been the difficulty in obtaining high-resolution spectra. This need not be an intrinsic problem. For the first row elements the line broadening should be less than 0.2 eV. Recently, Werme *et al.*(109) have been able to see even vibrational structure from the x-ray spectra originating from a K vacancy in N_2. More details on the use of x-ray fluorescence as a complementary tool to the study of molecular orbitals may be found elsewhere.(110)

4.2. Mass Spectroscopy

Appearance potentials obtained from mass spectroscopy are also valuable as a check against photoelectron spectra. The agreement between the adiabatic ionization potential obtained from PES for the first ionization band and the first appearance potential in mass spectroscopy is usually fairly close. In certain cases where there is a large change in the internuclear distances between the initial and final states, transitions to the ground state vibrational level will be too weak to be observed in photoelectron spectra, and photoelectron data will yield too high a binding energy. For example, note the difficulty in observing the ground state in methane.(111) In general, however, it is easier to obtain a more accurate value for the binding energy from PES than from mass spectroscopy.

The correlation is less clear between mass spectroscopy and photoelectron peaks of higher ionization potentials. A break in the ion intensity curve in mass spectroscopy corresponding to a more tightly bound orbital does not generally occur, in part because a molecular ion transmits its electronic excitation to vibrational excitation and molecular decomposition in the time of $\sim 10^{-5}$ sec that is required for its detection. Breaks in the ion intensity curve that are seen usually are due to vibrationally excited states and, if electron impact is used, to electronic states not excited by photons.

Difficulties in correlating PESOS and mass spectral data have been further explored by Innorta *et al.*(112)

Coincidence measurements between an ion fragment using a time-of-flight spectrometer and a photoelectron whose energy is determined by a parallel-plate analyzer have been described by Danby and Eland.(113) In addition, information was recorded on the kinetic energy of the ion. Figure 4.15 shows a PESOS spectrum in coincidence with $CH_2Cl_2^+$ and CH_2Cl^+. Chapter 7 in Eland's book(18) gives an excellent review of the roles played by fragment ions as the result of photoionization, with the particular aim of relating this field to photoelectron spectroscopy.

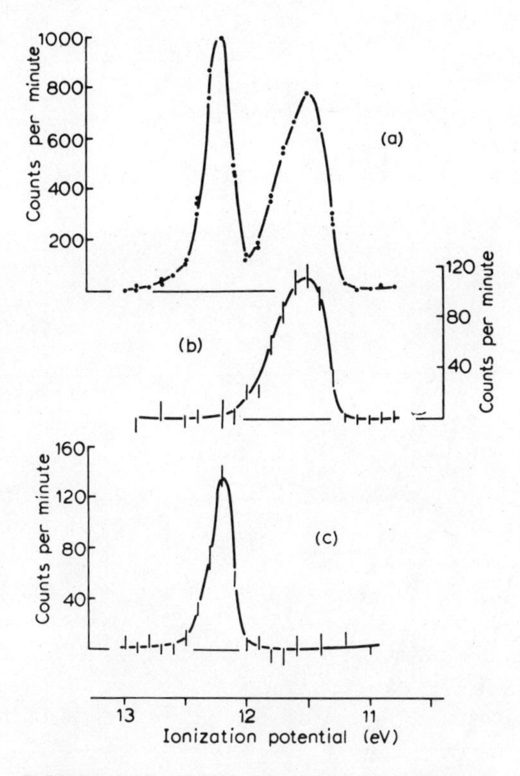

FIGURE 4.15. PESOS spectrum in coincidence with $CH_2Cl_2^+$ and CH_2Cl^+. Spectra show first two ionization bands of CH_2Cl_2. (a) All photoelectrons; (b) coincidences with $CH_2Cl_2^+$; (c) coincidences with CH_2Cl^+. [Reproduced from Danby and Eland,(113) Figure 4.]

5. SURVEY OF THE LITERATURE ON PESOS

In this section we shall review the literature for the specific atomic and molecular systems that have been studied by PESOS. For the most comprehensive compilation of data up to early 1973, the reader is directed to the CRC *Handbook of Spectroscopy*.(114) One table gives the ionization potentials, both vertical and adiabatic, together with molecular orbital assignments and vibrational spacings, when available. Approximately 900 compounds are listed, the data being taken from 236 references. A shorter table in the same handbook lists data on the valence shells using soft x-ray sources. Spectra of some 150 compounds using the 584-Å source are to be found in the book by Turner *et al.*(37) Two other books primarily on uv photoelectron spectra are by Baker

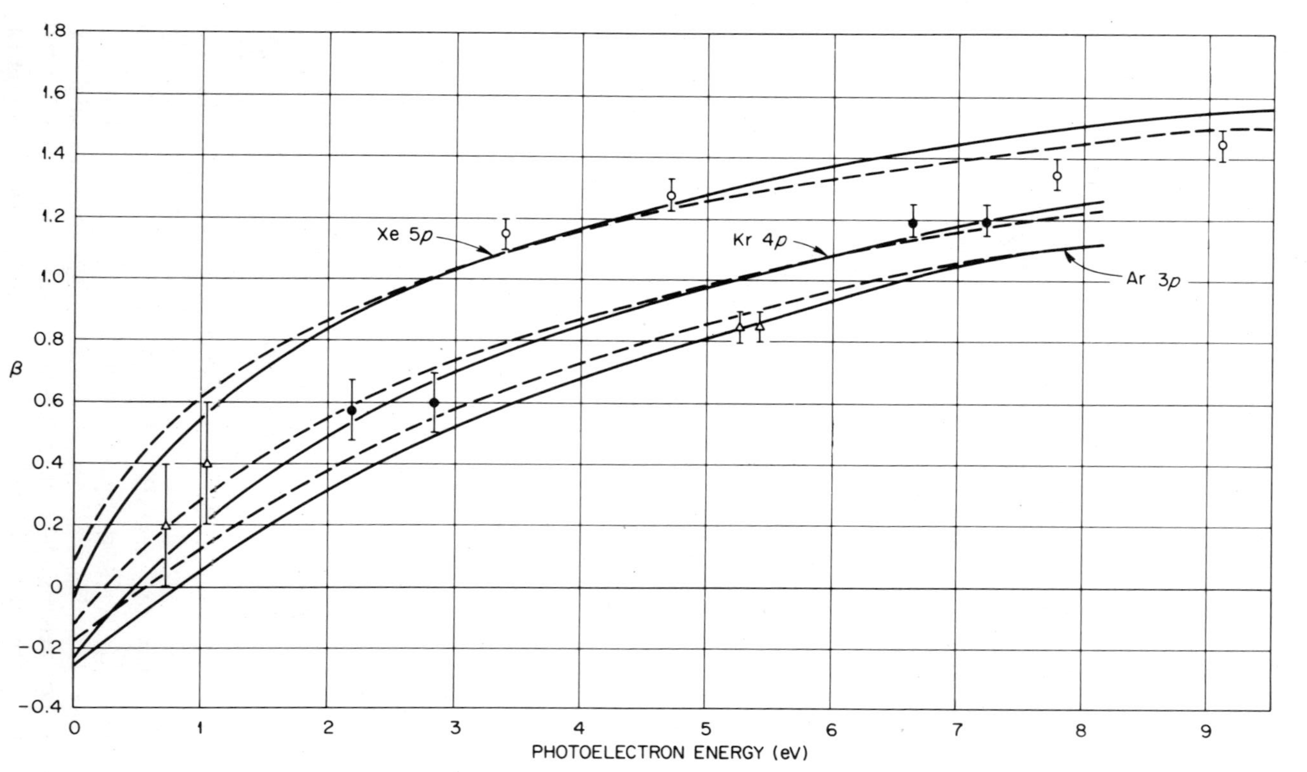

FIGURE 4.16. Comparison of experimental results with theory for the angular parameter β determined from the photoelectron spectra of the rare gases. Experimental data given as points. Solid lines are for calculations made in terms of dipole velocity and dashed lines for dipole length. [Reproduced from Carlson and Jonas,[84] Figure 5.]

and Betteridge[36] and by Eland.[18] The latter is quite recent and is an excellent textbook, but it covers the literature only up through 1972. Good comprehensive reviews of the literature have been written by Worley[115] and Betteridge.[116] There are in addition four excellent reviews in depth on inorganic compounds[117–120] and two by Heilbronner[48,121] on organic compounds.

For this book a review of *all* the literature will not be feasible, but it is hoped that the discussion will be sufficiently complete that the reader will have a good idea of the overall accomplishments of photoelectron spectroscopy.

We shall in this section primarily discuss the results and conclusions of PES using the He I (21.22 eV) and He II (40.8 eV) lines, and our discussion will generally restrict itself to work done with high-resolution dispersion spectrometers.

5.1. Atoms

Work on atomic systems forms the backbone of understanding the photoelectronic process itself. Thus, experimental studies and theoretical calculations on photoelectron cross sections have been made for the valence shells of atoms, not only with the more common photon sources (He I 584 Å and He II 304 Å), but also for a complete range of energies from uv to x rays.[122,123] In addition, results of calculations are available for angular distributions over a wide range of energy.[124] Figure 4.16 gives some results for the angular distribution parameter β. Reasonable satisfactory theoretical results have been obtained using Hartree–Fock wave functions, and we can in general assume that molecular physics rests on a firm basis of atomic behavior.

The photoelectron spectra of the outer shells of the rare gases have been of interest in the study of multicomponent structure arising from electron shake-up and configuration interaction. I shall defer discussion of these results, however, until Chapter 5, when a complete review of satellite line structure will be presented.

The accurately known binding energies of the orbitals of the rare gases and the essentially chemically inert behavior of these atoms make them an ideal source for calibration. Table 4.5 lists for reference the binding energies for the valence subshells of Ne, Ar, Kr, and Xe. Photoelectron studies using the He resonance lines have also been extended to some other atoms, e.g., Hg[125] and O.[126]

5.2. Diatomic Molecules

Like the rare gases, studies on the diatomic molecules are most important for the light they throw on the basic phenomenon of photoionization itself.

TABLE 4.5

Binding Energies of the Outer Shells of the Rare Gases[a]

Atom	Orbital	Final state M^+	Ionization energy, eV
Ne	$2s$	$2p^6(^2S)$	48.48
	$2p_{1/2}$	$2p^5(^2P^0); J=\frac{1}{2}$	21.67
	$2p_{3/2}$	$2p^5(^2P^0); J=\frac{3}{2}$	21.57
As	$3s$	$2p^6(^2S)$	29.24
	$3p_{1/2}$	$3p^5(^2P^0); J=\frac{1}{2}$	15.94
	$3p_{3/2}$	$3p^5(^2P^0); J=\frac{3}{2}$	15.76
Kr	$4s$	$4p^6(^2S)$	27.52
	$4p_{1/2}$	$4p^5(^2P^0); J=\frac{1}{2}$	14.67
	$4p_{3/2}$	$4p^5(^2P^0); J=\frac{3}{2}$	14.00
Xe	$5s$	$5p^6(^2S)$	23.40
	$5p_{1/2}$	$5p^5(^2P^0); J=\frac{1}{2}$	13.44
	$5p_{3/2}$	$5p^5(^2P^0); J=\frac{3}{2}$	12.13

[a] Taken from C. E. Moore, *Atomic Energy Levels*, Vols. I, II, III (NBS 467, U.S. Govt. Printing Office, Washington, D.C., 1949, 1952, 1958), using $1\ \text{cm}^{-1} = 1.240 \times 10^{-4}$ eV.

The presence of only one degree of vibrational freedom makes the analysis of vibrational structure particularly straightforward.

5.2.1. H_2

Molecular hydrogen has only one occupied orbital, and thus only one ionization band is observed in its photoelectron spectrum. However, it has been examined extensively. Because the orbital is highly bonding, the photoelectron spectrum shows a large number of vibrational bands, with the peak of intensity occurring at v_2'. The relative experimental intensities of the vibrational bands, together with calculated values, are given in Table 4.6. A clear picture of the photoionization process emerges in terms of the initial and final potential curves. (Cf. Figure 3.5.) The vibrational spacing in the molecular ion is considerably less than in the neutral molecule, because of the greater internuclear separation. The 0–1 vibrational spacing for H_2^+ is 2260 cm^{-1} while it is 4280 cm^{-1} for H_2. The decrease in vibrational spacing in the ionized molecule is endemic to bonding orbitals and is particularly notable when H is part of the bond. The calculated Franck–Condon factors (cf. Chapter 3, Section 3.7) give qualitative agreement with the measured intensities, but there is a steady deviation with energy due to the dependence of photoelectron cross section with photoelectron energy. Itikawa[(127)] has recently calculated these intensities without invoking the Born–Oppenheimer approximation (cf. Table 4.6). The spacing between the vibrational bands decreases as one approaches the disso-

TABLE 4.6

Relative Intensities of the Vibrational Bands in the Photoelectron Spectra of H_2 and D_2

	H_2			D_2		
v'	Exp.[a]	FC[b]	YI[c]	Exp.[a]	FC[b]	YI[c]
0	0.463	0.523	0.446	0.230	0.260	0.210
1	0.863	0.920	0.853	0.538	0.640	0.554
2	1.0	1.0	1.0	0.812	0.940	0.851
3	0.884	0.877	0.940	0.950	1.048	0.999
4	0.700	0.686	0.783	1.0	1.0	1.0
5	0.579	0.504	0.608	0.928	0.864	0.903
6	0.472	0.357	0.453	0.756	0.699	0.763
7	0.313	0.249	0.330	0.624	0.541	0.614
8	0.241	0.172	0.238	0.487	0.407	0.478
9	0.137	0.119	0.170	0.342	0.300	0.365
10	—	0.083	0.122	0.270	0.219	0.274

[a] Experiment [J. Berkowitz and R. Spohr, *J. Electron Spectros.* **2**, 143 (1973)].
[b] Franck–Condon factor.
[c] Calculations of Itikawa.[127]

ciation limit of H_2^+, and the point at which this limit is reached can be clearly seen (Figure 4.17).

Because of the small mass of H_2, the energy spacings of the rotational bands in H_2 are relatively large and can be observed (cf. Figure 4.18). The ob-

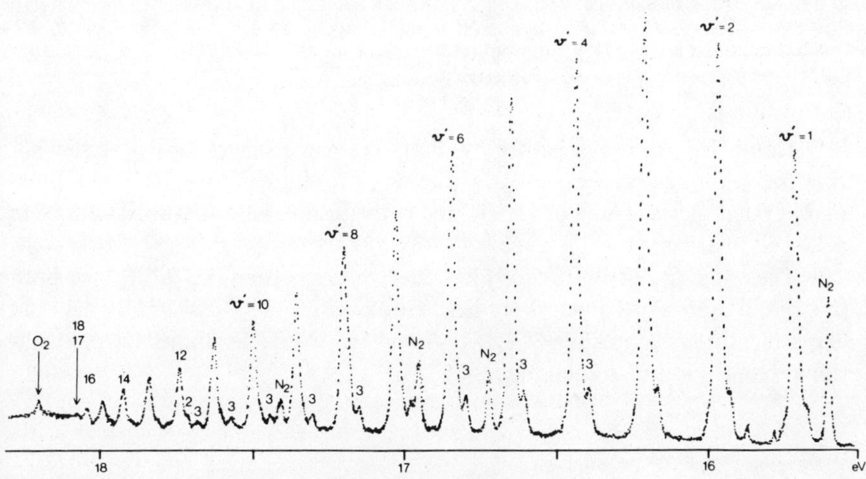

FIGURE 4.17. Photoelectron spectrum of H_2 using $h\nu = 584$ Å. v_0 is not shown in the figure. The dissociation limit occurs at 17, 18. [Reproduced from L. Åsbrink, *Chem. Phys. Lett.* **7**, 549 (1970). Figure 1.]

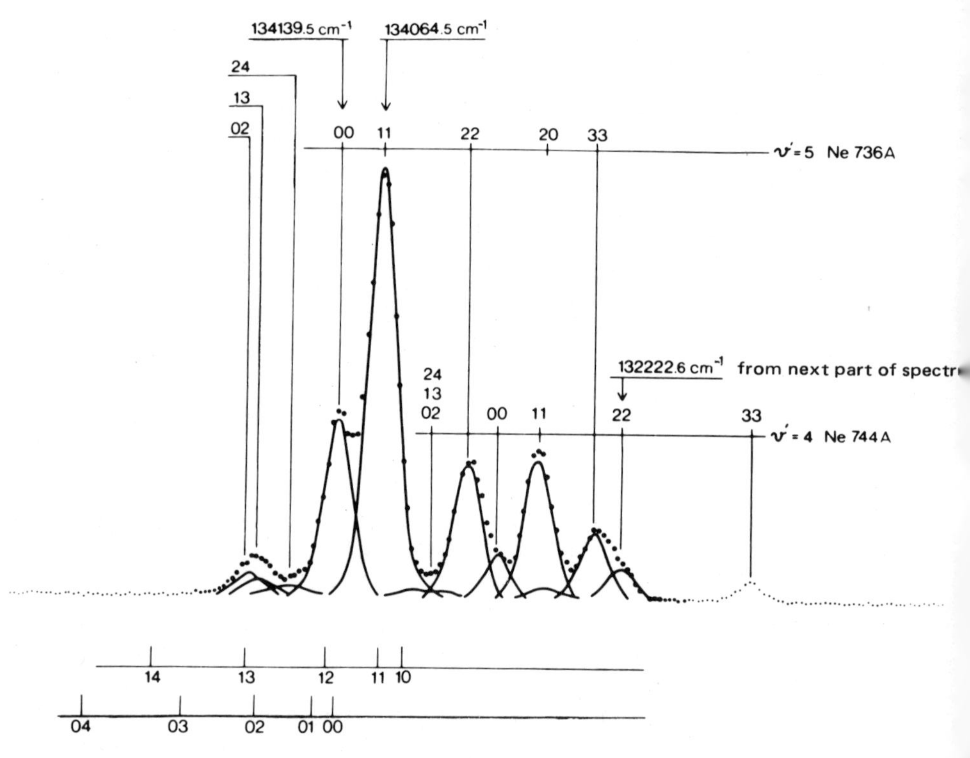

FIGURE 4.18. Detail of the photoelectron spectrum of H_2, showing rotational structure in th $v' = 4, 5$ bands. Neon radiation at 736, 744 Å has been used. The rotational lines have been desig nated by two numbers, J'' and J'. The lines corresponding to the selection rule $\Delta J = 0, \pm 2$ ha been indicated above the curve. [Reproduced from L. Åsbrink, *Chem. Phys. Lett.* **7**, 549 (1970 Figure 3.]

servation of rotational structure by photoelectron spectroscopy is somewhat of a tour de force, and systematic studies of rotational structure for a large number of molecules are not anticipated in the future, but that success has been achieved in at least one instance is gratifying.

The angular distribution of photoelectrons has been the subject for both theoretical and experimental studies. Good agreement between theory and experiment has been achieved (cf. Figure 4.19). Angular studies on the rotational bands have also been made.[88]

5.2.2. N_2 and CO

Nitrogen and carbon monoxide are isoelectronic and have the same orbital configuration. Of the first three ionization bands observed in photoelectron spectroscopy (cf. Figure 4.9), the first and third show most of the transitions going to the ground state vibrational level, indicative of the σ nonbonding orbitals, while the second band has a more complex spectrum, with the highest intensity being observed for an excited vibrational state, which is characteristic of a bonding orbital as is anticipated for the π_g orbital. In addition, the second band shows a decrease in vibrational spacing relative to that found for the ground state of the neutral molecule: namely 23% in the case of N_2 and 25% in the case of CO, which is likewise indicative of a bonding orbital. In addition, analysis of the angular distribution of photoelectrons for each of the three ionization bands also suggests, in agreement with the vibrational analysis, that the order of orbitals in terms of increasing binding energy is $2p\sigma_u$, $2p\pi_u$, $2s\sigma_g$ for both N_2 and CO.[84]

In apparent contradiction to experimental analysis, Hartree–Fock calculations on N_2 have the order in terms of eigenvalues π_u, σ_u, σ_g. (For CO the order of the eigenvalues agrees with experiment.) Presumably, Koopmans' rule fails in the case of nitrogen, and the eigenvalues or binding energies of the frozen orbitals are not in the same order as the experimental binding energies. (Cf. Chapter 3, Section 3.6 for a discussion of Koopmans' theorem.) Experiment should, however, agree with the theoretical ordering based on binding energies derived from the total energies for ions whose configurations correspond to a removal of a single electron from one of the given orbitals.

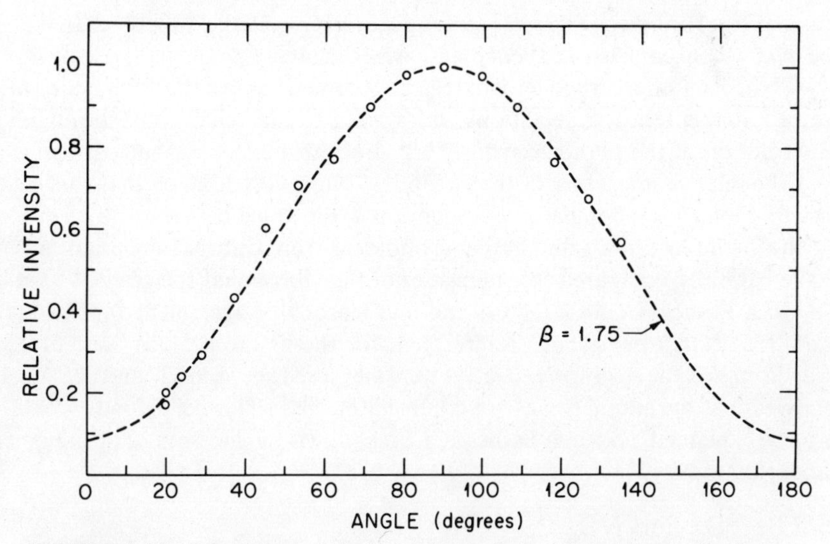

FIGURE 4.19. Comparison of theory and experiment of angular distribution of photoelectron spectrum of H_2 ($^1X_g^+$, v_2'). Circles are experimental data. Dashed line is theory. [Reproduced from Carlson and Jonas,[84] Figure 6.]

TABLE 4.7

Comparison of Theoretical and Experimental Binding Energies for N_2

Theory[a]	Δ(ionization potential) $(A\ ^2\Pi_u) - (X\ ^2\Sigma_g^+)$, eV
SCF orbitals of N_2 $(^1\Sigma_g^+)$	−0.78
Large basis set, very close to HF limit	−0.65
Variational calculations for each state; Nesbit basis set	−0.85
After limited configuration interaction	+0.58
After correction for difference in correlation energy	+1.17
Experiment	+1.12[b]

[a] See Table II, G. Verhaegen, W. G. Richards, and C. M. Moser, *J. Chem. Phys.* **47**, 2595 (1967).
[b] J. L. Franklin, J. G. Dillard, H. M. Rosenstock, J. T. Herron, K. Draxl, and F. H. Field, *Natl. Std. Ref. Data Ser. Natl. Bur. Std.* 26 (1969).

However, in the case of N_2 even using an *ab initio* Hartree–Fock treatment with an extensive basis set the difference in total energies between the $A\ ^2\Pi_u$, $^2\Sigma_g^+$ states are at variance with the experimental binding energies by 1.8 eV. Further improvements including corrections for correlation finally bring theory and experiment together (cf. Table 4.7). This would seem to augur poorly for correlating experimental binding energies with theory by use of Koopmans' theorem for more complex molecules. Fortunately, with more complex molecules Koopmans' theorem in general works as well as or better than in the case of N_2. However, Table 4.7 serves as a warning of the difficulties that can be met when comparing theory and experiment.

Nitrogen and carbon monoxide show somewhat peculiar behavior with respect to theoretical expectations with regard to the behavior of the angular distribution of the photoelectrons.[81, 84] In both molecules, but particularly N_2, the relative intensities of the v_0' and v_1' vibrational bands that make up the first ionization band are dependent on θ, the angle between the direction of the incoming photon and outgoing photoelectron. Gardner and Samson[100] have carefully compared the intensities of the vibrational spectra of CO and N_2 with Franck–Condon factors and find reasonable agreement for the $A\ ^2\Pi$ and $B\ ^2\Sigma^+$ states, but strong disagreement for the first ionization band of N_2. Furthermore, the $v = 1$ to $v = 0$ ratios were measured between 537 and 645 Å and showed that the anomaly was not confined to 584 Å. This behavior as yet cannot be explained, and may indicate a breakdown in the Born–Oppenheimer approximation.

5.2.3. O_2 and NO

Both oxygen and nitric oxide have unfilled molecular orbitals in the neutral molecule. These configurations are, respectively,

$$O_2: \ (\sigma_g 1s)^2(\sigma_u 1s)^2(\sigma_g 2s)^2(\sigma_u 2s)^2(\sigma_g 2p)^2(\pi_u 2p)^4(\pi_g 2p)^2 \ {}^3\Sigma_g^-$$
$$NO: \ (\sigma 1s_O)^2(\sigma 1s_N)^2 (\sigma_g 2s)^2(\sigma_u 2s)^2(\sigma_g 2p)^2 (\pi_u 2p)^4(\pi_g 2p)^1 \ {}^2\Pi$$

In the case of oxygen there are two electrons of unpaired spins in the π_g $2p$ orbital, while there is one unpaired spin for NO. If an electron is removed from the outermost π_g $2p$ orbital, the resulting ion has only one orbital with an unpaired spin in the case of O_2^+ and all filled shells for NO^+. Thus, only one final state occurs from photoionization in the outermost orbital; but removal of an electron from any other orbital will result in more than one final state, dependent on the coupling of the unpaired spins. For example, in oxygen both the ${}^4\Pi_u$ and ${}^2\Pi_u$ have been identified as arising from ionization in the π_u $2p$ orbital at 16.12 and 17.1 eV. Working with high resolution and sensitivity, Edqvist *et al.*(9) have identified all six states that can come from ionization of its π_u $2p$ orbital. (Fig. 4.20) A similar careful study has been made on NO.(10)

For a complete discussion of the number and variety of states possible see Herzberg.(106) Studies on the angular distribution of photoelectrons from O_2 and NO have also been made,(83) and it was determined that the angular parameter β did not depend sensitively on the final state resulting from the removal of an electron from a given orbital, but rather depended more on the nature of the initial orbital.

5.2.4. Diatomic Molecules Containing Halogen

The halogen acids reveal two ionization bands; the first, corresponding to the ${}^2\Pi$ state, derives from the nonbonding p electron of the halogen. There

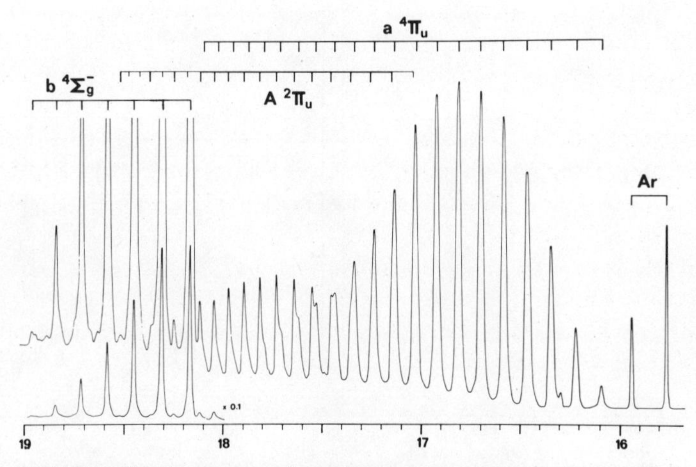

FIGURE 4.20. Portions of photoelectron spectrum of O_2 using 584-Å radiation. Energy scale in terms of ionization potential (eV). [Reproduced from Edqvist *et al.*(9)]

actually appear two final states due to spin–orbit coupling, $^2\Pi_{3/2}$ and $^2\Pi_{1/2}$, which correspond to the $p_{3/2}$ and $p_{1/2}$ orbitals of the halogen atom. The spin–orbit splittings for HCl, HBr, and HI are, respectively, 0.08, 0.33, and 0.66 eV, in good agreement with spectroscopic values and theoretical calculations. The second ionization band is due to the $^2\Sigma^+$ state and arises from ionization of the σ hydrogen–halogen bond. It is strongly bonding and thus presents a broad peak.

Vibrational structure is clearly seen for HF and HCl, but not for HI. In the second ionization band of HBr one has the intermediate situation in which the first portion of the band up to v_3' is clearly resolved, but in the remaining portion of the structure the vibrational structure becomes increasingly broad until one experiences a continuum spectrum. Lempka *et al.*(128) have demonstrated the cause of this phenomenon in terms of the potential curves of the ionized state in which the $^2\Sigma^+$ levels are being crossed by a $^4\Pi$ repulsion state, leading to $H + X^+$. Delwiche *et al.*(129) have compared the dissociation limit of HBr with DBr. (Cf. Figure 4.21.)

The halogens F_2, Br_2, and I_2 and the interhalogens ClF, BrF, ICl, and IBr have been studied.(130) The first ionization band is due to the removal of an electron from the π_g orbital. It is the only band of the first three seen with He I

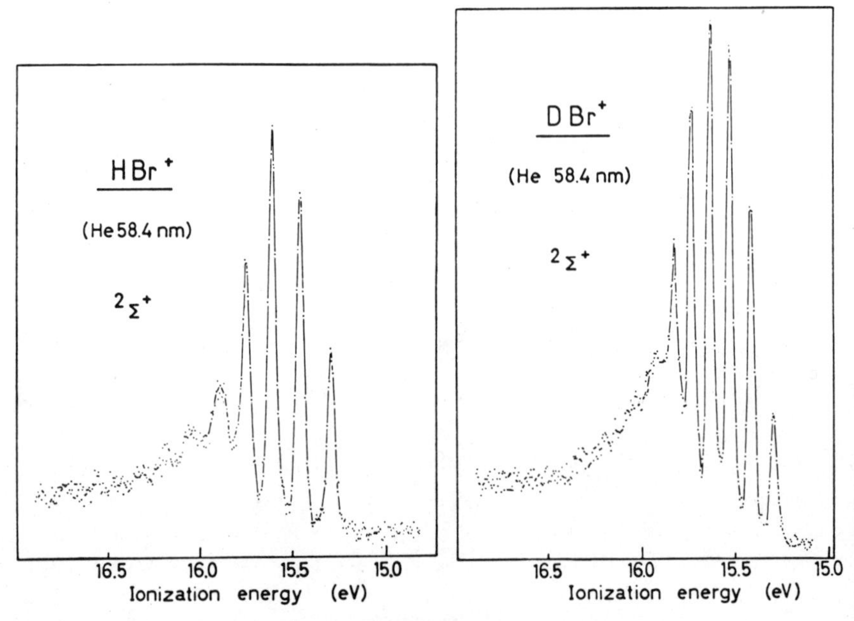

FIGURE 4.21. Photoelectron spectrum of second band in HBr and DBr. [Reproduced from Delwiche *et al.*,(129) Figures 3 and 4.]

(21.22 eV) that shows vibrational structure. It also shows spin–orbit splitting, which has been checked theoretically with good results.[1] The bromine spectra also give evidence for a hot band, i.e., transitions from the thermally excited v_1 vibrational level in the neutral molecule. At room temperature the probabilities, using a calculated Boltzmann factor, for populating the v_1 state is 21 %, in agreement with the experimentally observed hot band. For a mixed halide, e.g., ClF, the spin–orbit splitting lies between that for F_2 and Cl_2 but weighted toward the chlorine, the splitting for the three molecules being, respectively, 628, 337 and 645 cm^{-1}. Again experiment agrees nicely with theoretical predictions. (Cf. Table 4.1.)

5.3. Triatomic Molecules

5.3.1. Linear Triatomic Molecules

The most common linear triatomic molecules N_2O, CO_2, and CS_2 have been carefully studied. Three nonbonding orbitals are found among the first four ionization potentials as evidenced by the high transition probability to ground state vibrational level. In CO_2, COS, and CS_2 these bands correspond to the first, third, and fourth ionization states, but in N_2O the first, second, and fourth. These observations, together with data on the emission spectra of the ions, have enabled Brundle and Turner[131] to assign designations to the final states. Disagreement is found with the ordering as predicted from calculated eigenvalues, perhaps due to a failure in Koopmans' theorem.

Of particular interest for the triatomics is the vibrational structure. There are three degrees of vibrational freedom (actually four in the case of linear molecules) rather than the single degree of vibrational freedom that characterizes the diatomic molecules. One is curious as to which of the vibrational modes is activated in the photoionization process. Selection rules covering polyatomic molecules are as follows for transitions from the ground vibrational state of the neutral molecule:

1. The final state may have any number of quanta of a vibrational mode that is symmetric to all symmetry species in the molecule.

2. Only zero or an even number of quanta of a mode that is antisymmetric to symmetry species are allowed.

This means that ν_1 (symmetric stretching) can occur for all four linear molecules, ν_2 (bending mode) can have only even values of v', while ν_3 (antisymmetric mode) can occur in all values of v for N_2O and COS but only as even values for CO_2 and CS_2. Experimentally, vibrational levels corresponding to ν_1 and ν_3 are found, apparently consistent with the above selection rules; however, no vibrational peaks were found that would correspond to ν_2. Most difficult to understand is the substantial decrease in the vibrational frequency for ν_3 found in the molecular ions of CO_2 and CS_2 as compared to the neutral species.

Angular distribution studies[86] have been carried out on CO_2, COS, and CS_2. Most curious is the analysis of the angular distribution of the vibrational states of the fourth ionization band of COS, CS_2, and CO_2 ($\tilde{C}\,^2\Sigma^+$). Normally, within the framework of the Born–Oppenheimer approximation the angular distributions of the individual vibrational bands making up an electronic transition should be the same, as is usually found to be the case. But as seen in Table 4.8, the angular parameters β of the different vibrational lines vary extensively; in addition, this variation seems to follow a pattern. That is, in CO_2 and CS_2 the symmetric stretching mode has a high β, in agreement with the ground state transition, while the antisymmetric mode has a much lower β. In COS it is the antisymmetric mode that has the higher β and is consistent with the ν'_{000} transition, while the symmetric mode is much lower.

Another important linear triatomic is HCN. Some confusion exists in the location of the ionization band due to the 5σ orbital, which is essentially a nitrogen lone-pair orbital. It is believed[37] that the missing band overlaps the initial ionization band arising from the 1π level. This assignment is borne out when comparing the orbital assignments of the homologous compounds methyl cyanide and methyl isocyanide.

5.3.2. Bent Triatomic Molecules

Perhaps the most studied of the bent triatomic molecules is water. Often these studies are inadvertent since water is a most persistent contaminant and its first ionization band with the high intensity peak corresponding to the adiabatic ionization potential (12.61 eV) can be found in spectra even after extensive bakeout of the spectrometer. This band is due to the nonbonding lone-pair orbital. A detailed study of the first vibrational band with resolution pushed to 4 meV (FWHM) has revealed the presence of rotational structure.[132] One

TABLE 4.8

Angular Distribution Parameter[a] for the Fourth Ionization Band ($C\,^2\Sigma^+$) of CO_2, COS, and CS_2

Vibrational band[b]	CO_2	COS	CS_2
ν'_{000}	1.3 ± 0.1	0.55 ± 0.1	1.2 ± 0.1
ν'_{100}	1.4 ± 0.1	0.25 ± 0.07	1.1 ± 0.1
ν'_{001}	—	0.72 ± 0.08	—
ν'_{002}	$+0.2 \pm 0.1$	0.68 ± 0.10	$+0.1 \pm 0.3$

[a] Data taken from Carlson and McGuire.[86]
[b] Vibrational assignments taken from Turner *et al.*[37]

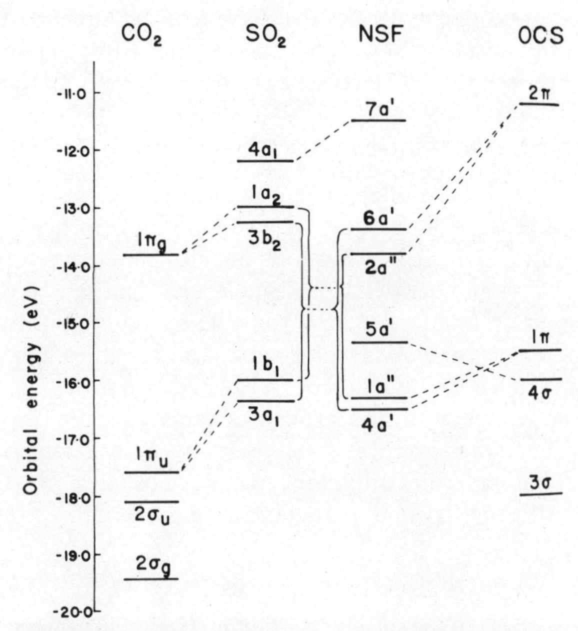

FIGURE 4.22. Correlation of ionization potentials for CO_2, SO_2, NSF, and OCS. [Reproduced from R. N. Dixon, G. Duxbury, G. R. Fleming, and J. M. V. Hugo, *Chem. Phys. Lett.* **14**, 60 (1972), Figure 3.]

feature of the spectrum of the second band of H_2O^+ is possible evidence for Renner–Teller splitting as seen from the irregularities of the individual vibrational peaks.

The study of water also affords us with an example of the use of isotopic substitution. Comparison of the photoelectron spectra of H_2O and D_2O shows little change in the ionization potentials but substantial change in vibrational frequency and Franck–Condon factors. Handled with care, isotopic labeling may be used to identify vibrational modes in the photoelectron spectrum, which in turn can lead to orbital identification. From the calculation of Botter and Rosenstock(133) it has been possible to use the Franck–Condon factor of the $^2B_1$ (first ionization band) to obtain the bond angle and bond distance for H_2O^+ and D_2O^+, for which Brundle and Turner(134) find $\theta = 109°$ and $r = 0.995$ Å. Bergmark *et al.*(135) have investigated the effect of isotopic substitution of ^{18}O for ^{16}O and find evidence of vibronic coupling.

Studies on the angular distribution of photoelectrons from H_2O and H_2S have been carried out.(86) They reveal that for a given orbital the sulfur compound has a substantially higher β than the oxygen counterpart, in confirmation of the observations on CO_2, COS, and CS_2.

Other bent triatomic molecules that have received attention from photoelectron spectroscopists are SO_2, NO_2, and O_3, and NSF. Figure 4.22 shows a correlation between the MO level schemes for SO_2, NSF, CO_2, and COS.

5.4. Organic Molecules

Polyatomic molecules offer more complexity in terms of their photoelectron spectra because of a greater variety of molecular orbitals and a larger number of vibrational degrees of freedom, but also because there is also an almost overwhelming number of different molecules than can be or have been studied. However, in numbers there is sometimes strength, for by studying a group of homologous compounds information can be gained that would elude even the most detailed investigation of an isolated system. This is particularly true of organic compounds. The groupings can take many forms, such as similarity in a functional group, a symmetry group, or a characteristic element (e.g., halide). Some molecules will belong to more than one group. In discussing the polyatomics we shall try to be particularly cognizant of this group behavior.

5.4.1. Methane, Alkanes, and Tetrahedral Symmetry

The energy levels of the saturated straight-chain hydrocarbons are shown in Figure 4.23. As the molecule grows in length, the number of orbitals as evidenced by the ionization bands increases. However, it is interesting to note that there appears a general grouping that falls into two energy sections, whose basic origins are the $2s$ a_1 and $2p$ t_2 orbitals of methane.

Methane is not only the simplest member of the paraffin family, but is also the simplest example of a symmetric tetrahedral compound. Methane is one of the best studied examples of Jahn–Teller splitting. Upon removal of an electron from the highly symmetric molecule, the degeneracy is removed and the resulting photoelectron band is split into two discernible peaks and a shoulder. The phenomenon has been discussed in detail by Grimm and Godoy[14] and Dixon[13]; Figure 4.24 shows the photoelectron spectrum of methane. Not only are the electronic bands distorted, but the vibrational structure shows an enhanced complexity. It is interesting to follow the Jahn–Teller splitting through a series of related symmetric compounds. When heavier atoms are used to replace the hydrogen in methane the Jahn–Teller splitting is greatly reduced. When halide atoms are used for replacement, the Jahn–Teller splitting is essentially eliminated.[38, 39] However, replacement of carbon by other members of the IVa group in the periodic table (namely the series CH_4, SiH_4, GeH_4, SnH_4) only reduces[16, 136] the splitting by the change in force constants (about 30%). In the series of $M(CH_3)_4$ compound where M is C, Si, Ge, Sn, and Pb, the Jahn–Teller splitting is increasingly apparent as one goes up in atomic number.[16, 17] Rabalais and Katrib[137] have made a careful study of ethane

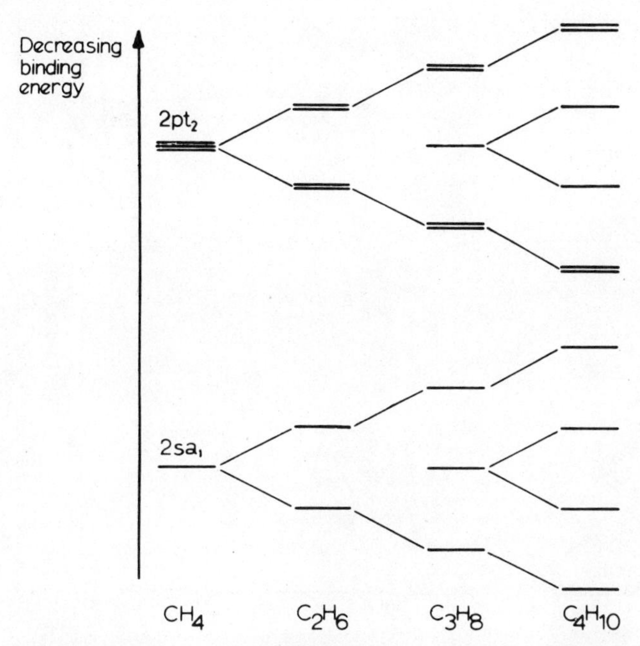

FIGURE 4.23. Schematic diagram showing relative orderings of C–H orbitals in alkanes. [Reproduced from A. D. Baker, D. Betteridge, M. R. Kemp, and R. E. Kirby, *J. Mol. Struc.* **8**, 75 (1971), Figure 1.]

and from the vibrational analysis assign the ground state to a Jahn–Teller-split ^2E_g state. Murrell and Schmidt(138) have studied the series $CH_{4-n}(CH_3)_n$. They find that formally degenerate pseudo-π CH orbitals of the methyl groups are split by "through space" interaction which amounts to 2.7 eV in the case of neopentane.

5.4.2. Unsaturated Aliphatics

The presence of double and triple bonds in an adiabatic compound adds their peculiar characteristics to the photoelectron spectrum. This comes in the form of the characteristic vibrational spacings and the presence of a π orbital that is nonbonding and is generally responsible for the lowest ionization band in the spectrum. Figure 4.25 shows how the spectra of aliphatic molecules with a single double bond are related to the spectrum of ethylene. From angular distribution studies on photoelectron spectra of unsaturated hydrocarbon it appears that the angular parameter β is characteristically large for the π orbitals associated with the carbon double bond structure. This is nicely demonstrated in the case of butadiene. Angular distribution studies(139) reveal that

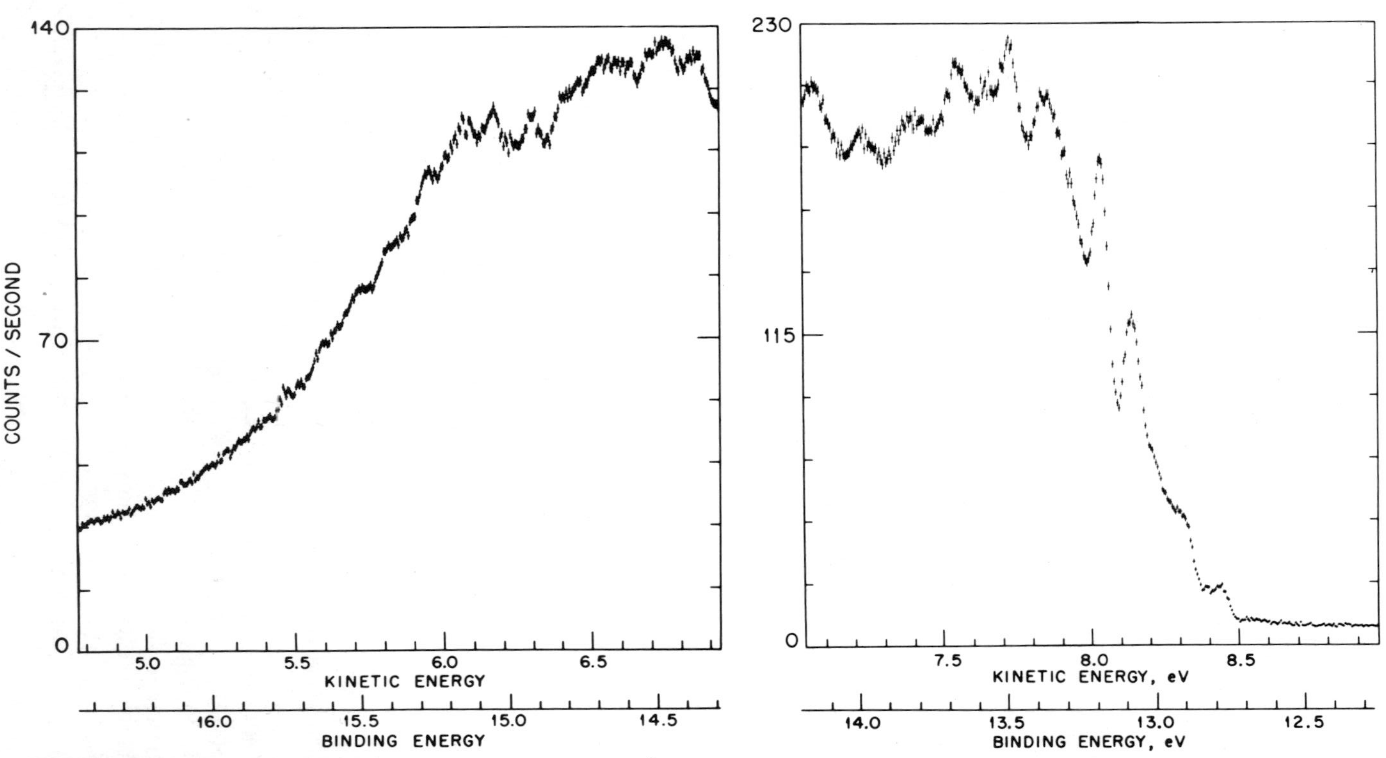

FIGURE 4.24. Photoelectron spectrum of methane ($h\nu = 584$ Å) showing complex vibrational structure and the three principal electronic bands that result from Jahn–Teller splitting, whose centers are located at the binding energies of 13.5, 14.4, and 15.1 eV. [Reproduced from Pullen *et al.*,[136] Figure 7.]

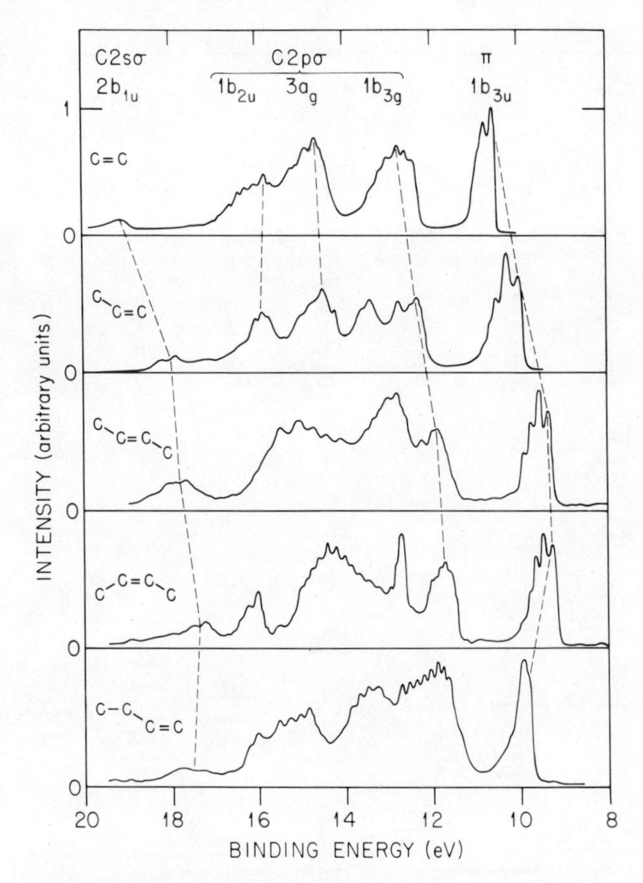

FIGURE 4.25. Comparison of the photoelectron spectra of HCH=CHY molecules. [Reproduced from White *et al.*,[139] Figure 7.]

the first two ionization bands have unusually large angular parameters β, in agreement with the analysis of Åsbrink *et al.*[140] that the first two bands are the carbon $2p$ π bands.

It is also useful to compare all the compounds containing ≡N with those of the isoelectronic structure ≡CH. Figure 4.26 shows the systematics of a series of triply bonded molecules as taken from Turner *et al.*[37]

5.4.3. Ring Compounds

The behavior of the orbitals studied by photoelectron spectroscopy of simple three-member ring compounds has been correlated by Turner *et al.*[37]

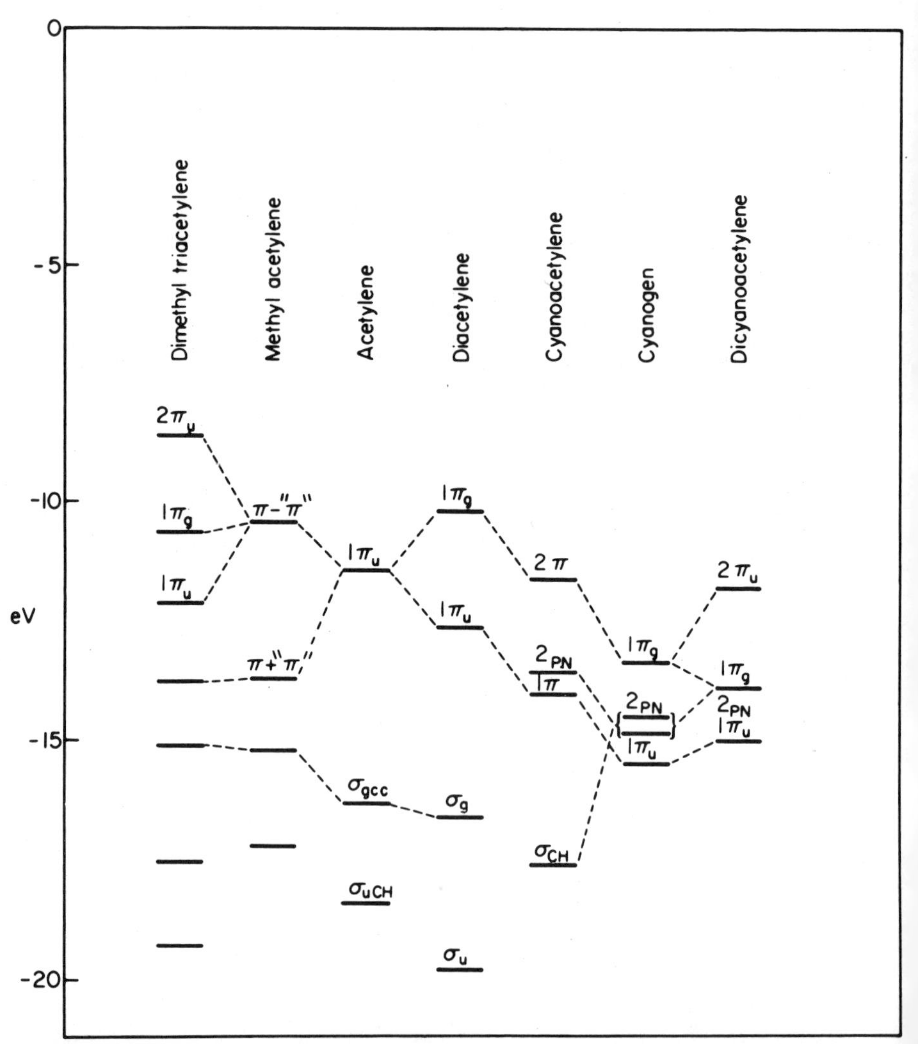

FIGURE 4.26. Correlation of energy levels in acetylene, diacetylene, and related compounds produced by replacing ≡C by ≡N. [Reproduced from Turner *et al.*,[37] Figure 6.34.]

using Walsh's and Sugden's description of the molecular orbitals as a starting point.

Probably the most important of the ring compounds is benzene, and it has been studied extensively by photoelectron spectroscopy. Using data from both the He I and He II radiation and data by Jonsson and Lindholm on mass spectroscopy and Rydberg transitions, Åsbrink *et al.*[141] have been able to assign orbitals to the different ionization peaks. Some confusion arose as to the possible presence of two σ orbitals at 10.3 and 10.8 eV as suggested from data from photoionization efficiency curves. However, it appears now that all the orbitals in benzene have been accounted for in the photoelectron spectrum and that each peak has the same order of magnitude of intensity. The suggestion that photoelectron peaks might be missed because of low intensity in the region 9.3–11.3 eV, that is, between the first two observed peaks, seems unlikely. Measurements[(82)] on the angular distribution of the photoelectron spectrum of benzene have been made. First, the overlapping bands can be distinguished from one another from the angular distribution data, and second, the angular parameter β obtained for the first six ionization bands is in qualitative agreement with expectations based on the orbital assignment of Åsbrink *et al.*[(141)] The ordering of some of the bands is still in dispute by Potts *et al.*,[(142)] and they have used evidence of resonance absorption in the spectrum of C_6D_6 to support their claims.

Many substituted benzene molecules have been studied by photoelectron spectroscopy.[(37, 142–145)] These data have helped clarify the nature of the benzene orbitals. Of great importance has been the information on the first

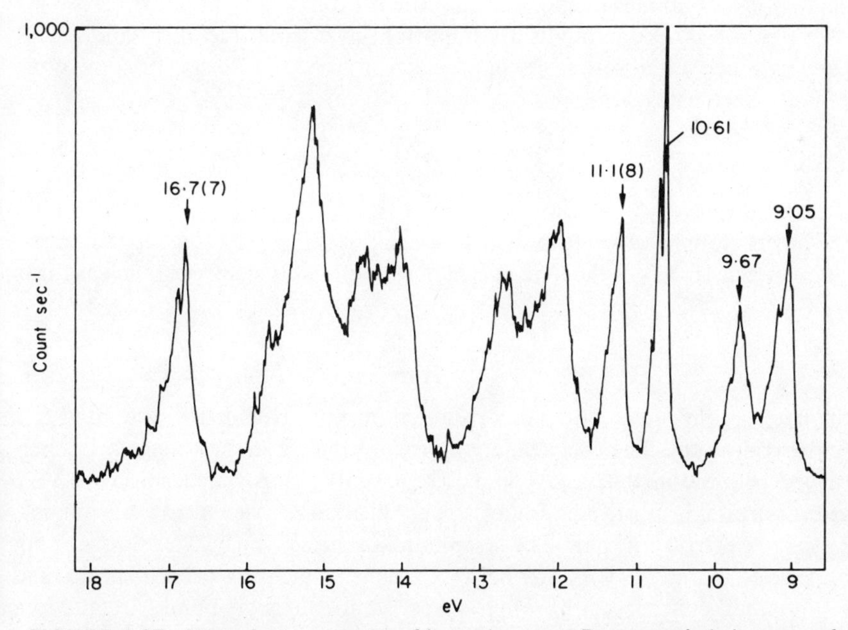

FIGURE 4.27. Photoelectron spectra of bromobenzene. Energy scale is in terms of ionization potential. Note splitting of the e_{1g} orbital of benzene into peaks at 9.05 and 9.67 eV. Note also lone-pair orbitals at 10.61 and 11.1 eV. [Reproduced from Turner *et al.*,[(37)] Figure 11.4.]

ionization band. This orbital, the doubly degenerate $(e_{1g})^4$ in benzene, splits on substitution into two orbitals (B_1 type, π_3) and (A_1 type, π_2). For example, see Figure 4.27. At the point of substitution the B_1 type has its maximum electron density, while the A_2 has a node. Thus, most molecules, such as the halobenzene, phenol, and aniline derivatives, show a splitting, with the π_2 orbital remaining the same in energy as the original e_{1g} benzene orbital, while the (B_1 type, π_3) orbital shows its ability to accept electrons donated by the mesomeric effect from the substituent. The splitting is further enhanced by double substitution, especially when the substitution occurs in the *para* position. Some cases, such as $C_6H_5CF_3$, C_6H_5CN, and C_6H_5CHO, show shifts in the A_2-type orbitals as well, indicating that sometimes there may be either efficient transmittance of the inductive effect to the *ortho* positions or a through-space polarization, either of which can effect the A_2 orbital. The study of photoelectron bands from the π_2 and π_3 orbitals in substituted benzene thus offers an excellent way in which to study inductive effects.

In halogen compounds it is also interesting to note the lone-pair bands (cf. Figure 4.27). The lower energy band is due to the "p_x" electrons and is narrower because of their nonbonding character, while the higher energy band is due to the "p_y" electrons, which, being perpendicular to the benzene ring, are slightly less nonbonding and therefore the band is slightly broader.

Heterocyclic compounds are important in organic chemistry and are receiving much attention from photoelectron spectroscopists. In some compounds, such as pyridine

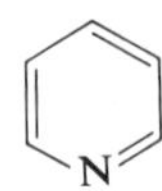

the heteroatom has a lone pair which is seen as a sharp peak (the second ionization band at 10.5 eV is the lone pair in pyridine). In other species, such as furan

C—C
‖ ‖
C O C

the heteroatom's lone pair is incorporated into the aromatic π system.

Pyridine and the other azaderivatives of benzene are best regarded as perturbed benzene in that a CH unit is replaced by nitrogen, whereby a N $2p$ atomic orbital is substituted for a C $2p$, but in addition each nitrogen atom possesses an electron lone pair. The systematics of this system has been worked out by Glutes *et al.*,[146] with the help of Hückel molecular orbital theory and Hoffmann's concept of "through space" and "through bond" interaction. Baker *et al.*[147] have also made systematic studies on the heterocycles. Rabalais *et al.*[148] have looked at thiophene and its derivatives. A particularly thorough investigation of the nitrogen heterocyclics has been carried out at the

Royal Institute of Technology in Stockholm.[104] In these studies high resolution PESOS has been combined with other spectroscopy methods to help in elucidating the nature of the molecular orbitals. Kobayashi and Nagakura[149] have intercompared members of the amino- and cyanopyridine families. For the most comprehensive review of the extensive application of photoelectron spectroscopy to heterocyclic compounds the reader is directed to Heilbronner *et al.*[48]

The photoelectron spectra of cyclooctateraene and its hydrogen derivatives have been studied by Batich *et al.*,[150] from which information about the twist angles between consecutive π bonds has been elicited.

5.4.4. Multiring Compounds

More and more complex molecular systems are being studied. A number of complex molecules were first studied by PESOS using a retarding grid electron spectrometer.[115] More recently, these data are being replaced by those taken with high resolution electron spectroscopy. For example, Van den Ham and Van der Meer[151] have investigated the diazonaphthalenes by PESOS, and Chadwick *et al.*[152] have investigated the norboranones and norbornenones. Boschi *et al.*[152a] have studied a series of ring compounds from anthracene (three rings) to ovalene (ten rings). Boekelheide *et al.*[153] measured *trans*-15,16-dimethyl-dihydropyrene, which gives an IP of 6.7 eV, the lowest value thus far measured for a closed shell hydrocarbon. Heilbronner *et al.*[154] have studied nonbornadiene and its derivatives, showing that through-space interaction between basis π orbitals can be more important than through-bond interaction. Through-space interaction has also been studied[155] for tris-bridged cyclophane. Allan *et al.*[156] have studied some 17 tropones. As molecular systems increase in molecular weight they become less volatile and harder to study in the gas phase. However, this has been circumvented by introducing samples into the target chamber on a heated probe through a vacuum lock, and keeping the whole chamber, including the electron energy analyzer, at elevated temperature.[157] Spectra have been taken with the analyzer as high as 300°C.[158]

5.4.5. Organic Halides

Many homologous series of organic compounds can be found in which the halide can be varied. The trend in behavior of these compounds often yields an important insight into the nature of the chemical bonding. Various comprehensive studies on organic halides have been made, as, for example, the halomethanes, the vinyl monohalides, halogen-substituted benzenes, the halodiacetylenes,[159] and fluoropyridines.[160] The electron lone pair of the halide can usually be characterized by its sharp peak. In addition, the orbitals associated with the halide will exhibit spin–orbit splitting, which can be characterized.

Carbon tetrabromide offers a good example of this and in addition demonstrates a situation where the Jahn–Teller effect interacts with the spin–orbit splitting.[37] The $^2T_1$ and $^2T_2$ states of CBr_4 undergo spin–orbit coupling to give respectively two doublets: $E_{1/2} + G_{3/2}$ and $E_{5/2} + G_{3/2}$. It was shown by Green *et al.*[161] that the spin–orbit splitting of T_1 and T_2 is expected to be, respectively, $\frac{1}{2}\rho$ and ρ, where ρ is the observed separation between the $^2P_{3/2}$ and $^2P_{1/2}$ states in the bromine atoms, or 0.46 eV. From Figure 4.28 we see a quartet of lines, the first of them having evidence of additional splitting. This suggests that the first and third bands are due to $G_{3/2}$ states since the additional splitting probably indicates a Jahn–Teller effect which can occur with the G states. The separation between the first and second ionization bands is approximately one-half that of the separation between the third and fourth bands; the first pair of bands is assigned to the $^2T_1$ states, while the second pair belongs to the $^2T_2$ states. Thus, the Jahn–Teller and spin–orbit splitting have made it possible to identify the first four bands in CBr_4.

The systematics of the methyl halides has been investigated by angular distribution measurements. Figure 4.29 shows the energies and ordering of the bands in terms of orbital assignments together with the angular parameter β (cf. Section 3.4). It is noteworthy that the ordering of the orbitals follows the

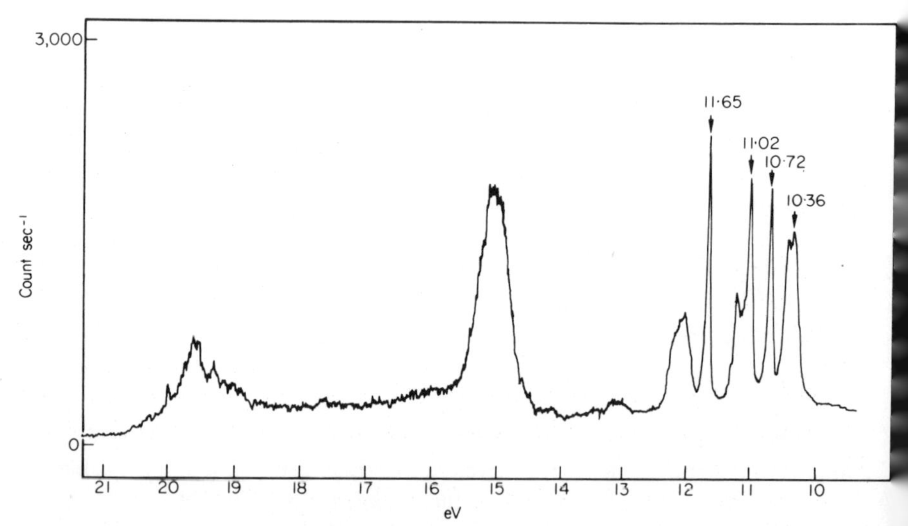

FIGURE 4.28. Photoelectron spectrum of carbon tetrabromide. Energy in terms of ionization potential. [Reproduced from Turner *et al.*,[37] Figure 8.26.]

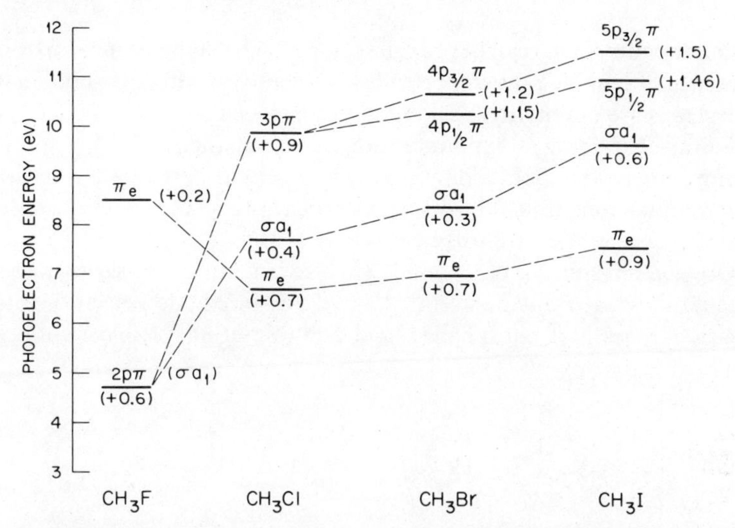

FIGURE 4.29. Energy level diagram of the methyl halides. The values for the angular parameter β are given in parentheses. [Data taken from Carlson and White.[85]]

magnitude of β for each molecule, i.e., $\beta(np\pi) > \beta(\pi_e) > \beta(\sigma a_1)$. In addition, the β in general rises with increasing Z of the halogen.

5.4.6. Miscellaneous Organic Compounds Containing Oxygen, Nitrogen, Sulfur, and Phosphorus

Organic compounds are often characterized by the inclusion of a nonhydrocarbon element, the most common being oxygen, nitrogen, or sulfur. The photoelectron spectra can likewise be characterized by these elements. The cyanide group C≡N is characterized by a weakly bound π orbital with its large vibrational stretching mode (~2000 cm^{-1}). Similarly, the carboyl group C═O is characterized by its vibrational spacings and behavior of its π bond. These characteristics are, however, not always clearly evident from analysis of a single molecule but can be deduced by following its behavior through a series of related compounds. A series of alcohols has been examined by Robin and Kuebler,[162] and Chadwick *et al.*[163] have looked at ketones. A correlation in the latter study was found between the carbonyl stretching frequency and oxygen ionization potential. A similar study was carried out on carboxylic acids and their derivatives.[163a]

Lone-pair electrons with their sharp peaks can sometimes be used to characterize the presence of the nonhydrocarbon. However, in nitrogen-containing compounds there often occurs no feature that can be described as entirely nonbonding. One can nevertheless trace the behavior of a band throughout a series of homologous compounds and identify it with an orbital in which the electrons are essentially localized about nitrogen.

Sulfur-containing compounds are often characterized by spin–orbit splitting for the lone-pair orbitals, and by *d* orbital participation by way of configuration interaction. A bird's-eye view of a number of nitrogen- and sulfur-containing compounds is illustrated in Ref. 147.

Organic phosphorus compounds are also beginning to receive their share of attention. Schäfer and Schweig[164] have compared photoelectron peaks of aromatic amines and phosphanes, and Schweig *et al.*[165] have studied the

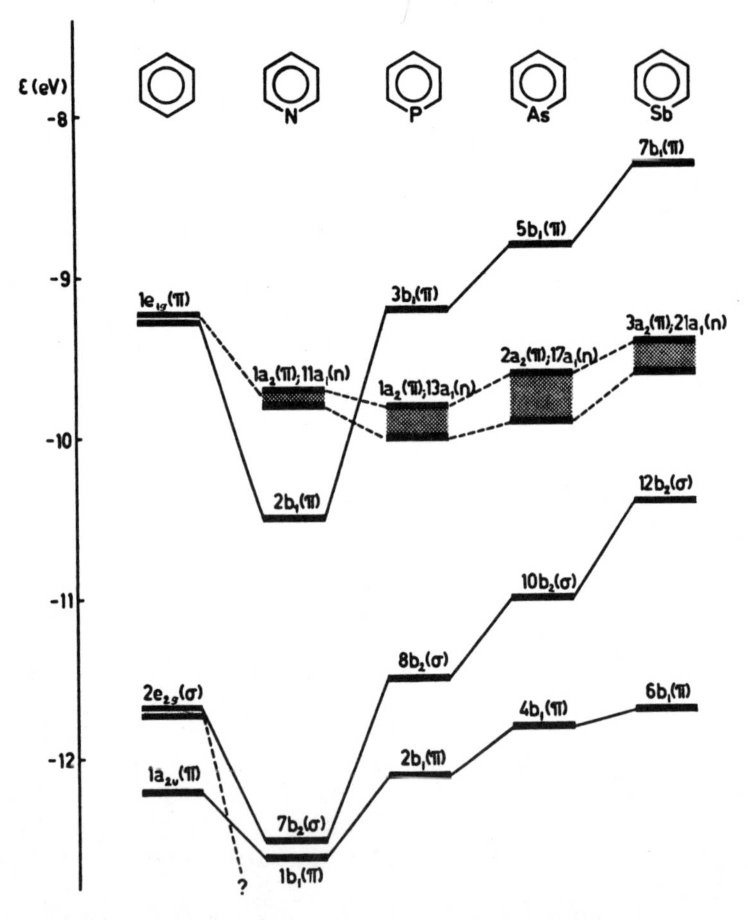

FIGURE 4.30. Orbital correlation diagram for (M-ring) where M is an element from the VB period of the periodic table. [Reproduced from Batich *et al.*,[166] Figure 2.]

phosphorin system. Batich *et al.*(166) have studied a series of compounds in which the nitrogen in pyridine is substituted, respectively, by P, As, and Sb (cf. Figure 4.30). Betteridge *et al.*(167) have looked at organophosphorus pesticides and related compounds.

5.5. Organometallics and Miscellaneous Inorganic Polyatomic Molecules

Although gaseous inorganic molecules are not available in numbers comparable to the hydrocarbons and their substituted derivatives, they nevertheless offer a large area for study. In addition, they give the photoelectron spectroscopist extra challenges in terms of different types of bonding and in the variety of elements that can be compared.

NH_3 has been studied by photoelectron spectroscopy in some detail. The first ionization band arises from a nonbonding orbital made up primarily of the $2p_z$ electrons from nitrogen. This is at first surprising since the band is very broad, with the vertical ionization peak occurring with vibrational band ν_6'. However, it is well known that the ground state of NH_3^+ is planar, in contrast to the puckered structure of the neutral molecule. The change in bond angle from 107 to 170° should cause a long series of bending vibrations as the result

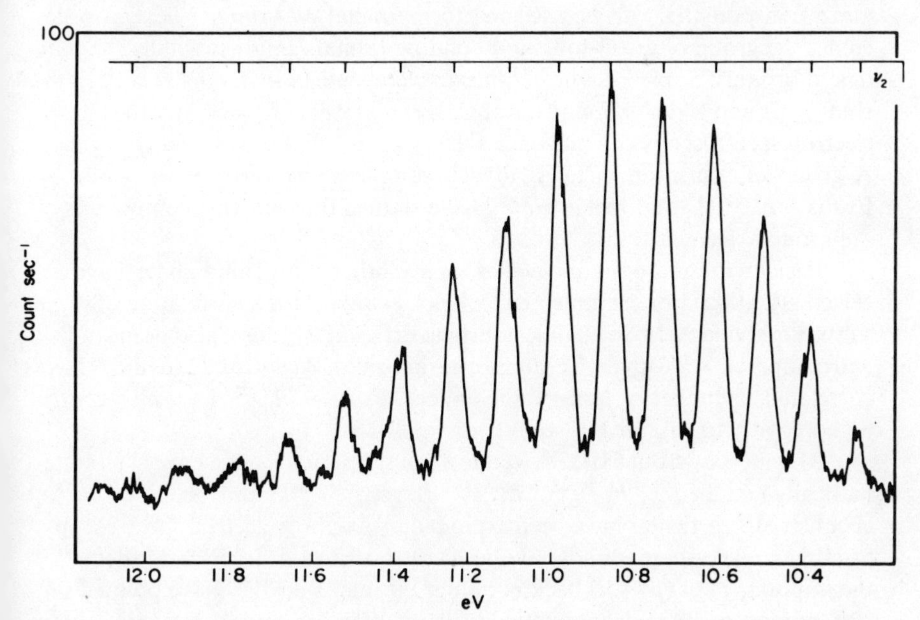

FIGURE 4.31. Photoelectron spectrum of first band of NH_3. Energy is given in terms of ionization potential. [Reproduced from Turner *et al.*,(37) Figure 14.2.]

of photoionization, as indeed it does (cf. Figure 4.31). Rabalais *et al.*[168] have analyzed the second band for Jahn–Teller effects.

Careful attention has also been given to the interesting molecules of P_4[169] and carbon suboxide.[170]

The metallocenes[37] form an interesting series (bicyclopentadienyl compounds of Cr, Fe, Co, and Ni). First, the 3*d* metal atomic orbitals are sensitive to the symmetry of the electrostatic field which the ligands produce. In iron the five unpaired 3*d* electrons are split into two groups, one nondegenerate and the other doubly degenerate, and in fact the first ionization band of ferrocene is split into two bands, one approximately twice the intensity of the other. Extra bands are found in the photoelectron spectra of cobaltocene and nickelocene since the metal atoms have respectively one and two extra electrons outside the ferrocene closed shell; likewise, chromocene has extra bands because it has one less electron. The first ionization potential for cobaltocene is unusually low (5.5 eV) due to the presence of the extra electron, which gives the compound alkali metal-like properties.

A variety of tris-chelates, ML_3, where the metal was respectively Al, Ga, Sc, Ti, V, Cr, Co, Mn, Fe, and Ru, have been studied by Evans *et al.*[171] From these studies details were obtained on the metal–ligand bonding, and in the case of the transition metal complexes, information was gained about the relative energies of the metal *d* and ligand orbitals. As a general rule the lowest ionization potential can be assigned to the metal 3*d* orbitals; next in order comes the group of metal-to-ligand bonding orbitals and then finally the pure ligand orbitals. Some π-arene complexes were studied by Evans *et al.*,[172] including Cr and Mn compounds. Rabalais *et al.*[11] have looked into the photoelectron spectroscopy of some open shell systems: Ni, Fe, Mn, and $Cr(C_5H_5)_2$. A group of transition metal sandwich complexes has been investigated by Evans *et al.*,[173] and Elbel *et al.*[174] have studied the trimethyl compounds of the group V elements.

Besides the diatomic molecules, many other inorganic halides have received attention. For example, see DeKock *et al.*[175] for a correlation of some sulfur hexahalides. Mixed halides, such as BrF_5 and IF_5 have also been studied and compared[175] with Se, Te, and Xe hexaflorides. Bassett and Lloyd[176] have correlated a number of tetraatomic halides. Frost *et al.*[177] have also contributed greatly to this area of study.

Members of groups III, IV, V, and VI of the periodic table have been systematically studied.[15–17, 39, 161, 166, 178] Boron[179, 180] and silicon[181,–184] chemistry in particular have been exploited by uv photoelectron spectroscopy. Studies on silicon compounds have been used both to affirm[185] and to deny[181] the importance of $(p \rightarrow d)$ back bonding. Pitt and Bock[182] have pointed out σ–π mixing in phenylpentamethyldisilane. Weidner and Schweig[184] have shown the existence of through-conjugation through the silicon atom.

Mercury[186] and rare gas[187] compounds have also received attention, and in fact nearly half of the elements in the periodic table have been studied by PESOS.

5.6. Ions, Transient Species, and Other Special Studies in PESOS

Not only stable molecules have been studied by PESOS, but also ions and transient species. Cairns *et al.*[188] have presented a review on photoelectron spectroscopy of beams, both ion and molecular.

Transient species produced by some form of preexcitation can be studied by PES. Jonathan *et al.*,[189] for example, have produced photoelectron spectra of the SO radial as produced by electrodeless discharge of inert gas–SO_2 mixtures or by reaction of atomic oxygen and CS_2. These sorts of studies are valuable not only for the determination of the nature of the molecular orbitals for transient species, but photoelectron spectroscopy can be used as an analytical tool for following the production and extinction of such species. Other studies of transient species have included atomic hydrogen, nitrogen and oxygen,[190] and CS.[191] Kroto *et al.*[192] have also examined species found under pyrolysis. Cornford *et al.*[193] have looked at a series of stable free radicals, e.g., NF_2, ClO_2, and $(CF_3)NO$, and SO_3F. As examples of organic free radicals that are stable, Morishima *et al.*[194] have studied several nitroxide radicals. Ozone[195] has recently received attention.

Daintith *et al.*[108] have reported on photoelectron spectroscopy in coincidence with ion fluorescence, and Ames *et al.*[196] have discussed the concurrent use of mass spectroscopy with He I photoelectron spectroscopy.

The use of PESOS on high-temperature vapors has been demonstrated by Berkowitz.[197] The interest in this area of research lies not only in establishing the molecular orbitals of these vapors, but in learning about the formation of dimers, trimers, and polymers. Allen *et al.*[198] have also begun a series of studies using a laser beam to obtain vapors of a wide variety of salts.

In principle, one ought to be able to follow chemical equilibrium through the use of uv photoelectron spectroscopy. This has indeed been achieved[199] by observing changes in the spectrum of acetylacetone as a function of temperature. From these changes the equilibrium constant for the keto $\rightleftharpoons$ enol reaction was determined. Compounds of 3-methylacetylacetone and 3.3 dimethylacetone were similarly studied.

6. STUDIES ON SOLIDS

The principal use of PESOS discussed in this book has been in the investigation of molecular orbitals for free molecules in the gas phase. In many solids there are no longer molecular orbitals as such for the outermost valence shells, but rather the interest resides in the study of band structure. There has been much work on this subject through the use of vacuum-uv radiation in which information is gained from photons over a continuum energy range up to 11.6 eV (LiF window cutoff).

Recently, higher energy photon sources from the He resonance lines at 21.2 and 40.8 eV to the Mg and Al $K\alpha$ soft x rays have been made. The study of the valence bands of metals using XPS has been made by several groups.(200) Though the resolution of soft x rays is limited, the interpretation of the bands is theoretically more straightforward at higher photoelectron energies. In addition, with the use of monochromators it is expected that this work will be repeated with substantially improved resolution. Such work is in fact underway at Berkeley and elsewhere(201) using Hewlett-Packard instruments with a resolution of about 0.4 eV. Figure 4.32 shows the valence band region for Pd through Te.

Eastman(202) has demonstrated the value of studying the conduction band

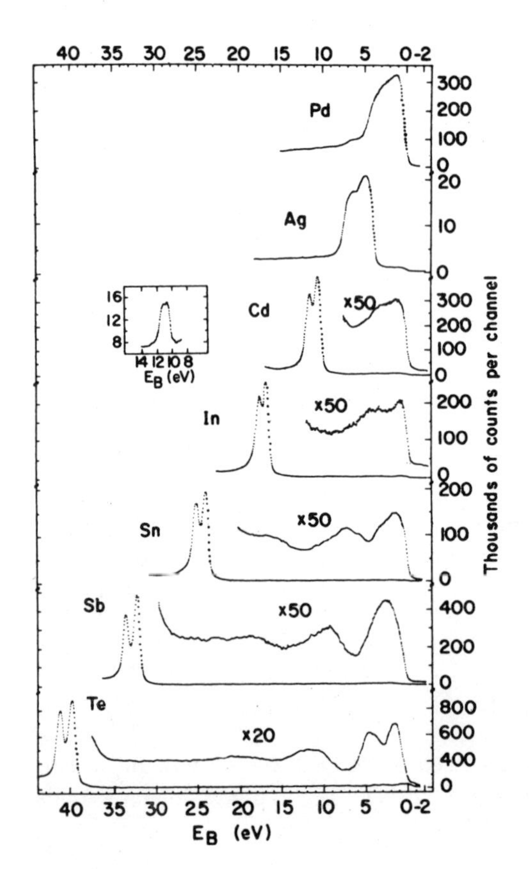

FIGURE 4.32. Valence band region for Pd through Te. It can be easily seen how the $4d$ orbital develops into a core shell. [Reproduced from Pollak *et al.*, *Phys. Rev. Lett.* **29**, 274 (1972), Figure 1.]

with photoelectron spectroscopy as a function of energy over a wide energy range from 10.2 to 40.8 eV. From these data it has been possible to ascertain the amounts of angular momentum character that make up the band. For example, in Figure 4.33 we see the emergence of the *f*-electron contribution and decrease of the *p*-electron contribution with increasing photoelectron

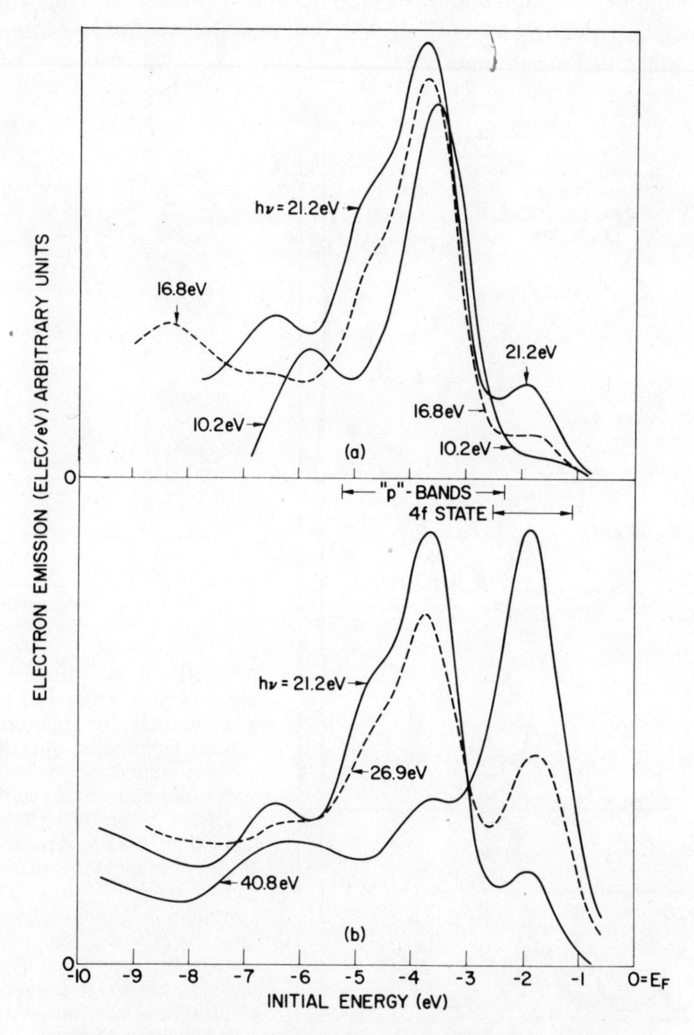

FIGURE 4.33. Study of correlation bands of EuS as a function of photon energy. Note the increased importance of the *f*-electron contributions with increasing photon energy. [Reproduced from Eastman,[202] Figure 1.]

energy. More recently Eastman has made use of synchrotron radiation for a continuous scan of energy up to 40 eV, with expectations of increasing this energy scan. Brundle *et al.*[203] have designed a source volume that allows a record of the photoelectron spectrum from the He I lamp and from Al $K\alpha$ x rays without disturbing the sample.

It has been found that the valence band structure can be very sensitive to the cleanliness of the surface, more so than might be anticipated from the analysis of impurities using data from core shell photoionization. Thus, the emergence of photoelectron spectroscopy under ultrahigh-vacuum conditions is a greatly welcomed development.

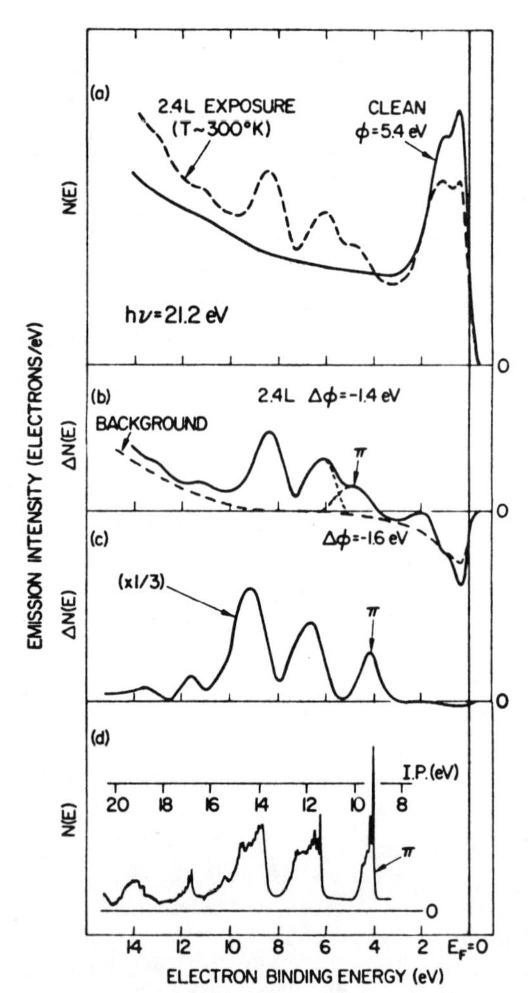

FIGURE 4.34. Comparison of photoelectron spectra of benzene taken with 21.2-eV photons under various conditions. Spectrum (d) shows the spectrum in the gaseous state. Spectrum (c) shows the spectrum of a thick layer condensed at 150°K. Spectrum (b) shows the spectrum of a chemisorbed layer on nickel after the contribution of the substrate has been subtracted. Spectrum (a) shows the contribution of the nickel to the spectrum (solid line) and the total spectrum of benzene plus substrate (dashed line). [Reproduced from Demuth and Eastman[206] Figure 1.]

Nonconductors and semiconductors as well as metals have been studied by photoelectron spectroscopy. The literature on photoelectron emission of solids for both experiment and theory is too vast to be covered in this book. However, the reader is directed to several reviews[204] for a partial exposure to the field.

Under many conditions solids yield photoelectron spectra whereby the analysis in terms of molecular orbitals is very profitable. Sehi *et al.*[205] have carried out a high-resolution photoelectron spectrum using 584-Å radiation on a naphthalene polycrystal. Of particular importance is the study of molecules adsorbed on the surface of a solid. Figure 4.34 shows the photoelectron spectrum of benzene under various conditions. A thick layer of condensed benzene has very much the same photoelectron spectrum as the gas phase, though the ionization bands are less well resolved and show no vibrational structure. There is also a general shift in energy due to the work function. When one views only a monolayer chemisorbed on nickel there is a slight shift of the ionization bands due to relaxation energy effects, with the one exception of the π, which has undergone substantial change. This behavior is expected because chemisorption is believed to involved bonding of the π orbital with the nickel $3d$ orbitals. In another experiment Demuth and Eastman[206] adsorbed ethylene on nickel at low temperature. When the temperature was allowed to rise, dehydrogenation took place and the photoelectron spectrum of acetylene appeared in place of that of ethylene. The behavior of adsorbed molecules would appear to be profitably studied by PESOS. For other studies of this nature, see Brundle and Roberts,[207] Atkinson *et al.*,[208] and Egelhoff *et al.*[209]

Another area of PESOS for solids that has gained increased interest is the study of complex anions using soft x rays. The resulting spectra can be analyzed in terms of molecular orbital theory as calculated for the free anion. These studies include a variety of oxyanions,[210–212] hexafluorides,[212] and $MoOS_3^{2-}$ and WOS_3^{2-}.[213]

7. ANALYTICAL APPLICATIONS OF PESOS

The use of photoelectron spectroscopy for the study of molecular orbitals has been primarily of benefit to basic research. Nevertheless, each molecule has a photoelectron spectrum that is unique. This, together with a relatively high sensitivity and reasonably good peak-to-background ratio, makes PESOS a candidate for use in gas analysis. Complex molecules have complex spectra; and, in general, PESOS would not be useful for identifying mixtures of a large number of such compounds. However, simple molecules such as CO could be easily identified even if present in small concentration. A partial gas pressure as low as 10^{-7} Torr could be detected with present-day equipment. Application to air pollution analysis would appear to be a good possibility.

However, to use PESOS as a highly sensitive technique for gas analysis, in parts per million or less, one would need to clean up the photon source with a monochromator, and preseparation by such means as gas chromatography would be desirable. Although the analytical potential of PESOS has been discussed,[214] as yet no serious demonstration of its usefulness or competitiveness with other techniques has been reported.

Chapter 5

Photoelectron Spectroscopy of the Inner Shells

The study of the inner shells of atoms is basically a study of atomic orbitals. The shifts in the binding energies of the core electrons as a function of chemical environment can be considered as a perturbation on the atomic system. Experimentally, PESIS started as a means to obtain information on atomic levels, the utilization of the data for chemical description coming as a by-product of better resolution. Thus, PESIS is in the first place an atomic measurement. It affords elemental identification and is another tool for qualitative and quantitative analysis.

The special enthusiasm for PESIS, however, has come from its usefulness in teaching us about chemical bonding. It not only identifies the different elements in a compound, but indicates the nature of the chemical environment surrounding each of the atoms. Though the bulk of the effort in eliciting chemical information from PESIS comes from core binding energy shifts (frequently called chemical shifts), it will be demonstrated in this chapter that a substantial amount of other information can also be gained from satellite lines and the study of surface phenomena; but let us start with the atoms themselves.

1. ATOMIC STRUCTURE

As mentioned previously, the initial impetus for PESIS was the desire to obtain atomic binding energies. Binding energies of the different atomic orbitals have been obtained previously from the study of x-ray spectra. However, because of dipole selection rules, the binding energies of the *s* orbitals with principal quantum number greater than 1 are known with less certainty. More important, as one goes to the outer shells, x-ray data are less and less dependable.

Data from optical spectroscopy can be used to determine with a high degree of accuracy the binding energy of the least bound orbital, but are only occasionally helpful for the other orbitals. Photoelectron spectroscopy fills the gaps left by x ray and optical spectroscopy admirably. Kai Siegbahn and his co-workers at Uppsala[1] have made a comprehensive study of atomic binding energies using x-ray sources ranging in energy from the Mg $K\alpha$ (1254 eV) to Ag $K\alpha_1$ (22,163 eV). Elements throughout the whole periodic table have been studied with x-ray photoelectron spectroscopy, including even the transuranic elements.[2]

Since the atomic binding energies are dependent on their chemical surroundings, it makes a difference in which material the energies were measured. Studies made on solids usually quote binding energies referenced to the Fermi level of the particular material employed. For example, the table of atomic binding energies given by Siegbahn *et al.*[1] is based on zero binding energy at the Fermi level. This table is based on data from XPS and also on data collated by Bearden and Burr[3] on x-ray emission. It is also desirable to have

TABLE 5.1

Comparison of Experimentally and Theoretically Determined Binding Energies for Xenon

	Experimental			Theory	
Subshell	Optical[a]	X ray[b]	XPS[c]	$-\varepsilon$[d]	$-(\varepsilon + R_E)$[e]
$5p_{3/2}$	12.13	—	12.1	12.0	11.4
$5p_{1/2}$	13.43	—	13.4	13.5	12.8
$5s$	—	—	23.3	27.6	26.3
$4d_{5/2}$	—	—	67.5	71.0	66.4
$4d_{3/2}$	—	—	69.5	73.1	68.5
$4p_{3/2}$	—	147	145.5	162	157
$4p_{1/2}$	—	—	—	175	170
$4s$	—	—	213.2	229	223
$3d_{5/2}$	—	672.3	676.4	693	677
$3d_{3/2}$	—	—	689.0	706	690
$3p_{3/2}$	—	937	940.6	959	944
$3p_{1/2}$	—	999	1002.1	1023	1008
$3s$	—	—	1148.7	1168	1143
$2p_{3/2}$	—	4782.2	4787.3	4833	4795
$2p_{1/2}$	—	5103.7	5107.2	5159	5120
$2s$	—	5452.8	5453.2	5507	5472
$1s$	—	34,564	—	34,689	34,622

[a] C. E. Moore, *Atomic Energy Levels*, Vol. II (NBS 467, U.S. Govt. Printing Office, Washington, D. C., 1952).
[b] Bearden and Burr.[3]
[c] Siegbahn *et al.*[48]
[d] Eigenvalues from optimized relativistic HFS wave functions.[48]
[e] Corrected for relaxation energy determined for iodine. A. Rosen and I. Lindgren, *Phys. Rev.* **176**, 114 (1968).

experimental binding energies from the free atom. This is a well-defined state and amenable to comparison with *ab initio* atomic structure calculations. Binding energy data on atomic gases are becoming more available. Besides the rare gases, binding energies have been measured in the free atomic state for Hg,[1] Bi,[3] and Pb.[3] Krause[4] and Lotz[5] have constructed comprehensive tables of binding energies based on free atoms. Appendix 1 lists binding energies for all elements up to $Z = 106$ based mainly on Lotz, but supplemented with recent experimental and theoretical results. Appendix 2 lists the differences of binding energies between the $n, l + \frac{1}{2}$ and $n, l - \frac{1}{2}$ subshells of all the atoms up to $Z = 106$. These values are taken from relativistic Hartree–Fock–Slater wave functions, and for splittings below 10 eV may often be considered more reliable than experimental values. These splittings in the core shell of atoms are quite characteristic of the subshell and can be of value in identifying the shell.

Table 5.1 lists experimentally and theoretically determined binding energies for xenon. The columns containing the experimental data give some idea of the effective range of application for x-ray analysis and optical and photoelectron spectroscopy. For a discussion of the relative merits and problems of atomic structure calculations, see Chapter 3.

2. THEORETICAL BASIS OF CHEMICAL SHIFTS OF CORE ELECTRONS

2.1. Valence Shell Potential Model

The most valuable aspect of PESIS with regard to molecular structure comes from the measurement of chemical shifts in the core electrons. A complete understanding of the chemical shifts requires solutions of the total wave functions for the neutral molecule and the singly ionized counterpart, corresponding to a vacancy in a given inner shell orbital. That is, the chemical shift ΔE_{nl} for orbital nl between the free atom A and a particular molecular state M is

$$\Delta E_{nl} = E_{nl}(\mathrm{M}) - E_{nl}(\mathrm{A}) = [T^{+}_{nl}(\mathrm{M}) - T(\mathrm{M})] - [T^{+}_{nl}(\mathrm{A}) - T(\mathrm{A})] \qquad (5.1)$$

where T and T^{+}_{nl} are the total energies, respectively, for the neutral species and for the ionized species having a single vacancy in the nl orbital. Such calculations for molecular systems have rarely been performed, but Koopmans' theorem can be invoked and eigenvalues for the core electrons can be taken from molecular and atomic Hartree–Fock solution and used as binding energies, i.e.,

$$-\Delta E_{nl} \simeq \varepsilon_{nl}(\mathrm{M}) - \varepsilon_{nl}(\mathrm{A}) \qquad (5.2)$$

Even so, Hartree–Fock calculations have been restricted to only simple molecules. The use of even a modified Hartree–Fock solution for solids has been limited. Fortunately, a rather simple model can be used to explain and predict chemical shifts. In fact, one of the main reasons for the success of PESIS has been the simple manner in which one can correlate chemical shift data and make predictions regarding the chemical nature of an unknown material without any knowledge of the material other than the inner shell binding energies of the different types of atoms making up the sample.

The simple model of chemical shifts is based on the idea that the core electrons feel an alteration in the chemical environment as a change in the potential of the valence shell as electrons are drawn away from or toward that shell. For example, if we envision the core electrons as being inside a hollow electrostatic sphere, then the potential seen by each of the core electrons due to the valence shell will be $q_v e^2/r_v$, where q_v and r_v are the charge and radius of the valence shell. The chemical shift would then be

$$\Delta E_{nl} = \frac{q_v e^2}{r_v}(\mathrm{M}) - \frac{q_v e^2}{r_v}(\mathrm{A}) \tag{5.3}$$

If q_v is given in units of electron charge, r_v in Bohr units, and e the electron charge $= 1$, then ΔE_{nl} will be in eV if multiplied by 27.2.

Verification of this model comes from atomic SCF free-ion calculations. Tables 5.2 and 5.3 contain calculated binding energy shifts in the core electron as the result of removing electrons from the valence shell of the atom. Calculations based on a relativistic Hartree–Fock–Slater atomic wave function program were used in Table 5.2. The Tm atom, $(4f)^{13}(6s)^2$, is the initial or atomic state A, while Tm^{2+}, $(4f)^{13}(6s)^0$, and Tm^{3+}, $(4f)^{12}(6s)^0$, are the final states. ΔE_{nl} was obtained from equation (5.2), where the ions represent a pseudochemical change and may be taken as the M state. Note that the core electrons *all* feel essentially identical binding energy shifts. ΔE_{nl} estimated from equation (5.3) using a calculated $\langle 1/r \rangle$ is in approximate agreement with the binding energy shift obtained from the eigenvalues. When an electron is removed from the $4f$ shell, the $5s$ and $5p$ orbitals do not exhibit binding energy shifts equal to the other core electrons. This is easily understood since the mean radius for the $4f$ orbital is smaller than those of the $5s$ and $5p$ shells, and in this case the $5s$ and $5p$ electrons cannot be considered to be core electrons. The principal effect on the $5s$ and $5p$ shells of removing a $4f$ electron is to alter the shielding. Table 5.3 reinforces the arguments made above and employs calculations based on a nonrelativistic Hartree–Fock code. Calculation of ΔE_{nl} have utilized both equations (5.1) and (5.2). The results are slightly different, but the binding energy shifts remain remarkably constant.

Arguments as to the nature of binding energy shifts in the core electron have received verification from molecular as well as atomic analysis. For

TABLE 5.2

Comparison of Changes in Binding Energies (in eV) as a Function of Removal of Valence Shell Electrons in Tm[a]

Shell	E, eV	$\langle r\rangle$, Bohr	$Tm^{2+} - Tm^0$	$Tm^{3+} - Tm^{2+}$
$1s$	59,503	0.020	15.85	20.50
$2s$	10,083	0.084	15.76	21.15
$2p_{1/2}$	9630	0.070	15.77	21.02
$2p_{3/2}$	8638	0.077	15.76	21.09
$3s$	2278	0.219	15.68	21.95
$3p_{1/2}$	2076	0.210	15.68	21.90
$3p_{3/2}$	1865	0.223	15.68	21.95
$3d_{3/2}$	1516	0.196	15.69	21.86
$3d_{5/2}$	1469	0.200	15.69	21.87
$4s$	454	0.495	15.69	21.50
$4p_{1/2}$	372	0.502	15.70	21.44
$4p_{3/2}$	324	0.534	15.70	21.28
$4d_{3/2}$	187	0.559	15.70	21.10
$4d_{5/2}$	178	0.570	15.70	21.05
$5s$	58.3	1.212	15.65	16.91
$5p_{1/2}$	35.5	1.362	15.58	16.13
$5p_{3/2}$	29.5	1.475	15.53	15.53
$4f_{5/2}$	10.9	0.745	15.68	20.02
$4f_{7/2}$	9.6	0.761	15.68	20.10
$6s$	6.1	3.853	—	—
Chem. shift $= \Delta q\langle 1/r\rangle 27.2$			18.6 eV	47.1 eV

[a] Obtained using equation (5.2) from eigenvalues of neutral atoms and ions of Tm as calculated by T. C. Tucker from relativistic HFS code reported by T. C. Tucker, L. D. Roberts, C. W. Nestor, Jr., T. A. Carlson, and F. B. Malik, *Phys. Rev.* **178**, 998 (1969). Configuration for Tm^0 is $4f^{13}6s^2$; for Tm^{2+}, $4f^{13}6s^0$; and for Tm^{3+}, $4f^{12}6s^0$.

example, Schwartz[7-9] and others[6, 10] have theoretically examined the core-level binding energy shift from molecular orbital theory. They found that *ab initio* calculations which treat the core electrons explicitly give core-binding energy shifts essentially in agreement with experiment. They also concluded that explicit core-hole calculations are unnecessary, and furthermore that only the potential due to the valence electron is important, so that MO schemes that do not include core electrons may be used.

To use Koopmans' rule and to explain binding energy shifts solely in terms of the electrostatic environment of the neutral molecule is to neglect the relaxation energy. The relaxation energy may be defined as the difference between Koopmans' energy and the adiabatic energy, or

$$R_E = -\varepsilon_{nl} - (T_{nl}^{+} - T) \tag{5.4}$$

The relaxation energy for core electons is not negligible. However, one

TABLE 5.3

Comparison of Changes in Binding Energies (eV) as a Function of Removal of Valence Shell Electrons for Sr → Sr^{2+} ($4s^2 \rightarrow 4s^0$)[a]

Shell	$-\varepsilon^b$	$\Delta\varepsilon^c$	ΔT^d	$\Delta E_{nl}{}^e$
1*s*	15,881	13.4	15,828	15.3
2*s*	2187	13.4	2160	15.3
2*p*	1986	13.4	1956	15.1
3*s*	367	13.3	357	14.9
3*p*	291	13.4	282	14.9
3*d*	155	13.4	145	15.0
4*s*	52	13.0	49	14.3
4*p*	30	13.1	28	14.3
		$\Delta q\langle 1/r\rangle 27.2 = 14.6$ eV[f]		

[a] Calculated by J. C. Carver from a code of C. Froese Fischer, *Can. J. Phys.* **41**, 1895 (1963).
[b] Eigenvalues.
[c] Binding energy shift from equation (5.2).
[d] Binding energies taken from differences in total energies.
[e] Binding energy shift from equation (5.1).
[f] Binding energy shift from equation (5.3). $\langle 1/r\rangle$ taken from C. Froese, Univ. of British Columbia Report, Vancouver, B.C. (1968).

usually assumes that for a given orbital energy the relaxation energy is constant with changes in the chemical environment. Table 5.3 shows that for free-ion calculations this assumption is approximately true, namely $-\varepsilon_{nl} \simeq \Delta T$. Shirley *et al.*[11–14] have critically evaluated the importance of the relaxation energy in chemical shifts. Davis and Shirley[13, 14] have suggested how to estimate the relaxation energy for free gas molecules. Let us follow their arguments. From the theoretical analysis of Hedin and Johansson[15] one can obtain the binding energy of a 4*s* electron by the following approximation:

$$E_B(1s) = -\tfrac{1}{2}[\varepsilon(1s) + \varepsilon(1s)^*] \tag{5.5}$$

where $\varepsilon(1s)$ and $\varepsilon(1s)^*$ are the core orbital energies for the ground and a 1*s* hole state (the same relationship will hold for any core orbital). In terms of binding energy shift between two molecules we have

$$\Delta E_B = \tfrac{1}{2}[\Delta\varepsilon(1s) + \Delta\varepsilon(1s)^*] \tag{5.6}$$

If we then replace $\Delta\varepsilon$ by ΔV_n or the change in potential at the nucleus [as given, for example, by (5.3)], we have

$$\Delta E_B \cong \tfrac{1}{2}[\Delta V_n + \Delta V_n^*] \tag{5.7}$$

To aid in evaluating V_n^*, the equivalent core approximation is made, which states that the potential of the valence shell V_n^* for the case of an ion having a core vacancy is the same as that when the nuclear charge is increased by one, or

$$\Delta E_B \cong \tfrac{1}{2}[\Delta V_n + \Delta V_n(Z+1)] \tag{5.8}$$

Adams and Clark[16] have examined the goodness of the equivalent core approximation and find it to be satisfactory for core electrons that are highly shielding. Further use of this approximation will be discussed in Section 2.6.

Equation (5.8) can also be written

$$\Delta E_B = -\Delta V_n - \Delta V_R \tag{5.9}$$

where

$$V_R = \tfrac{1}{2}[V_n(Z+1) - V_n] \tag{5.10}$$

That is, ΔV_n is the chemical shift in which the change in relaxation energy ΔV_R is neglected.

By including the effect of relaxation energy, Davis and Shirley[13, 14] were able to improve agreement between calculation and theory. In some series of compounds neglect of extraatomic relaxation is not serious when correlating chemical shifts, but if there is little change in the inductive effects of the ground state properties, variation in relaxation energy may dominate. This can be clearly seen in Figure 5.1 for a series of simple nitrogen compounds.

Shirley[11] pointed out that in addition to atomic and molecular relaxation, there may occur extraatomic relaxation accompanying photoelectron emission in the solid state. A manifestation of this solid state relaxation will be pointed out in Chapter 6 when we compare chemical shifts from Auger and photoelectron spectra. Ley *et al.*[12] have estimated the extraatomic relaxation energy in metals using a screening model for semilocalized excitons. Table 5.4 gives the results from this paper of $\Delta E_B(\text{expt}) = E_B(\text{atomic}) - E_B(\text{conductor})$,

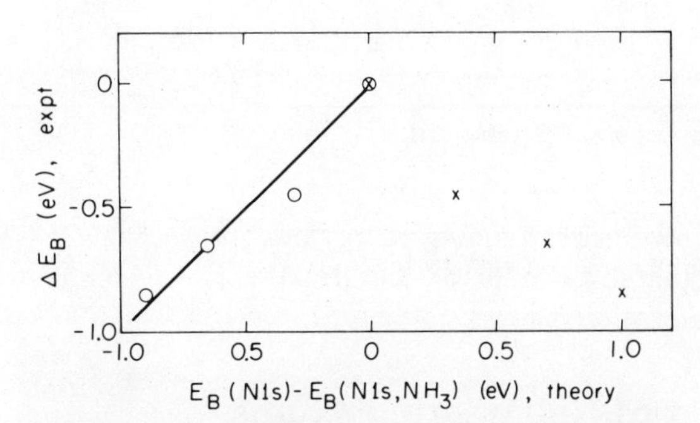

FIGURE 5.1. Relative experimental vs. theoretical N(1*s*) binding energies for the molecules (from top): NH_3, CH_3NH_2, $(CH_3)_2NH$, $(CH_3)_3N$. The theory that includes relaxation energies (open cirles) predicts the shifts very well (straight line is a perfect fit), while the theory that does not include relaxation (×'s) actually predicts the shift to be in the wrong direction. [Reproduced from D. A. Shirley, *J. Electron Spectros.* **5**, 135 (1974), Figure 5.]

TABLE 5.4

Experimental Relaxation of Binding Energies between Gaseous and Solid Targets ΔE_B(expt) and Theoretical Evaluation of Extraatomic Relaxation Energy ΔE_B(theor)[a]

Element	Core level	ΔE_B(theor), eV	ΔE_B(expt), eV
C (graphite)	$1s$	12.5	8.6
Ne (in Cu)	$1s$	4.0	3.0
Na	$2s$, $2p$	5.0	5.3
Mg	$2s$, $2p$	4.9	2.5
Al	$2s$, $2p$	6.0	5.1
Ar (in Al)	(average)	3.1	2.8
K	$2s$, $2p$	3.9	5.5
Ca	$2s$, $2p$	10.7	8.4
Sc	$2s$, $2p$	12.0	5.5
Ti	$2s$, $2p$	13.2	6.6
V	$2s$, $2p$	13.2	9
Cr	$2s$, $2p$	15.2	6.6
Mn	$2s$, $2p$	13.2	10.5
Fe	$2s$, $2p$	17.3	9.4
Co	$2s$, $2p$	18.3	12.3
Ni	$2s$, $2p$	18.4	12.5
Cu	(average)	5.0	3.6
Cu	$3d$	4.8	2.9
Zn	(average)	5.0	5.2
Zn	$3d$	4.8	3.4
Kr (in Cu)	(average)	2.9	2.5
Cd	$4d$	4.2	3.0
Xe (in Cu)	(average)	2.5	2.2
Pb	$4f_{7/2}$	5.2	3.5
Bi	$4f_{7/2}$	5.6	3.6

[a] Table taken from Ley *et al.*,[12] Table VIII.

which is the difference in binding energy of core electrons between that in a free atom and that in a conductor, and ΔE_B(theor), which is an estimate of the extraatomic relaxation energy.

2.2. Effect of Neighboring Atoms

Before approaching the problem of calculating the valence shell potential in order to evaluate the observed core-binding energy shifts, it is necessary to take into account the total electrostatic environment, which includes not only the valence shell but all the other atoms in the molecule, or, if it is a solid, all the atoms in the solid. We may regard each atom, other than the atom under study, as having a net charge q, which will act as a point charge located at the center of the atom. The core electron in atom i will thus see the effective potential of all the other electrons as

$$V = \sum_j (q_j e^2 / R_{ij}) \tag{5.11}$$

where R_{ij} is the distance between atoms i and j. The binding energy shift is then

$$\Delta E_B = \Delta(q_v e^2/r) - \Delta V \tag{5.12}$$

Since the change in the charge of the valence shell is essentially equivalent to the change in the net charge, calculations on the net charge can be applied equally to q_v and q_j. In a free molecule ΔV can be obtained by summing over a finite number of atoms and is commonly known as the molecular potential. In a solid or crystal we must carry out the calculation on V, which has been called in this context a crystal potential, in such a way that the sum will converge after taking a finite number of units. Such calculations are akin to Madelung potential calculations but are not identical to them. In the present use one always calculates the potential for a singly charged core electron. Special computer codes have been written for this use.[17] It might be noted here that the crystal potential for a surface atom will differ from that of an atom in the bulk. This can be substantial for the very first surface layer, but it has been demonstrated that the difference quickly becomes negligible as one goes to deeper layers.[18]

It is interesting to note that changes in the core binding energies can be affected by changing the temperature. In the case of KCl and LiF the shifts in the binding energies have been interpreted[19] as alterations in the crystal potential due to thermal expansion.

In correlating experimental binding energy shifts with theory it has been common practice to ignore ΔV. However, this is unjustified since ΔV is usually the same order as $\Delta q_v e^2/r$ and in some instances is larger. The only reason such neglect still yields a reasonable straight-line correlation is that V also often changes linearly with q, particularly if we compare molecules of similar structure. Take, for example, the carbon tetrahalides, CX_4. The calculated charge on carbon q_C is four times the charge on the halogen q_X.
Therefore

$$\Delta E_C = \left(\frac{\Delta q_C}{r} - \frac{4 \Delta q_X}{R} \right) e^2 = \Delta q_C \left(\frac{1}{r} - \frac{1}{R} \right) e^2 \tag{5.13}$$

Thus, a linear relationship will exist between the binding energy shifts in carbon ΔE_C and changes in the calculated net charge of carbon Δq_C for this series of compounds if a linear relationship also exists between the carbon–halogen internuclear distance R and q_X.

It has been often assumed that in a solid of a nonionic species it is sufficient to calculate only the molecular potential, that is, to neglect the contribution to V of the other atoms in the solid. That this assumption is not always valid has been pointed out by Hulett and Carlson[20] in their work on nitrogen with guanosine and related compounds. In these studies best agreement with experiments was obtained only after several molecular units were included in the calculation of V. Barber *et al.*[21] have pointed out a significant change in the relative binding energies of C(1*s*) for CH_3CN between the gas and solid phases, the cause being assigned to the effect of the atoms in the neighboring molecules.

2.3. Calculation of Net Charge from Electronegativity

Having demonstrated in Section 2.1 that the chemical shift is closely related to the electron density surrounding the atom, it seemed only reasonable to relate the core-binding energies of a given element to some chemical criterion reflecting this density. Siegbahn *et al.*[1] first suggested the use of a calculated charge based on Pauling's scale of electronegativity. The rules for obtaining the calculated charge for each atom of a given molecule are as follows.

1. Write down the formula for the molecule, including assignments for formal charge where applicable. The formal charge is defined as the net charge on an atom if all electrons shared in bonds are divided equally between the atoms. In certain cases there may be more than one resonance form.

2. For each atom determine the partial ionic character as contributed from each of the neighboring atoms. This value is derived from the electronegativity scale for elements (cf. Table 5.5) and is equal to

$$I = \frac{\chi_A - \chi_B}{|\chi_A - \chi_B|}\{1 - e^{[-0.25(\chi_A - \chi_B)^2]}\} \tag{5.14}$$

where χ_B and χ_A are the electronegativities, respectively, for the atom under study and its neighbor.

3. If a formal charge of +1 is present on the atom, the electronegativity is given by $\frac{2}{3}$ the value in Table 5.5 between Z and $Z + 1$, where Z is the atomic number of the atom with the formal charge. If the formal charge is −1, the electronegativity will be $\frac{2}{3}$ the value listed in Table 5.5 between Z and $Z - 1$, i.e.,

$$\chi(+1) = \chi(Z) + \tfrac{2}{3}[\chi(Z + 1) - \chi(Z)]$$

$$\chi(-1) = \chi(Z) + \tfrac{2}{3}[\chi(Z - 1) - \chi(Z)]$$

This rule is based on comparing electronegativities under conditions of equal effective nuclear charge.

4. To obtain the calculated charge, sum the contributions of the different partial ionic characters, counting each bond separately, and add the formal charge on the atom if present. That is,

$$q = \sum_i I + Q \tag{5.15}$$

TABLE 5.5

Electronegativity Scale for Elements[a]

Z	Element	χ	Z	Element	χ	Z	Element	χ
1	H	2.20[b]	20	Ca	1.0	49	In	1.7
3	Li	1.0	21	Sc	1.3	50	Sn	1.8
4	Be	1.76[b]	22	Ti	1.5	51	Sb	1.9
5	B^-	1.98[b]	23	V	1.6	52	Te	2.1
5	B	2.20[b]	24	Cr	1.6	53	I	2.5
6	C^-	2.33[b]	25	Mn	1.5	55	Cs	0.7
6	C	2.45[b]	26	Fe	1.8	56	Ba	0.9
6	C^+	2.80[b]	27	Co	1.8	57–71	La–Lu	1.1–1.2
7	N^-	2.80[b]	28	Ni	1.8	72	Hf	1.3
7	N	3.15[b]	29	Cu	1.9	73	Ta	1.5
7	N^+	3.40[b]	30	Zn	1.6	74	W	1.7
8	O^-	3.40[b]	31	Ga	1.6	75	Re	1.9
8	O	3.65[b]	32	Ge	1.8	76	Os	2.2
8	O^+	3.82[b]	33	As	2.0	77	Ir	2.2
9	F^-	3.82[b]	34	Se^-	2.37[b]	78	Pt	2.2
9	F	4.00[b]	34	Se	2.4	79	Au	2.4
9	F^+	4.25[b]	35	Br	3.05[b]	80	Hg	1.9
11	Na	0.9	37	Rb	0.8	81	Tl	1.8
12	Mg	1.2	38	Sr	1.0	82	Pb	1.8
13	Al	1.5	39	Y	1.2	83	Bi	1.9
14	Si	1.95[b]	40	Zr	1.4	84	Po	2.0
15	P	2.47[b]	41	Nb	1.6	85	At	2.2
15	P^+	2.47[b]	42	Mo	1.8	87	Fr	0.7
15	P^+	2.47[b]	43	Tc	1.9	88	Ra	0.9
16	S^-	2.47[b]	44	Ru	2.2	89	Ac	1.1
16	S	2.75[b]	45	Rh	2.2	90	Th	1.3
16	S^+	3.00[b]	46	Pd	2.2	91	Pa	1.5
17	Cl	3.25[b]	47	Ag	1.9	92	U	1.7
19	K	0.8	48	Cd	1.7	93–102	Np–Mo	1.3

[a] As given by Siegbahn *et al.*[1] from L. Pauling, *The Nature of the Chemical Bond*, 2nd ed. (New York, 1945), except where noted otherwise.
[b] Taken from W. L. Jolly, *J. Am. Chem. Soc.* **92**, 3260 (1970).

5. In the case of more than one resonance form, average the results according to the weight given each form.

There is a great deal of arbitrariness in obtaining the calculated charge. In particular, there is the choice of the most probable resonance form, or if more than one form seems important, what the relative weights should be. For complex ions and organic molecules it is well to use group electronegativities.

Illustration of the determination of some calculated charges might serve as the best description of the method.

Example 1. Calculate q for carbon in CH_3F.

Bond	$\chi_A - \chi_B$	I_i	Number of bonds	$\sum_i I_i$
C–H	2.20 – 2.45	–0.02	3	–0.06
C–F	4.00 – 2.45	+0.45	1	+0.45
				$q = +0.39$

Example 2. Calculate q for nitrogen in aniline.

Resonance form 1:

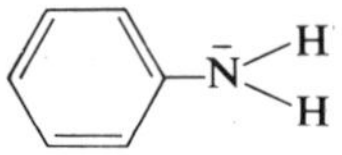

Bond	$\chi_A - \chi_B$	I_i	Number of bonds	$\sum_i I_i$
C–N	2.45 – 3.15	–0.12	1	–0.12
H–N	2.20 – 3.15	–0.20	2	–0.40
				$q = -0.52$

Resonance form 2:

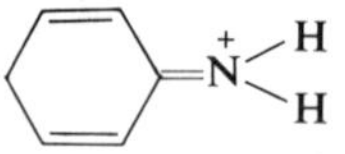

Bond	$\chi_A - \chi_B$	I_i	Number of bonds		$\sum_i I_i$
$C{=}N^+$	2.45 – 3.40	–0.20	2		0.40
$H{-}N^+$	2.20 – 3.40	–0.30	2		–0.60
				$Q = +1$	+1.00
					$q = 0.00$

Siegbahn *et al.*[1] used the measured binding energy of nitrogen in aniline to estimate the relative importance of the two resonance forms given above by the following expression:

$$E_B = \sum_n C_n E_B(q)$$

where E_B is the experimental binding energy, $E_B(q)$ is the binding energy obtained from a plot of q vs. E_B, and C_n is the percentage of resonance form n. They found $C_1 = 46\%$ and $C_2 = 54\%$.

The use of Pauling's electronegativity scale is very simple, applies to all the elements, and works surprisingly well. There is, however, some arbitrariness in determining the proper electronic structural formula on which the calculation is to be performed, and if resonance occurs in the molecule there is no easy *a priori* method for assigning weights to different resonance structures (a rather arbitrary description in itself).

Bus and de Jong[22] have tested the calculated charge as obtained from electronegativity using a form of equation (5.12), namely

$$E_B - V = kq + l \tag{5.16}$$

where k and l are adjustable parameters. Ideally, l would be the binding energy where $q = 0$, and k would be e^2/r. Bus and de Jong, using carbon $1s$ binding energy data, found that electronegatives from Jolly and Johnson[23] gave a better fit than Pauling's. (It might be pointed out here that data from PESIS itself might be used to establish an electronegativity scale.) However, the correlation in either case was poor, and in fact neglect of V gave a better fit. This observation has also been made by Gelius *et al.*,[24] who concluded that charges calculated by means of electronegativity incorporate the essential part of the molecular potential. However, Bus and de Jong[22] have suggested an alternative procedure to improve the correlation. They argued that the calculated charge be modified somewhat when a given carbon atom attached to a highly electronegative group has as a neighbor another carbon also attached to a highly electronegative group, for example, hexafluorobenzene. When such occurs, the partial ionic character is not strictly additive as suggested by equation (5.15), but is somewhat diminished:

$$I' = I(1 - nc) \tag{5.17}$$

where n is the number of adjacent neighbors with highly electronegative groups, and c is a correction factor, which was found to give best correlation at 0.35. Figure 5.2 gives the results of correlation obtained by Bus and de Jong using the above arguments.

Though such parameterization is a dangerous game, it does suggest that electronegativities can provide a useful tool for correlating binding energy shifts in the absence of more sophisticated calculations.

Jolly[25] has determined that since one needs to consider the final ionized state following photoionization as well as the initial state of the neutral molecule in determining binding energies (that is to say, the relaxation energy), the calculated charge should fall approximately half-way between the charge calculated for the neutral molecule and the case where a core electron has been removed. Equation (5.16) becomes

$$E_B = \tfrac{1}{2}k(q_i - q_f - 1) + V_{\text{av}} + l \tag{5.18}$$

where q_i and q_f are the charges on the atom before and after losing the core electron. V_{av} is the crystal or molecular potential using the average charge $(q_i - q_f - 1)/2$ rather than q. To further aid in this concept, Jolly has devised

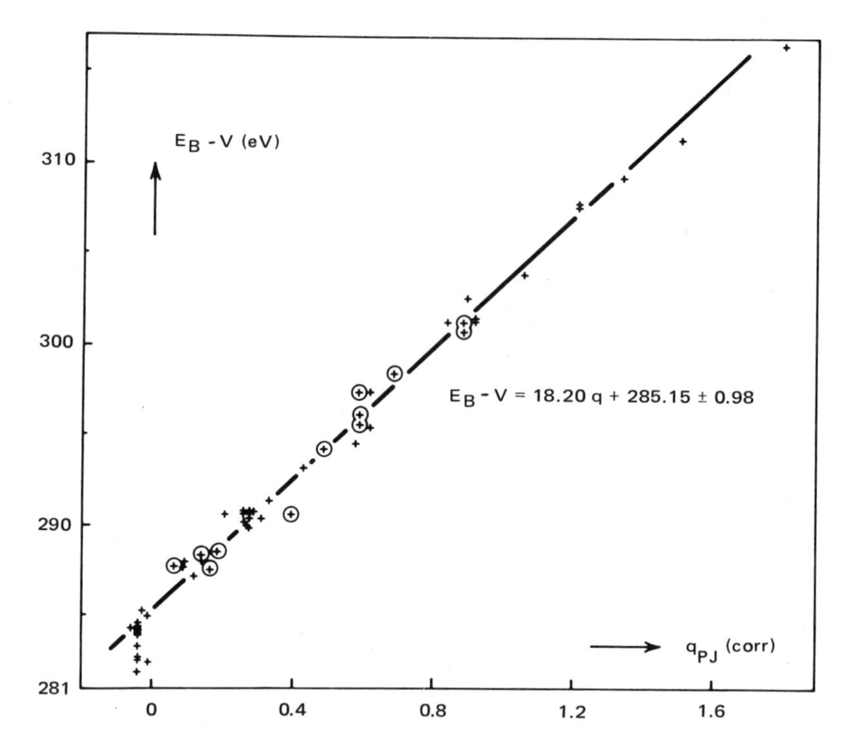

FIGURE 5.2. Plot of $E_b - V$ for photoionization in the C(1*s*) shell vs a corrected charge based on equation (5.16). [Reproduced from Bus and de Jong,[22] Figure 2.]

an alternative model for estimating atomic charges. Jolly *et al.*[26] in fact explored various ways to obtain both calculated charges from the concept of electronegativity by adding extra parameters. Account is taken of the polarizability and the amount of *s* and *p* character in a given orbital. Calibration of the parameters has been made both with experimental core binding energies and charges calculated from *ab initio* molecular orbital calculations.

Finally, it needs to be pointed out that rather than using electronegativities only for single atoms, one may determine electronegativities for a whole group.[24] In such a treatment the molecular potential or V is dropped from consideration and we take

$$\Delta E = \sum_{\text{gr}} E_{\text{gr}} + l \tag{5.19}$$

Group shifts ΔE_{gr} are obtained from a least squares fit of the experimental data to equation (5.19). The group shifts can also be used to derive more generally applicable group electronegativities by means of the relation

$$\Delta E_{\text{gr}} = \text{const} \times [1 - \exp -|0.25(\chi_{\text{gr}} - \chi_{\text{X}})|^2] \tag{5.20}$$

TABLE 5.6

Group Chemical Shifts and Electronegativities for Carbon, Phosphorus, Nitrogen, and Arsenic Obtained from X-Ray Photoelectron Spectroscopy

Group	ΔE_{gr}, eV	χ_{gr}
Carbon[a]		
—CH_2NH_2	−0.52	1.9
—$CH(CH_3)OC$	−0.52	1.9
—$CH(OC)_2$	−0.52	1.9
—CH_3	−0.32	2.0
—CH_2OH	−0.42	2.0
—CH_2Cl	−0.32	2.0
—CH_2OCH_2C	−0.23	2.1
═C(OH)═C[c]	−0.25	2.1
—$C(O)OCH_2X$	−0.35	2.1
—CH═C[c]	−0.18	2.2
—C(O)C	−0.17	2.2
—CH_2C	−0.06	2.3
—C(O)OH	−0.15	2.3
—H	0.01	2.6
—$CH_2OC(O)X$	0.08	2.7
—NH_2	0.25	2.9
—CCl_2F	0.46	3.0
$CClF_2$	0.48	3.1
CF_3	0.59	3.1
—$N(CH_2)_2$	0.79	3.2
═S	0.85	3.3
—Br	0.89	3.3
—OH	1.51	3.5
—OCH_2C	1.37	3.5
—OCH_3	1.46	3.5
—$OCH(CH_3)O$	1.31	3.5
—$OCH(OCH_3)_2$	1.68	3.6
—Cl	1.56	3.6
—OC(O)C	1.91	3.7
—OC(O)X	2.14	3.8
—F	2.79	4.0
Arsenic[d]		
—Ar	0.21	2.6
—*ϕ*	0.38	2.8
—ONa	0.50	3.0
—CH_3	0.48	3.0
—SAs	0.60	3.1
—AsS_2	0.90	3.4
—OH	1.02	3.5
—NR_2	1.00	3.5
—F_{hexa}	0.98	3.5
—Br	1.60	4.1
═O	2 × 1.00	4.8
Phosphorus[b]		
—O	0.24	2.7
—	0.30	2.8
—C	0.33	2.9
—H	0.34	2.9
—S	0.57	3.1
—N	0.60	3.2
—ONH_4	0.70	3.3
—S_3	0.70	3.3
—OH	0.87	3.4
—Br	0.87	3.4
—OC	0.89	3.5
—OP	1.12	3.7
—Cl	1.22	3.8
—S_2	1.38	3.9
—F	1.43	4.0
═O	1.58	4.1
═S	1.87	4.5
Nitrogen[d]		
≡Acr	−(3 × 0.76)	2.3
≡Quiniz	−(4 × 0.58)	2.3
≡Azol	−(3 × 0.59)	2.4
≡Py	−(3 × 0.46)	2.5
≡CR	−(3 × 0.30)	2.6
—*ϕ*	−0.56	2.7
—H	−0.29	2.7
—CHR_2	−0.22	2.7
—*n*-Bu	−0.22	2.8
—CH_3	−1.19	2.8
—Ar	−0.17	2.8
—PO_2Na	−0.25	2.8
═PCl_2═	1.5 × 0.03	3.1
≡N(O)Ar	2 × 0.11	3.2
—S—N	0.25	3.2
—SO_2Ar	0.38	3.3
—SAr	0.39	3.3
═NAr	2 × 0.43	3.4
—N^+H_3	0.80	3.4
—S—S	0.59	3.4
—SO_3^-	0.63	3.4
═S═N		3.5
—ONa		3.5
—OR		3.5
—OH		3.5
═O		3.9
≡O_2Na		4.1

[a] Cf. Gelius *et al.*,[24] Table V.
[b] Cf. Hedman *et al.*[27]
[c] Aromatic compounds.
[d] Lindberg and Hedman.[28]

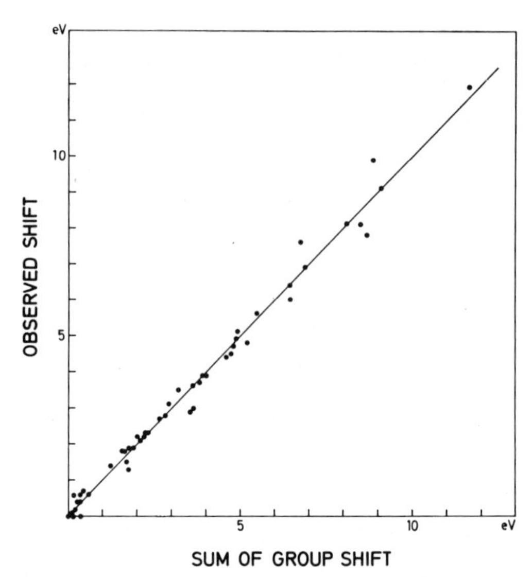

FIGURE 5.3. Plot of ΔE [group shift energy; cf. equation (5.19)] vs observed binding energy shift in C(1*s*). [Reproduced from Gelius *et al.*,[24] Figure 12.]

where X is the element whose core binding energy shifts are being studied. Table 5.6 and Figure 5.3 have been derived from data on carbon.[24] Also included in Table 5.6 are group electronegativities for constituents attached to phosphorus, nitrogen, and arsenic.[27,28] Riggs[29] has suggested that group electronegativities be used whereby the arithmetic mean of the electronegativities of the substituent atoms is employed. The sum is limited, however, to the central atom of the group and the atoms attached directly to it.

A modified Sanderson method has been devised[30] which uses the concept of the stability ratio enunciated by Sanderson,[31] but in which group behavior is substituted for atomic contributions. The authors feel their method retains the simplicity of Sanderson's model, i.e., no parameterization and no resonance form, while providing a generally applicable empirical means of calculating charges.

2.4. Calculation of Net Charge from Semiempirical MO

Wave function programs for molecules based on semiempirical methods (cf. Chapter 3, Section 5) such as CNDO, INDO, and MINDO treat the inner shell electrons as a fixed core and thus cannot be used directly to yield shifts in the inner shell binding energies. However, such calculations can be used to give estimates of the charge density by means of, for example, Mulliken's population number[32]:

$$q_{\mu} = \sum_{\nu} P_{\mu\nu} S_{\mu\nu} \tag{5.21}$$

where q_{μ} is the charge density on atom μ for a given molecular orbital. The molecular orbital is expressed in terms of a linear combination of atomic orbitals, i.e.,

$$\psi_i = \sum_{\nu} \phi_{\nu} c_{\nu i} \tag{5.22}$$

$S_{\mu\nu}$ are the overlap integrals between ϕ_{ν} and ϕ_{μ}, and $P_{\mu\nu}$ is the bond order, or

$$P_{\mu\nu} = 2 \sum_{k} c_{k\mu} c_{k\nu} \tag{5.23}$$

A simple plot of charge density as derived from CNDO calculations gives a rather large scatter. However, good correlation has been obtained using equation (5.16). Equation (5.16) can be derived from molecular orbital theory, and has been successfully applied to CNDO calculations by Gelius *et al.*[24] (cf. Figure 5.4). Ellison and Larcom[10] have extended the method of calculating the effects of neighboring atoms by including not only monopole charges centered at the nuclei of the atoms, but also electronic dipoles and quadrupoles. They find, however, that though these refinements substantially improve some nitrogen data, results on carbon and oxygen were rather mediocre. Ellison and Larcom[10] also suggested a two-parameter fit in which

$$kq \quad \text{becomes} \quad k_s q_s + k_p q_p \tag{5.24}$$

the subscripts corresponding to the atomic s or p character of the molecular orbitals. That is, the valence shell can be thought of as two subshells with two slightly different radii, just as in the case of the s and p orbitals in an atom. The two-parameter fit generally gives better agreement and in particular explains the odd behavior of compounds like CO.

Schwartz *et al.*[8] have critically evaluated the use of the CNDO scheme in determining binding energy shifts and have found it lacking, suggesting the exploration of other semiempirical methods which employ some less drastic differential overlap neglect. They also suggest that equation (5.16) be amended to

$$E_B = \alpha + \beta q + \gamma V \tag{5.25}$$

where α, β, and γ are adjustable parameters. Davis and Shirley[33] have given perhaps the "definitive" treatment of binding energy shifts within the framework of CNDO calculations. Using equation (5.16) as their basis, they also consider (1) the effect of relaxation energy and (2) the replacement of point charges by $\langle 1/r \rangle$ integrals which are sensitive to bond directions of p orbitals.

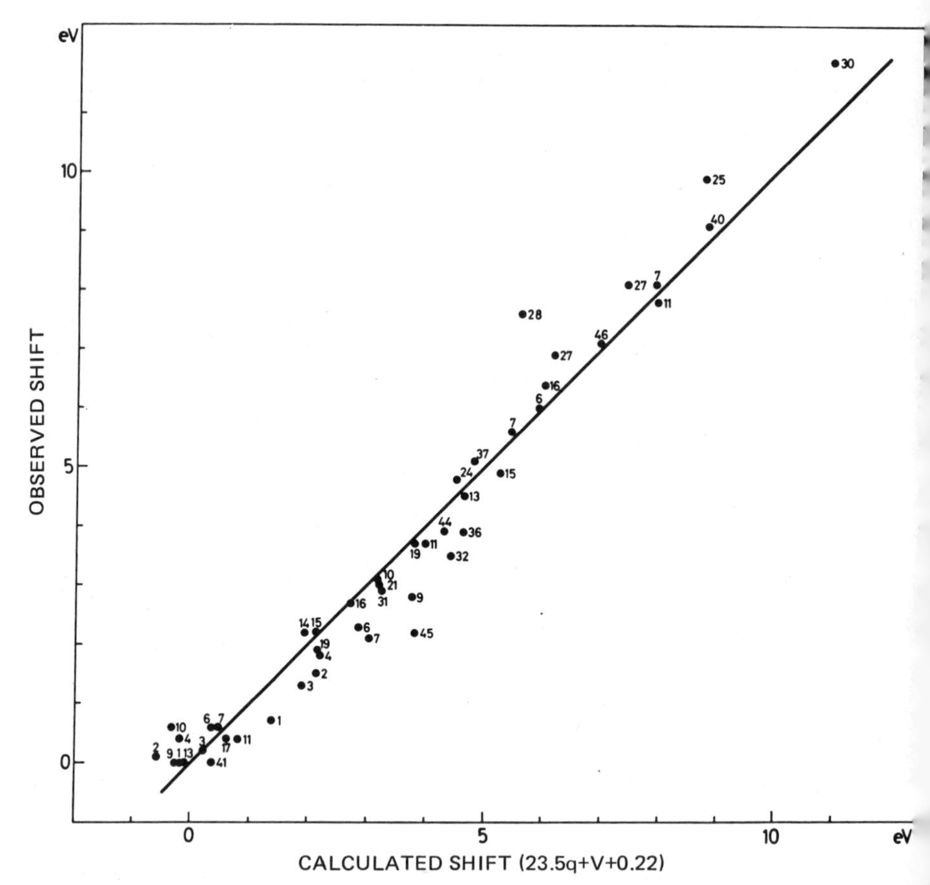

FIGURE 5.4. Comparison between measured shifts on C(1*s*) and shifts calculated with equation (5.16) using charges obtained from CNDO/2 calculations. [Reproduced from Gelius *et al.*,[24] Figure 10.]

Although there still remain problems, it would seem that use of semiempirical MO wave functions for predicting chemical shifts of at least free molecules is on very sound ground.

One limitation of the semiempirical methods is that they have not been perfected much beyond the lighter elements, although this problem is being worked on (cf. Chapter 3, Section 5.1.).

2.5. Use of *Ab Initio* Calculations for Chemical Shifts

The most direct method for comparing experimental changes in the core binding energies with theory is through a Hartree–Fock calculation in which all electrons, including core electrons, are treated in a self-consistent fashion. Chemical shifts for orbital n are obtained from the eigenvalues by means of equation (5.2). Correlations of *ab initio* calculations with experimental data are given in Figure 5.5. The calculations give absolute binding energies that are too high, due in part to the approximation contained in Koopmans' theorem. The slope should be unity, and in the case shown for carbon it is. The orbital wave functions can also be used to assign a value of q to each atom in the molecule. The limitation of the *ab initio* method is the fact that such calculations still require prohibitive computer time except for relatively small molecules made up of first row elements, and again such calculations are for free molecules, not solids.

2.6. Correlation of Chemical Shift with Thermochemical Data

Jolly and his co-workers[34] have suggested an alternative model to correlating chemical shifts of core electrons using thermochemical data. The basic assumption is as follows: Removal of a core electron is chemically equivalent to the increase of nuclear charge by one. The effective charge felt by a valence electron Z_{eff} is

$$Z_{\text{eff}} = Z - \alpha \tag{5.26}$$

where Z is the nuclear charge and α is the screening of the inner shell electron. With regard to the effective charge, the removal of one screening electron is thus equivalent to the increase in nuclear charge by one. For example, the total chemical energy E_T for $C^*O_2^+$ is approximately equal to that for NO_2, where an asterisk indicates a core vacancy. In this way one estimates E_T for molecular states following photoionization from available thermochemical data and thereby the chemical shift. Take the case of the chemical shift in C(1s) binding energy between CH_4 and CO_2. The binding energies are respectively represented by

$$CH_4 \rightarrow C^*H_4^+ + e^- \qquad E_B(CH_4) \tag{5.27}$$

$$CO_2 \rightarrow C^*O_2^+ + e^- \qquad E_B(CO_2) \tag{5.28}$$

From arguments given above on equivalent central potentials

$$C^*H_4^+ \equiv NH_4^+ \tag{5.29}$$

$$C^*O_2^+ \equiv NO_2^+ \tag{5.30}$$

Thus, substituting (5.29) and (5.30) into (5.27) and (5.28) and subtracting yields

$$CO_2 + NH_4^+ \rightarrow NO_2^+ + CH_4, \qquad E_B(CO_2) - E_B(CH_4) \tag{5.31}$$

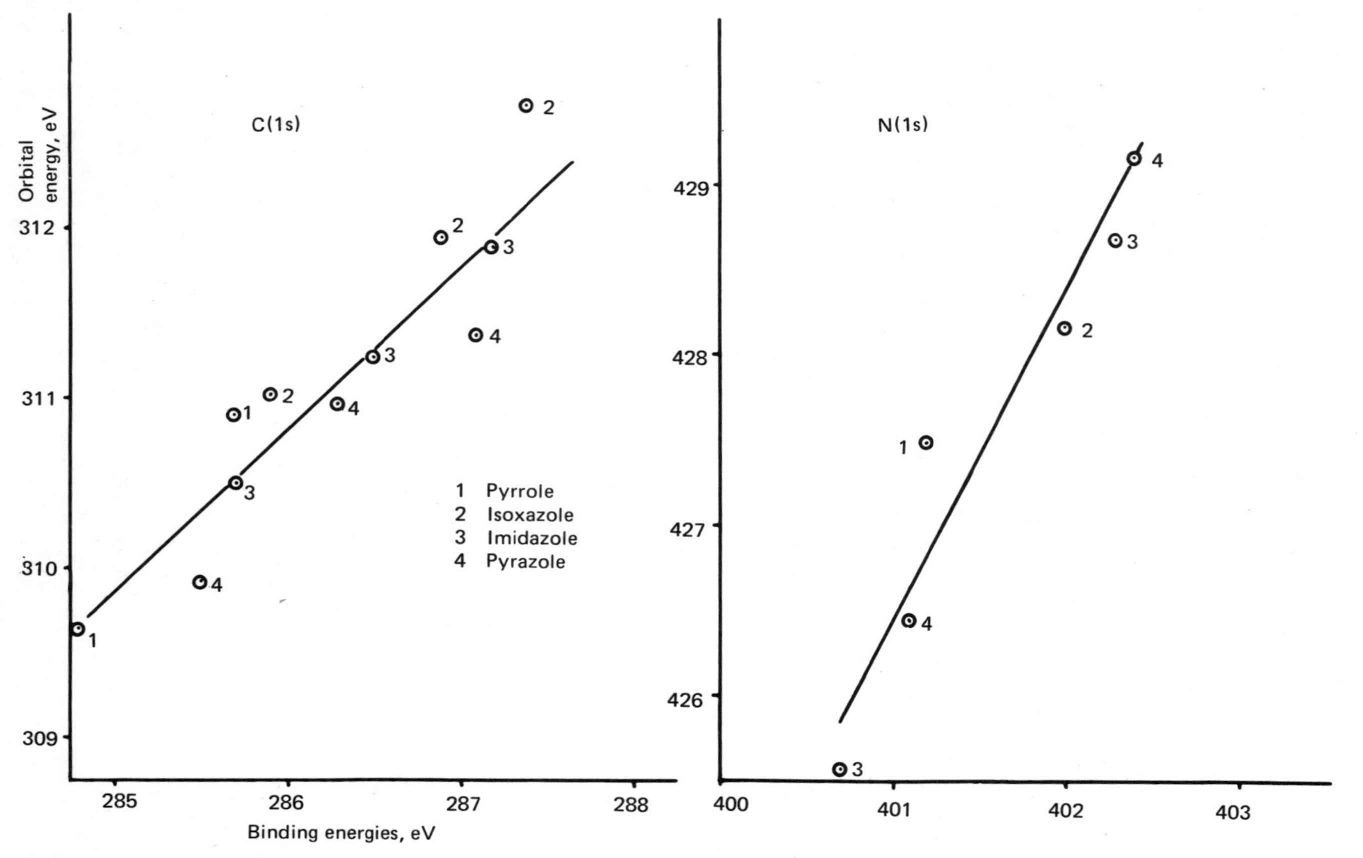

FIGURE 5.5 Plot of C(1*s*) and N(1*s*) binding energies vs. eigenvalues derived from *ab initio* molecular orbital calculations. [Reproduced from Clark and Lilley,(44) Figure 1.]

The difference in core binding energies should be closely related to the difference in thermochemical energies represented by the reaction shown in (5.31). They in fact both yield[34] 6.8 eV. Figure 5.6 shows a plot of relative (C1*s*) binding energies versus the thermodynamically estimated values for a series of gaseous compounds.

The principal value of using thermochemical data rather than calculated charge is that it circumvents the problem of computing the molecular or crystal potential V. Its weakness lies in the limited availability of needed thermal data and its requirements regarding the use of data from species having effective charge equivalence.

Jolly[34] has extended his arguments so as to be able to use Pauling's electronegativity. He states that, based on empirical observation, covalent contributions to the bonds in a species are equal to those in any isoelectronic species, and thus one can equate a difference in dissociation energy to differences in the sum of ionic contributions to the bonds. From these arguments he derives the dissociation energy:

$$\Delta = \sum_i [(\chi_A{}^2 - \chi_B{}^2) + 2(\chi_B - \chi_A)\chi_i] + [10.5 \sum_j 1/(1+k)]C, \qquad (5.32)$$

Here χ_A and χ_B are the electronegativities of the transmutable atoms A and B

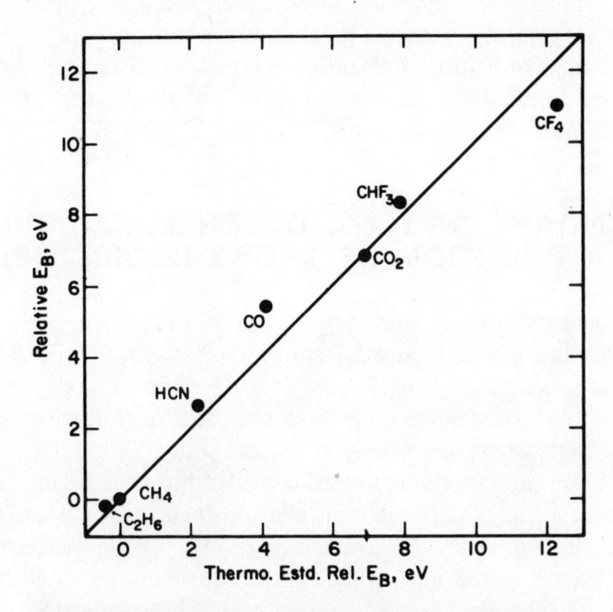

FIGURE 5.6. Plot of carbon 1*s* binding energies vs thermodynamically estimated energies. [Reproduced from Jolly,[34c] Figure 1.]

(the atomic number of atom A is one less than that of atom B); χ_i is the electronegativity of an atom directly bonded to atom A (or B); and C_j is the formal charge of an atom separated by k atoms from A (or B). The sum $\sum_i$ is carried out over the i atom directly bonded to atom A (or B) and the sum $\sum_j$ is carried out over all the atoms in the species, except atom A (or B). The equations are valid only when the transmutable atoms and the atoms to which they are directly bonded have formal charges -1 to $+1$.

Heats of reaction can also be calculated using molecular orbital theory, these values then being used to calculate binding energy shifts via the method of Jolly. This approach has been made by Clark and Adams[35] with CNDO/2 calculations and by Frost *et al.*[36] applying MINDO/1.

Another valuable relationship that has been explored[37] is that between core binding energy shifts and proton affinity. The relationship exists because on the one hand PESIS measures the effect of change in nuclear charge at a given atomic site due to a removal of a shielding electron, while on the other proton affinity measures the effect of nuclear charge by the addition of a positive charge or proton. The proton affinity is defined as $-\Delta H_f = E_0 - E_{\text{ion}}$. For example, the proton affinity for a series of alcohols as determined by the reaction $ROH + H^+ \rightarrow ROH_2{}^+$ should be related to the chemical shifts in O(1s) photoionization: $ROH \rightarrow RO^*H$ (1s hole) $+ e^-$. This relationship works best for a homologous set of compounds where there is a lone-pair set of electrons at the protonation site. The model breaks down when there are large changes in geometry upon protonation. Relaxation energies are associated with protonation, but are affected by substituents less than in the case of inner shell photoionization.

3. SUMMARY OF DATA ON CHEMICAL SHIFTS AS A FUNCTION OF THE PERIODIC TABLE

PESIS can be applied to any element having an inner shell, which is to say any element with $Z > 2$. Figure 5.7 shows the extent of work done on core binding energy shifts. At least 78 elements have been investigated for these shifts. More than 100 studies have been reported for a number of elements. The CRC *Handbook of Spectroscopy*[38] lists about 3000 studies.

The amount of data has increased considerably since Figure 5.7 was produced, but the influx of papers based solely on chemical shifts and correlation of chemical shifts has subsided as the interest of x-ray photoelectron spectroscopy has been diverted into other applications, such as satellite lines and surface studies. The maximum observed shift has been indicated in Figure 5.7 by a bar. The more thoroughly an element has been investigated, the more likely the maximum shift will be observed, but investigators will normally seek

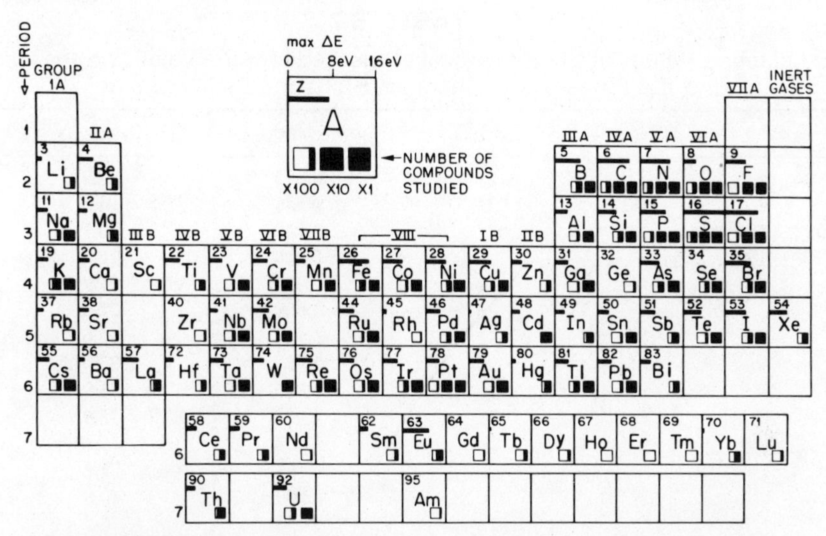

FIGURE 5.7. Chart showing the extent of studies of core binding energy shifts by PESIS throughout the periodic table. A bar indicates the maximum observed shift. [Chart prepared by N. Fernelius on data available as of July 1972.]

out comparison between compounds with large chemical shifts, and the maximum observed shift offers a crude criterion for evaluating the magnitude of the shift for a given element.

Appendix 3 lists a compendium of data in an attempt to illustrate the extent of the chemical shifts that are encountered for each of the elements. Jørgensen and Berthou[39] have been particularly industrious in terms of obtaining an overall view, looking at photoelectron spectra for 77 different elements. Care must be taken when comparing their values with other workers since they calibrate against a carbon 1*s* binding energy from hydrocarbon contamination on a double-stick tape backing, taken to be 290.0 eV rather than the commonly used value of about 285.0 eV. Even when this correction is made, agreement between Jørgensen and Berthou and other authors for similar compounds is often rather poor; as a matter of fact, in general, agreement between different investigators is rather unsatisfactory. Usually, relative binding energies for a given set of data taken by one author are much more dependable than a comparison of binding energies obtained at different laboratories. This is probably due for the most part to the way in which the calibration and the problem of charging are handled. The elements that appear best suited for study are the light elements most commonly found in different chemical combinations: C, N, S, and P. As one goes to elements of higher Z, at least for a given group, the chemical shifts in general decrease. This can in part be understood by the fact that the radius of the valence shell increases with Z for a given group

TABLE 5.7

Expectation Values of the Reciprocal of the Radii for the Valence Subshells and the Slope of the Chemical Shifts for Each Element

Z	Element	Shell	$\langle 1/r \rangle$[a]	$\langle 1/r \rangle e^2$, eV[b]
3	Li	$2s$	0.365	9.9
4	Be	$2s$	0.548	14.9
5	B	$2s$	0.760	20.7
		$2p_{1/2}$	0.619	16.8
6	C	$2s$	0.953	25.9
		$2p_{1/2}$	0.809	22.0
7	N	$2s$	1.139	31.0
		$2p_{1/2}$	0.989	26.9
		$2p_{3/2}$	0.987	26.9
8	O	$2s$	1.321	35.9
		$2p_{1/2}$	1.163	31.6
		$2p_{3/2}$	1.161	31.6
9	F	$2s$	1.500	40.8
		$2p_{1/2}$	1.335	36.3
		$2p_{3/2}$	1.332	36.2
10	Ne	$2s$	1.679	45.7
		$2p_{1/2}$	1.506	41.0
		$2p_{3/2}$	1.500	40.8
11	Na	$3s$	0.325	8.8
12	Mg	$3s$	0.432	11.8
13	Al	$3s$	0.557	15.2
		$3p_{1/2}$	0.398	10.8
14	Si	$3s$	0.660	18.0
		$3p_{1/2}$	0.507	13.8
15	P	$3s$	0.754	20.5
		$3p_{1/2}$	0.605	16.5
		$3p_{3/2}$	0.602	16.4
16	S	$3s$	0.845	23.0
		$3p_{1/2}$	0.698	19.0
		$3p_{3/2}$	0.694	18.9
17	Cl	$3s$	0.934	25.4
		$3p_{1/2}$	0.786	21.4
		$3p_{3/2}$	0.781	21.3
18	Ar	$3s$	0.992	27.0
		$3p_{1/2}$	0.812	22.1
		$3p_{3/2}$	0.805	21.9
19	K	$4s$	0.259	7.0
20	Ca	$4s$	0.331	9.0
21	Sc	$3d_{3/2}$	0.869	23.6
		$4s$	0.355	9.7

[a] In Bohr units (1 Å = 0.529 Bohr units) from relativistic Hartree–Fock–Slater wave functions. See C. C. Lu, T. A. Carlson, F. B. Malik, T. C. Tucker, and C. W. Nestor, Jr., *Atomic Data* **3**, 1 (1971).

[b] Slope of chemical shift. Value should be approximately equal to k given in equation (5.16), and is in terms of eV/unit charge.

TABLE 5.7 (*continued*)

Z	Element	Shell	$\langle 1/r \rangle$[a]	$\langle 1/r \rangle e^2$, eV[b]
22	Ti	$3d_{3/2}$	0.961	26.1
		$4s$	0.376	10.2
23	V	$3d_{3/2}$	1.045	28.4
		$4s$	0.394	10.7
24	Cr	$3d_{3/2}$	1.053	28.7
		$3d_{5/2}$	1.048	28.5
		$4s$	0.368	10.0
25	Mn	$3d_{3/2}$	1.203	32.7
		$3d_{5/2}$	1.209	32.9
		$4s$	0.428	11.6
26	Fe	$3d_{3/2}$	1.287	35.0
		$3d_{5/2}$	1.281	34.9
		$4s$	0.444	12.1
27	Co	$3d_{3/2}$	1.364	37.1
		$3d_{5/2}$	1.357	36.9
		$4s$	0.489	13.3
28	Ni	$3d_{3/2}$	1.440	39.2
		$3d_{5/2}$	1.432	39.0
		$4s$	0.474	12.9
29	Cu	$3d_{3/2}$	1.452	39.5
		$3d_{5/2}$	1.441	39.2
		$4s$	0.438	11.9
30	Zn	$3d_{3/2}$	1.590	43.3
		$3d_{5/2}$	1.579	43.0
		$4s$	0.502	13.7
31	Ga	$4s$	0.586	15.9
		$4p_{1/2}$	0.398	10.8
32	Ge	$4s$	0.651	17.7
		$4p_{1/2}$	0.480	13.1
33	As	$4s$	0.712	19.4
		$4p_{1/2}$	0.551	15.0
		$4p_{3/2}$	0.537	14.6
34	Se	$4s$	0.770	21.0
		$4p_{1/2}$	0.614	16.7
		$4p_{3/2}$	0.599	16.3
35	Br	$4s$	0.826	22.5
		$4p_{1/2}$	0.674	18.3
		$4p_{3/2}$	0.656	17.8
36	Kr	$4s$	0.881	24.0
		$4p_{1/2}$	0.731	19.9
		$4p_{3/2}$	0.712	19.4
37	Rb	$5s$	0.243	6.6
38	Sr	$5s$	0.302	8.2
39	Y	$4d_{3/2}$	0.593	16.1
		$5s$	0.330	9.0
40	Zr	$4d_{3/2}$	0.666	18.1

TABLE 5.7 (*continued*)

Z	Element	Shell	$\langle 1/r \rangle$[a]	$\langle 1/r \rangle e^2$, eV[b]
40	Zr	$5s$	0.351	9.6
41	Nb	$4d_{3/2}$	0.692	18.8
		$5s$	0.340	9.3
42	Mo	$4d_{3/2}$	0.755	20.5
		$4d_{5/2}$	0.745	20.3
		$5s$	0.854	9.6
43	Tc	$4d_{3/2}$	0.850	23.1
		$4d_{5/2}$	0.840	22.9
		$5s$	0.398	10.8
44	Ru	$4d_{3/2}$	0.871	23.7
		$4d_{5/2}$	0.859	23.4
		$5s$	0.378	10.3
45	Rh	$4d_{3/2}$	0.926	25.2
		$4d_{5/2}$	0.912	24.8
		$5s$	0.389	10.6
46	Pd	$4d_{3/2}$	0.943	25.7
		$4d_{5/2}$	0.926	25.2
47	Ag	$4d_{3/2}$	1.032	28.1
		$4d_{5/2}$	1.015	27.6
		$5s$	0.409	11.1
48	Cd	$4d_{3/2}$	1.114	30.3
		$4d_{5/2}$	1.097	29.8
		$5s$	0.458	12.5
49	In	$5s$	0.519	14.1
		$5p_{1/2}$	0.366	10.0
50	Sn	$5s$	0.566	15.4
		$5p_{1/2}$	0.427	11.6
51	Sb	$5s$	0.611	16.6
		$5p_{1/2}$	0.481	13.1
		$5p_{3/2}$	0.454	12.4
52	Te	$5s$	0.653	17.8
		$5p_{1/2}$	0.529	14.4
		$5p_{3/2}$	0.499	13.6
53	I	$5s$	0.693	18.9
		$5p_{1/2}$	0.573	15.6
		$5p_{3/2}$	0.541	14.7
54	Xe	$5s$	0.733	19.9
		$5p_{1/2}$	0.615	16.7
		$5p_{3/2}$	0.580	15.6
55	Cs	$6s$	0.221	6.0
56	Ba	$6s$	0.270	7.3
57	La	$5d_{3/2}$	0.496	13.5
		$4s$	0.292	7.9
58	Ce	$4f_{5/2}$	1.312	35.7
		$5d_{3/2}$	0.504	13.7
		$6s$	0.297	8.1

TABLE 5.7 (*continued*)

Z	Element	Shell	$\langle 1/r \rangle^a$	$\langle 1/r \rangle e^2$, eV[b]
59	Pr	$4f_{5/2}$	1.365	37.1
		$5d_{3/2}$	0.511	13.9
		$6s$	0.301	8.2
60	Nd	$4f_{5/2}$	1.350	36.7
		$6s$	0.288	7.8
61	Pm	$4f_{5/2}$	1.401	38.1
		$6s$	0.292	7.9
62	Sm	$4f_{5/2}$	1.451	39.5
		$6s$	0.296	8.1
63	Eu	$4f_{5/2}$	1.484	40.4
		$4f_{7/2}$	1.465	39.9
		$6s$	0.292	7.9
64	Gd	$4f_{5/2}$	1.589	43.2
		$4f_{7/2}$	1.571	42.7
		$5d_{3/2}$	0.520	14.1
		$6s$	0.315	8.6
65	Tb	$4f_{5/2}$	1.580	43.0
		$4f_{7/2}$	1.557	42.4
		$6s$	0.299	8.1
66	Dy	$4f_{5/2}$	1.626	44.2
		$4f_{7/2}$	1.602	43.6
		$6s$	0.303	8.2
67	Ho	$4f_{5/2}$	1.687	45.9
		$4f_{7/2}$	1.662	45.2
		$6s$	0.314	8.5
68	Er	$4f_{5/2}$	1.732	47.1
		$4f_{7/2}$	1.704	46.4
		$6s$	0.318	8.7
69	Tm	$4f_{5/2}$	1.761	47.9
		$4f_{7/2}$	1.731	47.1
		$6s$	0.341	9.3
70	Yb	$4f_{5/2}$	1.820	49.5
		$4f_{7/2}$	1.788	48.7
		$6s$	0.326	8.9
71	Lu	$4f_{5/2}$	1.910	52.0
		$4f_{7/2}$	1.880	51.2
		$5d_{3/2}$	0.548	14.9
		$6s$	0.353	9.6
72	Hf	$5d_{3/2}$	0.608	16.5
		$6s$	0.373	10.1
73	Ta	$5d_{3/2}$	0.660	18.0
		$6s$	0.391	10.6
74	W	$5d_{3/2}$	0.706	19.2
		$6s$	0.407	11.1
75	Re	$5d_{3/2}$	0.752	20.5
		$5d_{5/2}$	0.721	19.6

TABLE 5.7 (*continued*)

Z	Element	Shell	$\langle 1/r \rangle$[a]	$\langle 1/r \rangle e^2$, eV[b]
75	Re	$6s$	0.423	11.5
76	Os	$5d_{3/2}$	0.795	21.6
		$5d_{5/2}$	0.762	20.7
		$6s$	0.437	11.9
77	Ir	$5d_{3/2}$	0.836	22.7
		$5d_{5/2}$	0.802	21.8
		$6s$	0.452	12.3
78	Pt	$5d_{3/2}$	0.855	23.3
		$5d_{5/2}$	0.816	22.2
		$6s$	0.442	12.0
79	Au	$6s$	0.454	12.4
80	Hg	$6s$	0.490	13.3
81	Tl	$6s$	0.539	14.7
		$6p_{1/2}$	0.384	10.4
82	Pb	$6s$	0.578	15.7
		$6p_{1/2}$	0.437	11.9
83	Bi	$6s$	0.617	16.8
		$6p_{1/2}$	0.488	13.3
		$6p_{3/2}$	0.407	11.1
84	Po	$6s$	0.654	17.8
		$6p_{1/2}$	0.532	14.5
		$6p_{3/2}$	0.448	12.2
85	At	$6s$	0.690	18.8
		$6p_{1/2}$	0.572	15.6
		$6p_{3/2}$	0.485	13.2
86	Rn	$6s$	0.724	19.7
		$6p_{1/2}$	0.609	16.6
		$6p_{3/2}$	0.518	14.1
87	Fr	$7s$	0.226	6.1
88	Ra	$7s$	0.271	7.4
89	Ac	$6d_{3/2}$	0.415	11.3
		$7s$	0.297	8.1
90	Th	$6d_{3/2}$	0.458	12.5
		$7s$	0.317	8.6
91	Pa	$5f_{5/2}$	0.942	25.6
		$6d_{3/2}$	0.438	11.9
		$7s$	0.309	8.4
92	U	$5f_{5/2}$	0.990	26.9
		$6d_{3/2}$	0.447	12.2
		$7s$	0.315	8.6
93	Np	$5f_{5/2}$	1.035	28.2
		$6d_{3/2}$	0.454	12.4
		$7s$	0.320	8.7
94	Pu	$5f_{5/2}$	1.041	28.3
		$7s$	0.307	8.4
95	Am	$5f_{5/2}$	1.085	29.5

TABLE 5.7 (*continued*)

Z	Element	Shell	$\langle 1/r \rangle$[a]	$\langle 1/r \rangle e^2$, eV[b]
95	Am	$5f_{7/2}$	1.053	28.7
		$7s$	0.311	8.5
96	Cu	$5f_{5/2}$	1.145	31.2
		$5f_{7/2}$	1.128	30.7
		$6d_{3/2}$	0.470	12.8
		$7s$	0.334	9.1
97	Bk	$5f_{5/2}$	1.198	32.6
		$5f_{7/2}$	1.165	31.7
		$6d_{3/2}$	0.474	12.9
		$7s$	0.339	9.2
98	Cf	$5f_{5/2}$	1.207	32.8
		$5f_{7.2}$	1.170	31.8
		$7s$	0.324	8.8
99	Es	$5f_{5/2}$	1.245	33.9
		$5f_{7/2}$	1.206	32.8
		$7s$	0.328	8.9
100	Fm	$5f_{5/2}$	1.309	35.6
		$5f_{7/2}$	1.272	34.6
		$6d_{3/2}$	0.483	13.1
		$7s$	0.354	9.6

in the periodic table, and thus the k in equation (5.16) decreases. Table 5.7 lists the reciprocals of outer shell atomic radii and k atomic radii as a function of the periodic table. Apart from the radius of the valence shell, the crystal field potential and the ability for an element to engage in a variety of different types of bonding determine the magnitude of chemical shifts for a given element. Although a wide variety of behavior occurs with regard to the extent of the chemical shift for the different elements, there is essentially no part of the periodic table that cannot be studied with some value.

In this section we shall first examine more closely the data obtained on chemical shifts for the more exhaustively studied elements, and bring out some special consideration associated with these elements. Second, we shall examine the behavior of core binding energy shifts for different groups in the periodic table, trying to draw some general rationale out of the relative behavior.

3.1. Carbon

Carbon, the most versatile of all elements, has been extensively examined for chemical shifts in the $1s$ shell. Good correlation was obtained between Pauling's calculated charge and ΔE_B for a wide variety of compounds.[24] Comparison was made between the binding energies observed in the solid phase versus those in the gaseous phase. Although, because of differences in the reference levels, compounds in the gaseous phase have C 1*s* binding energies of about 5 eV greater than in the solid phase, the *chemical shifts* in the two phases are about the same (cf. Figure 5.8).

By the use of equation (5.16) and charges obtained from *ab initio* and CNDO/2 calculations, even better fits to the experimental binding energy shifts have been obtained.[24] The empirical slopes (18.3 and 23.5, respectively, for *ab initio* and CNDO/2 calculations) may be compared with atomic Hartree–Fock calculations. The value of $\langle 1/r \rangle e^2$ is 22.0 eV from Table 5.7, and ΔE_B due to a loss of a 2*p* electron in carbon is 21.2 eV as calculated from differences of eigenvalues[40] for the neutral and singly charged ion.

When enough data have been obtained, it is possible to assign chemical shifts to groups. That is, each group attached to carbon is assigned a characteristic chemical shift. In this model one assumes that the groups behave independently, which is only partially valid. Only neighboring groups are considered rather than the whole solid, so that the treatment is restricted to nonionic compounds. Within these limitations the treatment is quite successful and easy to apply. (Cf. Section 2, together with Table 5.6 and Figure 5.3, for further discussion.)

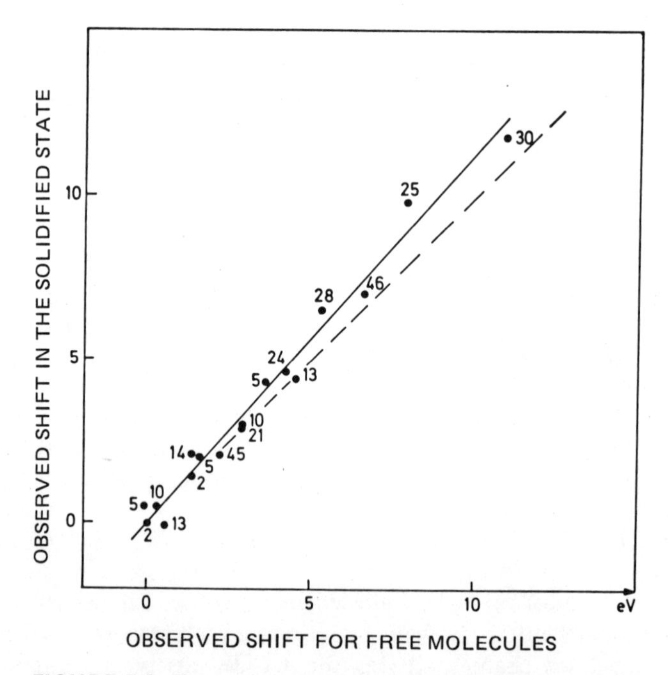

FIGURE 5.8. Comparison between carbon 1*s* binding energy shifts in the gaseous and solidified states. [Reproduced from Gelius *et al.*,[24] Figure 5.]

Gelius *et al.*[24] also pointed out with their carbon data that, on neglecting V in equation (5.16), ionic compounds correlate with q just as well as nonionic compounds, which implies that the detailed effects of the crystal potential to some extent average out for ionic crystals as well as for nonionic compounds. They also note that on calculating V using q_P (Pauling's calculated charge) the expression

$$E_B = kq + \gamma V + l \tag{5.33}$$

gives the best correlation if $\gamma = 0$. In contrast, the CNDO analysis gave the best fit to $\gamma = 1$. This imples that q obtained from the CNDO/2 calculation can be taken as a fair estimate of the net charge on each atom, while q_P already contains a crude estimation of the molecular potential. However, note objections to this assertion given by Bus and de Jong (Section 2.3). Davis *et al.*[41] have correlated binding energies for a large number of carbon compounds. In particular, they have discussed the effects of relaxation. They have used the chemical shift data to derive charge distribution and obtain good agreement with values calculated from CNDO/2 molecular orbital calculations.

A number of other authors have also correlated data on carbon 1s binding energies with molecular orbital calculations: e.g., Clark and Kilcast[42] have compared CNDO/2 calculations with results for halogenated methanes. Adams and Clark[43] have compared results for the same molecules with *ab initio* calculations and have examined different types of calculations for predicting chemical shifts and their sensitivities with regard to the size of the basis set. For a minimal basis set the equivalent core method gives better results than either Koopmans' theorem or hole-state calculations. Nonempirical LCAO-MO calculations using contracted Gaussian basis sets have been employed by Clark and Lilley[44] for examining several heterocyclic compounds; Nelson and Frost[45] have employed a floating spherical Gaussian orbital (FSGO) model to make calculations on some small hydrocarbons. A large number of organic compounds have been studied for their chemical shifts in carbon and some of this material will be covered in Section 4.3. Cubic carbides of groups IVb and Vb have received attention from Ramqvist *et al.*[46] Of particular note were the chemical shifts as a variation of carbon content for different phases. These results were correlated with heats of formation and atomic ratios and are summarized in Figure 5.9.

Carbon is one of the most difficult elements to study with regard to contamination. One of the most common contaminants on the surface of solid targets is absorbed hydrocarbons from pump oil, *O*-rings, and stopcock grease. In fact, the contamination is often used for calibration purposes. From personal experience I have found that when working under "normal" vacuum conditions (10^{-6}–10^{-7} Torr) the carbon from the contamination layer accounts for about 10–20% of what might be expected for a pure carbon sample. A

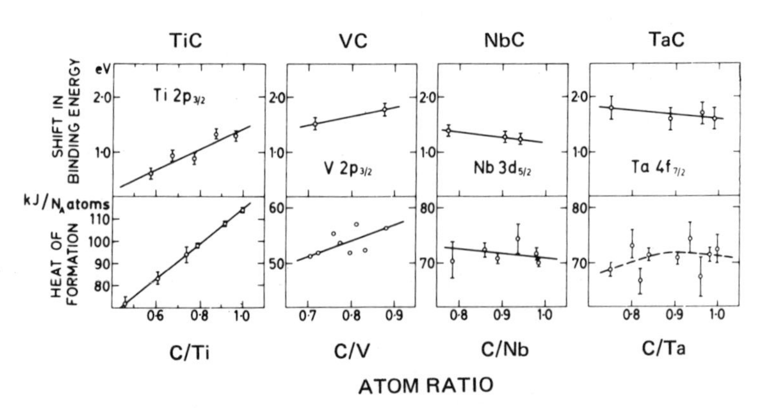

FIGURE 5.9. Binding energy shifts and heats of formation vs atom ratio x for TiC_x, VC_x, NbC_x, and TaC_x within the homogeneity ranges of the carbides. [Reproduced from Ramqvist *et al.*,[46b] Figure 1.]

liquid nitrogen trap in the source volume reduces this contamination considerably. To completely remove uncertainty from $1s$ carbon analysis, the source volume should be bakeable and free from any but metal seals. Quite often contamination comes from the sample itself and the nature of the contamination depends on how the sample has been prepared. Studies carried out in the gas phase, or those by which a volatile material is continuously condensed on a cold finger serving as the target, are nearly free from contamination worries.

3.2. Nitrogen and Phosphorus

Nitrogen and phosphorus both have been studied extensively for their core-electron chemical shifts. In nitrogen the $1s$ electron is the core electron measured, while in phosphorus it is the $2p$ shell. The correlation of chemical shifts with Pauling's charge q_P for nitrogen gives a monotonic but not a linear dependence.[1] Application[47] of CNDO/2 and extended Hückel molecular orbital calculations were found to give a linear dependence but with rather bad scattering, and compounds formed subgroups depending on whether the compounds were neutral or ionic, and if ionic, whether the nitrogens were part of an anion or cation group. The above comparisons have *not* been corrected for molecular potential, i.e., V in equation (5.16). When this was done for a few simple gaseous molecules[48] a good correlation was obtained with a value of $k = 21.5$ eV, which may be compared with $\langle 1/r \rangle e^2$ from Table 5.7 for the nitrogen $2p$ shell (26.9 eV). *Ab initio* calculations on nitrogen compounds[49] also gave a good correlation with $k = 22.4$ eV. It was shown, however, that a minimal basis set was not adequate, but that a double-zeta basis close to that obtained for a wave function near the Hartree–Fock limit was necessary. Also, inclusion of the lattice potential (i.e., V) is necessary for both ionic and non-ionic compounds to be accommodated by the same straight line.

Jack and Hercules[50] have studied a large number of quaternary nitrogen compounds. The order found in terms of decreasing binding energy was R_4N^+, pyridinium > isoquinolinium > quinolinium > acridinium, benzoquinolizinium. Single-atom correlations were found to be of only limited value, and account of the effects of both the counterion (anion) and substituent is needed. Studies on a series of nitrogen bases have been reported by Cox *et al.*[51] and the effects of lattice potentials were evaluated for pyridine hydrochloride and pyridine hydrogen nitrate. Sharma *et al.*[52] have studied closely the core and valence electronic levels of the alkali azides.

Nitrogen has proven to be one of the most useful elements for applying PESIS to chemical problems. The behavior of nitrogen in organic molecules is most helpful in shedding light on the electronic structure of the whole molecule. This is particularly true when nitrogen is part of the ring in heterocyclic compounds. The oxidation states of nitrogen, as evidenced by its binding energies, have given clues for protein analysis and pollution monitoring. All these applications will be discussed later in this chapter.

An initial study on phosphorus compounds gave rather poor correlation[53] between theory and chemical shifts, but it was shown by Hedman *et* al.[27] that if only nonionic compounds were compared using calculated charges from electronegativity and computing the molecular potential V via equation (5.16), good correlation could be obtained. Hedman *et al.*[27] also determined for phosphorus the group shifts and electronegativities according to the scheme given in equations (5.19) and (5.20) (cf. Table 5.6).

Problems with the correlation of ionic compounds have also been cited for nitrogen.[47,50] Presumably, if crystal field calculations were made, better correlation could be obtained for both phosphorus and nitrogen.

Swartz and Hercules[54] have determined $2p$ phosphorus binding energies for a series of quaternary phosphonium salts, $R(C_6H_5)_3P^+Y^+$. They found poor correlation with the electronegativity of R substituents, but were able to correlate the data with ^{31}P NMR chemical shifts despite the fact that NMR data were taken on salts in solution, while the PESIS data were from solids. Morgan *et al.*[55] and Blackburn *et al.*[56] have studied a wide variety of phosphorus compounds, in which they interpret the absence of chemical shifts in terms of backbonding through the π bond. A number of phosphorus compounds have also been studied by Stec *et al.*[57] and a comparison made with ^{31}P NMR data. They have also discussed the effects of diasterioisomerism and of substituents in cyclic structures involving phosphorus in the ring. From a study of P, PCl_3, and $POCl_3$ Barber *et al.*[58] suggested that the results from PESIS be directly interpreted in terms of the phosphine lone pair and the nature of the additional bonds which are formed. Swartz *et al.*[59] have looked at some bis(triphenylphosphine) iminium salts. From the P(2*p*) and N(1*s*) binding energies they concluded that phosphorus atoms bear some degree of positive charge, while the nitrogen atom has some negative character. Swartz *et al.*[60] have extended their studies to diphenylphosphino ethane and related compounds and have correlated binding energy shifts of phosphorus with charges calculated with the help of Sanderson electronegativities.

3.3. Sulfur and Oxygen

Sulfur has been studied for its chemical shifts probably more than any element in the periodic table, with the possible exception of carbon. Shifts of up to 16 eV have been reported and surprisingly good correlation has been achieved with the calculated charge as obtained from Pauling's electronegativities. (cf. Figure 5.10). Part of the problem in assigning a calculated charge is the presence of a resonance form. In sulfur this problem is particularly acute in explaining the >S=O bond. Lindberg *et al.*[61] examined this question closely and reached the conclusion that the >S=O has a polarity about halfway between that of a purely semipolar and a purely double-bonded structure.

CNDO calculations as well as *ab initio* calculations have been carried out on a few molecules containing sulfur. From these calculations[62] one finds a value for k [cf. equation (5.16)] of 14.1 eV, compared with 18.9 for $\langle 1/r \rangle e^2$ (Table 5.7). *Ab initio* calculations[63] showed that a small amount of S(3*d*) polarization in the basis set improves agreement with experiment considerably.

Lindberg *et al.*[61] have done extensive work on correlating various functional groups containing sulfur. Figure 5.11 illustrates some of their work. Like nitrogen, sulfur has been invaluable in evaluating the behavior of organic molecules. For example, Lindberg and Hamrin[64] have studied the influence of conjugation on chemical shifts of core electrons for a large group of sulfur-substituted nitrobenzene compounds. Thiathiophene compounds have received attention from Clark *et al.*[65] and Gleiter *et al.*[66] The question examined was principally whether the sulfurs labeled 1 and 6 in the compound

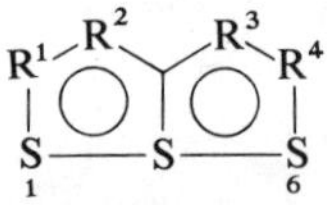

R^1 R^2 R^3 R^4
S—S—S
1 6

are different, depending on the substitutes R^n. This question was answered nicely for a number of compounds and the results examined with the help of SCF-MO calculations[65] and He I photoelectron spectra.[66]

Kramer and Klein[67–69] have devoted much attention to core binding energy chemical shifts in sulfur and iron in a number of iron complexes. Their

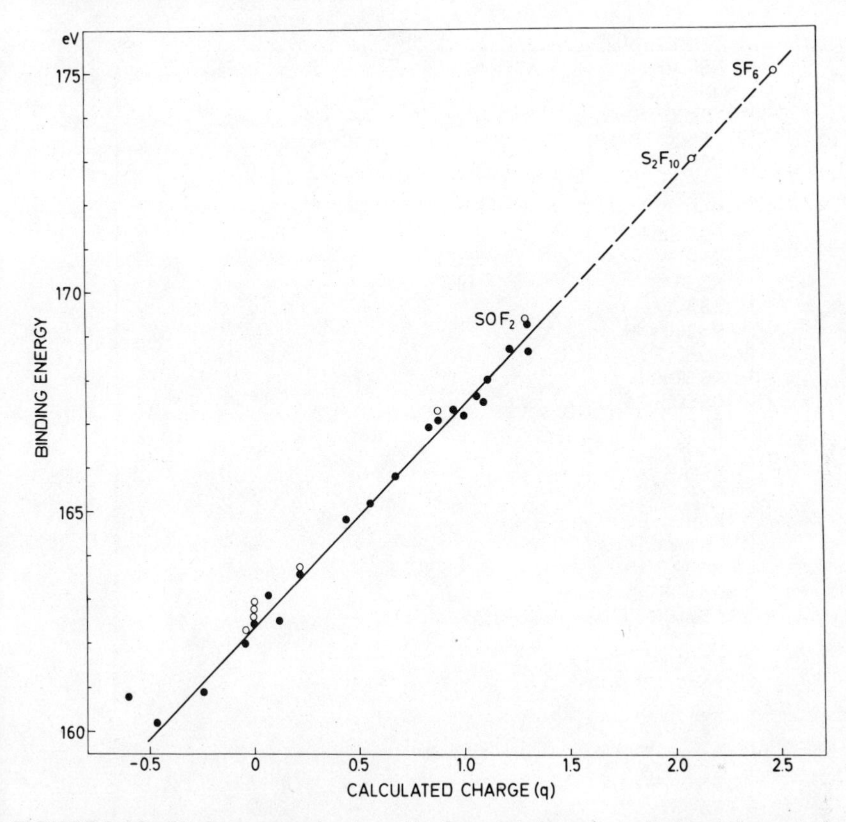

FIGURE 5.10. Binding energy for the sulfur 2*p* subshell versus calculated charge based on Pauling's electronegativities. Filled points indicate averages from compounds studied in the solid phase at room temperature. Open points represent liquids and gases studied by freezing. [Reproduced from Siegbahn *et al.*,[1] Figure 5.39.]

interest lies with the biological importance of these compounds. In fact, there have been a number of investigations on sulfur 2*p* binding energies for biological and environmental problems, and discussion of these studies will be found in Section 6 of this chapter.

The value of $\langle 1/r \rangle e^2$ for oxygen atoms is 31.6 eV (Table 5.7), one of the highest values in the periodic table, and nearly twice that for sulfur. However, the maximum observed shift so far has been only 6 eV. Oxygen and fluorine have the highest electronegativities of all the elements and tend to always assume the same chemical form. Thus, in spite of the high value of k, the maximum chemical shift is held to a low value. Charges based on CNDO/2 calculations[48] give only a fair fit of the data to equation (5.16) and a k of 25.8 eV.

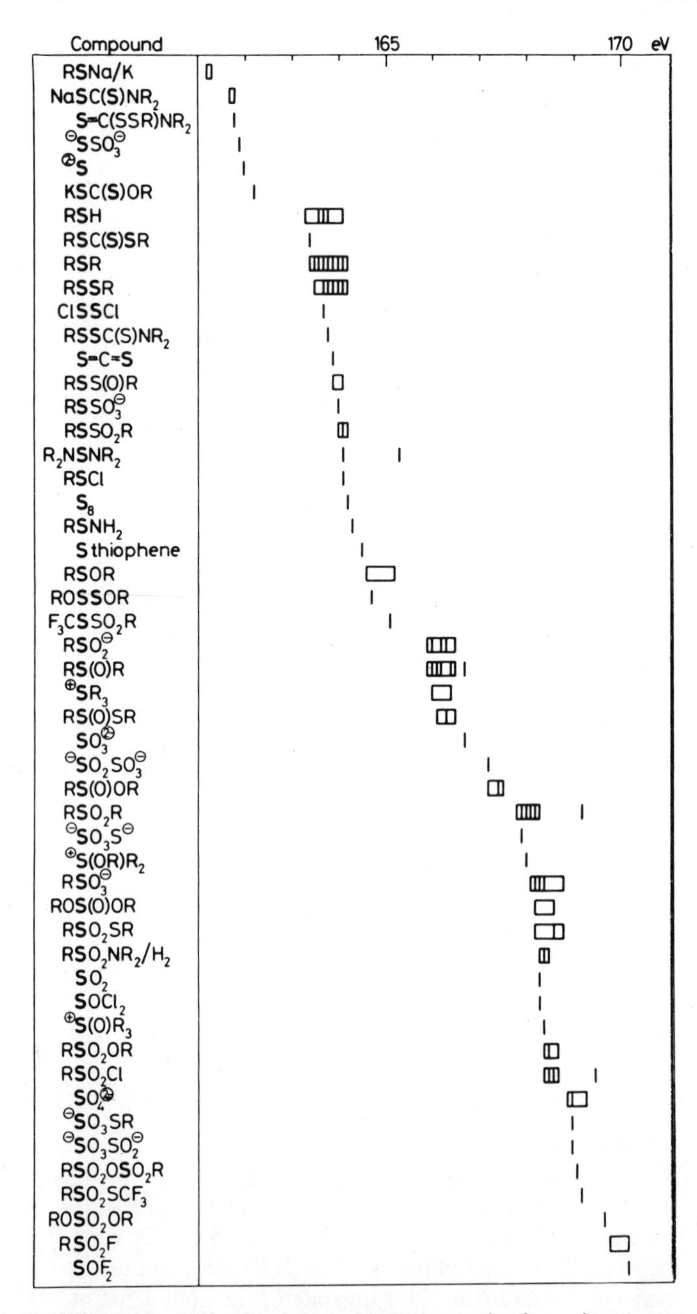

FIGURE 5.11. Sulfur 2*p* electron binding energies for various types of sulfur compounds. R signifies alkyl or aryl carbon. [Reproduced from Lindberg *et al.*,[61] Figure 3.]

Ellison and Larcom[10] obtained 26.4 eV. Davis and Shirley[14] have examined binding energy shifts in oxygen theoretically, including relaxation effects.

Some chemical shift studies with oxygen have proved valuable. Yin *et al.*,[70] for example, have detected energy differences between bridging and nonbridging oxygen atoms in a silicate chain. Bus[71] has resolved two distinct peaks in hexanitrosobenzene.

Together with carbon, oxygen is one of the most common contaminants. It is virtually impossible to keep the surface layer of a metal free from oxidation unless the metal is cleaned *in situ* under high vacuum. In addition, water, CO, and CO_2 are common adsorbents to the surfaces of all solids. PESIS can, of course, be of immense aid in the study of oxygen on the surface.

Both O_2 and NO are subject to multiplet splitting. This phenomenon and electron shakeoff in molecules containing oxygen are of considerable interest and their consideration will be deferred until Section 5.

3.4. Group IIIA, IVA, VA, and VIA Elements

As one traverses the periodic table from group I to VIII, the radius of the valence shell shrinks and thus the slope [k in equation (5.16)] of the chemical shift equation increases. As one goes to lighter elements the radius also becomes smaller and, hence, k becomes larger. Examining the periodic table as a whole, one further expects and generally finds a wide variety of oxidation states for the center groups. Thus, it is not surprising that the elements that have been most successfully exploited for their chemical shifts are the light elements of the periodic groups IIIA–VIA such as B, C, N, Si, P, and S.

3.4.1. Group IIIA: B, Al, Ga, In, and Tl

This group includes the elements B, Al, Ga, In, and Tl. Boron, as indicated above, has received the most attention. Data on B(1*s*) have been compared with calculated charges obtained from MO calculations.[72] Employing equation (5.16), Allison *et al.*[72] found 19.1 for the value of k. In B_5H_9 two peaks are observed in the ratio 4:1. The odd boron has a distinctively lower binding energy, which is in support of chemical evidence for an electrophilic attack on the apical boron. (Cf. Figure 5.12.) No correlation was found between core binding energy shifts and ^{11}B NMR data.[72b, 73] Good agreement was obtained between PESIS data on boron and thermochemical shifts.[73] (Cf. Table 5.8.)

McGuire *et al.*[74] have studied a series of Al, Ga, In, and Tl salts and oxides and have compared chemical shifts for related compounds. They have also demonstrated the importance of the crystal potential in these comparisons. Müller *et al.*[75] have studied a number of thalium compounds, in which they

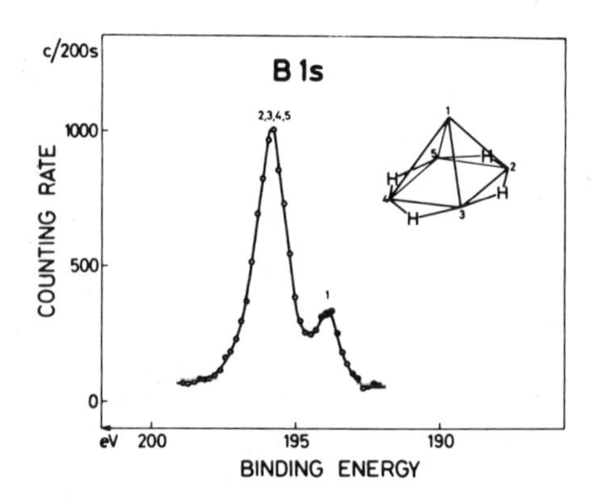

FIGURE 5.12. The B(1*s*) spectrum of B_5H_9 demonstrating the presence of the odd boron atom. [Reproduced from Allison *et al.*,[72b] Figure 7.]

have shown the strong polarizing effect of Tl(I) on the anion. Aluminum has been studied with regard to its behavior in minerals.[76]

3.4.2. Group IVA: C, Si, Ge, Sn, and Pb

The lightest member of group IVA, carbon, is of course the most important and has been discussed in detail separately. Silicon compounds have been investigated by Nordberg *et al.*[77] The results were discussed in terms of back-bonding and a comparison made between silicon and phosphorus and sulfur compounds of related structure. The behavior of silicon in minerals has also been studied.[76] Perry and Jolly[78] have studied chemical shifts in the tetrahedral compounds of carbon, silicon, and germanium, finding that the results required orbital participation in the silicon and germanium compounds.

TABLE 5.8

Boron 1*s* Chemical Shifts, Estimated Boron Charges, ^{11}B NMR Chemical Shifts, and Thermochemically Estimated Chemical Shifts[a]

Compound	Relative binding energy, eV	Pauling charge	CNDO charge	Extended Hückel charge	^{11}B NMR chemical shifts, ppm	Thermochemical energy, eV
BF_3	0	0.92	0.70	1.65	6.6	0
BCl_3	−2.3	−0.22	0.25	0.81	−29.2	−2.42
$B(OCH_3)_3$	−4.4	0.36	0.49	1.13	0.0	−3.02
B_2H_6	−6.3	−0.29	−0.03	−0.01	0.5	−5.42
$B(CH_3)_3$	−6.4	−0.62	0.13	0.31	−68.2	−6.86
BH_3CO	−7.6	−0.78	−0.43	0.15		−8.03
$BH_3N(CH_3)_3$	−9.1	−0.60	−0.08	−0.10	24.9	−9.46

[a] Taken from Finn and Jolly,[73] Table I.

Germanium has been studied for its effect on doping in semiconductors.[79] Hollinger *et al.*[80] followed the Ge 3*d* binding energies for a series of molecular glasses. Perry and Jolly measured a number of silicon and germanium compounds in the gas phase[81] and compared[82] Auger and core binding energy shifts to ascertain the effects of relaxation energy for a variety of germanium compounds. Tin has received more attention. Barber *et al.*[83] successfully correlated a series of tin compounds [$Y_2Sn(ox)_2$, where ox = 8-quinolinolate and Y = ethyl, phenyl, Cl, Br and I] with Mössbauer data. Some scattered data on lead compounds have been reported by Jørgensen and Berthou[39] and others.[85–87] The importance and identification of lead in pollution samples have been reported by several groups.[88, 89]

Morgan and Van Wazer[86] have made a systematic study of chemical shifts for the group IVA elements. Figure 5.13 shows a plot of the relative shifts for compounds of similar structure. The relative shifts decrease with decreasing *Z*. Such plots as Figure 5.13, though quite useful, depend strongly, however, on a uniform behavior of crystal potential.[74]

3.4.3. Group VA: N, P, As, Sb, and Bi

Group VA is one of the more interesting groups in the periodic table for the study of core binding energy shifts, because its elements can occur in such a wide variety of compounds with different oxidation states. Nitrogen and phosphorus have already been discussed in some detail. Some 32 compounds of arsenic have been studied by Stec *et al.*[90] with examples of compounds having coordination numbers of 3, 4, 5, and 6. They have also examined the relative shifts of N, P, and As compounds having the same basic structure. Lindberg and Hedman[28] and Hedman *et al.*[27] have developed group shifts for nitrogen, phosphorus, and arsenic compounds, and Morgan *et al.*[91, 92] have made a systematic study of chemical shifts in all the elements of group VA. Nefedov *et al.*[93] have examined antimony and bismuth. Hulett and Carlson[94] studied a series of arsenic compounds in order to demonstrate the usefulness of XPS in monitoring the decomposition of cacodylic acid, $(CH_3)_2AsO(OH)$, in soil.

3.4.4. Group VIA: O, S, Se, and Te

Group VIA is essentially a one-element group as far as PESIS is concerned. Sulfur, which has been discussed separately, is probably the most studied element for core binding energy shifts. Oxygen has only one valence state. Its valence shell electron density, and consequently its 1*s* binding energy, remains fairly constant throughout a series of compounds. Polonium does not occur naturally, and selenium and tellurium have as yet not been extensively studied.

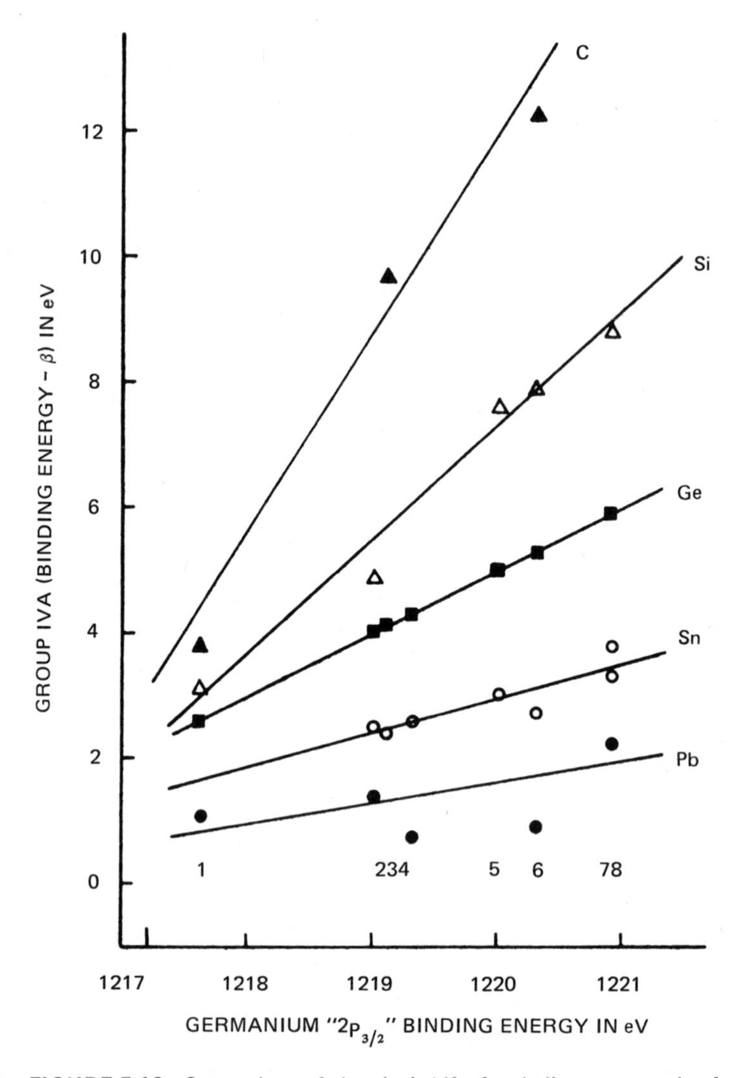

FIGURE 5.13. Comparison of chemical shifts for similar compounds of the group IVA elements. [Reproduced from Morgan and Van Wazer,[86] Figure 1.]

Swartz *et al.*[95] have compared the binding energy shifts in Se and Te for the oxidation states –2, 0, +4, and +6. Comparison is made with HFS semi-empirical MO calculations for the group VI oxyanions. Table 5.9 shows in general a smaller binding energy shift as one goes to a higher Z in accord with

TABLE 5.9

Comparison of Binding Energies for the Group VIA Elements

Compound	Oxidation state	E_B[a]
Na_2S	−2	0
S	0	1.4
Na_2SO_3	+4	5.0
Na_2SO_4	+6	6.9
KSeCN	−2	0
Se	0	0.8
Na_2SeO_3	+4	4.5
Na_2SeO_4	+6	5.0
Na_2Te	−2	0
Te	0	0.6
K_2TeO_3	+4	3.5
K_2TeO_4	+6	4.2

[a] From data of Swartz *et al.*[95]
[b] Normalized to −2 state.

calculations for $\langle 1/r \rangle$. A more extensive investigation of selenide compounds has been carried out by Malmsten *et al.*[96] and a correlation made with calculated charge based on electronegativity. Norbury *et al.*[97] have examined the behavior of the thiocyanate, selenocyanate, and tellurocyanate ions.

3.5. Halides and Rare Gases

The group VII and VIII elements have the smallest valence shell radii of the periodic table. However, the elements are so electronegative that they do not tend to share their electrons in different chemical combinations, and the halides usually occur in the valence state of −1, while the rare gases maintain their closed shell configuration. No compounds of neon or argon have yet been reported and fluorine exhibits only one oxidation state and thus has a maximum shift of only 5 eV. However, Davis *et al.*[41] have done a good job in correlating observed chemical shifts in F 1*s* with theory. The zenon fluoride compounds[48] have been studied by PESIS and it was found that each fluorine draws at least 0.3 electron from the xenon atom, and Carroll *et al.*[98] have established evidence for "orbital independence" in the xenon–fluorine bonding.

Chlorine, bromine, and iodine, in contrast to fluorine, do form a number of oxidation states and exhibit large chemical shifts, as expected. Figure 5.14 shows results for the oxyhalides together with calculations based on atomic Hartree–Fock calculations of the nominal oxidation states of the halides. The atomic calculations give unrealistically large values, but when normalized to

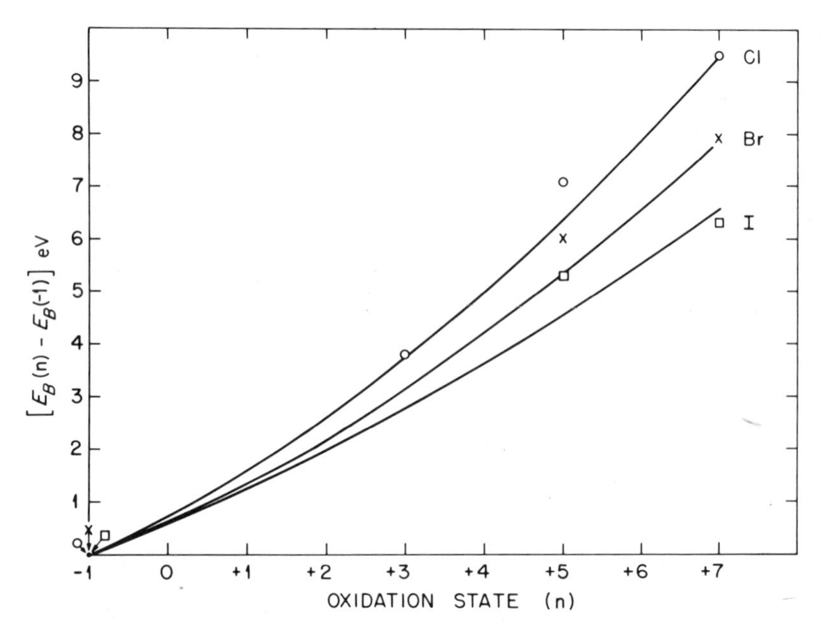

FIGURE 5.14. Comparison of binding energy shifts in the halogens as oxidation state increases. The quantity $[E_B(n) - E_B(-1)]$ for a given halogen represents the difference in binding energy for its oxidized state n and its halide state. The curves are from Hartree–Fock calculations normalized to the data. Solid line = theory × 0.067. [Reproduced from Hulett and Carlson,[94] Figure 2.]

one point, the value of ClO_4^-, the relative dependence of the three halides is reproduced quite well.

Except for the oxyhalides, halogens usually find themselves in the −1 oxidation state and there is generally little variety in the core binding energies as a function of compound. However, their large electronegativities give rise to substantial changes in the neighboring atoms. In addition, the electronegativity varies substantially from fluorine to iodine. Thus, studies on binding energy shifts of a given element usually include a variety of halide compounds. Clark *et al.*[35, 42, 99–106] have made copious studies on the effect of substituting halogens, particularly fluorine, for hydrogen in organic molecules, some of which will be discussed in Section 4.3. Figure 5.15 is a plot of carbon 1*s* binding energies for a series of halogen-substituted compounds. When the data for hydro, chloro, and fluoro substitution were plotted separately, the slopes obtained were 25.5 + 2.3 (hydro series), 24.3 + 1.4 (chloro series), and 20.9 + 1.0 (fluoro series). Hayes and Edelstein[107] have examined the point charge model for various metal-ion fluorides, and Jørgensen *et al.*[108] reported on a large number of fluorine compounds. Hashmall *et al.*[109] have correlated binding energy shifts in the $3d_{5/2}$ shell of iodine for alkyl iodides with PESOS data to help in

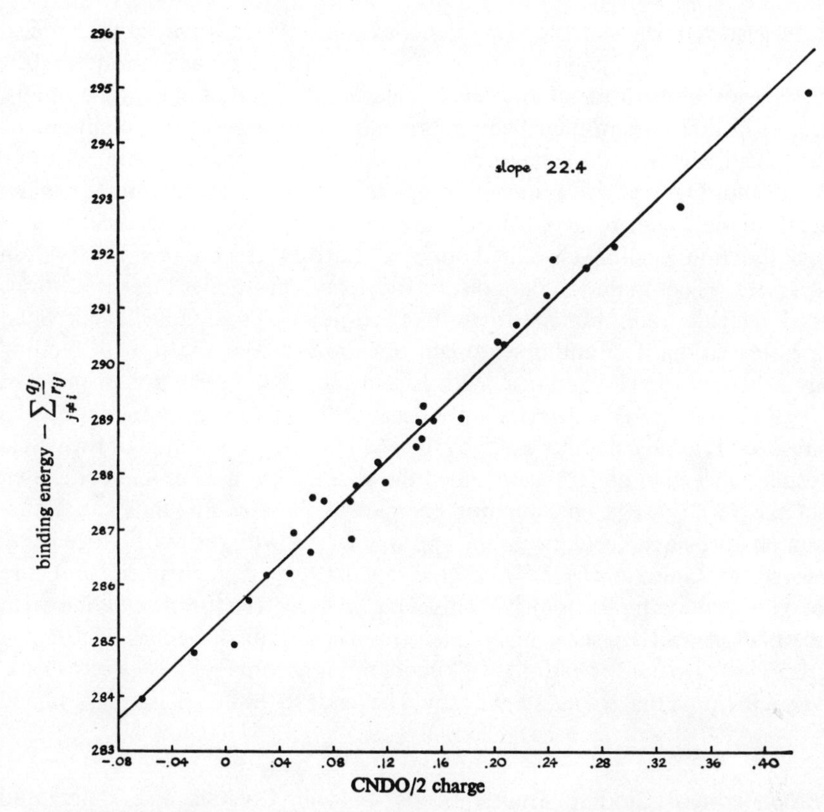

FIGURE 5.15. Plot of measured binding energies (corrected for molecular potential) vs charge from CNDO calculations for the C(1s) shell. All data come from halogen-substituted compounds. [Reproduced from Clark *et al.*,(104) Figure 4.]

estimating the magnitude of hyperconjugation in these compounds. Hamer and Walton(110) were able to show that the Cl 2p binding in a series of metal halide clusters of rhenium and molybdenum was higher for the bridging than for the terminal metal–chlorine bond and were then able to use this fact in characterizing structures of amorphous solids not amenable to x-ray diffraction analysis.

3.6. Alkali Metals and Alkaline Earths

Groups IA and IIA of the periodic table are probably the least interesting of all the elements in terms of measurable chemical shifts. They tend to have only one oxidation state, and the radii of the valence shells are large. Chemical shifts of modest size, however, are noted for one alkaline earth, namely Be. This element has the smallest radius of any element in groups I and II and thus the largest chemical shift slope k. Nicholls *et al.*[111] have studied the chemical shifts for Mg and Al compounds as a function of their coordination number.

Since the alkali metal and alkaline earths form principally ionic bonds, interest in them lies in their being representative of purely ionic compounds. The alkali metal halides have been particularly subject to close scrutiny. Citrin and Thomas[112] assumed complete ionicity in evaluating the alkali metal halides, and, to avoid problems concerning charges and definition of work functions, mainly evaluated only the relative shifts between the levels in the crystal as compared to data on the free ions. Even under these conditions improvements were obtained when the fixed-point-charge model is corrected for polarization. The authors brought out the fact that the shift in binding energy between an electron in a neutral atom and the same electron in an ion in a crystal is small. This arises simply because the change in potential due to ionization is almost compensated by the Madelung-like potential of the whole crystal. Morgan *et al.*[113] also studied the alkali metal halides and concluded that the variations in core binding energies of these compounds can be accounted for by a molecular-binding approximation, with the ionic lattice being ignored. McGuire *et al.*[114, 85] pointed out that observed chemical shifts are not consistent with pure ionicity. This is particularly true for the alkaline earth salts and group IIIA salts, but is even true for the alkali halides. (Cf. Table 5.10.) They also used equation (5.16) in obtaining q from Pauling's electronegativity and found that a consistent fit could be made to the data if it were assumed

TABLE 5.10

Comparison of Binding Energy Shifts in Ionic Crystals and Calculated Shifts Based on Complete Ionicity

		$(-)V$,[a] eV	ΔE_{calc}[b]	ΔE_{exp}[c]
NaF	$q=1$	10.9	0.0	0.0
NaCl		9.0	1.9	0.6
NaBr		8.5	2.4	0.9
NaI		7.7	3.2	1.0
MgF_2	$q=2$	22.1	0.0	0.0
$MgCl_2$		19.1	3.0	1.5
$MgBr_2$		14.9	7.2	1.7
MgI_2		13.3	8.8	2.0
AlF_3	$q=3$	33.0	0.0	0.0
$AlCl_3$		23.0	10.0	−1.6
$AlBr_3$		20.7	12.3	−1.1

[a] Calculation of crystal potential derived from McGuire,[85] assuming complete ionicity.
[b] Changes in metal ion binding energies taken from crystal potentials relative to fluoride for a given metal ion. $\{E_B(\mathrm{MF}) - E_B(\mathrm{MX})\}$.
[c] Changes in metal ion binding energies relative to fluoride from data given in Ref. 85 and from R. Albridge and W. E. Moddeman (private communication) for alkaline earth halides.

that k is given more in terms of the reciprocal of the ionic radii than the atomic radii.

3.7. Transition Metals

Transition metals have received a fair amount of attention from electron spectroscopists because of the large variety of interesting compounds that can be studied, including numerous complexes and organometallics. The behavior of the transition metals depends largely on the outer unfilled d orbital. This orbital is also responsible for the presence of satellite lines from electron shakeup and multiplet splitting. The multicomponent structure is as valuable to the understanding of transition metal compounds as are the chemical shifts and will be discussed in Section 5.1.

The trends in the chemical shifts of the transition metals are among the most difficult to interpret, since (1) they tend to form complex ionic salts whose crystal potentials are hard to evaluate and (2) the valence shell involves both the outer primary shell s electrons and the d electrons of the penultimate principal shell. Caution should be taken in assigning binding energy shifts from the experimental data. First, if the d shell is unfilled, multiplet splitting will occur, and more than one final state in the photoelectron spectrum will be observed as the result of ejecting an electron from a given subshell. A comparison of core binding energy shifts requires a knowledge of this splitting. The problem will be discussed more fully in Section 5.1. Second, the removal of the valence d electrons may cause a slightly different shift for orbitals in the penultimate shell than for the remaining core electrons because of differences in shielding. (Cf. Table 5.2.) Both these problems will be reduced if a deeper principal shell is studied. For example, the chemical shift data should be more reliable for studies done on the $2p$ subshells of the first transition metal series than for the $3p$ subshell.

As seen from expectation values of $\langle 1/r \rangle$ taken from free atom calculations (Table 5.7), the effects on the chemical shifts would seem to be much larger when d electrons are involved then when the outer s electrons are removed. In actual molecular bonding the d electrons are already fairly well removed from the metal ion regardless of the compound (cf. Section 5.1 on multiplet splitting). This may account for why the net chemical shifts are not as spectacular as one might expect from the removal of electrons of such small atomic radii.

3.7.1. First Transition Metal Series: Sc, Ti, V, Cr, Mn, Fe, Co, Ni

Scandium, titanium, and vanadium have not yet received much attention, although Groenenboom *et al.*[115] have looked at some titanium and vanadium organometallic compounds. Hamrin *et al.*[116] obtained PESIS data on several tungsten vanadium oxide compounds and were able to identify the oxidation states of vanadium, confirming previous suggestions based on structural data.

Hendrickson *et al.*[72] have studied a group of chromium compounds, relating them to extended Hückel calculations. Kramer and Klein[68] did the same for iron compounds. Carver *et al.*[117] compared compounds of chromium, iron, manganese, and cobalt with calculated charges from electronegativity. They found the slope of $\Delta E_B/\Delta q$ to be the same for Cr^{3+}, Fe^{3+}, and Co^{3+} compounds but somewhat less for Fe^{2+} and Mn^{2+} compounds, in agreement with the larger ionic radii for the latter ions. (The slopc is dependent on $1/r$.) None of these studies explicitly calculated the effects of the crystal potential, although Carver *et al.* explained qualitatively some of their data in terms of changes in the crystal potential. Helmer[118] has looked at chromium oxide in different oxidation states. Barber *et al.*[119] have investigated core shifts of chromium carbonyl compounds.

Jørgensen[120] has studied a number of nickel compounds and Tolman *et al.*[121] and Matienzo *et al.*[122] have measured over 100. Grim *et al.*[123] looked at nickel dithiolate complexes, coming to the conclusion from the data that these compounds have an indeterminate charge on nickel with electrons encompassing the whole π-bonding system. The metal carbonyls of Cr, Mn, Fe, and Ni have been studied by Barber *et al.*[124] The metal carbonyls and the pentadienyls were investigated by Clark and Adams,[125] and from these studies the net effect was evaluated of σ donation of the CO ligand and the back π donation from the metal ion. They found from observing the C($1s$) and O($1s$) and the inner shell binding energies of the metal that a net donation from the metal to the ligand took place. The results were compared with various MO calculations. Ferrocene compounds were studied by Cowan *et al.*[126] Presence of Fe II and Fe III in stochiometric amounts was noted. The mixed valence compounds of K_xFeF_3[127] and Prussian blue[128] have also been determined. Studies on a number of iron-containing compounds of biological interest have been made and results will be presented in Section 6.

3.7.2. Second Transition Metal Series: Y, Zr, Nb, Mo, Tc, Ru, Rh, Pd

Extensive studies have yet to be made on yttrium, rhodium, and technetium, the last named because it does not exist as a stable nucleus in nature. However, the others have received at least some special attention. Kharitonova *et al.*[129] have looked at some zirconium and hafnium complexes.

Novakov and Geballe[130] have measured the $5d_{5/2}$ core electron of niobium in a series of rocksalt and beta-tungsten structure. McGuire *et al.*[74] have studied niobium compounds as well as tungsten and tantalum compounds and made a comparison of compounds containing these elements and having similar structure.

Molybdenum was studied for its binding energy shifts by Swartz and Hercules[87] and an attempt was made to correlate the binding energy with oxidation state. Development of a quantitative analytical procedure for bulk analysis of MoO_2–MoO_3 mixtures was given. Lane *et al.*[131] applied XPS to the determination of the structure of deprotonated ethylenediamine ruthenium complex by studying this complex and other ruthenium compounds of oxidation state from zero to six. From the data they proposed Ru II. Prather and Zatko[132] have examined 40 different compounds of ruthenium, relating their chemical shifts to Mössbauer isomer shifts.

Palladium has received the most attention of all the elements of the second transition metal series. Hedman *et al.*[133] have studied some palladium alloys (including rhodium). Kumar *et al.*[134] have studied a number of palladium complexes. For those complexes containing only halide ligands that are not π acceptors, they observed that the Pd binding energy shifts were proportional to q calculated from electronegativity. For those complexes containing cyano-ligands, which are known to be effective π-acceptor ligands, the Pd core energies were considerably larger than afforded by electronegativity calculations. Detailed comparisons[135] have been made between Pd and Pt complexes. Clark *et al.*[136] have studied Pt and Pd complexes with the view to evaluating the "*trans*" influence in square-planar complexes and the distinction between terminal and bridging chlorine in *trans*-$L_2M_2Cl_4$ ($L = R_3P$ or CH_3CHCH_2; $M =$ Pd or Pt).

3.7.3. Third Transition Metal Series: Hf, Ta, W, Re, Os, Ir, Pt

As might be expected, the third transition metal series has been the least studied, but there has recently appeared a fair number of investigations. In particular, the core binding energies of Pt compounds have been measured for a wide variety of compounds.

As mentioned before, McGuire *et al.*[74] have looked at Ta and W, making comparisons with Ni. The relative intensities of tungsten metal and tungsten oxide as a function of surface oxide thickness have been ascertained.[137] Some rhenium salts were measured by Cain[138] and shown to have a linear fit to q as calculated from Pauling's electronegativity; some results from each element were reported by Jørgensen and Berthou.[39] Nefedov *et al.*[139–141] and Mason *et al.*[142] have looked at a number of rhenium and iridium (III) complexes. Holsboer and Beck[143] have compared Mössbauer and XPS data on five-coordinated iridium complexes.

Besides the studies mentioned before on comparing chemical shifts in Pt and Pd compounds, studies on Pt have been carried out by Cook *et al.*[144]

and Nefedov *et al.*[145] Kim *et al.*[146] have used PESIS to study platinum–oxygen surfaces. Riggs[29,147] has demonstrated that the oxidation state of Pt—0, +2, or +4—can be determined from the core binding energy. Riggs has also demonstrated a linear plot between binding energies and modified Pauling electronegativities. Butler *et al.*[148] have studied one-dimensional platinum compounds.

3.8. Groups IB and IIB: Cu, Ag, Au, Zn, Cd, Hg

Groups IB and IIB share with the alkali metals and alkaline earths the double disadvantage that the elements occur primarily only in one oxidation form and that the valence shell has a large radius; thus, the chemical shifts in core binding energies are relatively small.

Copper has two oxidation states, I and II, and some simple compounds illustrating both states have been studied by Novakov and Prias.[149] More interesting than the chemical shifts is the large degree of multicomponent structure found with copper compounds. Frost *et al.*[150] have made an extensive study of copper compounds and found that while substantial satellite structure is seen with Cu(II) compounds, none is found with Cu(I) compounds. A more complete discussion of this phenomenon will be given in Section 5.1. Murtha and Walton[151] compared a number of Ag(I) and (II) complexes and found the chemical shifts to be rather small, but Zatko and Prather[152] noted a 3-eV shift between Ag(II) and Ag(III) compounds. Shifts in the core shells for gold compounds have received little attention. However, it is fair to say that the $4f$ lines of metallic gold have been measured more than any other photoelectron peaks. Since gold is a conductor and is inert to oxidation in the air, and the $4f$ shell has a high photoelectron cross section, yielding lines with high resolution, the gold $4f_{5/2,7/2}$ doublet is an excellent standard. It might be said that PESIS is on the gold standard. However, even gold as a standard is not free from criticism.[153]

Vesely and Langer[154] have looked at the binding energies of some simple Zn, Cd, and Hg compounds. In general, the chemical shifts are fairly small. Also the electronegativity for Hg is equal to that of hydrogen, so that substitution of Hg for hydrogen in organic compounds is not easy to detect in terms of chemical shifts. On the other hand, the $4f$ shell of Hg is very sensitive to XPS analysis, which should be of interest for pollution detection.

3.9. Rare Earths and Actinides

As the amount of chemical shift for the transition metals depends on the behavior of the unfilled d shell, so the behavior of the rare earths and actinides depends on the f shell. The same problems of multiplet splitting and shielding noted for the transition metals also apply here. Further complications arise since the valence shell is made up of sometimes partially filled *ns*, $(n-1)d$, and $(n-2)f$ electrons, where *n* is the quantum number of the outermost principal shell.

The shifts are fairly small for the rare earths, largely because the oxidation state is usually only +3, in which two *s* electrons and a single *d* electron participate in the bonding. In the core of Eu, however, we have a half-filled 4*f* shell which has added stability, and Eu^{2+} compounds are also found. The chemical shift is remarkably large,[155] one of the largest noted for a change of only one oxidation state. This large shift in binding energy is probably due to the removal of an electron from the 5*d* subshell, having a relatively small radius. Jørgensen and Berthou[39] have recently recorded other examples of large chemical shifts as the result of different oxidation states for the rare earths.

Some comprehensive studies have been made on the rare earths, but the emphasis has not been on shifts in the core electrons. Rather, Bonnelle *et al.*[156] and Wertheim *et al.*[157] have discussed the nature of the photoelectron spectra of the 4*f* shell, which, of course, is part of the valence shell. Cohen *et al.*[158] have looked at the multiplet splitting in the 4*s* and 5*s* orbitals and Jørgensen and Berthou[159] have studied multicomponent structures in a series of lanthanide compounds. Signorelli and Hayes[160] have examined the multicomponent structure found in the 4*d* shell of the lanthanides.

None of the elements in the actinide series is radioactively stable, and thus only the long-lived elements thorium and uranium have thus far been studied for their chemical shifts (although PESIS has been most useful in obtaining atomic binding energies, e.g., americium[2]). Several groups[161,162] have looked at the various oxides of uranium, and Jørgensen[34,163] has reported on several compounds of both thorium and uranium. Adams *et al.*[164] investigated some uranium (VI) complexes.

4. SPECIAL TOPICS ON SHIFTS IN CORE BINDING ENERGIES

In this section we shall bring together some topics that affect the phenomenon of chemical shifts in general and have not as yet been discussed in previous sections or would benefit from being united here.

First, a summary will be made of some problems encountered in measuring core binding energies and in comparing them for their chemical shifts.

Second, we shall consider topics that are characteristic of inorganic and organic compounds as a whole. Specifically, we shall consider special problems found with different types of bonding. Much of the data encountered with core binding energies can be interpreted solely in terms of electronegativity in which electrons are extracted from or pushed into the valence shell. However, other bonding phenomena must be considered, too, e.g., backbonding in inorganic complexes and resonance in organic molecules. In this section we shall also include in the discussion of these special topics other experimental data not covered in Section 3.

Finally, we shall discuss how the behavior of chemical shifts derived from PESIS compares with chemical shifts obtained from other physical measurements.

4.1. Experimental and Interpretive Problems in PESIS

Considering the natural width of the Al and Mg x-ray lines and the normal spectrometer resolution and intensity, one may expect to locate the center of a photoelectron peak to ± 0.1 eV. A number of standards for calibrating binding energies have been carefully determined for the core levels in gaseous atoms and molecules.[165, 166] (For example, the neon $1s$ ionization potential has been measured[166] to be 870.312 ± 0.017 eV.) Similarly, core binding energies in conducting solids relative to the Fermi level have been determined[165] for calibration purposes. For example, the Au $4f_{7/2}$ has been measured to be 83.8 ± 0.2 eV and the spin splitting between $4f_{7/2}$ and $4f_{5/2}$ levels to be 3.67 ± 0.03 eV. Improved resolution through use of a monochromator and sensitivity through employment of a position-sensitive detector make one hope to be able to establish the relative binding energy to ± 0.01 eV. However, as observed from comparison of binding energies from different laboratories, or even as measured on the same machine, the actual reproducibility is far worse. A number of experimental problems need to be overcome before binding energies by PESIS can be obtained with the desired precision. Let us now discuss some of these problems.

4.1.1. Comparative Problems in the Gas and Solid Phases

The problems that beset the study of PESIS in the gas and solid states are often quite different. For gases the signal strength is much weaker and the variety of compounds suitable for study is smaller. However, there are strong advantages for studying free molecules, which become more apparent as time passes. With gases the problem of secondary electrons can be eliminated or controlled. The resolution of photoelectron peaks from PESIS will eventually be pushed to 0.2 eV for most gaseous samples. Because of charging and other factors, the ultimate resolution for solid samples appears to be much worse. The interpretive problems faced in calculating chemical shifts in solids are much greater than for gases, particularly as one goes to more sophisticated models. Thus, I predict that in the future subtle chemical shifts will be measured with gaseous molecules and interpreted in terms of molecular orbital theory. With solids, however, I predict that while PESIS will be utilized to solve a large number of practical problems connected with product control and surfaces, and will be applied with continuing success for elucidating special questions on chemical structure, there will be severe limitations on interpreting details of chemical bonding from small shifts in the core binding energies.

4.1.2. Charging

One of the largest difficulties encountered with nonconducting solids (which make up most of the samples studied by PESIS) is charging. These problems were discussed in Chapter 2, Section 1.1. Peaks are broadened because of the nonhomogeneity of the charged surface, and unless one establishes good electrical contact with a standard, binding energies could be quite erroneous. Such electrical contact depends on charge centers formed by radiation. The larger the separation between bands in a nonconductor, the harder it becomes to establish electrical contact. It is particularly difficult, for example, to obtain reproducible binding energies on the alkali metal fluorides, which have an unusually large band separation. The problem of charging can sometimes be turned into an asset, for the study of the degree of charging can help characterize a given material or surface.*

4.1.3. Definition of Binding Energy for Insulators

Even when a reproducible binding energy is obtained for an insulator, its theoretical meaning is obscure. It represents a binding energy referenced to somewhere between the Fermi level and the conduction band. When comparing the core binding energies of two insulators, one must be concerned with changes in the reference levels. With this problem in mind, studies on doping were made by Sharma *et al.*[79] The results showed no effect, suggesting that the zero of binding energy in XPS does not in general vary with the bulk Fermi level of the sample. Hedman *et al.*,[169] however, found that differences between the binding energies of heavily *n*- and *p*-doped silicon do occur.

In interpreting binding energy shifts one also needs to be concerned with extraatomic relaxation energy, a problem which was discussed in Section 2.1. Calculations of the relaxation energy for free molecules and for metals are possible, if difficult. For insulators, the problem of interpreting the effect of relaxation energy is again more complicated.

* For example, the relative thicknesses of two silver oxide layers were evaluated for the relative ability of the surface layers to charge up.[167]

4.1.4. Binding Energy of Surface Atoms

PESIS is a surface or near surface measurement and consideration of the differences in core binding energies for the surface and bulk atoms must be made. First, the crystal potential or Madelung-like energy will differ, particularly for the first monolayer. Any potential field will be felt differently for those atoms close to the surface. For this reason, it is unwise to have the target area at a different voltage than its surroundings for purpose of deceleration. Adsorbed contaminants will affect the surface atoms. Changes in binding energy might under certain conditions be dependent on particle size and surface condition.

4.1.5. Radiation Effects

Solid compounds sensitive to radiation damage will change their photoelectron spectrum during the measurement. This is particularly true of compounds of biological interest, but examples may be found throughout the whole literature. If damage is suspected, the spectrum should be examined as a function of time and the data extrapolated to $t = 0$. Increased counting rates achieved without increasing the strength of the radiation source, such as with a position-sensitive detector, ensure that in the future PESIS studies can be made with negligible effects from radiation damage. Monochromators which will cut out all unneeded radiation also will be of aid in this problem.

4.1.6. Linewidths

When an atom occurs in more than one chemical form, the difference in core binding energies is often smaller than the widths of the photoelectron peaks, and one may wish to deconvolute the spectra. This requires a knowledge of linewidth and shape. These quantities are determined in part by the instrument and the natural widths of the x-ray source, which can be determined by a separate experiment on a pure material, e.g., Au. In addition, we have the natural width of the subshell being studied and the broadening in the solid itself due to charging, etc. The former may be obtained from a photoelectron study of another compound containing only one chemical form of the element or from x-ray data, and the latter may be ascertained by measuring a photoelectron peak corresponding to a shell of another element that is known to have a small natural width. Also, the material used for a calibration standard mixed into the sample being studied may reflect the same broadening due to charging, although this is not always certain.

If prior knowledge indicates the number of chemical forms expected and the relative intensities, one may deconvolute using an assumed shape (e.g.,

Gaussian or Lorentzian) and a width which is assumed to be identical for each photoelectron peak making up the band. Even here problems can arise. Relative intensities can be affected by electron shakeoff and shakeup (cf. Section 5). Widths may change due to the lifetimes of states, which are governed by Auger transitions that are in turn influenced by the nature of the valence shell. Friedman *et al.*[169] and Shaw and Thomas[170] have studied broadening of photoelectron peaks as a function of chemical environment for solids and gases respectively, interpreting their results in terms of variation of the lifetime of a core vacancy as a function of chemical bonding.

Broadening of the photoelectron peak that arises from core shell ionization may be also due to vibrational excitation by means of Franck–Condon transitions. In free molecules this is sometimes reflected by a slight asymmetry of the photoelectron peak. With the high resolution afforded by an x-ray monochromator, Gelius[171] was able to discern vibrational peaks in the C 1*s* spectrum of methane and to obtain a vibrational spacing that was in agreement with theory. In solids the analogous phenomenon is that of lattice vibrations or phonon excitation. This was discussed by Citrin *et al.*[172] and Matthew and Devey.[173] Citrin *et al.*[172] suggested that only for polar materials will phonon broadening be a dominant factor in PESIS linewidth measurements. Hüfner *et al.*[174] pointed out that core lines of metals can have asymmetric distortion due to coupling of the final hole state to the conduction electrons.

Finally, in deconvoluting, background must be subtracted, and the dependence of the background as a function of photoelectron energy can be difficult to ascertain at times. All these factors can be dealt with, but they require careful consideration.

4.2. Inorganic Compounds

Because PESIS can be applied to any element in the periodic table above $Z = 2$, it makes an ideal tool for comparative inorganic chemistry. There are difficulties, however. One of the principal ones is that most inorganic solids are ionic and some evaluation of the crystal potential is necessary. The crystal potential may in fact completely countermand the changes in electron density in the valence shell. This problem is discussed in some detail in Section 2. Also note comments made with regard to the alkali metal halides in Section 3.6.

It is possible to have an element having essentially the same chemical bonding but different core binding energies due to differences in the lattice potential.[175] Changes in the core binding energies due to lattice expansion from heat have been noted by Butler *et al.*[19]

Below are some further considerations of problems faced with the study of core binding energy shifts in inorganic compounds.

4.2.1. Multiple Chemical Environment

Unlike organic compounds, a given element in inorganic compounds often appears in only one chemical environment. There are numerous exceptions, however, particularly if an element can exist in both anionic and cationic groups, e.g., NH_4NO_3. The study of such compounds is an advantage since chemical shifts between two atoms in the same molecule can be determined with greater confidence than if they are measured in separate compounds. Shaw *et al.*[176] studied ClF_3, SF_4, and PF_5, showing the presence of unequal fluorines, demonstrating the occurrence of chemical shifts between atoms that differ only in their geometric orientation with respect to the same central atom. Most interesting are those compounds that exist in two oxidation states (spinels), such as prussian blue. This compound also illustrates the problems encountered when assigning two chemical forms in those compounds that also possess multicomponent structure, which is particularly true for the transition metal compounds (cf. Section 5).

4.2.2. Coordination Complexes

Coordination complexes form an important area of inorganic chemistry, which comes under the domain of ligand field theory. Description of core binding energy shifts solely in terms of electronegativity is no longer warranted. Stated simply,[135] a coordination compound forms when orbitals of the metal and those of the ligands overlap to bring about a net decrease in the energy of the system. When the symmetries of the orbitals allow for one point of beneficial overlap, a sigma bond is formed; two points of contact create a pi-type bond. By means of one or both of these bonding modes the ligand (L) and the metal (M) transfer electron density between each other. According to one theory,[177,178] the sigma bond results from an overlap of a filled ligand orbital with an empty metal one. This overlap gives rise to an $L \rightarrow M$ charge transfer, in which case the ligand is acting as a Lewis base and the metal as a Lewis acid. The formation of a pi bond can result in either an $L \rightarrow M$ or $M \rightarrow L$ charge transfer. In the former case the metal will possess pi-acceptor orbitals and will act as a Lewis acid insofar as pi bonding is concerned. In the latter case the ligand will possess pi-acceptor orbitals and will act as a Lewis base in relation to pi bonding. This latter case is known as back bonding. Thus, the transfer of charge in pi bonding can tend either to augment or to compensate the transfer due to sigma bonding.

The electron spectroscopy group at Vanderbilt has studied[55,56,135] the nature of back bonding in the triphenyl phosphorus compounds of nickel, palladium, and cadmium as well as in platinum and palladium complexes. For example, Figure 5.16 shows a plot of the comparative chemical shift data for analogous Pt and Pd complexes. For the halogen ligands, which do not pi bond, the charge distribution and, hence, chemical shift are determined by the electronegativity differences, which for Pt and Pd complexes are the same (the electronegativity is 2.2 for both Pd and Pt). The cyanide and nitrate complexes show a deviation from this straight line, corresponding to Pd having a relatively higher binding energy. This is explained by the fact that in those compounds that do pi-bond, M → L charge transfer can occur, and because palladium has the smaller ionic radius, there is more pi-orbital overlap and thus more M → L charge transfer.

Further studies of inorganic complexes were discussed in Section 3. Cox and Hercules,[179] in their investigation of potassium hexachlorometallate

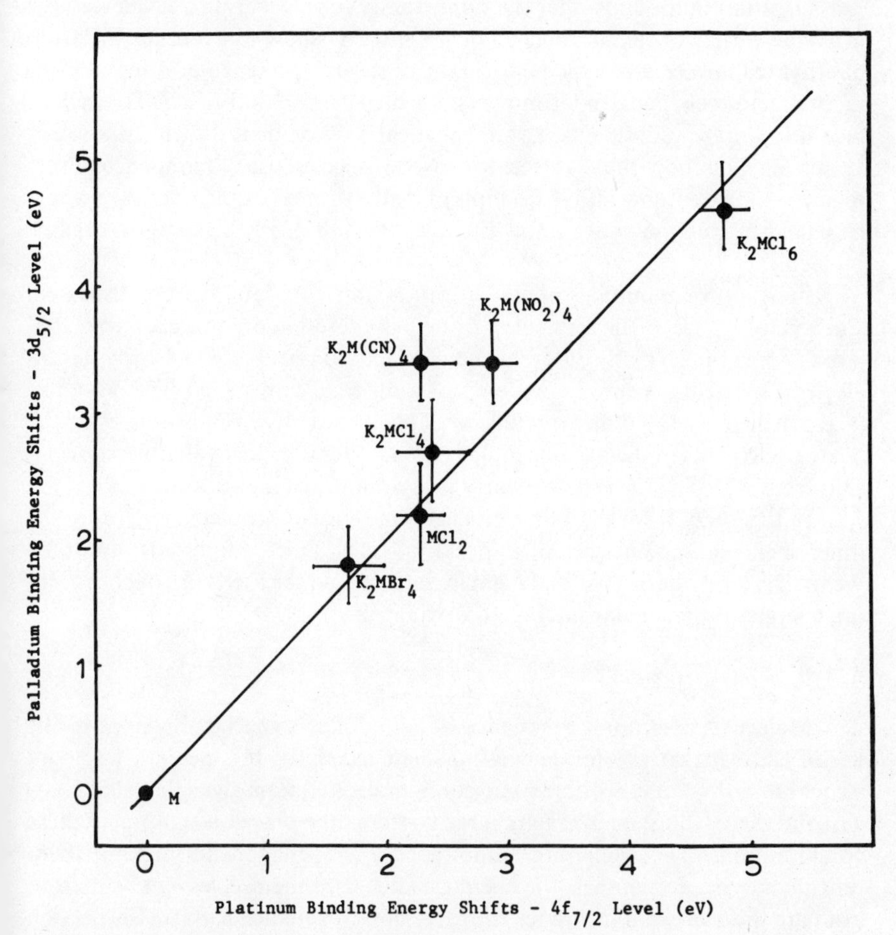

FIGURE 5.16. Platinum vs palladium binding energy shifts for corresponding compounds. [Reproduced from Moddeman *et al.*,[135] Figure 1.]

compounds, afford an example of the variety of information that can be gleaned from an x-ray photoelectron spectroscopy study. First, they were able to explain their results on the chlorine $2p$ binding energies with the help of crystal potential calculations. Second, from the binding energy shifts of the metal ion in the complex relative to the pure metal, Cox and Hercules were able to rationalize the change in the populations of the $5d$ and $6s$ subshells of Re, Os, Ir, and Pt. Third, they were able to correlate three photoelectron bands observed in the valence region with MO calculations.

4.3. Organic Compounds

Organic compounds offer the opportunity for measuring a given element in a wide variety of chemical behavior. Figure 5.17 shows the famous spectrum of ethyl trifluoracetate, which clearly illustrates the presence of four substantially different chemical environments in which carbon finds itself. The lack of resolution makes subtle changes in chemical environment difficult to discern at this time, but hopefully, at least for gaseous species, use of monochromators to narrow the photon width to a couple of tenths of an eV could open up a whole new field of study. Yet even now there is a wealth of PESIS data on organic molecules.

In Section 3 a number of studies on organic compounds were discussed. Carbon, of course, is the element of principal interest in organic chemistry, but it may appear in so many different chemical environments that its x-ray photoelectron spectrum is simply made up of a series of unresolved photoelectron peaks, which are too difficult to deconvolute and analyze. Thus, it is often the "extra" elements in the organic molecule that offer the most valuable information from PESIS. This is particularly true of nitrogen and sulfur.

In this section we shall first discuss some special problems regarding the inner-shell binding energies of importance to organic compounds, and then we shall extend our discussion begun in Section 3 on the literature of core binding energies of particular interest to organic chemistry.

4.3.1. Resonance

Molecules containing resonance bonding offer a special challenge to the use of electronegativity for calculating chemical shifts. It is necessary to construct more than one possible ionic configuration, obtain the calculated charges, and then deduce an average charge based on the proper weighting of these configurations. This weighting cannot generally be made without some theoretical analysis. Sometimes, the chemical shift data themselves can be used to evaluate the relative importance of the various resonance forms. (Cf. Section 2.3 for an example.) In general, it is necessary to use *ab initio* or semiempirical molecular orbital calculations to properly predict chemical shifts for aromatic

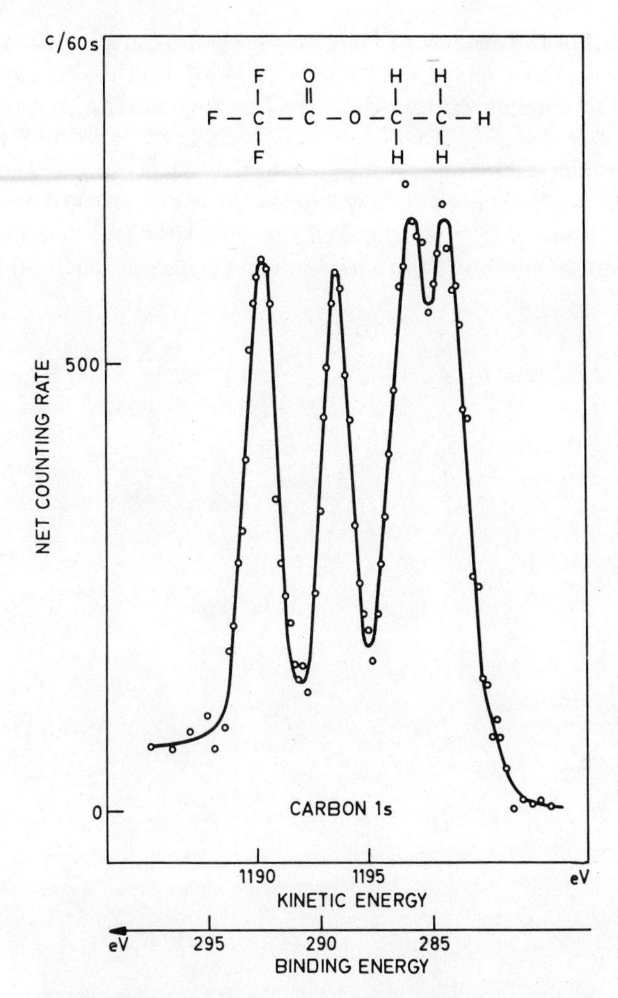

FIGURE 5.17. Photoelectron spectrum of the C(1*s*) shell of ethyl trifluoroacetate using Al *Kα* irradiation. [Reproduced from Siegbahn *et al.*,[1] Figure 1.16.]

compounds. Extended Hückel calculations are particularly valuable for organic compounds. Use can also be made of examining the behavior of core binding energies in a series of homologous compounds.

4.3.2. Substituent Effects

The chemical shift of an atom in a molecule is predominantly influenced by its nearest neighbor. However, substituent effects due to the influence of atoms or groups at a distance of more than one atom can also be shown to play a role. For example, Clark *et al.*[35,42] showed that in the series CCl_3X the relative C(1*s*) binding energy shifts are 0.0, 1.1, and 1.6, respectively, for $X = CH_3$, CCl_3, and CF_3. Clark and his co-workers[99–101,103–106] have made extensive studies on fluorosubstitution in benzene, heterocyclic, and multiring compounds. Additional studies on fluorinated hydrocarbons have been carried out by Davis *et al.*[180] The large electronegativity of fluorine allows one to see substitution effects in PESIS normally not observed. The C(1*s*) spectrum

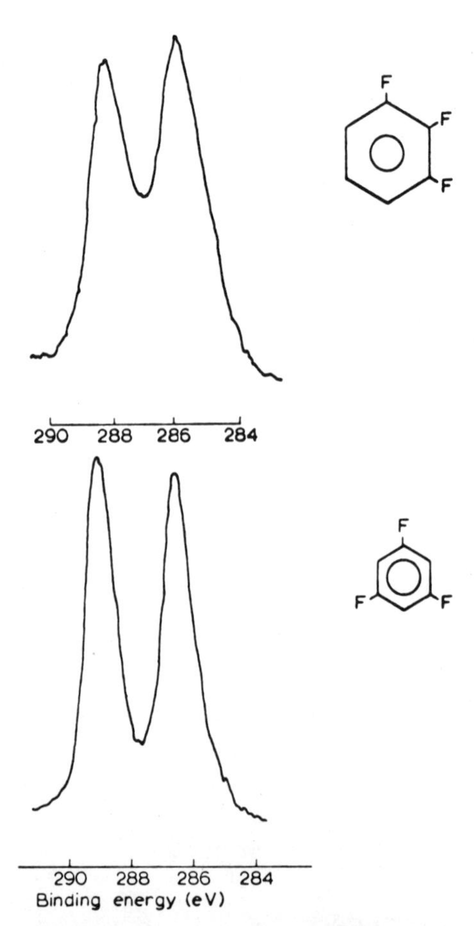

FIGURE 5.18. Comparison of C(1*s*) photoelectron spectra for 1,3,5- and 1,2,3-trifluorobenzene. [Reproduced from Clark *et al.*,[106] Figures 1 and 5.]

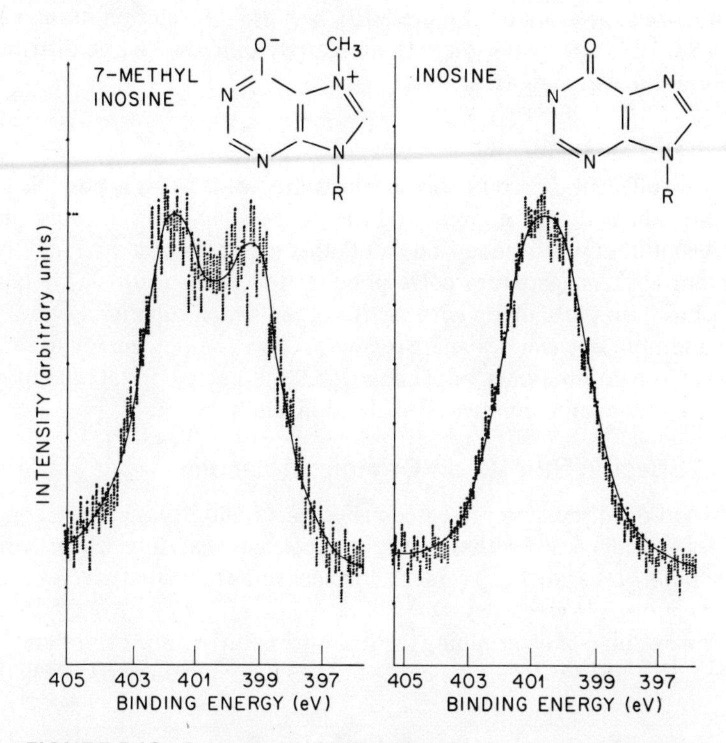

FIGURE 5.19. Comparison of N(1*s*) photoelectron spectra for inosine and 7-methyl inosine. [L. D. Hulett and T. A. Carlson, unpublished data. See also Ref. 297.]

given in Figure 5.18 shows the effects of fluorine substitution in benzene. There are two carbon peaks, clearly separated, their 1*s* binding energies dependent on whether carbon is bound directly to H or F. However, when fluorine is substituted in the 1,2,3 positions the photoelectron peaks are noticeably broader than for 1,3,5-fluorobenzene. In the latter compound there are only two types of carbon with different chemical environments. In 1,2,3-fluorobenzene, carbon 2 is different from carbons 1 and 3, not by its nearest neighbor but by its next nearest neighbor; and a similar situation holds for carbon 4 compared with carbons 3 and 5. With the advent of high resolution (~0.2 eV) the extra unresolved peaks such as illustrated above should be easily separated in the gas phase, and subtle inductive effects should be more amenable for close study.

A dramatic effect arising from substitution is found in compounds where the substitution changes the whole nature of the resonance structure. For example, comparison of 7-methyl inosine and inosine shows an entirely different nitrogen spectrum (cf. Figure 5.19), because the addition of the methyl group gives rise to a zwitterion with an entirely different charge distribution throughout the whole molecule.

4.3.3. Group Analysis

When sufficient data on chemical shifts are available, it is possible to assign group shifts; that is, a group attached to a given atom can be assigned a chemical shift that will be independent of other groups. From this analysis one can obtain electronegativities corresponding to given groups. This type of analysis has been particularly effective for organic molecules (cf. Section 2.3).

In addition, one can designate regions of core binding energy for an element in a given group. For example, Figure 5.20 shows regions of binding energies found with a large number of oxidized disulfides.

4.3.4. Specific Studies on Organic Molecules

Most of the discussion presented in Sections 2 and 3 that involved organic molecules was concerned with the correlation of chemical shifts with calculated charges for its own sake. Here we shall consider some cases that involve specific problems in organic chemistry.

As an example of determining the structure of an organic compound let us take the case quoted by Hedman *et al.*[181] A certain reaction product had one

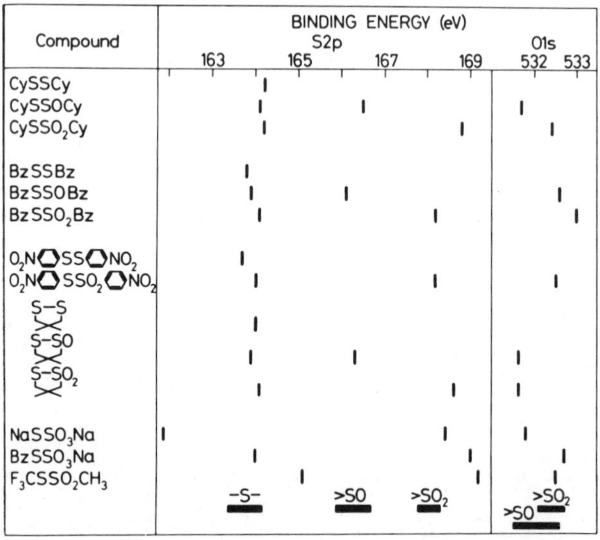

FIGURE 5.20. The S(2*p*) and O(1*s*) electron binding energies for oxidized disulfides. [Reproduced from Lindberg *et al.*,[61] Figure 8.]

of the two possible isomeric structures (I) and (II):

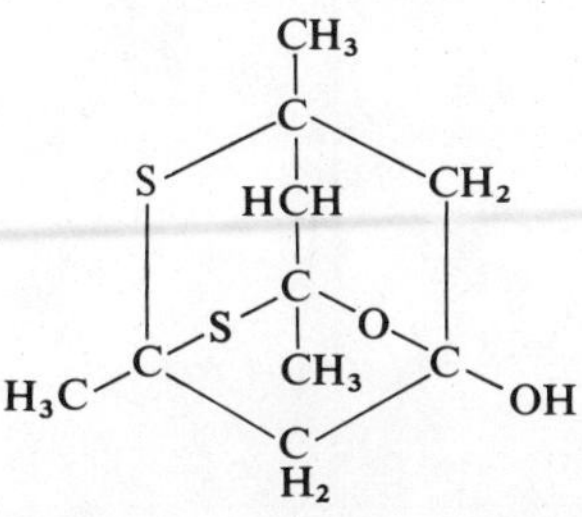

I 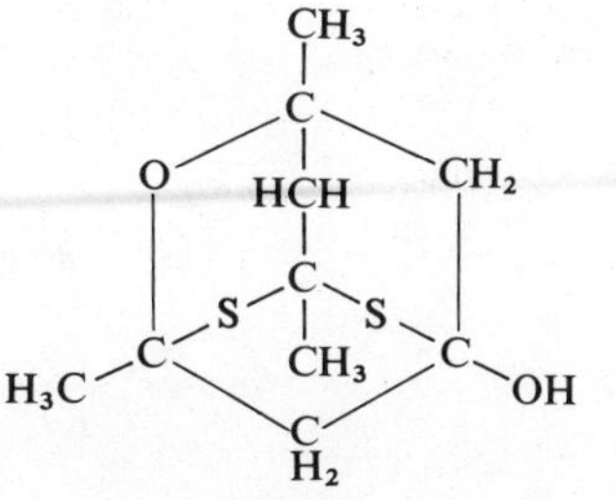 II

From binding energies corresponding to the calculated charge, one can construct a theoretical spectrum for the two structures (cf. Figure 5.21). From experiment it is obvious that structure I is the correct one.

Clark *et al.*[182] demonstrated the use of PESIS in determining the structure of allyl pentachlorocyclopentadiene from simple inspection of the C(1*s*) spectrum. In another case,[99] two isomers of C_9HF_7 were distinguished by comparing x-ray photoelectron spectral data with theoretical spectra derived from equation (5.16) using CNDO-calculated charges. Patsch and Thieme[183] have studied a series of mesoionic compounds. From the core binding energies of nitrogen and sulfur they were able to confirm Katritzky's view that a betaine structure predominates in the ground state of mesoionic 1,3,4-thiadiazolethiones and similar compounds.

Meteescu and co-workers[184–186] have studied a number of stable organic ions derived from hydrocarbons, and their hydroxy and carbonyl derivatives. Positive charge localization on a carbon atom (specifically the carbenium ion center of tert-, butyl-, and 1-adamantyl cations) increases the C(1*s*) electron binding energy by 4–5 eV. Most interesting was the study on the question of whether the norbornyl cation possesses a "classical" or "nonclassical" electronic structure. Figure 5.22 shows the results for the norbornyl cation and the adamantyl cation. In the latter case we see a high energy C(1*s*) peak clearly separated from the rest of the spectrum with an intensity ratio of approximately 1:9. The positive charge is clearly localized on a single carbon in a "classical" structure. The norborane shows a more complex spectrum indicating a "nonclassical" electronic structure. Palmer and Findlay[187] have studied norbornadiene.

4.4. Comparison of Core Electron Binding Energy Shifts with Other Physical Quantities

Since shifts in the inner-shell binding energies can be correlated with changes in the chemical environment, it seems only natural to compare chemi-

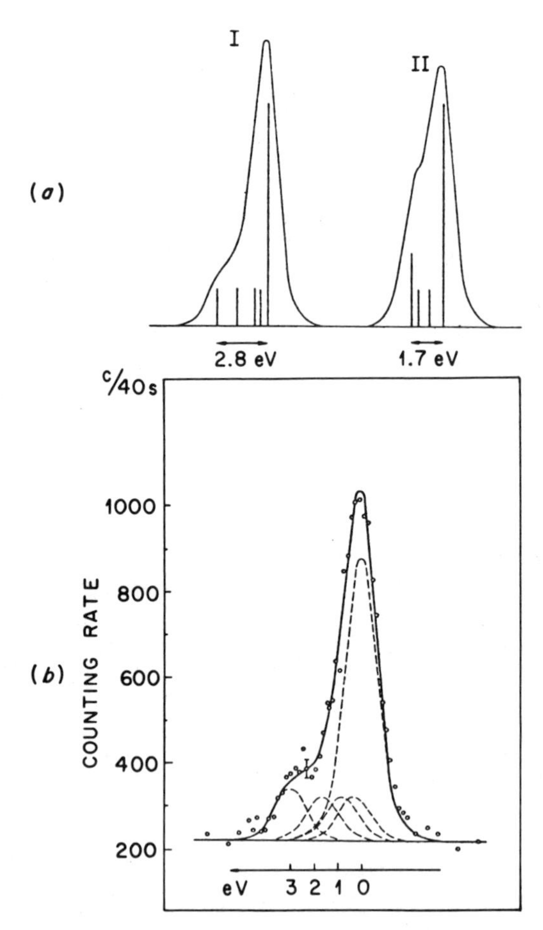

FIGURE 5.21. Theoretically calculated shapes of S(2*p*) photoelectron spectra for two possible structures of oxadithia-adamantanol, and fit of experimental data to structure I. (a) Predicted spectra from I and II. Positions and relative intensities of the 1*s* electron peaks from the different carbon atoms are indicated. (b) Carbon 1*s* spectrum from I. The solid line shows the sum of the deconvoluted components, which are given as dashed lines. This PESIS spectrum gives conclusive evidence of structure I. [Reproduced from Hedman *et al.*,(181) Figures 1 and 2.]

cal shifts from PESIS with other physical measurements that reflect changes in the nature of the electronic structure of molecules. In this section we shall review the results of such comparisons. Primarily, we shall concentrate our attention on the Mössbauer isomer shift and the chemical shifts in NMR.

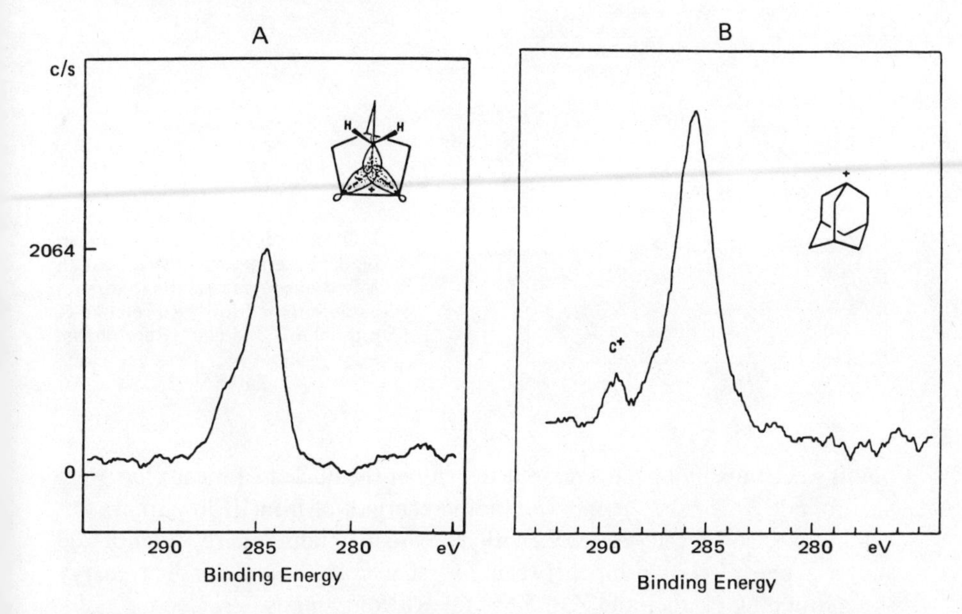

FIGURE 5.22. The C(1*s*) photoelectron spectra of (A) norbornyl cation 5 and (B) adamantyl cation. [Reproduced from Mateescu and Riemenschneider,[184] Figures 5 and 6.]

4.4.1. Mössbauer Isomer Shift

The Mössbauer isomer shift reflects changes in the electron density near the nucleus, while PESIS reflects changes in the electron density in the valence shell. The electron density near the nucleus is made up mostly of contributions from *s* orbitals: the lower the principal quantum number, the higher the contribution. However, changes come mainly from alterations in the valence shell *s* electrons, either directly or through changes in the screening as other orbital electrons with $l > 0$ are involved in bonding. If electrons from the *d* or *f* subshells with principal quantum numbers one less than the outermost shell are used in bonding, changes in screening may also substantially affect the electron density of *s* electrons in the penultimate shell.

Though the Mössbauer isomer shift and core electron binding energy shifts are not related in a simple way, the electron density at the nucleus in general decreases as the electron density of the valence shell decreases. Deviations from this rule may reflect special changes in the bonding. For example, if electrons with $l > 0$ are drawn away from a given atom, the loss of shielding will increase the density of *s* electrons at the nucleus as the overall density of the valence shell decreases.

Barber *et al.*[83] compared isomer shifts with binding energy shifts for some tin compounds. A linear relationship was found, with an increase of

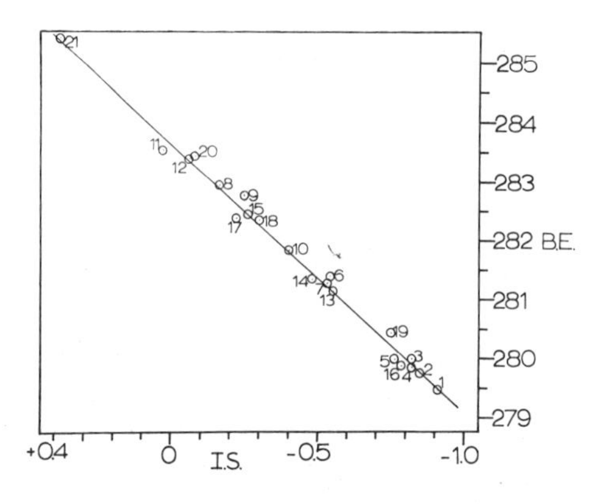

FIGURE 5.23. The Ru($3d_{5/2}$) binding energies (eV) versus Mössbauer isomer shifts (mm/sec) as determined for ^{99}Ru relative to ruthenium metal. [Reproduced from data of Prather and Zatko.[132]]

binding accompanying a decrease in density at the nucleus. Linear plots were also found[188] for the $2p$ and $3p$ binding energies of iron(II) low spins. But here the electron density *increased* with increase in bonding energy. Figure 5.23 shows a linear relationship between Mössbauer shifts and binding energy shifts found by Prather and Zatko[132] for Ru compounds.

The Mössbauer splitting in iron was used together with the classical shifts from PESIS to help characterize the structure in some iron compounds.[126, 189, 190] Holsboer *et al.*[143, 191] have used the combination of Mössbauer and PESIS to examine metal ligand bonding effects in both gold and iridium complexes.

4.4.2. NMR

Nuclear magnetic resonance has proven to be a very powerful and versatile tool for the study of the nature of the electronic structure of molecules, particularly in organic chemistry. It then seems only natural to attempt to correlate chemical shifts found in NMR with those from PESIS.

It has been generally acknowledged that an overall correlation between NMR and PESIS does not exist. However, correlations have been found for special groups of related compounds, such as the quaternary phosphonium salts,[54] some methyl halides,[184, 192] some carbocations,[184] and molecules with similar carbon–oxygen linkage.[184]

The chemical shifts from NMR come from differences in the magnetic shielding at a nucleus due to changes in the chemical environment. The average screening constant may be considered to be composed of two parts, diamagnetic and paramagnetic. Basch[193] showed that part of the diamagnetic term is related to an electrostatic potential, which in turn describes the chemical shifts in PESIS. Gelius *et al.*[194] closely reviewed the problem and concluded that changes in the average diamagnetic screening constant $\Delta\sigma_A(\text{av})$ can be expressed as

$$\Delta\sigma_A(\text{av}) = -0.652\,\Delta E_A + \frac{0.3646}{g_A}\,\Delta\sum_\alpha C_{A\alpha}I_\alpha \tag{5.34}$$

where ΔE_A is the change in core binding energy, g_A is the g-factor for nucleus A, $C_{A\alpha}$ is the spin-rotation constant for nucleus A along the αth molecular inertial axis, and I_α is the corresponding principal moment of inertia of the molecule. Gelius *et al.*[194] further pointed out that the portion of equation (5.34) that reflects changes in the binding energy is an order of magnitude lower than the total NMR chemical shift, so that correlation between NMR and PESIS data cannot generally be expected. They did point out, however, that equation (5.34) relates NMR and PESIS data with chemical shifts of molecular spin-rotation constants obtained from molecular beam magnetic resonance or microwave spectroscopy.

To support their contention of lack of direct correlation between NMR and PESIS, Gelius *et al.*[194] compared data for boron compounds and for the halomethanes. Zeroka[195] also indicated poor correlation with fluoromethanes.

Lindberg[196] examined the correlation between NMR and core binding energy shifts pragmatically in a rather comprehensive review and developed a number of rules for imposing limitations on which compounds would be suitable for comparison.

4.4.3. Other Physical Data

As indicated in Chapter 1, comparisons of inner-shell binding energies with other physical methods have almost no limit, since a phenomenon that measures a quantity proportional to electron density is at the heart of describing chemical structure. However, we shall restrict ourselves here to other cases not yet mentioned where specific correlation between binding energy shifts and other physical measurements has been described.

First, the reader should be reminded of the correlation with thermal data described in Section 2.7. Another important area to be discussed elsewhere (Chapter 6, Section 2) concerns the correlations among photoelectron data, x-ray fluorescence, and Auger spectra. Stucky *et al.*[197] have compared the electron density as obtained by core binding energies and derived from x-ray and neutron diffraction data. Verbist *et al.*[162] have correlated shifts in the core shells of uranium with the half-life for internal conversion for a very low energy isomeric state of ^{235}U.

Lindberg and Schröder[198] discussed the correlation of polarographic half-wave potentials with the sulfur $2p$ electron binding energies for a series of substituted nitrobenzenes. *Para*-substituted compounds gave a good linear correlation, while *meta* compounds seemed to give a second correlation with a different slope. The same series of compounds, again looking at S(2*p*), has been correlated with the asymmetric NO stretching vibration. *Ortho*-substituted sulfur groups showed the largest shift in vibrational frequencies and *meta*-substituted groups the least. Folkesson[199] has combined the uses of XPS and infrared spectroscopy.

5. OTHER APPLICATIONS OF PESIS

The information from PESIS that is generally of most interest and of widest applicability is related to chemical shifts. However, the photoelectron spectra of core electrons yield still other data, which form a body of knowledge that is extremely important in itself. First, there is often more than one photoline per atomic orbital, due to various phenomena such as multiplet splitting, electron shakeoff, configuration interaction, and characteristic energy losses. It is important to understand the nature of these "extra" photoelectron peaks, both because we cannot properly interpret photoelectron spectra without this understanding and because the phenomena are in themselves of fundamental interest. In addition, these phenomena can be of extreme importance in learning about the nature of chemical bonding.

A field that will most certainly take on increasing importance is the use of PESIS for the study of surfaces. PESIS is an analytical tool that is highly sensitive to the surface, yielding an analysis of the elements in the first few monolayers as well as an indication of their chemical bonding. PESIS can also reveal the changes in chemical environment between the surface layer and the bulk material. It may also be possible to use PESIS to study how far a given element or compound lies beneath the surface.

Finally, angular studies have been coupled with PESIS on crystals to yield structural information.

5.1. Multicomponent Structure

Normally, one expects a single photoelectron peak associated with each atomic or molecular orbital, governed by the familiar equation

$$E_e = h\nu - E_B \tag{5.35}$$

The binding energy E_B can be taken as the difference in total energy for the initial and final states:

$$E_B = T_f - T_i \tag{5.36}$$

If there is more than one final state resulting from removal of a core electron, then more than one photoelectron peak will occur. The satellite line that represents the excited state whose energy is T^* is related to the energy of the normal photoelectron peak by

$$\Delta E = E_e(\text{normal}) - E_e(\text{satellite}) = T_f^* - T_f \qquad (5.37)$$

That this situation occurs quite often will be apparent from the discussion in this section. Most of the earlier sections of this chapter concentrated on the main photoelectron peak representing transitions to the ground state of a species in which a single electron has been removed from one of the orbitals, but where no other excitation has occurred. We shall now discuss transitions to excited states represented in photoelectron spectra as satellite lines or multicomponent structure. Not only is it important to evaluate this multicomponent structure in order to understand multiple processes that can be involved in photoionization, but the nature of the transitions to excite states offers another important tool to PESIS for the exploration of the electronic structure of molecules.

We shall discuss, in order, multiplet splitting, electron shakeoff, configuration interaction, and characteristic energy losses. Finally, a section will be given on unraveling the nature of the observed satellite lines from a comprehensive study of the photoelectron spectra.

5.1.1. Multiplet or Exchange Splitting

As described in Chapter 3, Section 3.2, when there are one or more electrons in the valence shell with unpaired spins, the act of photoionization in another shell of the same atom can lead to more than one final state, depending on the ways in which the unfilled shells couple. For free ions or atoms the total energies of the various final states can be calculated using solutions from Hartree–Fock wave functions, assuming L–S coupling. The differences in the total energies should be reflected in the differences in the photoelectron energies. The first photoelectron spectra of core electrons to illustrate this phenomenon were taken on O_2 and NO in the gaseous phase by Hedman *et al.*[49, 200] and for transition metal salts of Mn and Fe by Fadley *et al.*[201] A very careful study by Davis and Shirley[202] was made on NO, yielding a multiplet splitting of 1.41 ± 0.02 eV for a $1s$ vacancy in nitrogen and 0.53 ± 0.02 eV for a $1s$ vacancy in oxygen. This is to be compared with molecular Hartree–Fock calculations by Schwartz of the total initial and final state energies, which gave 1.35 and 0.48 eV, respectively. Bagus *et al.*[203] have extended this study on O_2 and NO by examining the intensity ratios in the multiplet splitting and come to the conclusion that substantial multiple excitation involving charge transfer must be present to obtain agreement between theory and experiment.

Studies on other paramagnetic gas molecules, such as N_2F, N_2F_4, NO_2, and $(CF_3)_2NO$, have been carried out by Davis *et al.*[204] Multiplet splitting has also been measured[205] in the organic free radical diphenylpicrylhydrazyl (DPPH). However, by far the most important areas of chemical research displaying multiplet splitting are the transition metal compounds, where there is an unfilled *d* shell, and the rare earths and actinides, where there are unfilled *f* and *d* shells.

Experimental values on molecules give a smaller splitting than calculated from free ion calculations. This arises in part because in a molecular bond the unpaired electrons in the valence shell are drawn away from the parent atom, decreasing the chance for overlap between the unfilled orbitals of the core and valence shells. Errors in the free ion calculation may also arise from the failure to include electron correlation. Hartree–Fock calculations by Bagus *et al.*[206] on multiplet splitting using wave functions with configuration interaction indicated that neglect of antiparallel correlation may be an important factor in the discrepancy between earlier calculated splittings and experiment. In addition, the calculations by Bagus *et al.* explain the discrepancy between the observed relative intensities of the multiplet peaks and those expected from multiplicity alone. Finally, the calculations including correlation predict higher energy satellites, which have in fact been observed.[207] Correlation effects can be expected to decrease as one goes deeper into the core. Hüfner and Wertheim[208] found that splitting following photoionization in the 2*s* subshell is in better agreement with results from uncorrelated wave functions than is the case for the 3*s* subshell. Ellis and Freeman[209] performed specific molecular Hartree–Fock calculations of the extent of multiplet splitting for $[MnF_6]^{4-}$, and agreement with experiment is good.[201] Most important is the fact that multiplet splitting can give a measurement of the bonding character of the unfilled valence shell.

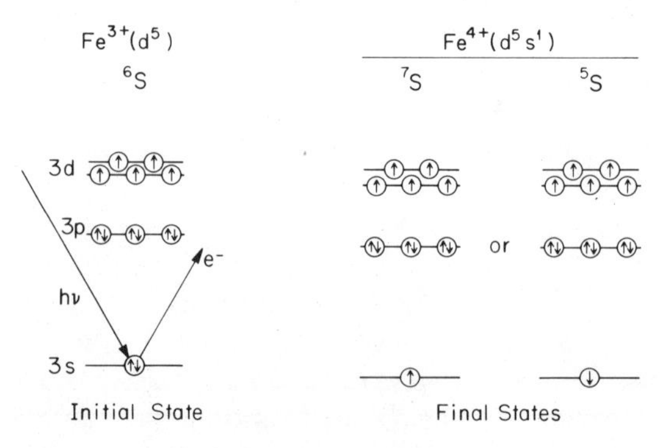

FIGURE 5.24. Schematic of multiplet splitting following photoionization in Fe^{3+}.

For example, systematic studies of the nature of bonding in metallic salts have been made by Carver *et al.*[117] and Wertheim *et al.*[210] In these studies multiplet splitting yields information on the nature of the $3d$ bonding. Photoelectrons from the $3s$ shell are studied, because (1) neglecting the effects of configuration interaction, a vacancy in an s shell leaves only one unpaired spin, which leads to only two possible final states (cf. Figure 5.24), and (2) the overlap is greatest when the two subshells are in the same principal shell. (For example, see Table 5.11 for calculated multiplet splitting in Fe^{4+} and Mn^{3+}.) Variations in multiplet splitting as a function of the element are shown in Figure 5.25. For a given ligand (in this case fluorine) the experimental values for multiplet splitting follow closely the number of unpaired spins (Figure 5.26). In this example the number of d electrons changes from zero to ten. The greatest splitting occurs with five unpaired electrons. The experimental results are also compared with free ion calculations, which are always larger by a constant factor. If the two sets of results are normalized, the calculations show the proper experimental trend.

TABLE 5.11

Calculation[a] of Multiplet Splitting for Fe and Mn Ions with Initial Configurations of $3d^5 4s^0$ as a Function of Inner Shell Vacancy

Inner shell vacancy	Final state	Intensity, %		Relative energy, eV	
		Mn^{3+}	Fe^{4+}	Mn^{3+}	Fe^{4+}
$3s$	^7S	58	58	0.0	0.0
$3s$	^5S	42	42	14.2	15.7
$2s$	^7S	58	58	0.0	0.0
$2s$	^5S	42	42	6.1	7.2
$1s$	^7S	58	58	0.0	0.0
$1s$	^5S	42	42	0.08	0.10
$3p$	^7P	58	58	0.0	0.0
$3p$	^5P	42	42	17.3	18.9
$3p$	^5P (1)[b]	28	28	4.0	4.6
$3p$	^5P (2)[b]	0	0	9.4	10.5
$3p$	^5P (3)[b]	14	14	24.0	26.4
$2p$	^7P	58	58	0.0	0.0
$2p$	^5P	42	42	6.1	7.2
$2p$	^5P (1)[b]	24	23	3.2	3.6
$2p$	^5P (2)[b]	0	0	7.7	8.7
$2p$	^5P (3)[b]	19	18	10.0	11.7

[a] Calculated by C. W. Nestor, Jr., J. C. Carver, and T. A. Carlson from nonrelativistic Hartree–Fock code of C. Froese Fischer. Energies are taken from the total energy of the ion for the given designation. For further results of similar calculations, see A. J. Freeman, P. S. Bagus, and J. V. Mallow, *Int. J. Mag.* **4**, 35 (1973).
[b] With configuration interaction.

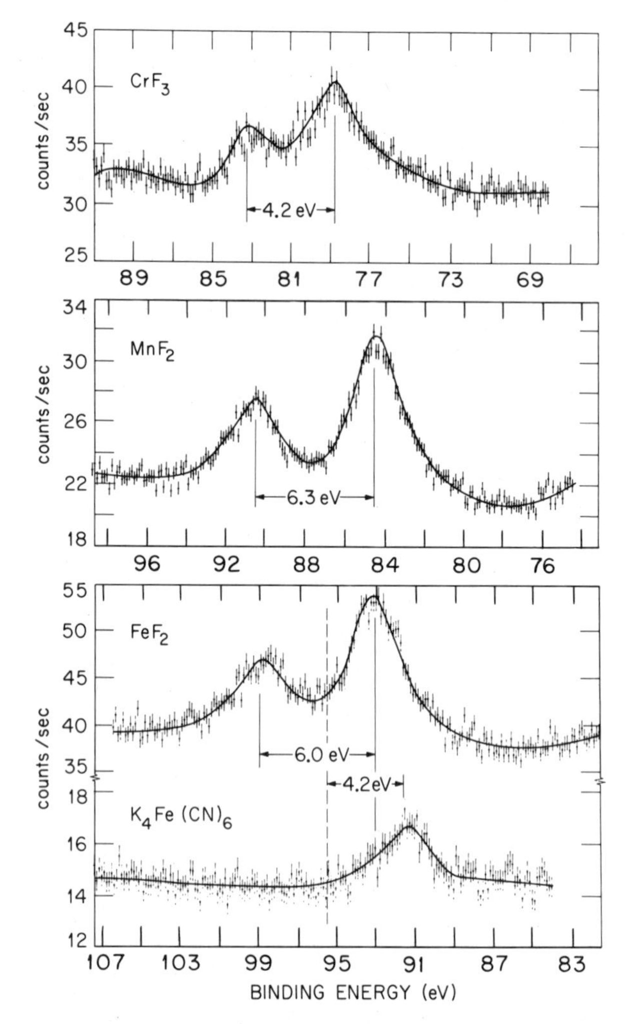

FIGURE 5.25. Photoelectron spectra of 3*s* shell in some transition metal compounds showing effect of multiplet splitting. [Reproduced from Carver *et al.*,[117] Figure 2.]

For a given transition metal ion the multiplet splitting in a chemical compound is determined by three major factors: (1) the extent of decoupling in the *d* orbital due to strong field ligand bonding, (2) the extent that the *d* electrons are delocalized due to the nature of the chemical bond, and (3) correlation effects. See Table 5.12 for results on multiplet splitting as a function of ligand. Compounds such as $K_4Fe(CN)_6$ are known from magnetic susceptibility measurements to have all the unpaired spins in the *d* orbital of iron coupled.

Thus, one expects no multiplet splitting, and indeed such is the case (cf. Figure 5.25). For $K_3Cr(CN)_6$ the *d* electrons still have unpaired spins, but the electrons are so highly delocalized that little or no splitting is observed. The splitting also follows the degree of covalency: The more covalency, the more delocalized are the electrons, and the smaller is the splitting. Thus, on comparing compounds of the same transition metal ion, the exchange splitting for oxides is greater than for sulfides, and for halides the order is F > Cl > Br > I. Wertheim *et al.*(210) found for the manganese oxides that the splitting did not change as much as predicted solely on the basis of the number of unpaired spins, which they attributed to an increase in *s–d* overlap with increasing oxidation state. A correlation(208) has been established between multiplet splitting and the contact term in hyperfine splitting. This has been used by Santibáñez and Carlson(211) to interpret Mössbauer data on some spinel compounds, $NiFe_xCr_{2-x}O_4$. Tricker(212) has used the satellite structure in the 2*p* shell of iron to follow the thermal equilibrium between the $^6A_1$ and $^2T_2$ states of iron(III) tris(*N*, *N'*-dialkyldithiocarbamate). In this case the satellite structure probably arises more from electron shakeup than multiplet splitting, but the structure is still intensified by the presence of a high spin iron.

Clark and Adams(213) have studied the 3*s* shells of chromium in $Cr(CO)_6$, $Cr(\pi C_5H_5)_2$, and $Cr(hfa)_3$ (hfa = hexafluoroacetonylacetonate). These compounds have, respectively, zero, two, and three unpaired electrons in the valence shell, and correspondingly the multiplet splitting was found to be 0.0, 3.1, and 4.5 eV, respectively.

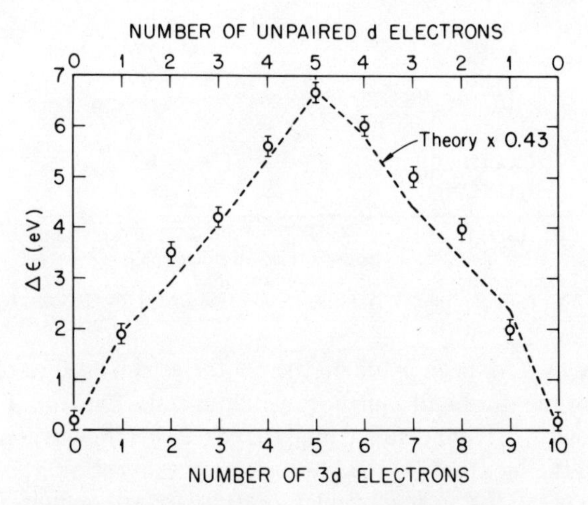

FIGURE 5.26. Variation in multiplet splitting as a function of unpaired 3*d* electrons. Experimental values taken from 3*s* data of transition metal fluorides. Theoretical calculations are from Hartree–Fock solutions for the free ions. [Data taken from Carver *et al.*(117)]

TABLE 5.12

Multiplet Splitting Observed from the Photoelectron Spectra of the Penultimate *s* Shell in Transition Metal Compounds[a]

Compound	Shell	$\Delta\varepsilon$, eV
CrF_3	3*s*	4.2
$CrCl_3$	3*s*	3.8
$CrBr_3$	3*s*	3.1
Cr_2O_3	3*s*	4.1, 3.3[b]
Cr_2S_3	3*s*	3.2
$K_3Cr(SCN)_6$	3*s*	<0.2
$K_3Cr(CN)_6$	3*s*	<0.2
MnF_2	3*s*	6.3, 6.5[c]
$MnCl_2$	3*s*	6.0
$MnBr_2$	3*s*	4.8
MnO	3*s*	5.5, 5.7[c]
MnS	3*s*	5.3
Mn_2O_3	3*s*	5.4
MnN	3*s*	5.5
MnF_3	3*s*	5.6
FeF_2	3*s*	6.0
$FeCl_2$	3*s*	5.6
$FeBr_2$	3*s*	4.2
FeS	3*s*	6.3
FeF_3	3*s*	6.7, 7.0[c]
$FeCl_3$	3*s*	6.2
$FeBr_3$	3*s*	4.9
Fe_2O_3	3*s*	7.3
K_3FeF_6	3*s*	7.0
Na_3FeF_6	3*s*	7.0
$K_4Fe(CN)_6$	3*s*	<0.2, <0.1[c]
CoF_3	3*s*	6.0
$K_3Co(C_2O_4)_3$	3*s*	<0.2
$K_3Co(CN)_6$	3*s*	<0.2

[a] Cf. Table II of Carver *et al.*[(117)] Values of $\Delta\varepsilon$ are from Carver *et al.*,[(117)] except where noted otherwise.
[b] J. C. Helmer (personal communication).
[c] C. S. Fadley and D. A. Shirley, *Phys. Rev. A* **2**, 1109 (1970).

Some studies have been made on the photoelectron spectra of the penultimate *s* shell of the third and fourth transition metals. Evidence for multiplet splitting was found,[(214)] but detailed analysis of the spectra was experimentally difficult due to the fact that (1) the photoelectron cross section for the *s* electrons is relatively small, (2) the multiplet splitting is somewhat less for higher *Z* (for example, from theoretical calculations* on free ions, the exchange splitting

* Nestor and Carlson,[(215)] using C. Froese Fischer's nonrelativistic Hartree–Fock code.

for Ru^{4+} $4s^14d^5$ is 74% of that for Fe^{4+} $3s^13d^5$, both elements belonging to the same group in the periodic table), and, most important, (3) the line broadening for the 4s and 5s subshells of the second and third transition metal series is broader (4–5 eV) than that for the corresponding 3s shell (3 eV) of the first transition metal series. In all three cases line broadening occurs because of the short lifetime of s-shell vacancies which are subject to being filled by Coster–Kronig transitions ($ns - np$, $(n+1)\,l$), but the problem seems even more severe for the heavier elements.

Experimentally, one might be advised to study multiplet splitting as seen in the photoelectron spectra of p, d, and f core subshells. The lines are sharp due to the absence of Coster–Kronig transitions, the intensities are greater, and for a given principal quantum number the multiplet splitting is generally greater. However, the spectra become much more complex when a core vacancy is made in a subshell with $l > 0$ (cf. Table 3.4). Ekstig *et al.*[216] made free ion calculations including configuration interaction for the various neutral transition metal atoms with one to eight d electrons and a single vacancy in the 3p subshell. These calculations give the final term value, the energy separation, and relative intensities. Similar calculations on Mn^{3+} and Fe^{4+} appear in Table 5.11. In these calculations L–S coupling is assumed and configuration interaction is introduced into the Hamiltonian of a SCF Hartree–Fock calculation. The mixing coefficients are allowed to vary to achieve a minimum total energy in a self-consistent fashion.

The number of possible states is independent of the principal quantum number of the core level in which photoionization takes place. Also the same number of states occurs whether the core vacancy is in a p, d, or f subshell. The intensities of the lines are proportional to the multiplicity, i.e.,

$$I \propto |C_{ba}|^2(2S' + 1)\,(2L' + 1) \qquad (5.38)$$

where S' and L' are the total spins and angular momentum of the final state following photoionization. C_{ba} is the mixing coefficient. If only one configuration is allowed, the intensities are proportional only to the multiplicity and not to the detailed nature of the wave function. For example, photoionization in a 3s shell of high spin Fe^{3+} (cf. Figure 5.24) results in two states with total spins of 2 and 3 and $L = 0$. From equation (5.32) the multiplicities are respectively 5 and 7. The experimentally observed peaks[117,201,210] generally follow the theoretically predicted ratios qualitatively. (But not exactly! As was mentioned before, inclusion of correlation can lead to a reduction of this ratio.) If configuration interaction occurs, C_{ba} is dependent on the radial wave function and thus is sensitive to changes in the chemical environment. The amount of splitting in terms of energy is sensitive to both the principal and angular momentum quantum numbers of the subshell in which photoionization occurs and to the nature of the d orbital.

The simplest case one may have for multiplet splitting with the transition metals when photoionization occurs in the p shell is with Mn^{2+} and Fe^{3+}, whereby there are five unpaired electrons in the $3d$ shell. Only two final states occur without configuration interaction, ^7P and ^5P. With configuration interaction ^5P can divide again into three states. Fadley *et al.*(201) have studied the $3p$ spectrum of FeF_3 and identified what they believe to be satellite lines from multiplet splitting. Data from the $3p$ shell has thus far been difficult to analyze and it has been hard to identify any multiplet structure associated with the $3p$ shell in some compounds where it should exist. Identification is complicated by the fact that one of the ^5P lines is close in energy to the ^7P state and is seen only as a shoulder, the second line is of negligible intensity, while the third is so separated in energy that it appears on top of a plasmon loss peak.

The nature of multiplet splitting in the p shells of transition metal compounds has been clarified by calculations of Gupta and Sen.(217) In these calculations the authors have included effects of spin–orbit coupling and crystal field. Their results are shown in Figure 5.27. The experimental spectrum should

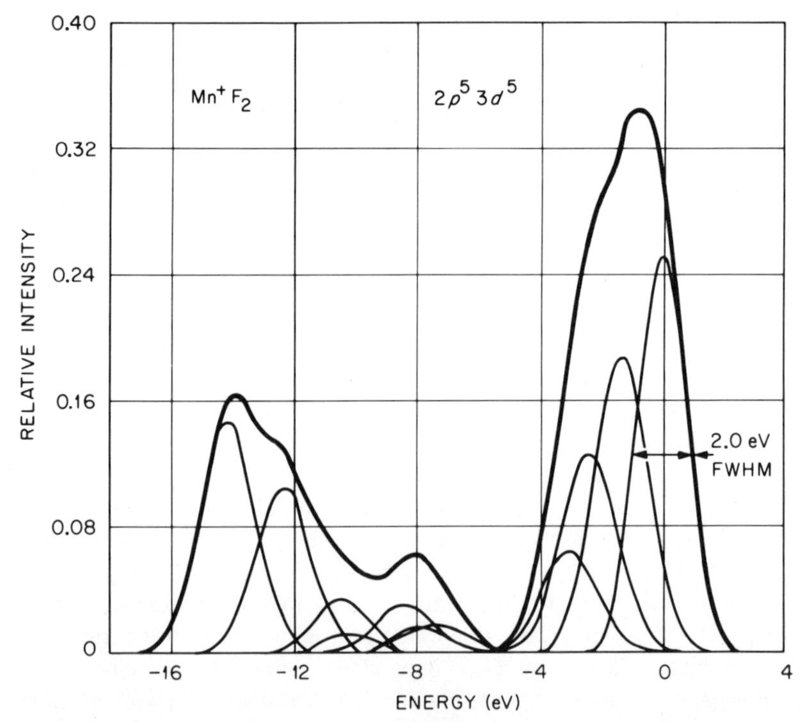

FIGURE 5.27. Convolution of states calculated from multiplet splitting for MnF_2 for ionization in the $2p$ subshells. Full-width half-maximum is arbitrarily chosen to be 2 eV and the energy scale is read in terms of kinetic energy relative to the largest peak set at zero. [Reproduced from Carlson *et al.*,(263) Figure 3, as based on calculation of Gupta and Sen.(217)]

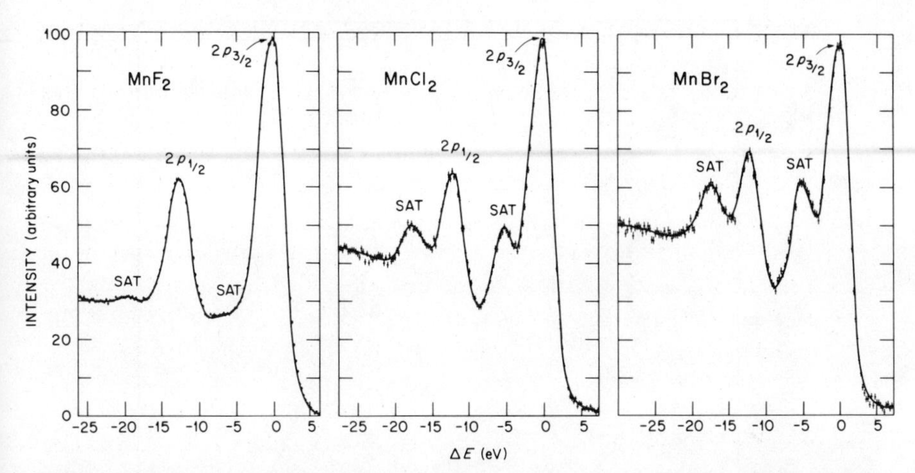

FIGURE 5.28. Photoelectron spectra of the $2p$ shell of some Mn halides showing dependence of satellite structure on the nature of the ligand. SAT indicates satellite structure. [Reproduced from Carlson *et al.*,[255] Figure 1.]

be affected by an asymmetric broadening of the $2p_{3/2}$ peak toward lower energies, while the $2p_{1/2}$ peak will slope toward higher energies. A small contribution is also seen between the peaks. These observations have been confirmed experimentally.[218] Comparison between the theory and experiment for the $3p$ shell shows that the multiplet splitting is considerably distorted by correlation effects.

Large satellite structure in the $2p$ photoelectron spectrum is found with other manganese compounds and with many other transition metal compounds. Most of this structure is believed to be due to electron shakeup and not multiplet splitting. For example, see Figure 5.28. Little satellite structure is apparently present in MnF_2, but, as stated above, close examination of the spectrum[218] does reveal unresolved structure consistent with multiplet splitting. For $MnCl_2$ and $MnBr_2$ large satellite structure is seen with the structure on the low-energy side of the $2p_{1/2}$ photoelectron peak mirroring that seen with the $2p_{3/2}$ photo peak. Frost *et al.*[219] have interpreted the increased separation of the $2p_{1/2}$, $2p_{3/2}$ levels in paramagnetic complexes of cobalt as opposed to low-spin diamagnetic compounds as due to multiplet splitting. However, they attributed the strong satellite lines found with the paramagnetic compound as due to electron shakeup. Multicomponent structure found with the $2p$ shell will be further discussed in Section 5.1.2. It should suffice to mention here that satellite lines have been seen with many photoelectron spectra of the $2p_{1/2,3/2}$ shells of metal compounds and should prove to be a very important area of study for PESIS. In Section 5.1.5 we shall discuss experimental methods for ascertaining the true source of these satellites.

Other important series of elements that can be studied for their multiplet

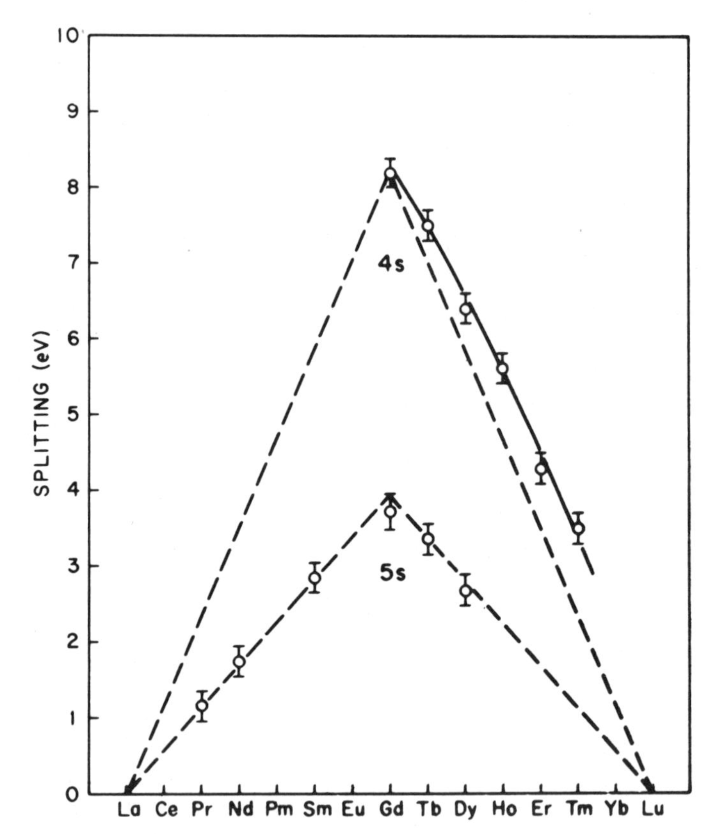

FIGURE 5.29. Plot of multiplet splitting in the 4*s* and 5*s* shells of the rare earths as obtained from x-ray photoelectron spectroscopy. [Reproduced from Wertheim and Cohen,[4] Figure 4.]

or exchange splitting are the rare earths and actinides in which the outermost *f* orbital is partially filled. Wertheim *et al.*[186] have studied a group of trivalent fluorides and oxides of the rare earths. The splitting of the 4*s* and 5*s* shells follows closely the number of unpaired spins (cf. Figure 5.29). The splitting in the 4*s* shell is larger because of the larger overlap with the 4*f* orbital. Splitting was also observed[157] in the 4*f* shell. Two principal peaks were observed and were associated with a spin-up and a spin-down final state following photoionization.

One question that naturally arises is, "How does one evaluate the chemical shifts for an element which is susceptible to multiplet splitting?" The question can be easily answered for the case of photoionization in the *s* shells. From the exchange integral [cf. equation (3.10)] one may regard the "unperturbed" binding energy as the average energy of the two multiplet states weighted by the Clebsch–Gordan coefficients. That is,

$$E_B = \frac{E_B(\mathrm{I})C^2(\mathrm{I}) + E_B(\mathrm{II})\,C^2(\mathrm{II})}{C^2(\mathrm{I}) + C^2(\mathrm{II})} \tag{5.39}$$

where $C^2(\mathrm{I})$ and $C^2(\mathrm{II})$ are the Clebsch–Gordan coefficients associated with the two final multiplet states whose binding energies correspond to $E_B(\mathrm{I})$ and $E_B(\mathrm{II})$. For the case of s vacancies C^2 will be simply proportional to the multiplicity.

For photoionization in shells whose angular momentum quantum number is greater than zero, the spectrum is generally too complex to average correctly, and the most intense peak is taken to represent the binding energy. This could lead to substantial error in evaluation of chemical shifts if multiplet splitting is not negligible. For example, results of Kramer and Klein[68] on chemical shifts in iron compounds using data on the $3p$ binding energies without corrections were considerably at variance with data taken by Carver *et al.*,[117] in which the multiplet effects were taken into account.

5.1.2. Electron Shakeoff and Shakeup

For more details on the theoretical nature of electron shakeoff and shakeup the reader is directed to Chapter 3, Section 3.5. When a core electron is removed by x-ray photoionization, there is a sudden change in effective charge due to the loss of a shielding electron. This sudden change in effective charge gives rise to the possibility for monopole excitation or ionization. That is, an outer electron can be transferred from an original orbital described by a set of quantum numbers to an excited orbital in which only the principal quantum number has been changed. That is, $\Delta l = \Delta s = \Delta j = 0$. If a transition is to an excited, but bound, state, this will appear as a discrete satellite photoelectron line at a kinetic energy lower than the main line by the difference between the ground and excited states of the ion with a core vacancy. This process is known as monopole excitation or electron shakeup. If the electron is excited into the continuum, the photoelectron spectrum will evidence a continuous spectrum rising smoothly from a lower kinetic energy to a threshold whose energy difference is equal to the ionization potential for the ground state of the ion with a core vacancy. This process is called monopole ionization or electron shakeoff. See Figure 5.30 for a diagrammatic representation in the case of photoionization in the K shell of neon.

Since there is no change in spin or angular momentum of the individual electrons in monopole excitation, the whole process of photoionization and monopole excitation maintains the selection rules for photoionization. Thus, the final excited state of the ion must differ from the initial state of the neutral atom by $\Delta L = 1$, $\Delta S = 0$.

The probabilities for electron shakeoff and shakeup can be calculated using the sudden approximation. (Again, see Chapter 3 for more details.)

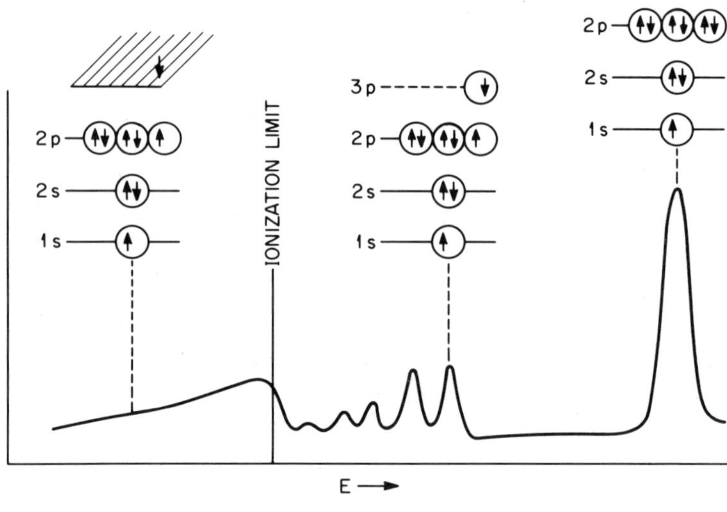

$$\Delta J = \Delta L = \Delta S = 0 \;;\; \Delta \ell = \Delta s = 0$$

$$P_{n\ell j - n'\ell j} = \left| \int \psi^{*}_{n'\ell j}\, \psi_{n\ell j}\, d\tau \right|^{2}$$

FIGURE 5.30. Schematic of satellite structure arising in photoelectron spectrum of core shell of neon (monopole excitation and ionization). The shape of the shakeoff spectrum has been exaggerated for viewing purposes. In reality it would be expanded over a greater energy range and would merge smoothly with the shakeup peaks at the ionization limit. *E* is energy of the photoelectrons. Final state configurations of Ne^+ are shown above the spectrum.

This probability is greatest for the outermost shells and amounts to about 10–20% for free atoms. For an assessment of shakeoff plus shakeup probabilities in photoionization, see Table 3.5. When an electron is excited from an inner shell it generally goes into the continuum; but if excitation occurs in the valence shells, electron shakeoff and electron shakeup have the same order of importance. The probability for shakeup is increased as the overlap between the ground state orbital and excited state orbital increases. Thus the presence of low-lying states that can be reached by monopole excitation favors the presence of electron shakeup.

Studies of electron shakeoff following photoionization have been made by Carlson and Krause for the rare gases using charge spectrometry(221, 222) and photoelectron spectroscopy.(222, 223) When vacancies were formed in the core shell, the experimental results were in reasonable agreement with calculations made using single-electron wave functions. The distribution of the kinetic energy of shakeoff electrons was determined(223) for neon and argon and compared with theoretical calculations of Levinger(224) (cf. Figure 5.31). The probability of electron shakeoff was experimentally found[221] to be essentially independent of photoelectron energy (except near the threshold), as predicted from the sudden approximation. When photoionization takes place in the valence shell the observed[222, 225] electron shakeoff probability is several times that predicted from the use of single-electron wave functions. However, in the case of He, calculations[226] were carried out using a Hylleras-type wave function for the bound state which included electron correlation explicitly, and the calculations gave good agreement with experiment. Åberg[227] showed that at high photoelectron energy the sudden approximation would also give agreement with experimental data on He if a correlated wave function was used.

With the use of high-resolution x-ray photoelectron spectroscopy, the effects of electron shakeoff are not readily seen, the broad spectrum merging into the background. Rather, one sees the contributions of the discrete lines from electron shakeup. (Cf. Figure 5.32.) In photoelectron spectroscopy one expects generally to observe only shakeup lines from the valence shell. Contributions of excitation from core electrons are not visible because (1) the probability for excitation and ionization is smaller and (2) when they do occur, transitions go to a broad continuum, separated from the main peak by the

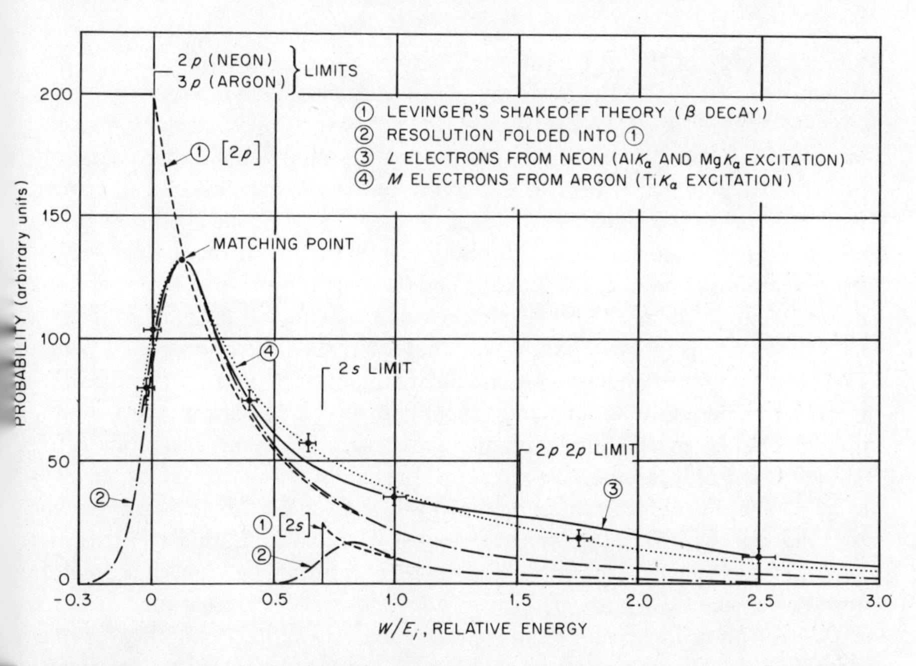

FIGURE 5.31. Comparison of distribution of kinetic energy of electrons which have been shaken off from the valence shell as the result of photoionization in the K shell of neon and L shell of argon compared with theory. W is the kinetic energy of electrons, E_i the ionization energy required for shakeoff. [Reproduced from Krause *et al.*,[223] Figure 5.]

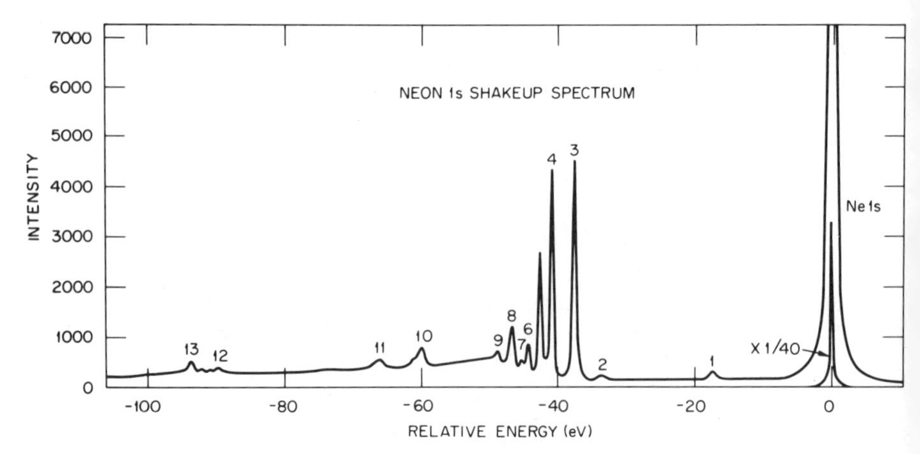

FIGURE 5.32. Shakeup and shakeoff spectrum arising from *K* shell ionization in neon. Peak 1 is due to collision losses. Remaining peaks, except 2, are from "normal" monopole transitions. Shakeoff spectrum is seen as high background gradually decreasing from about —45 eV down. [Reproduced from U. Gelius, E. Basilier, S. Svensson, and K. Siegbahn, Univ. of Uppsala Institute of Phys. report, UUIP-817 (1973).

binding energy of the core electron. The total effect of electron shakeoff and shakeup can be felt by the lessening of intensity of the main peak.

Siegbahn *et al.*[48] and Carlson *et al.*[228] have studied electron shakeup in neon. Based on Hartree–Fock solutions and assuming monopole excitation, both the calculated intensities and energies were found to be in good agreement with experiment (cf. Table 5.13). Spears *et al.*[229] studied electron shakeup in argon, krypton, and xenon, and found, in agreement with theoretical predictions, that the probability for shakeup and the excitation energy are essentially independent of where the initial photoelectron vacancy is created, so long as it is at least one principal quantum number lower than the valence shell. Slight variation in intensity occurs, in part due to slight differences in the shielding between the penultimate shell and deeper core levels. The deeper the shell, the greater change in shielding and the greater the shakeoff probability (cf. Table 3.5). For example, the shakeup probability determined experimentally[229] for photoionization using the Mg $K\alpha$ x ray in the $3d$ subshell of Xe was 8.4% as opposed to 5.1% for the $4d$ subshell. Slight variation in excitation energy is also seen because of the differences in coupling between the core hole and the valence shell, but again this is a second-order perturbation.

When photoionization occurs in the valence shell, the probability for electron shakeup, as with electron shakeoff, is no longer amenable to single-electron wave-function calculations. However, Carlson[222] pointed out that agreement occurred between measured shakeup in He and calculations based

TABLE 5.13

Photoelectron Spectrum of Neon Ionized in the *K* Shell (Monopole-Excited States)[a]

Designation	Energy (eV relative to unexcited transition)				Intensity (relative to unexcited transition)			
	Run A[b]	Run B[c]	S[d]	Theory[e]	Run A	Run B	S[d]	Theory[f]
$1s2s^22p^6\ ^2S$	0	0	0	0(868.6)	100	100	100	100
$1s2s^22p^53p\ ^2S$ lower	37.3(2)	37.1(3)	37.3	35.6	3.0(2)	3.2(3)	2.4(4)	2.3
$1s2s^22p^53p\ ^2S$ upper	40.7(2)	40.5(3)	40.7	39.5	3.1(2)	3.0(3)	2.6(4)	2.9
$1s2s^22p^54p\ ^2S$ lower	42.4(2)	42.0(3)	42.3	40.5	1.9	1.7(3)	1.5(4)	—
$1s2s^22p^55p\ ^2\mathrm{S}$ lower	44.4(2)	—	44.2	42.4	0.6(2)	—	0.5(2)	—
$1s2s^22p^54p\ ^2S$ upper	—	—	46.4	44.6	—	—	0.6(2)	—
$1s2s^22p^5np$ (remainder)	~46	~46	—	—	~1	0.8(4)	—	—
Ionization limit	~47.4	~48	~47	45.2	—	—	—	—
$1s2s2p^63s$	59.5(5)	59.1(7)	60	61	~1	—	—	—
$1s2s2p^64s$	68(2)	—	—	—	~0.2	—	—	—

[a] Table taken from Carlson *et al.*(228)
[b] Mg $K\alpha$ x rays used.
[c] Al $K\alpha$ x-rays used.
[d] Siegbahn *et al.*(48)
[e] Energies were obtained from Hartree–Fock wave functions.
[f] Intensities were calculated using the sudden approximation.

on correlated wave functions. Recently, Krause and Wuilleumier[230] measured electron shakeup in He near the threshold (Figure 5.33). Their measurements agree with the theory of Jacobs and Burke,[231] who included the possibility that since electrons are indistinguishable, it is permissible to have the outgoing electron in the *s* continuum with the excited electron going into an *np* orbital, rather than the normally expected situation where the ejected electron from the 1*s* shell of He goes into a *p* continuum while the remaining electron is excited into a discrete *ns* state. At high photoelectron energy the "normal" process predominates, but closer to threshold and even out to about 100 eV above threshold the "reverse" process plays a dominant role. The probability for electron shakeup remains constant with photoelectron energy except near threshold, where it may first rise and then decrease. Studies on electron shakeup in neon, argon, krypton, and xenon have also been carried out. Wuilleumier

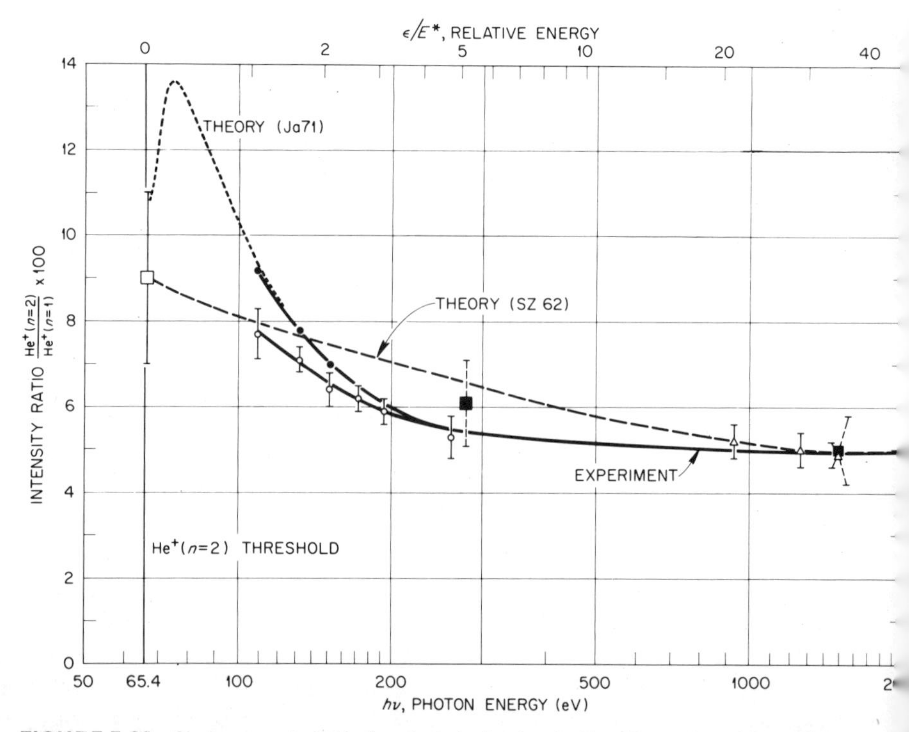

FIGURE 5.33. Shakeup probability for photoionization in He. [Reproduced from Krause and Wuilleumier (private communication).] Theory (Ja71) is Jacobs and Burke.[231] Theory (SZ62) is from E. E. Salpeter and M. H. Zaidi, *Phys. Rev.* **125**, 248 (1962). Experimental data from Krause and Wuilleumier,[230] ○ (90°), ● (54.7°); Wuilleumier and Krause (unpublished), ×; Carlson *et al.*,[222, 228] ■; and J. A. R. Samson, *Phys. Rev. Lett.* **22**, 693 (1969), □

and Krause[232] carefully studied satellite structure in the photoionization of the L shell of neon. The found their data consistent with monopole excitation. In contrast, Spears *et al.*[229] found that the principal contribution to the multicomponent structure found in the photoelectron spectra of the valence shell of Ar, Kr, and Xe came from configuration interaction (cf. Section 5.1.4).

Electron shakeup studies have been carried out on a number of simple gas molecules. Although the counting rates are lower, the background encountered with gas studies is also much lower, and corrections for characteristic energy losses can be determined. Thus, in gases the electron shakeup spectrum can be obtained with greater certainty, and the interpretation is also more straightforward. In solids the shakeup spectrum is often swallowed up by the characteristic energy loss spectrum. Figure 5.34 shows a typical spectrum with gaseous N_2O. Note the differences in the electron shakeup spectra, dependent on whether the core vacancy is created in the nitrogen or the oxygen. If the same valence states were excited, the two spectra would be very similar. By localizing the sudden change in charge to a given atom, one may differentially excite different molecular orbitals. The orbital whose density is localized near the atom where the initial hole is formed is most likely to be excited. For example, Gelius *et al.*[233] showed that in the photoelectron spectra of carbon suboxide (O═C═C═C═O) the more intense carbon $1s$ peak had a strong shakeup peak in agreement with the O($1s$) spectrum, while the other carbon (the central carbon) has no visible shakeup spectrum. The ratio of the main carbon peaks, 1:1.8 rather than 1:2, also supports the idea that shakeup (and shakeoff) take place more readily with the outer carbons. From the above they concluded that the shakeup transition was $1\pi_g \rightarrow 2\pi_g{}^*$, a monopole transition involving an orbital that has a node at the central carbon, and, in addition, the $2\pi_g$ level is the lowest unoccupied orbital. Recently Spears *et al.*[234] have been able to rationalize their data on nitric oxide and nitrous oxide on the basis of Mulliken population analysis.

The selection rules for shakeup in molecules must follow the general concept of monopole excitation. There should be no change in spin and parity, and the general symmetry properties ought to hold constant. In other words, one expects transitions from a given orbital to occur in one of its Rydberg states. Unfortunately, neither experimental nor theoretical data exist for excited states of molecules with core vacancies. However, using arguments similar to Jolly's with respect to equivalent charge (cf. Section 2.7), one may equate $N_2{}^+(1s^1)$ with NO^+, the excited states of the latter having been obtained by Gilmore.[235] In this way it was suggested[228] that the energy needed to transport an electron by monopole excitation for the $2p\pi^b\ {}^1\Sigma^+$ to $2p\pi^*\ {}^1\Sigma^+$ state was 15 eV, in good agreement with the main shakeup peak in the $1s$ photoelectron spectrum of N_2. Gelius[171] has reexamined shakeup in nitrogen, using X_α

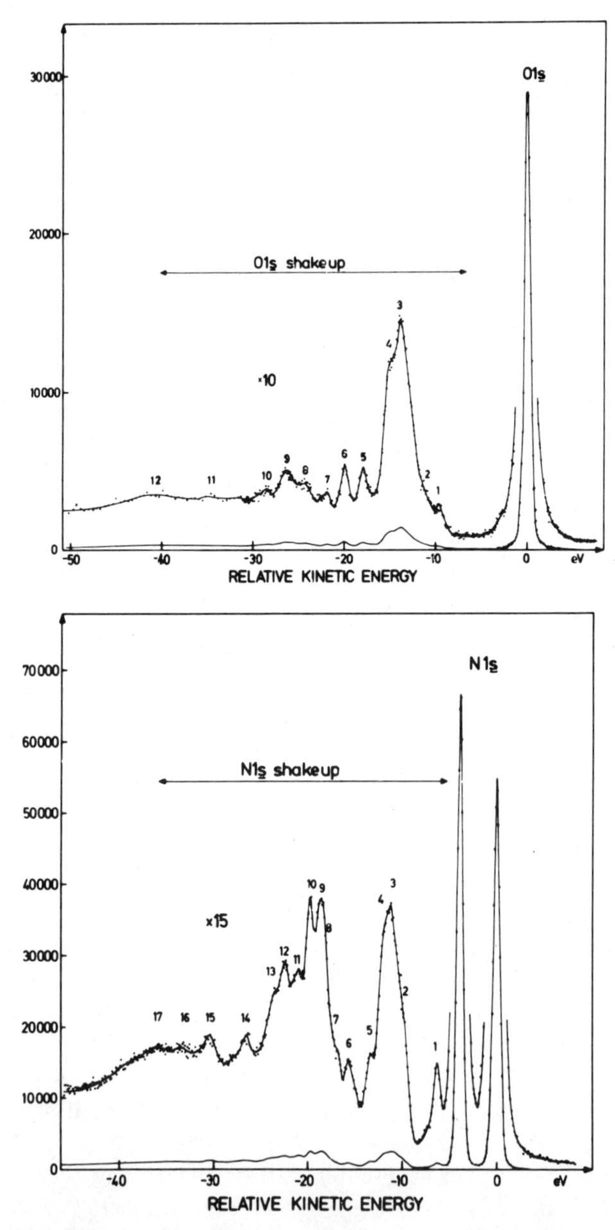

FIGURE 5.34. The O(1*s*) and N(1*s*) shakeup spectra from N_2O excited by monochromatic Al $K\alpha$ x rays. The numbers refer to distinguishable shakeup peaks, except for contributions around 11.3 eV, which may in part be due to inelastic scattering. [Reproduced from Gelius,[171] Figure 6.]

TABLE 5.14

Comparison of Energy Spacings[a] in Neutral Molecules vs. Molecules with a Core Vacancy

Transition	ΔE, eV Neutral molecule	ΔE, eV Isoelectronic molecular ion with pseudo-inner-shell vacancy
$X\,^2\Pi \rightarrow a\,^4\Pi_u$	NO (4.7)	O_2^+ (4.0)
$X\,^2\Pi \rightarrow b\,^4\Sigma^-$	NO (5.8)	O_2^+ (6.1)
Ionization potential	NO (9.3)	O_2^+ (24.4)
$X\,^1\Sigma^+ \rightarrow A\,^1\Pi$	CO (8.0)	NO^+ (8.0)
Ionization potential	CO (14.0)	NO^+(30.5)

[a] Data from Gilmore, *J. Quant. Spectra. Radiat. Transfer* **5**, 369 (1965); J. L. Franklin, J. G. Dillard, H. M. Rosenstock, J. T. Herron, K. Draxl, and F. H. Field, *Natl. Std. Ref. Data Ser., Natl. Bur. Std.* 26 (1969).

calculations for the energies, and has assigned a number of lines at variance with the above suggestion. He has also made a comparison of the CO and N_2 shakeup spectra, since by the equivalent charge arguments $C(1s^1)O^+$ should also have the same final valence shell as NO^+. The energy spacings in the two shakeup spectra (N 1*s* for N_2 and C 1*s* for CO) are indeed similar. However, the intensities vary because the ground states differ, even though the final states are similar, and hence the overlap integrals differ.

In their analysis of the triatomics Allan *et al.*[236] have compared their data on electron shakeoff assuming that the level spacings in the neutral molecule were the same as with an inner shell vacancy and found agreement with transitions to monopole states. Their assumption conflicts with atomic data. For example, the transition $2p \rightarrow 3p$ in neon is 18.3 eV for the neutral atom but 37.3 eV in $Ne^{+1}(1s^1)$. Based on the technique of equivalent nuclear charge, Table 5.14 compares energy spacings in neutral molecules vs. molecules with a core vacancy. The energies for transitions to the continuum are much larger for molecules with a core vacancy, but *surprisingly*, the energies required for transitions into lower lying unoccupied orbitals are rather similar. It would appear that transitions into low-lying excited orbitals do not effectively sense the inner shell vacancy. Thus, neutral molecules and molecular ions with core vacancies may be compared, but with caution.

Aarons *et al.*[237] have proposed that molecular orbitals for the final state excited ion can be derived using a semiempirical method such as INDO, if the inner shell vacancy is represented by an effective nuclear charge $Z + 1$. With the use of these molecular orbitals, both the energy of excitation and probability of electron shakeup can be calculated.

The probability for a transition from an occupied orbital i of the molecule to a virtual orbital j of the ion under perturbation of a core ionization in atom α is given as

$$P_{j\alpha\leftarrow i} = N|\sum_u C_{iu}C'_{ju}\langle\phi_u|\phi'_u\rangle|^2 \tag{5.40}$$

where C_{iu} and C'_{ju} are LCAO coefficients of the MO's and N is the number of electrons in orbital i. In view of the crudeness of the molecular wave functions, the agreement with experiment is quite good. Aarons *et al.*[238] have also explored the use of Hartree–Fock wave functions in calculating shakeup. Basch[239] has made multiconfiguration self-consistent field calculations on formaldehyde, which give mixed agreement with experiment.[240]

Electron shakeup peaks ought to occur with every molecule, and in the gas phase some structure in the photoelectron spectra can almost always be recognized as electron shakeup. However, in solids the large general background and in particular the characteristic energy losses of the main peak obscure the shakeup peaks. Thus, only when the satellite structure is unusually prominent has electron shakeup been discussed. Large satellite structure due to monopole excitation seems to occur primarily with transition metal and rare earth compounds. Barber *et al.*[124] have reported large satellite peaks associated with the 1s shells in carbon and oxygen for a group of carbonyl transition metal compounds. They appear at energies 5.4–6.0 eV less than the main peaks. Pignataro[241] has noted the same situation for chromium, molybdenum, and tungsten hexacarbonyls. Carbon monoxide studied in the gas phase[228, 242] does not exhibit this structure, so that it must be the property of the complex and it has been attributed to electron shakeoff. Rosencwaig *et al.*[243] have studied the 2p shells for a number of first row transition metal difluorides, finding multicomponent structure when the d shell is unfilled. Frost *et al.*[150] have studied a large number of copper compounds, finding satellite peaks with the Cu(II) d^9 compounds but not Cu(I) d^{10}. They also report[219] that low-spin diamagnetic compounds of cobalt show little satellite structure with photoionization in the 2p shell, but extensive satellites with high-spin paramagnetic complexes (cf. Figure 5.35). Similar results were reported[244, 245] on nickel compounds, and the appearance or absence of satellite structure was used to identify whether nickel compounds were paramagnetic tetrahedral complexes (satellites) or diamagnetic planar complexes (no satellites).

Multiplet splitting also occurs with paramagnetic compounds, but in the case of the 2p shells of the first transition metal compounds, Frost *et al.*[150] argued that the satellite structure is primarily due to electron shakeup because (1) the separation between the main peaks and satellite peaks is too large for multiplet splitting and (2) the intensity of the satellite peaks does not follow the number of unpaired spins in the valence shell. As seen in the discussion on

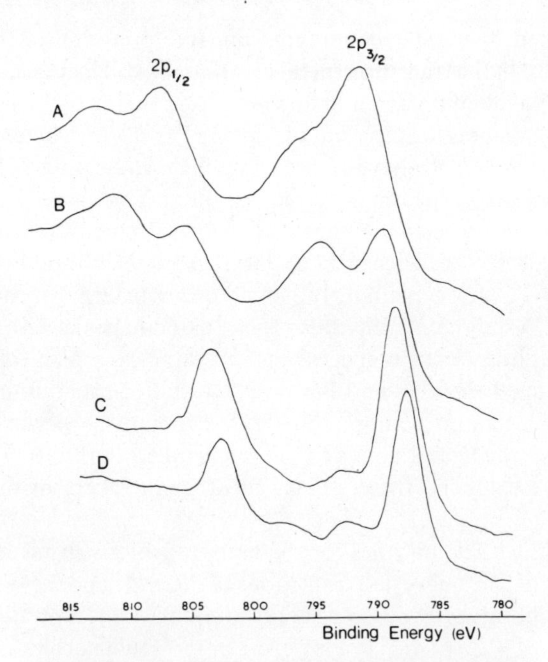

FIGURE 5.35. Photoelectron spectra of some Co compounds showing the $2p_{1/2}$, $2p_{3/2}$ lines and associated satellite structure. (A) CoF_2, (B) $CoCl_2$, (C) $K_3Co(CN)_6$, (D) $Co(en)_3Cl_3$. [Reproduced from Frost *et al.*,[219] Figure 1.]

multiplet splitting in the previous section, these arguments are not definitive. Frost *et al.*[150] also found satellite lines with the 3*s* shell of copper. The splitting here also seems too large to be multiplet splitting. Single-electron wave functions calculations carried out by Nestor and Carlson[215] predict the shake-off probability of *d* electrons as the result of holes in the 3*s* and 3*p* shells to be 36% and 42% of that found as the result of photoionization in the 2*p* subshell, while the energy of splitting is the same or slightly less for photoionization in the *M* shell versus the *L* shell. The theoretical predictions are hampered by the fact that when photoionization occurs in the same principal shell where shake-off takes place, extensive electron correlation may be involved. Some of the *M*-shell data from Ref. 150 fit the description suggested by single-electron calculations and some do not.

Novakov and Prins[149] claimed to find satellite structure in Cu_2O and CuCl which was enhanced by surface oxidation, and Castle[246] suggested that this effect could serve as a help in analyzing oxide layers on metals, but Frost *et al.*[150] felt that the satellite structure was not due to Cu(I) compounds but to Cu(II) formed by surface contamination. Interpretation of the multicomponent structure in the transition metal compounds still leaves many questions open and will be taken up again in Section 5.1.5.

Shakeup satellites have been observed[247,248] for heterocyclic compounds. Ikemoto *et al.*[249] have observed satellites indicating 2–4 V excitation in x-ray photoelectron spectra of tetracyanoquinodimethane (TCNQ) and its complexes. They interpreted their results as due to electron shakeup transitions between the π-electron valence orbitals. Ginnard *et al.*[250] and Butler *et al.*[251] examined TTF-TCNQ, explaining the satellite lines in terms of charge transfer. Pignataro and Distefano[252] have also found satellite peaks with nitrogen and oxygen in the photoelectron spectra of nitroanilines, which they ascribe to electron shakeup. Jørgensen and Berthou[159] studied satellite lines in the $4d$ spectrum of lanthanum compounds. They claim these peaks arise out of transitions from the ligand bonds to the unoccupied f orbitals. Signorelli and Hayes[160] have extended these studies to other members of the lanthanide series.

In discussing selection rules for shakeup transitions in metal complexes, Rosencwaig *et al.*[243] suggested transitions of $3d \rightarrow 4s$ and $3d \rightarrow 4p$. From an atomic viewpoint this presents some confusion since they contradict monopole selection rules $\Delta l = \Delta s = \Delta j = 0$. However, in molecular orbitals angular momentum is no longer a good quantum number and in the conduction band of solids there is a great deal of mixing of s, p, and d states. Matienzo *et al.*[245,246] have suggested that the reason paramagnetic compounds usually show satellite structure in the $2p$ shell, but not diamagnetic compounds, is that for transitions involving orbitals derived from the $3d$ metal ion, $\Delta L = 0$. Thus $3d \rightarrow 4s$ transitions are possible so long as the total angular momentum remains constant. This situation can only occur when L for the initial state is not zero, which in turn would prevent transitions from completely filled shells such as with the diamagnetic nickel and copper compounds. However, Matienzo's explanation does not explain the presence of satellite when the d shell is completely empty (e.g., Ti and Sc[253–256]. Alternatively, Kim[257] has examined which transitions are allowed in NiO and CuO, arriving at

$$(\mathrm{O}\,2p)\,e_g{}^b \rightarrow (\mathrm{Ni}\,3d)\,e_g{}^a \tag{5.41}$$

which is a charge transfer between the ligand orbital and the unfilled metal d orbital. The transition is consistent with a monopole transition and nicely explains why electron shakeup occurs in Cu(II) d^9 compounds and not in Cu(I) d^{10} compounds. In the former compounds the $e_g{}^a$ is partially filled, and transition can occur, while in the latter case the orbital is completely filled and transition cannot take place. Also, from Jørgensen,[258] the transition energy for charge exchange is about 4–8 eV, in reasonable agreement with the experimental separation of the satellite and main peaks. Vernon *et al.*[259,255] made a

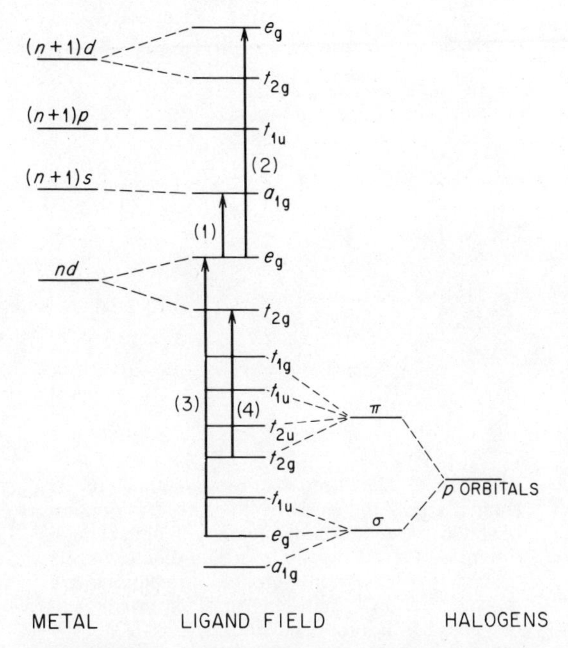

FIGURE 5.36. Schematics of molecular orbitals in a hexahalide complex (octahedral structure). Arrows show possible monopole transitions. Transitions 1 and 2 involve only metal ion orbitals, while transitions 3 and 4 are charge transfer transitions between ligand and metal ion orbitals. [Reproduced from Carlson *et al.*,[255] Figure 4.]

comprehensive study of satellite structure for a wide variety of transition metal compounds. The types of transitions possible are illustrated in Figure 5.36. Comparison with X_α calculations favor those transitions from the ligand orbitals to the unfilled 3*d* orbitals as suggested by Kim.[257]

5.1.3. Configuration Interaction

If an ion that results following photoionization can be described by a wave function that includes excited configurations, then it is possible that the ground state may interact with these excited states, and the final state may end up as one of these excited states.

For example, in the photoionization of the valence shell of argon a 3*s* electron can be photoejected, yielding the state of the ion as $3s^1 3p^6\,{}^2S$. One possible excited configuration that would maintain the same designation ${}^2S$ is $3s^2 3p^4 3d^1\,{}^2S$, which involves a two-electron excitation. (Incidentally, one could

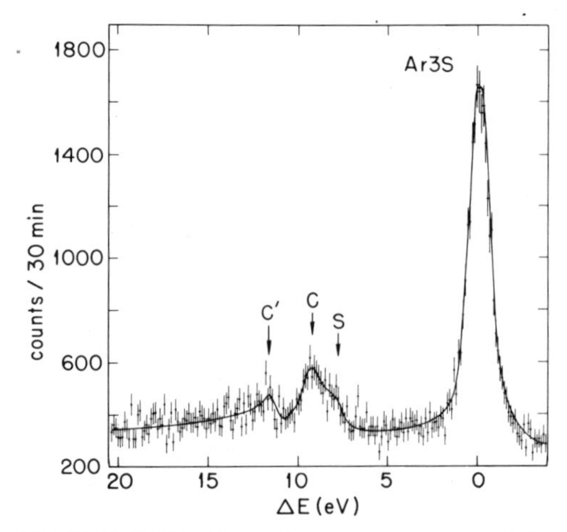

FIGURE 5.37. Photoelectron spectrum of $3s$ shell of argon and satellite structure. Designation of states assisted by comparison with optical data. Energies of states arising from two-electron excitations involving configuration interactions are marked C and C′. State due to electron shakeup is marked S. Reproduced from Spears *et al.*,[229] Figure 4.]

construe electron shakeup as a special case of a one-electron excitation to a configuration interaction state.) The large overlap between the $3s$, $3p$, and $3d$ orbitals makes the transition to this state highly probable. Figure 5.37[229] shows that the excitation energy observed with the main peak in satellite structure of the photoelectron and spectrum of argon matches the energy difference between the $3s^1 3p^6\,{}^2S$ and $3s^2 3p^4 3d^1\,{}^2S$ term values as obtained from optical data. A contribution due to electron shakeup is also noted, but is of secondary importance. Similar conclusions were reached as to the relative importance of configuration interaction with Kr and Xe. In contrast, Wuilleumier and Krause[232] found in neon that configuration interaction states played only a minor role in the photoionization of the L shell. Configuration interaction states such as $2s^2 2p^4 3d^1\,{}^2S$ or $2s^2 2p^4 3s^1\,{}^2S$ are possible but the smaller overlap between orbitals with different principal quantum numbers diminishes the probability for transition to these states.

Wertheim and Rosencwaig[260] have explained the presence of large satellite peaks in the x-ray valence shell photoionization of the alkali metal halides as due to configuration interaction. Since the alkli metal ions are isoelectronic with the rare gases, one might expect from atomic considerations similar behavior with regard to the production of satellite structure. Figure 5.38 shows a comparison of Kr and RbCl data. The large enhancement of satellite structure

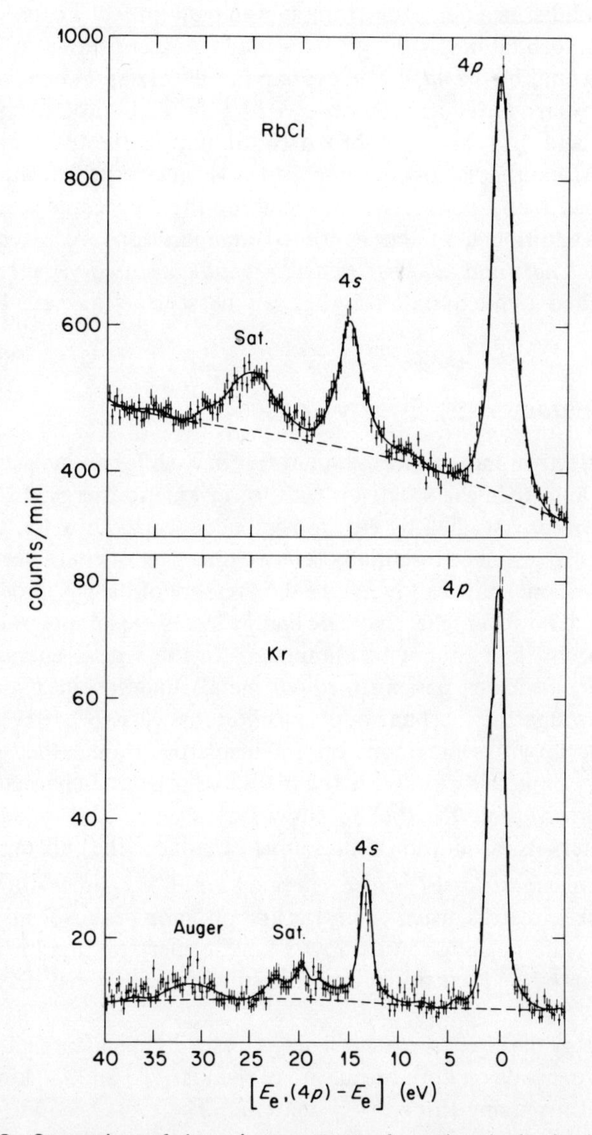

FIGURE 5.38. Comparison of photoelectron spectra from photoionization in the valence shell of Kr and RbCl. Photoelectron energies are given relative to the photoelectron peak corresponding to the 4*p* subshell, i.e., $E_e(4p) - E_e$. Satellite structure is marked Sat. The portion of the Kr spectrum marked Auger comes from some weak *LMM* Auger processes. Dashed lines give approximate backgrounds. [Reproduced from Spears *et al.*,[229] Figure 7.]

suggests a solid state effect; for example, the band in RbCl corresponding to the excited $4d$ orbital in Rb^+ may have a much greater overlap with the $4s$, $4p$ shells than would be the $4d$, $4s$, $4p$ overlap in the corresponding free ion.

Complex structure has been observed[171] in the photoelectron spectrum of Xe $4p_{3/2}$ and $4p_{1/2}$ shells. A broad continuum is due to a super-Coster–Kronig transition that is possible with a $4p_{1/2}$ vacancy. The transition is so fast that one has a many-body problem including the Auger and ejected photoelectrons. In addition, a number of excited configurations are mixed in, such as $4d^{-2}\,nf^*$, $4d^{-2}\,np^*$, and $4d^{-2}\,ns^*$ (where -2 indicates two vacancies, and the asterisk excited unfilled state), which can be seen as discrete lines in the spectrum.

5.1.4. Characteristic Energy Losses

As an electron moves through matter it may suffer an inelastic collision. It will then appear in the spectrum with lower kinetic energy. In gases these losses are carefully studied by electron impact spectroscopy (cf. Chapter 1). If one wishes to lessen or essentially remove the effect of characteristic energy losses in gases, one may simply reduce the pressure of the gas under investigation. In solids, however, characteristic energy losses are an integral part of the observed spectra and cannot be eliminated. In the kinetic energy region of 100–1500 eV, electrons passing through metals undergo inelastic collisions primarily through the mechanism of plasmon loss. It is believed that similar losses also occur with semiconductors and insulators, though they are less well understood. For metals we observe the main loss peak at a characteristic plasmon frequency (usually 5–20 eV) followed by a series of secondary peaks at harmonic intervals. In addition to plasmon losses due to the bulk material, there are surface plasmon losses whose energies are $1/\sqrt{2}$ times those of bulk material. For insulators, usually only the first plasmon peak is seen, with a long tail gradually diminishing toward lower kinetic energies. In addition to plasmon losses, interband transitions and molecular excitation may play important roles.

The energy of the characteristic energy losses is dependent on the material. Its intensity depends on both the nature of the material and the kinetic energy of the electron passing through the material. The cross section for inelastic collision due to plasmon loss decreases monotonically with increasing energy according to[261]

$$\sigma = \frac{h\omega_\rho/2r_0}{E_e\, ln\{[(1+y_\rho)^{1/2}-1]/[x-(x^2-y_\rho)^{1/2}]\}} \tag{5.42}$$

where $x=(E_e/E_F)^{1/2}$; $y_\rho = \mathrm{h}\omega_\rho/E_F$; E_e is the kinetic energy of the electron; r_0 is the Bohr radius; ω_ρ is the plasmon frequency; and E_F is the Fermi energy.

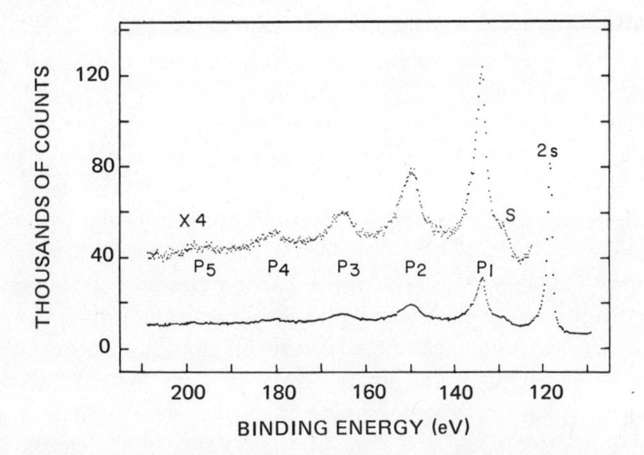

FIGURE 5.39. Photoelectron spectrum of Al(2*s*) subshell from clean Al metal surface, showing characteristic energy losses. P_1–P_5 are due to bulk plasmon losses. S is due to surface plasmon loss. [Reproduced from Pollak *et al.*,[262] Figure 2(b).]

In the energy region of interest to x-ray photoelectron spectroscopy, the energy dependence of the cross sections can be taken as approximately proportional to $E^{-1/2}$.

Characteristic energy losses in photoelectron spectroscopy need not be only a nuisance factor, but can be used to learn about the nature of the material being studied. For example, Pollak *et al.*[262] have made a study of plasmon losses in nine metals, using x-ray photoelectron spectroscopy. Their results on Al, for example, are shown in Figure 5.39. To take another example, electrons arising from a surface layer of an inhomogeneous material, e.g., C(1*s*) photoelectrons from a hydrocarbon film on a non-carbon-containing substrate, will show substantially reduced characteristic energy losses, and such layers may be identified by this behavior.

5.1.5. Determining the Nature of Multicomponent Structure

In the preceding sections it is obvious that confusion often is present in determining the nature of the satellite structure. In principle, however, if a careful study of the photoelectron spectra of all the subshells is made and the results intercompared, a complete understanding as to the source of the satellite structure can be achieved. Let us now consider the general behavior for each of the sources of multicomponent structure: (1) multiplet splitting, (2) electron shakeup, and (3) configuration interaction, as well as two other important sources for extra photoelectron peaks, (4) characteristic energy losses and (5) chemical shifts.

Multiplet Splitting. Normally, this will occur only in the element in which there is an unfilled valence shell. Photoionization in each subshell in that element will display a different satellite structure, with regard to both the energy separation and relative intensity. The relative intensities of the s peaks, which appear as a doublet, will be the same for all shells and can be predicted from the number of unpaired spins in the valence shell. The degree of energy splitting decreases as the principal quantum number differs from that of the unfilled shell.

Electron Shakeup. This will vary according to the chemical environment of the atom undergoing photoionization. However, each subshell of that atom displays nearly the same satellite structure on photoionization, with regard to both relative intensity and energy separation, if the subshell is at least one principal quantum number lower than the shell undergoing electron shakeup. If photoionization occurs in the valence shell, electron shakeup occurs, but its intensity and energy may differ than when a hole is created in the core. The extent of shakeup in the valence shell is usually less, but not necessarily.

Configuration Interaction. Satellite structure from configuration interaction, involving two-electron excitation, will usually occur only when photoionization takes place in the valence shell.

Characteristic Energy Losses. Characteristic energy losses are independent of which subshell or in which atom or element photoionization occurs, since any electron of the same energy will have the same losses. The intensity relation to the main peak will vary roughly as $\sqrt{E_e}$, and the spectrum is usually broad without sharp peaks (except for metals).

Chemical Shifts. Chemical shifts are usually predictable from knowledge of the molecular structure, but sometimes radiation damage or impurities will produce unexpected peaks. Every subshell in a given element will have the same intensity for the satellite peak and essentially the same energy separation.

If we turn our attention once again to the problem of the satellite structure in the $2p$ shell of transition metal compounds, we can eliminate all but multiplet splitting and electron shakeup. Since electron shakeup is associated with the $3d$ shell, it is not clear whether creating a vacancy in the $3s$ or $3p$ subshell will have the same effect as creating one in a shell with a lower principal quantum number. A study of the $2s$ shell would be desirable, but the $2s$ level is extensively broadened due to Coster–Kronig transitions. Nevertheless, studies on the $2s$ subshell have been carried out by Kim[257] for CuO and NiO and show similar multicomponent structure as the $2p$ shell. Photoionization studies in the $1s$ shell of the metal ions would be more valuable. Multiplet splitting in this shell as calculated from Hartree–Fock wave functions should have a negligible energy spread. Electron shakeup peaks, however, should be almost identical to those found in the $2p$ shell. One would need hard x-rays, e.g., Cu $K\alpha_1$ (8.048 keV) to promote K vacancies in the transition metals. The widths of the peaks from the photon source and natural width of the $1s$ shell are nearly 4 eV, but the resolution is still sufficient for distinguishing the satellite peaks now reported for the $2p$ shell. Carlson *et al.*[263] studied the satellite structure in the photoelectron spectrum of $FeCl_3$ and $FeBr_3$ and found it to be nearly identical with that observed in the $2p$ shell, showing that electron shakeup is the principal cause for the multicomponent structure for the $2p$ shell. Another method for distinguishing multiplet splitting from electron shakeup is to examine the $K\alpha$ x-ray spectrum ($1s$–$2p$). The final state of such an x-ray transition is the same as photoejecting an electron from the $2p$ shell. Thus, the multiplet splitting should show itself to be the same in the $K\alpha$ x-ray fluorescence spectrum as in the photoelectron spectrum. However, electron shakeup should not affect the x-ray spectrum substantially. For example, Asada *et al.*[264] found for a series of nickel compounds similar multicomponent structure between $K\beta$ emission ($1s$–$3p$) and the $3p$ photoelectron spectrum, demonstrating multiplet splitting to be the dominant cause for the multicomponent structure. However, in the $K\alpha$ x-ray spectrum the large multicomponent structure that is present in the $2p$ photoelectron spectrum is not seen, showing that the latter arises from electron shakeup.

TABLE 5.15
Summary of Data on Mean Free Path for Electrons of Energies from 100 to 1500 eV

Material	E_e, eV	λ,[a] Å	Reference
Be	60	4.7	Seah[b]
Be	355	10.0	Seah[b]
Be	935	12.7	Seah[b]
C	920	15	Steinhardt *et al.*[c]
C	1169	18	Steinhardt *et al.*[c]
Si	321	13	Klasson *et al.*[269]
Si	554	23	Klasson *et al.*[269]
Si	1178	39	Klasson *et al.*[269]
Ag	110	9.4	Seah[b]
Ag	362	8	Palmberg and Rhodin[d]
GeO_2	234	6.1	Todd and Heckingbottom[271]
GeO_2	266	6.8	Todd and Heckingbotton[271]
W	1455	13	Carlson and McGuire[137]
WO_3	1450	26	Carlson and McGuire[137]
Au	1405	22	Baer *et al.*[e]

[a] "Effective" mean free path for inelastic scattering as obtained from Auger and photoelectron spectroscopy.
[b] M. P. Seah, *Surf. Sci.* **32**, 703 (1972).
[c] R. G. Steinhardt, J. Hudis, and M. L. Perlman, *Phys. Rev. B* **5**, 1016 (1972).
[d] P. W. Palmberg and T. N. Rhodin, *J. Appl. Phys.* **39**, 2425 (1968).
[e] Y. Baer, P. F. Heden, J. Hedman, M. Klasson, and C. Nordling, *Solid State Comm.* **8**, 1479 (1970).

5.2. PESIS for Surface Studies

The photoelectron spectroscopy of a solid is basically a study of atoms on or close to the surface since the mean free path for inelastic collisions of photoelectrons normally encountered is the order of 10–20 Å. Table 5.15 lists experimentally determined values for average escape depths. The first surface layer generally makes up about 10% of the observed spectrum. Though we have usually discussed the results of photoelectron spectra in terms of the bulk material, it is obvious that we are in fact dealing with the surface layers. It is indeed surprising that so many analyses relating to bulk properties can be carried out without special precautions taken toward surface conditions. I shall often

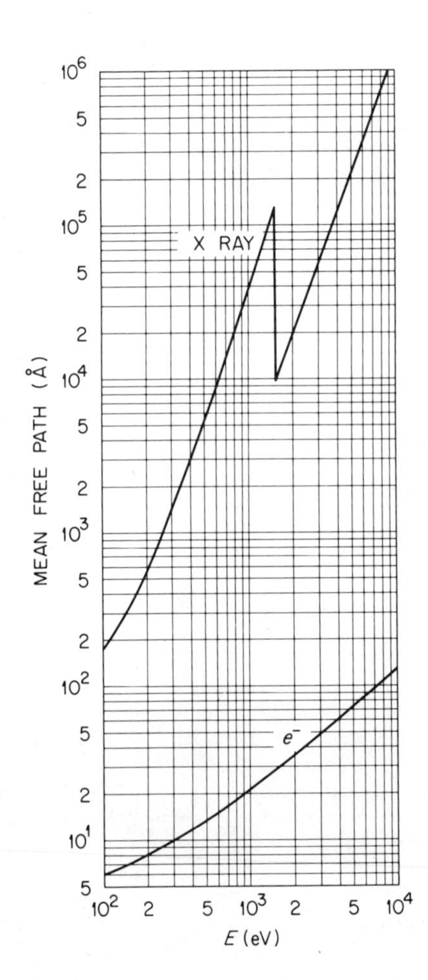

FIGURE 5.40. Comparison of mean free paths for electrons and x rays in Al.

refer to XPS of solids as a surface study, although most of the signal is commonly from below the outermost first layer of atoms. Nevertheless, we are so close to the surface that we must always be conscious of its possible effects.

Several phenomena determine the fraction of observed photoelectrons that originate at a given depth in the sample. Let us consider some of the equations related to these phenomena. First, we shall assume that the x-ray beam will be unattenuated over the depth from which photoelectrons can be effectively removed. This is a good assumption since the mean free path for x rays encountered is always several orders of magnitude higher than that for the associated photoelectron. (Cf. Figure 5.40.) Thus, for a one-component system we have for the number of photoelectrons escaping without inelastic scattering following photoelectron ejection from a given shell by a given photon

$$dN = N_0 e^{-\sigma x}\, dx \tag{5.43}$$

where N_0 is the number of photoelectron emissions per unit area for the given shell and given photon energy, σ is the inelastic cross section per unit length, and x is the depth below the surface.

The assumption that the x-ray beam is unaltered over the depth of interest implies that N_0 is also constant over that region and

$$N_0 = \alpha n F \tag{5.44}$$

where α is the photoelectron cross section for a given subshell, n is the concentration per area of those atoms that are undergoing photoelectron ejection, and F is the flux of x rays. Integrating equation (5.43) from zero to infinity, we have

$$N = \frac{\alpha n}{\sigma} F \tag{5.45}$$

If we compare the relative intensities for two compounds containing a common element, we obtain the relative inelastic cross section for the two compounds at the given photoelectron energy, i.e.,

$$\frac{\sigma_1}{\sigma_2} = \frac{N_2 n_1}{N_1 n_2} \tag{5.46}$$

The mean free path λ_m is related to σ by

$$\lambda_m = \frac{1}{I} \int_0^\infty dI = \frac{1}{\sigma} \tag{5.47}$$

Also useful is the escape depth at half-intensity, $\lambda_{1/2}$, or the depth below the surface from which half of the photoelectrons are observed:

$$\lambda_{1/2} = \frac{\ln 2}{\sigma} = \frac{0.693}{\sigma} \tag{5.48}$$

The value σ as determined experimentally from escape depth measurements is not the true attenuation coefficient for inelastic scattering, since we have neglected elastic scattering. Elastic scattering will have the net effect of lengthening the average time that it takes a photoelectron to emerge from the solid. Ideally, one should treat the escape of photoelectrons from a solid as a diffusion problem. It can be shown, however, that the total cross section is the cross section for inelastic scattering times some factor that is related to the elastic cross section.[263] That is, σ as determined in PESIS is the inelastic cross section times some constant slightly larger than one.

Let us next consider a two-component system, such as might occur with an oxide film on a metal surface. If the oxide layer is uniform, the relative intensity for the different peaks corresponding to the metal (1) and the oxide (2) is

$$\frac{N_1}{N_2} = \frac{n_1\sigma_2}{n_2\sigma_1} \frac{e^{-\sigma_2 x_o}}{1 - e^{-\sigma_2 x_o}} \tag{5.49}$$

where x_0 is the thickness of the oxide layer. For example, in the case of tungsten–tungsten oxide these ratios were determined as a function of the thickness of the oxide film x_0, which was obtained in a separate experiment. Figure 5.41 shows the experimental ratios plotted against x. The data, which stretch over three orders of magnitude, can be fitted to a single parameter σ (for WO_3) by means of equation (5.49). The excellence of the overall fit gives good evidence that the assumption of a uniform film of WO_3 is correct. This method of examining a two-component system by PESIS can thus be used to ascertain the depth of the surface film and its uniformity.

The mean free path or escape depth is a function of the photoelectron energy, and results from a variety of studies appear to fit a universal curve (cf. Figure 5.42). More recent results[266] show a much wider variation. In the case of hydrocarbons it is believed[267, 1] that λ_m may be as high as 100 Å for 1200-eV electrons. The energy dependence is close to that predicted from theory. For example, Tung and Richie[268] have obtained good agreement between calculation and experimental values of λ_m for aluminum. Escape depths as a function of electron energy from 320 to 3.6 keV have been obtained for silicon using electron spectroscopy.[269] Inelastic scattering can result from (1) plasmon losses (5–30 eV), (2) single-electron interaction, (3) interband transitions, which for valence states to higher bands may be from 1 to 8 eV and for deeper-lying levels to higher states are usually above 15 eV, and (4) phonon excitation, less than 1 eV. It has been shown by Thomas[270] that for electrons with kinetic energies above $E_F + h\omega_p$, where E_F is the Fermi level and ω_p is the plasmon frequency, plasmon production is the dominant process. In the range from 200 to 2000 eV, which is the range that is of particular interest to PESIS, λ varies roughly as $E_e^{1/2}$. At very low energies (below 10 eV) the mean free path rises rapidly as the mechanisms for inelastic scattering are eliminated.

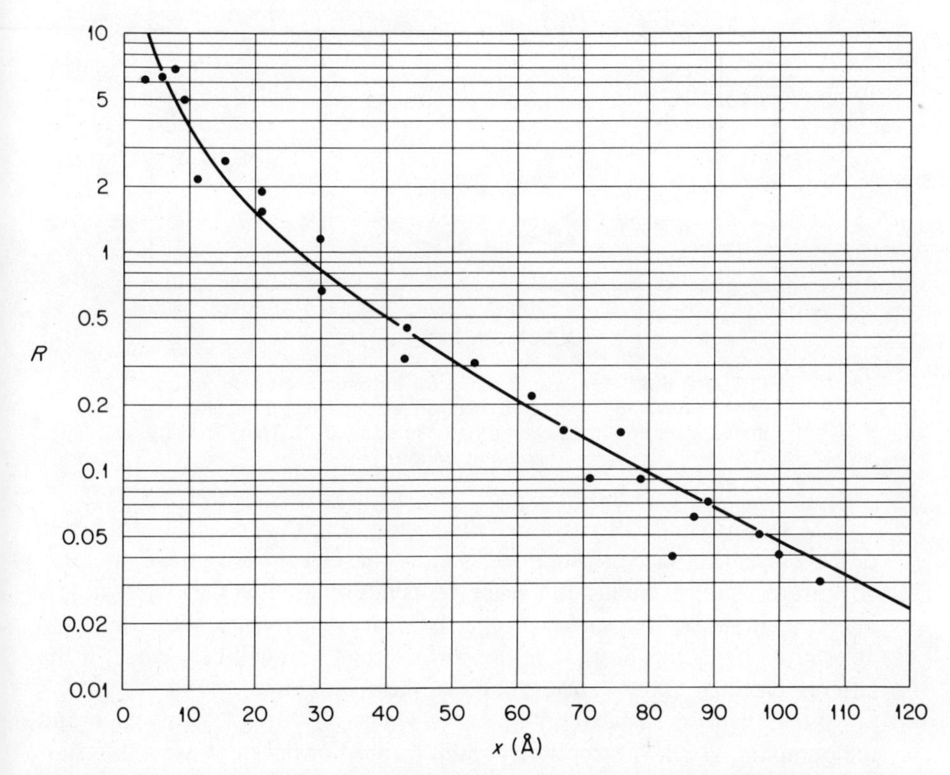

FIGURE 5.41. Plot of the ratio of intensities of photoelectrons arising from W relative to those arising from WO_3 as a function of the depth d of WO_3 film. $x = 2^{1/2}d$. Solid line is theoretical curve derived from equation (5.49) using $\sigma_1 = 0.038$ Å^{-1}. [Reproduced from Carlson and McGuire,(137) Figure 2.]

The dependence of λ_m on the photoelectron energy can be employed to locate atoms below the surface when they are distributed inhomogeneously. The photoelectron spectrum of a given atom generally consists of a variety of photoelectron peaks and Auger electron peaks of substantially different energies. The intensity ratio of these electrons will depend on the depth to which the atom is buried in the surface. For example, if for a sodium atom using Mg $K\alpha$ radiation the ratio of the number of photoelectrons ejected from the $1s$ shell (~180 eV) to the number of Auger electrons arising from the K–LL process (~990 eV) is unity when sodium is present on the surface of Al, then at a depth of 20 Å the ratio will be 1/25 using λ from Figure 5.40. Use has been made of this technique in determining escape depths.(271) Clark *et al.*(272) have used the ratios of photoelectron intensities of F(2s) to F(1s) to estimate the position of fluorine atoms in surface-fluorinated polyethylene film.

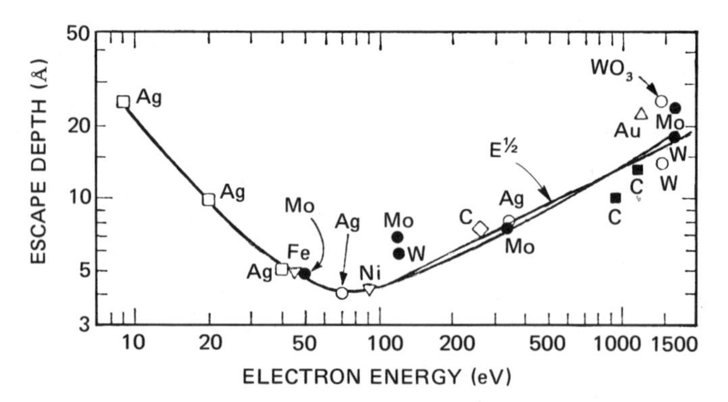

FIGURE 5.42. Mean free paths for inelastic scattering as obtained from Auger and photoelectron spectroscopy data. [Reproduced from original plot of data collated by J. C. Tracy (private communication).]

As set forth in the discussion of this section, PESIS of a solid is, for better or for worse, a surface measurement. Let us consider the advantages and disadvantages that this entails. In a sense it depends on whether the investigator is interested in the bulk or surface properties. If it is the former, the added variables given by PESIS because of the surface condition will be a curse; if the latter, a blessing. Let us consider some of these variables.

First, there are surface contaminants. Unless all *O*-ring seals are metal and the pumping system is exceptionally well trapped or does not use a diffusion pump, some hydrocarbon vapor will be adsorbed on the surface of the sample. The layer seems to quickly equilibrate with the residual pressure, and the extent of carbon contamination depends to a large degree on the nature of the surface of the sample target. Sometimes contamination will continue to increase if the carbon is "baked on" as the result of radiation damage. Usually the carbon layer does not interfere with the general analysis unless, of course, carbon is one of the elements of interest in the compound being studied. Also, there is some loss in the signal. Often the carbon contamination can be used as a reference standard, but other standards such as gold films are generally more reliable. Water vapor and oxygen are two other common contaminants. The only sure way to study surfaces completely free of contamination is to operate with a vacuum of 10^{-9} Torr or better and to provide *in situ* cleaning. Besides cutting down on intensity and adding unwanted peaks to the photoelectron spectra, contaminants may also react with the surface layer of the compound being studied, causing chemical shifts for the elements in that compound.

Even when there is no contamination on the surface, the surface atoms see a different chemical environment than those in the bulk material. For example, in calculating[18] the crystal potential as seen by a core electron, the

Madelung-like constant for a surface atom in NaCl is only 80% of that for an atom in the bulk crystal. One monolayer below the surface the Madelung-like constant rises rapidly to 99% of its full value. Changes in the work function also occur for the surface layer and the influence of potential fields is also sensitively felt by the surface atoms.[1]

Looking at the surface problems from a positive viewpoint, photoelectron spectroscopy is a sensitive tool for studying the surface. Approximately 10% of the total number of discrete photoelectrons observed come from the first monomolecular layer when Al $K\alpha$ x rays are used. Use of lower-energy soft x rays, such as the Zr and Y $M\zeta$ x rays, would substantially increase that figure. Also, studying photoelectron emission at a grazing angle will concentrate the attention on the surface layer. Such studies are currently being carried out by Fraser *et al.*[273] and Brunner and Zogg.[274] Figure 5.43 plots the fraction of signal that arises from the surface layer as opposed to the total signal, based on calculations for a flat surface with negligible x-ray refraction and reflection, which is a reasonable assumption for θ above 10°. The effects of angular studies on surface analysis have been examined thoroughly by Fadley *et al.*[275] They considered the various effects of x-ray refraction and reflection, nonuniform surface layer thickness, and surface roughness. Although these complications

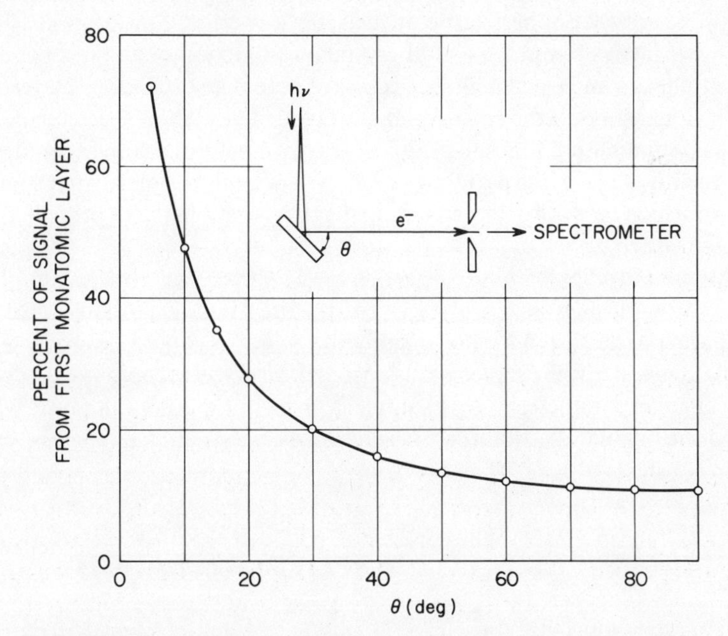

FIGURE 5.43. Calculation of percentage of signal arising from surface layer as a function of takeoff angle θ. Calculations based on $\lambda_m = 26$ Å and surface layer 2.9 Å as might occur for Al metal with x rays of 1.5 keV.

need to be kept in mind, the use of angular studies to obtain a semiquantitative estimate of the surface to bulk properties is a valuable and practical procedure.

It is not unreasonable to expect to observe contributions from 1% or less of the atoms making up the first monomolecular layer, and as usual with PESIS measurements, one obtains not only elemental analysis but chemical bonding information as well. As will be discussed in Chapter 6, Auger spectroscopy has been used extensively for surface analysis. The relative merits of Auger electrons and photoelectrons for surface study will be discussed there. Suffice it to say here, that although Auger spectroscopy has been used almost exclusively by surface scientists, photoelectron spectroscopy has a great deal to be said for it, and undoubtedly will be used more in the future. Some practical applications of surface measurements with PESIS will be discussed in Section 6 of this chapter.

5.3. Angular Studies with PESIS

Several investigators[48, 276, 277] have studied the photoelectron spectra of crystals with x rays and vacuum uv as a function of angle. The interest here resides in the fact that the mean free path is dictated by the crystal structure. That is, along certain crystal planes there is channeling, the electrons finding less opportunity for either elastic or inelastic scattering. For example, Figure 5.44 shows the spectrum for a gold crystal as a function of angle between the incident photon and ejected photoelectron. There is an oscillatory pattern that reflects at least in part the crystal planes of gold. The carbon spectrum derived from a contamination film shows, in contrast, no pattern. In one crystal alloy, 56% Au and 44% Ag, the pattern as a function of angle for silver was found[278] to be similar to that for gold, indicating that the Ag atoms were present as part of an fcc lattice.

Angular studies of photoelectrons ejected from the core shells of free atoms and molecules have not received the attention given the valence shell (cf. Chapter 4, Section 3.4). Presumably the behavior should be characteristic of the atom and not the molecule. Manson[279] has made copious calculations of the angular parameter β. An example is given in Figure 3.1. Krause[280] has studied the angular distribution of the photoelectrons ejected from the 3*s*, 3*p*, and 3*d* subshells of Kr.

6. USE OF PESIS FOR APPLIED RESEARCH

Although much of the effort in photoelectron spectroscopy currently being pursued falls into the category of basic research, the potential of PESIS for practical application is extensive. These applications fall into the areas of

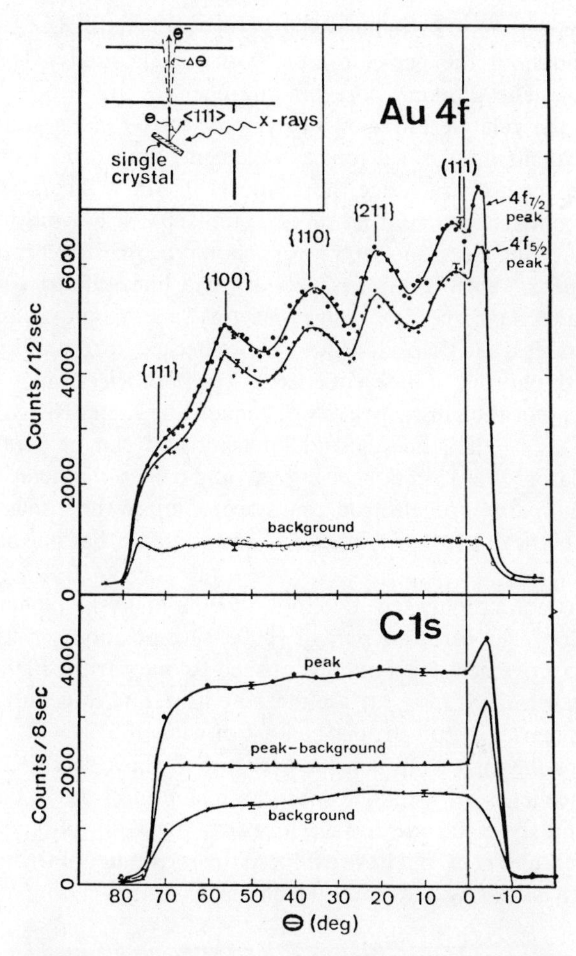

FIGURE 5.44. Intensity of photoelectron peak from gold as a function of angle between gold crystal and ejected electron. Lower portion of figure shows C(1*s*) peak from surface contamination layer. [Reproduced from Fadley and Bergström,[276] Figure 1.]

analytical chemistry, biological systems, geology, polymers and alloys, environmental studies, surface state studies, radiation damage studies, and industrial control.

6.1. PESIS as an Analytical Tool

As has been mentioned previously, PESIS can be used as a "triple threat" tool for chemical analysis: qualitative, quantitative, and diagnostic. For elemental analysis PESIS can be used for every element of $Z > 2$ and it can be applied throughout the periodic table with equal facility. The analysis is unambiguous. The different core binding energies are widely separated in energy and the relative energies and photoelectron cross sections for the different subshells are known for each element. Spin orbital doublets, e.g., $4f_{7/2,5/2}$, also help characterize the element. PESIS makes a good tool for analyzing a complex spectrum *in situ*, e.g., smoke particles, soil, and individual rocks. In analyzing electron spectra, care should be made to properly identify extraneous peaks, such as Auger spectra, extra lines due to x-ray satellites, and multicomponent structure. Auger spectral data are known and have been compiled in tables, and suspected lines can be checked by rerunning the electron spectrum with photons of different energy. The photoelectron peaks will shift, but the energy of the Auger peaks will remain the same. X-ray satellites are well known as to their energy and intensity and can be removed with a monochromator. Multicomponent lines should have a characteristic behavior related to the main photoelectron peaks according to the discussion given in Section 5.1. None of these problems is insurmountable, but one must be aware of them.

PESIS is not a highly sensitive tool for bulk analysis. Elemental analysis is generally only good to one part in 1000. With position-sensitive detectors and other improvements this may be pushed to one part in 10,000 or slightly better. However, PESIS is a surface measurement, and thus only a small portion of the material is studied; one can detect with PESIS as little as 10^{-12} g if it is located on the surface. Brinen and McClure[(281)] have described a procedure in which trace levels of metals in solution in solution were electrochemically deposited and the electrode surface measured by x-ray photoelectron spectroscopy. Hercules *et al.*[(282)] have explored the possibility of treating fiberglass surfaces with silylating reagents, which then can react with metal ions in solution.

PESIS can provide quantitative as well as qualitative analysis. The intensity of a photoelectron peak corresponding to an element in a homogeneous material is, from equation (5.45),

$$I = \frac{K\alpha n}{\sigma} F \tag{5.50}$$

K is a spectrometer constant involving the geometry and collection efficiency. It is independent of the atom but depends on the photoelectron energy. If electrons are not preaccelerated, their collection efficiency (or window width of the analyzer) should be proportional to their kinetic energy. Thus, to compare peaks in a photoelectron spectrum taken without preacceleration, one may simply take the areas and divide by E_e. For analyzers that maintain a fixed potential, no window width correction is needed. However, preaccelera-

TABLE 5.16

Comparison of Relative Intensities for Photoelectron Peaks in Gases ($h\nu = \text{Mg } K_\alpha$)

Compound	Selected ratio	Theoretical[a] σ ratio	Experimental[b] value
NN*O	N 1*s*/O 1*s*	0.622	0.62 ± 0.05
N*NO	N 1*s*/O 1*s*	0.622	0.57 ± 0.05
NO	N 1*s*/O 1*s*	0.622	0.69 ± 0.03
CO_2	O 1*s*/C 1*s*	2.85	2.3 ± 0.1
CO	O 1*s*/C 1*s*	2.85	2.5 ± 0.2
CH_3Cl	Cl 2*p*/C 1*s*	2.36	2.5 ± 0.2
CH_3I	I $3d_{5/2}$/C 1*s*	19.3	16.2 ± 1.1
			17.4 ± 1.2
C_6H_5F	F 1*s*/C 1*s*	4.26	3.9 ± 0.2
CH_2F_2	F 1*s*/C 1*s*	4.26	3.7 ± 0.2
CS_2	S 2*p*/C 1*s*	1.74	3.1 ± 0.8
			3.5 ± 0.5
CS_2	S 2*s*/C 1*s*	1.25	1.8 ± 0.5
			1.8 ± 0.3
CS_2	S 2*p*/S 2*s*	1.39	1.7 ± 0.2
			2.0 ± 0.2

[a] The theoretical σ ratios were obtained using the subshell photoelectron cross sections of Scofield, Ref. 287.
[b] From Carter *et al.*, Ref. 283.

tion or deceleration may alter the collection efficiency. This could, in principle, be calibrated. Equation (5.50) can be solved using values for the photoelectron cross section α obtained from other measurements or theory. For example, the relative subshell cross sections have been determined[(283)] from gas-phase XPS and are compared with theory in Table 5.16. The values of σ can be obtained from some sort of universal curve such as given in Figure 5.42. The value of K is a function of E_e only and is fixed for a given spectrometer. Since we are interested primarily in relative intensities, it is most profitable to compare intensities from homogeneous compounds of known concentration. This has been done by Wagner,[(284)] Jørgensen and Berthou,[(285] and Nevedov *et al.*[(286)] for the Varian instrument. By comparing data with those from a spectrometer having a known $K(E)$ (for example, one using no preacceleration) K can be ascertained.[(283)] In this way intercomparison between different spectrometers is possible. From (5.50)

$$\frac{I_1}{I_2} = \frac{\alpha_1 \sigma_2}{\alpha_2 \sigma_1} \tag{5.51}$$

Since in the region of interest σ is roughly proportional to $1/\sqrt{E_e}$, we have

$$\sigma_2/\sigma_1 = (E_1/E_2)^{1/2} \tag{5.52}$$

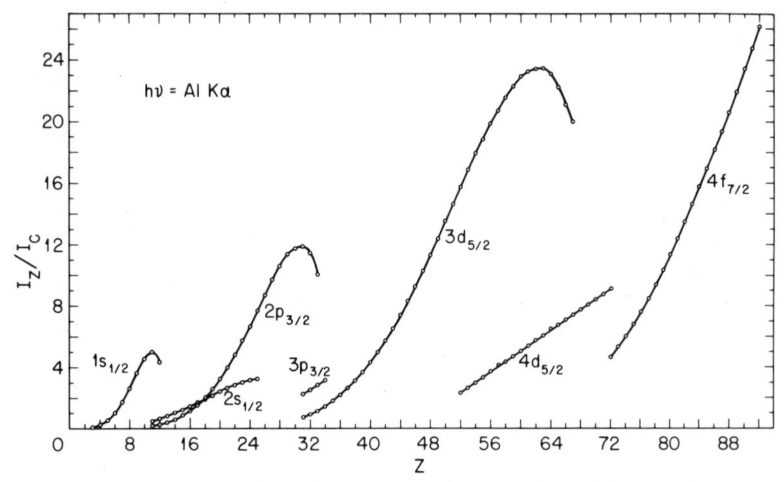

FIGURE 5.45. Relative intensities of photoelectrons ejected from various core shells for solids using Al $K\alpha$ x rays as a function of Z. Intensities are related to C(1s) = 1.00. [Reproduced from calculations of Carter *et al.*,[288] Figure 2.]

Scofield has made comprehensive calculations of subshell cross sections* and Carter *et al.*,[283, 288] using cross sections at 1254 and 1487 eV, determined the relative intensities expected for XPS of solids for all elements from helium to uranium. Results for Al $K\alpha$ x rays are given in Figure 5.45. Comparison of theory with experiment for quantitative analysis of XPS is given in Table 5.17. The agreement is satisfactory, the average deviation being 12%.

With further improvements in experiment and theory it should be possible to obtain from PESIS a quantitative analysis good to 5 or 10%. But a number of problems prevent more accurate determination. First, it is difficult for one to accurately subtract background substructure. Second, angular distribution may play a role. If data are taken in the gas phase at $\theta = 90°$, one may make a correction by multiplying by the factor $[1 + (\beta/4)]$. The values of the angular parameter β for inner shell orbitals can be taken from atomic calculations of Manson.[279] One may avoid the need for a correction for angular distribution by taking data at the magic angle of 54.7°, at which angle the intensity is always proportional to the total intensity regardless of β. In solids, channeling and elastic collisions make angular prognosis more difficult. Third, intensities may depend on the chemical bonding. This may be due, for example, to differences in electron shakeoff. Swingle has examined this problem and the general problem of energy loss with regard to their effects on quantitative analysis.[289] Finally and most importantly, any nonhomogeneity in the surface depth will

* Supplied to the author by J. H. Scofield for photon energies 1254 and 1487 eV. For a general report see Ref. 287.

TABLE 5.17

Comparison of Relative Intensities for Photoelectron Peaks in Solids ($h\nu$ = Al K_α)

	Experiment[a]				
Ratio	I_W	I_C	I_N	$I_{Average}$	I_{Theory}
C 1*s*/F 1*s*	0.24	0.29	0.24	0.26	0.277
O 1*s*/Na 1*s*	0.61	0.53	0.35	0.50	0.522
Na 1*s*/F 1*s*	2.09[c]	1.44	1.89	1.80	1.39
Si $2p_{3/2}$/F 1*s*	0.17	0.23	0.15	0.18	0.161
P $2p_{3/2}$/Na 1*s*	0.26	0.18	0.12	0.19	0.167
S $2p_{3/2}$/Na 1*s*	0.33	0.30	0.18	0.27	0.232
Cl $2p_{3/2}$/Na 1*s*	0.46	0.43	0.25	0.38	0.312
K $2p_{3/2}$/F 1*s*	0.85[c]	1.03	0.83	0.90	0.723
Ca $2p_{3/2}$/F 1*s*	1.01	1.06	0.98	1.02	0.903
Pb $4f_{7/2}$F 1*s*	4.10	4.12	—	4.11	3.74
Na 2*s*/Na 1*s*	0.065[c]	0.145[c]	0.077[c]	0.096	0.0919

[a] The I_W are from Wagner, Ref. 284; the I_C are experimental data from Carter *et al.*, Ref. 288, corrected for the window width of the spectrometer; the I_N are from Nefedov *et al.*, Ref. 286.
[b] Calculated from equations (5.51) and (5.52).
[c] Result is the average for several different compounds.

cause a difference in intensity. If the two photoelectron peaks being compared have large differences in their kinetic energies, a layer of contamination may affect the relative intensities. When two materials are mixed, one may be preferentially located on the surface. To avoid such a problem, Hercules[(290)] has suggested the use of a common matrix. For lead compounds a bisulfate fusion was made so that lead in an unknown sample could be compared under identical conditions. Phillips has investigated[(291)] the use of XPS in quantitative analyses for fluorochemicals, studying the effects of surface roughness, molecular orientation, surface contamination, and sample matrix type. He felt that by exercising proper caution, XPS could provide quantitative data on surface composition accurate to within 1–3%.

PESIS can provide sufficiently accurate information to approximately ascertain a compound's formula. Its unique value is that it gives an appraisal of the oxidation state or degree of ionicity for every atom that is measured, and this information is available for even the most complex substance. For example, we can thus study a protein *in situ* and establish the different types of amino acids from a determination of the elements present, and, most important, from their chemical shifts. This latter information can tell the number of types of groups such as carbonyl, amino, nitro, sulfonyl, etc. Swartz and Hercules[(87)] have used the chemical shift in MoO_2 and MoO_3 to develop an analytical procedure for bulk analysis of MoO_2–MoO_3 mixtures, which showed a relative standard deviation of 2%.

The sample size needed for analysis in PESIS is quite small, only a few milligrams. If thin films can be obtained, only one monomolecular layer is required. Usually, the sample is ground into powder, but objects of any irregular dimension can, in principle, be investigated. (Spectrometers with no restrictions regarding the target position, such as those employing spherical sector plates, are especially well adapted to study large, odd-shaped objects.)

PESIS can be also used as an analytical tool for gases. The peak-to-background ratio is much more favorable than with solids (for neon, ratios of 30,000/1 have been obtained) and one can make measurements for a given element under favorable conditions to as low as one part per million. The sensitivity depends more on the photoelectron cross section than the inelastic collision cross section for the ejected photoelectron. Thus, a choice of x-ray energies to give the maximum cross section is desirable.

6.2. Biological Systems

PESIS has been used to study several aspects of proteins. Klein and Kramer[292] have used ESCA to study powders made from various species of grain. The main result of their work is that quantitative estimates of the total protein content of the sample can be made by measuring the intensities of the nitrogen and sulfur peaks. In addition, for proteins rich in lysine and arginine the amine nitrogen may be distinguished from amino nitrogen.

Thus, photoelectron spectroscopy appears to be a convenient and rapid method for determining the quantity and quality of grain proteins, which could also be used in mass screening operations. Kramer and Klein[67–69] have also studied non-heme iron proteins and have interpreted the data with respect to hypothetical and synthesized iron–sulfur complexes. When comparisons could be made, the data agreed with results from other spectroscopic techniques. The results on rubredoxin, with high iron and sulfur binding energies, were compatible with one of several proposed models for its structure. Millard and Masri[293] have carried out a program of detection and analysis of modified proteins with the aid of XPS. Leibfritz[294] has studied the valence of iron in a series of ferredoxins. Lauher and Lester[295] have carried out x-ray photoelectron spectroscopy on complexes related to coboglobins (hemoglobin with the iron replaced by cobalt). Zeller and Hayes[296] have made an extensive study of porphyrin and phthalocyanine compounds.

In the field of biological structure determinations, Hulett and Carlson[20, 297] have made studies of several nucleosides and *t*-RNA. Of particular interest was evidence of a drastic change in the nitrogen spectrum of the nucleoside in which a methyl group has been substituted into the seventh position of the ten-membered ring that makes up the base. This change indicated the strong effect the methyl group had in altering the charge density of the molecule. What caused these findings to be especially exciting was the fact that these seven methyl nucleosides are believed to be involved in chemical mutagenesis of nucleic acid. Saethre *et al.*[298] have measured a group of narcotics and narcotic antagonists, in which they were able to correlate the nitrogen 1*s* binding energies with CNDO/2 calculations. To accomplish this, the crystal potentials were evaluated. Both calculations and experiment confirmed that the charge density on the nitrogen atom of the protonated species remains nearly invariant for a variety of species, in contrast to earlier speculation based on differences in pharmacological activity.

Many problems in clinical analysis can, in principle, also be solved by PESIS. A substantial program along these lines has in fact begun at a hospital in Gothenburg, Sweden.

6.3. Geology

Huntress and Wilson[299] have studied a number of lunar samples with x-ray photoelectron spectroscopy. It offered a rapid, nondestructive elemental analysis and in addition gave results on the oxidation of iron, which was found as Fe^{2+}. Comparison of fayalite and quartz show that the oxygen 1*s* binding energy is 0.5 eV smaller in fayalite.

Vinogradov *et al.*[300] studied specimens of lunar regolith. The PESIS spectra of O(1*s*), Si(2*p*), Al(2*p*), and Mg(2*p*) were much the same as the spectra of stony meteorites and of terrestrial materials, suggesting that Mg, Al, and Si are usually oxidized. From the aluminum 2*p* binding energy they concluded the coordination number is four and from the Si 2*p* line that there was an absence of extreme acid and alkali minerals in the regolith. The iron spectrum indicated a predominance of metallic iron in lunar regolith. There was also a small contribution from sulfur.

6.4. Environmental Studies

Hulett and Carlson[94] have measured chemical shifts in the binding energies of arsenic by PESIS in a number of compounds and in a soil sample inoculated with a commercial herbicide containing arsenic, in the form of cacodyllic acid, as its active ingredient. Using the soil sample directly as a sample, they were able to estimate the charge state of the arsenic *in situ*.

Another pollution study by Hulett *et al.*[301] has been concerned with the charge state of sulfur oxides adsorbed on the surface of fly ash and coal smoke particles. Both sample types showed a broad peak, indicating two species of higher oxidation state, probably sulfates, while the coal smoke particles showed a second peak, indicating a sulfide. Conventional chemical analysis showed the sulfur content of the fly ash to be only about 0.15%, but the strong signal for sulfur in the PESIS study indicated that this amount was located primarily on the surface.

PESIS has proved to be an ideal tool for examining particulate matter carried in the air. The material can be collected on filter paper giving a total sample cross section or with devices like a Lundgren cascade impactor, which spatially separates different size particles. A piece of the collected sample can

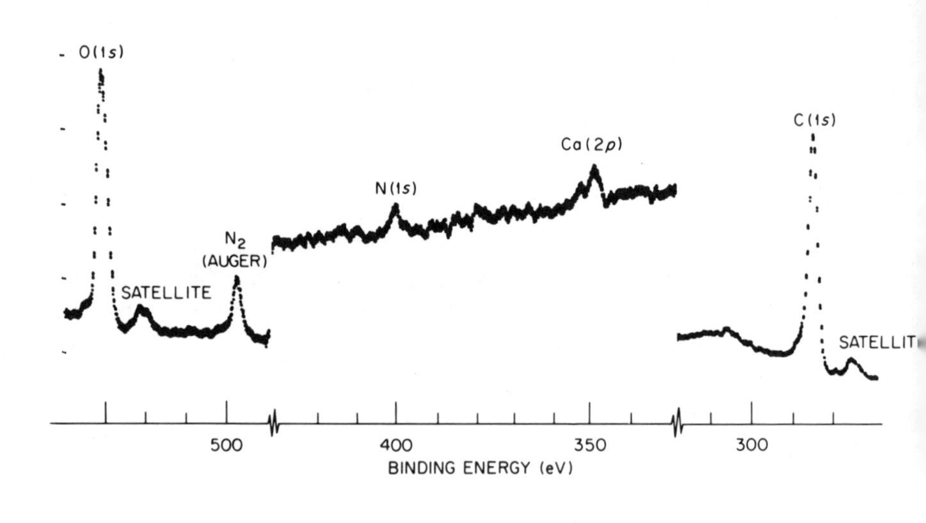

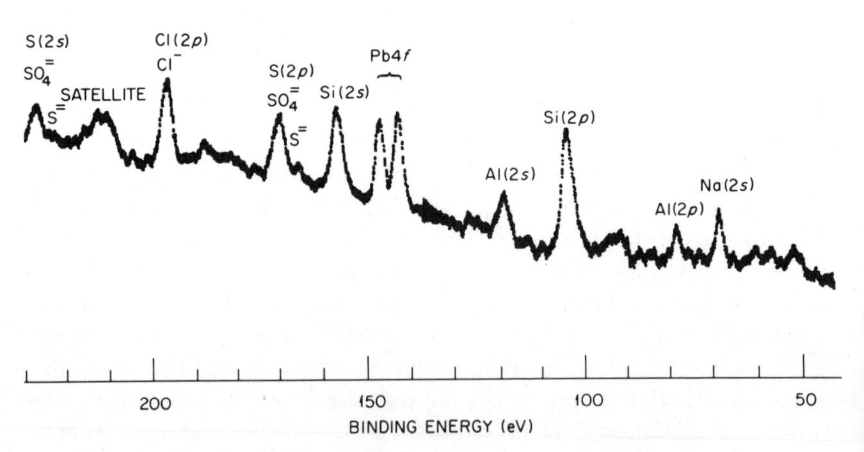

FIGURE 5.46. Portions of photoelectron spectrum of particulate matter collected in Knoxville, Tennessee. [Reproduced from Carter *et al.*[303]]

be measured directly by PESIS without further preparation. A total spectrum is taken with an analysis of all the elements and their oxidation states. See, for example, Figure 5.46. PESIS, of course, concentrates its attention on the surface. It is interesting to note the presence of lead in most of the samples thus far studied. Araktingi *et al.*[(302)] have made studies in the Baton Rouge, Louisiana, area, Novakov *et al.*[(89)] have studied aerosols in Pasadena, California, and Carter *et al.*[(303)] have investigated particulate matter from Knoxville, Tennessee. Novakov[(89)] found changes in the nitrogen spectrum during a 48-hr period. Of special interest was the behavior of N(1*s*) peaks assigned to peridino and amino groups which are known to be associated with gasoline additives. Studied as a function of particle size and time of day, their peak concentration occurs at early morning for small particles and at noon for larger particles, consistent with traffic conditions and a coagulation of smaller particles. Sulfur occurs in both the +4 and +6 states. The diurnal and size variations were found to be consistent with atmospheric oxidation that proceeds first by chemisorption of SO_2 followed by the oxidation to SO_4^{2-}. Novakov *et al.*[(304)] have suggested a strong relationship between the oxidation state of sulfur and the particle size of carbon particulates: the smaller the particulate, the more active the surface leading to sulfate. They have also shown that, in opposition to general belief, sulfate does not occur as NH_4SO_4 in particulate matter, but rather when ammonium ions are detected, they appear in equal molar concentration with nitrate. Carter *et al.*[(303)] found a lower oxidation of sulfur in the Knoxville environment, resulting either from the incomplete burning of soft coal or from a plant in the vicinity producing mercaptans.

A study of coal itself, particularly the oxidation states of sulfur, with a view to understanding its potential danger through pollution and possible means of removing that danger, has been the subject of several ESCA investigations.[(305)]

6.5. Surface Studies

For solid samples photoelectron spectroscopy provides a technique for surface studies. The interest in using XPS for surface studies has grown rapidly in the last few years. For a review of this subject, see Brundle.[(305a)] It is a tool that can be used in a number of areas. One is in the field of catalysis. Delgass *et al.*[(306)] have made a PESIS study of several commonly used catalysts. Brinen and Melera[(307)] used XPS for several rhodium catalysts and found that catalysts with high activity are characterized by a metal oxide ratio of less than unity. The interest in using PESIS for the study of catalysts and catalytic behavior has continued to grow.[(308)] In metallurgy, corrosion of metals occurs at the surface and a number of techniques have been developed to provide protective coatings. The major effects of various treatments, such as electroplating, electropolishing, anodizing, and acid pickling, occur at surfaces. The understanding of the bonding of adhesives is primarily concerned with the surface.

Kim *et al.*[146] have studied the formation of oxide film on Pt by adsorption and by electrochemical treatment using PESIS. Kim and Winogard[309] studied the adsorption of oxygen on lead, identifying two crystalline forms of PbO and a chemisorbed layer of atomic oxygen. Two groups[310, 311] have investigated the adsorption of oxygen onto the surface of nickel. Dianis and Lester[312] have studied the behavior of nitric oxide adsorbed on nickel oxide, cobalt oxide, and graphite as a function of temperature. Fischmeister and Olefjord[313] studied the surface of stainless steel. Hulett *et al.*[314] found that when a copper–nickel 70:30 alloy was passified in a sodium chloride solution

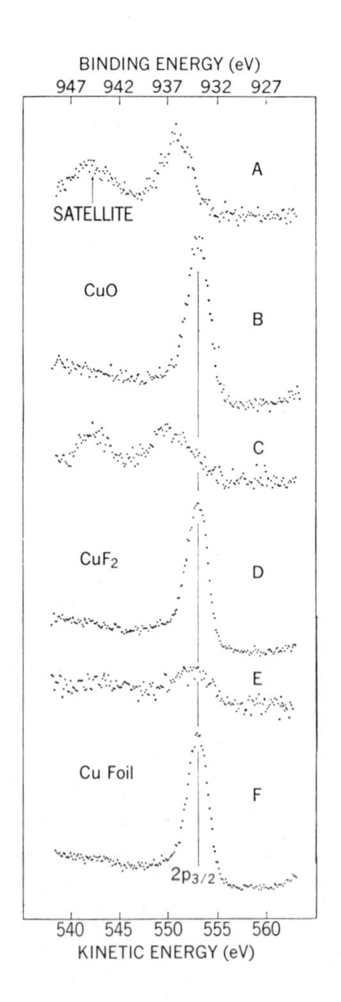

FIGURE 5.47. Dependence of x-ray photoelectron spectrum on sputtering. Spectra are from the Cu $2p_{3/2}$ level in CuO, CuF_2, and Cu foil. A, CuO prior to sputtering. B, CuO after 3 min of sputtering. C, CuF_2 prior to sputtering. D, CuF_2 after 3 min of sputtering. E, Cu foil prior to sputtering. F, Cu foil after 1.5 min of sputtering. The effect of sputtering is to clean off the surface and, in the case of CuO and CuF_2, to reduce the materials to their metallic form. [Reproduced from Yin *et al.*,[315] Figure 2.]

at anodic potentials above –0.3 eV, the surface was primarily nickel oxide with little or no copper.

Yin *et al.*[315] have studied the formation of metallic copper and iron on the surface of copper and iron compounds by means of reduction produced by ion sputtering (cf. Figure 5.47). Whan *et al.*[316] studied a catalyst prepared by condensing molybdenum hexacarbonyl onto alumina and found that the molybdenum in an activated disproportionation catalyst does not exist as the hexacarbonyl but rather in a higher oxidation form. Loss of activity through air oxidation shows molybdenum in a still higher oxidation state. Thomas *et al.*[317] reported on stress-annealed pyrolytic graphite, indicating two types of chemisorbed oxygen on the prismatic faces of the solid but no chemisorbed oxygen on the (001) face.

Yates *et al.*[318] have studied in detail the nature of adsorbed species on tungsten. A variety of chemical environments are detected through chemical shifts in the core binding energies and these are followed as a function of various adsorption temperatures. The data are interpreted in terms of the different types of adsorption sites in tungsten, whether the molecules are strongly or weakly chemisorbed, and whether they chemisorbed dissociatively. Brundle[305a] discussed the nature of chemisorption as revealed by electron spectroscopy, and pointed out that whereas chemical shifts can be large for the adsorbed molecules, they are generally rather small for the core electrons of the substrate.

6.6. Polymers and Alloys

Chemical information from PESIS is not restricted to simple molecular units, but is equally useful for less well-defined systems such as one finds with alloys and polymers. Studies on alloys have been mentioned previously in this chapter. Larson[320] has performed quantitative measurements on gold–silver alloys using x-ray photoelectron spectroscopy.

Fluoropolymers have received attention from both Ginnard and Riggs[321] and Clark *et al.*[102, 322, 323] One advantage of these studies is the thin films that can be prepared for study. With the homopolymer PTFE, Clark and Kilcost[102] showed that going from the monomer to the polymer widened the core levels only by 20%, and that the valence shell of the polymer was also relatively well resolved. From a study of copolymers[323] information regarding linkage was obtained. Some structural irregularities were noted for copolymers with $CF_2{=}CFCl$ and with $CF_2{=}CFH$. The reaction mechanism for polymerization was also discussed with help from PESIS and MO calculations. Ginnard and Riggs[321] were able to correlate the carbon 1*s* binding energies of the polymers with a method based on Pauling's electronegativities, modified according to Riggs.

6.7. Radiation Studies

Chemical changes due to x radiation can often be an unwanted problem in PESIS. However, PESIS can be a most useful tool for studying radiation effects. Cheng *et al.*[324] have reported the effects of radiation in EDTA, and Klein[325] has alluded to similar studies in biological compounds. Brinen and Wilson[326] have studied radiation damage in synthetic sodalites. They found that permanent coloration produced by electron bombardment results in a substantial decrease of surface halogen and sodium. Aita *et al.*[327] have evaluated the effects of neutron irradiation on copper oxide and Kim *et al.*[328] have studied the alteration of the surfaces by ion bombardment.

6.8. Industrial Uses

Brinen[329] has given a fine summary of applications of x-ray photoelectron spectroscopy to industrial chemistry. In particular, there should be a substantial use for PESIS in industrial control. As an example, corona discharge and low-temperature plasma treatments of wool yarn and fabric have been found to improve many properties of the yarn, e.g., reduce shrinkage and increase tensile strength. Millard[330] has made a PESIS study of the binding energies of sulfur, nitrogen, carbon, and oxygen atoms in treated and untreated wool. Only the sulfur line ($2p$) showed much change. By comparison with samples with known sulfur oxidation states, it appears that wool surface sulfur atoms have become oxidized to the +6 state. Similar effects have been observed in artificial fibers. Thus a useful technique for spot testing an industrial process has been found.

The foregoing examples are only a few cases of how PESIS has been put to work on applied problems. Actually, there are many more. At least several hundred electron spectrometers have been purchased for industrial and application uses. Unfortunately, most work on applied research does not reach the open literature, but as it does begin to trickle out, one may expect to see an added growth of interest in electron spectroscopy.

Chapter 6

Auger Electron Spectroscopy

When a vacancy is formed in one of the inner shells of an atom, it may be filled by either a radiative (x ray) or nonradiative (Auger) process. In most instances nature chooses the Auger process. Only when the transition energy exceeds roughly 10 keV is x-ray emission predominant. Thus, the fluorescence yield is greater than 0.5 only for vacancies in the K shells of atoms whose $Z > 30$, and in a few cases in the L shell for very heavy elements. (The fluorescence yield ω is defined as equal to the number of times a vacancy in a given shell is filled by a radiative process divided by the total number of times that hole is filled.) Why, then, are x rays well known even to the man in the street, while the understanding of the Auger processes has been until recently restricted to the specialist? The answer lies simply in the ease of measurement. X rays are a highly penetrating radiation, while Auger electrons have only a small mean free path in solids.

The Auger process was discovered by Pierre Auger[1] (pronounced *oh jay'*) using a Wilson cloud chamber. Tracks corresponding to ejected electrons could be seen along a beam of x rays. An Auger event could be detected by the presence of a double track emanating from a single source. One track was due to photoelectron ejection, while the second was due to the ejected Auger electron. The Auger effect has been of interest to the x-ray spectroscopist because of its profound influence in altering the fluorescence yield, or in broadening the natural width of the x-ray lines by shortening the lifetime of the vacancy state. For a review of x-ray fluorescence yields, together with Auger and Coster–Kronig transition probabilities, see Bambynek *et al.*[2] Nuclear spectroscopists have also needed to study the Auger process in order to better account for their data on electron capture and internal conversion in which inner shell vacancies are being promoted. If, for example, the probability for K capture for a given element is being monitored by the number of x rays observed, then the fluorescence yield needs to be known. In the study of internally converted electrons, beta spectroscopists also measure the discrete-energy Auger electrons.

For the most part, the early studies of the Auger process were studies of a nuisance effect. One needed to know about it in order to correct one's data, but the effect was not of too much intrinsic value. The Auger effect did shed light on atomic processes, such as the type of coupling (LS, j–j, or intermediate) throughout the periodic table; but, in general, relatively little constructive use was made of the phenomenon. In the last few years this situation has changed drastically. Two media in which the collision cross section does not prevent the detection of even low-energy Auger electrons are gases and surfaces. The study of Auger electrons under these conditions has proven to be most profitable.

Studies of Auger electrons in gases have been useful in learning about the basic phenomenon of ionization and about the nature of molecular ions, while in the study of surfaces, Auger spectroscopy has been invaluable as an analytical tool. In this chapter we shall discuss in detail some of the more recent applications of Auger spectroscopy in both gases and solids. But before doing so, we must pass on to a more thorough discussion of the phenomenon itself.

1. THEORY OF THE AUGER PROCESS

The first book to discuss the basic phenomenon of the Auger effect was by Burhop.[3] Other excellent reviews are by Mehlhorn,[4] Sevier,[5] Parilis,[6] and Burhop and Asaad.[7]

A nonradiative readjustment to an inner shell may take place by having one electron from a less tightly bound orbital fill the hole, while a second electron is ejected into the continuum with an energy equal to the difference in total energies of the initial and final states. For example, take the K–L_IL_{II} Auger transition. The initial state has a hole in the K shell, and the final state has two vacancies in the L shell. Symbolically, we represent the Auger process by the naming of the orbitals or shells in which vacancies occur, both in the initial state (in our example, the K shell) and the final state (the L_I and L_{II} subshells). Energetically, this is equivalent to producing a virtual K–L_{II} x ray that photoejects an electron from the L_I shell. However, this is not the way an Auger process occurs. It is rather a two-electron Coulombic readjustment to the initial hole. This is shown by the presence of the $1s$–$2s2s$ Auger process. A $1s$–$2s$ radiation process is not allowed in the dipole approximation.

The energy of an Auger electron E_A can be estimated from the binding energies, e.g., for a K–L_IL_{II} Auger process

$$E_A = E_K - (E_{L_I} + E'_{L_{II}}) \tag{6.1}$$

where E_K and E_{L_I} are the binding energies of the K and L_I shells of a neutral atom and $E_{L_{II}}'$ is the binding energy of an electron in the L_{II} shell of an ion having a single vacancy in the L shell. The binding energy for a given shell in an ion having a single-hole configuration is slightly larger than that for its atomic counterpart. Bergström and Hill[8] suggested $E_L' = E_L(Z + \Delta Z)$, where $E_L(Z + \Delta Z)$ is the binding energy of an electron in shell L lying between those for atoms with atomic numbers Z and $Z + 1$, where $0 < \Delta Z < 1$. For example, E_L' for Z around 80 has $\Delta Z = 0.55$. Atomic binding energies are available both experimentally and theoretically, so that equation (6.1) may be used to obtain an estimate of the Auger transition energy. Coghlan and Clausing[9] have published a complete listing of Auger energies for elements up to $Z = 92$, using the approximation

$$E_A = E_X - \{\tfrac{1}{2}[E_X(Z) + E_X(Z + 1)] + \tfrac{1}{2}[E_Y(Z) + E_Y(Z + 1)]\} \qquad (6.2)$$

Packer and Wilson[10] also have a comprehensive listing, but based on a less reliable approximation. Shirley[11] has considerably improved the estimates of Auger energies by consideration of relaxation effects, and Nicolaides and Beck[12] have examined the requirements for an "exact" calculation of *KLL* Auger energies. Comprehensive listings of experimental Auger energies are given by Siegbahn *et al.*,[13] in a chart put out by Varian,[14] and in a collection of spectra measured by Palmberg *et al.*[15]

A more detailed treatment of the Auger energies requires knowledge of a coupling in the final state which occurs between the two unfilled shells. For light elements the coupling scheme is pure L–S, for heavy atoms j–j, and for elements in the middle of the periodic table intermediate coupling needs to be invoked. Figure 6.1 shows the Auger lines that can arise from K–LL Auger transitions. For the light elements there are five lines (the $1s$–$1s2p\ ^3P$ transition is forbidden according to the L–S coupling scheme). Pure j–j coupling yields six lines, while in the intermediate region as many as nine lines are allowed and can be discerned.

Let us next turn our attention to the calculations of the Auger transition probability, which is given as

$$P_{i\to f} = \frac{2\pi}{\hbar}\left| \iint \chi_f^*(\mathbf{r}_1)\psi_f^*(\mathbf{r}_2)\, \frac{e^2}{|\mathbf{r}_1 - \mathbf{r}_2|}\, \chi_i(\mathbf{r}_1)\psi_i(\mathbf{r}_2)\, d\mathbf{r}_1\, d\mathbf{r}_2 \right|^2 \qquad (6.3)$$

The two electrons involved are initially in bound orbitals represented by χ_i and ψ_i. Following the nonradiative Coulombic readjustment, the final states of the electron are represented by a bound orbital χ_f^* (the filled vacancy) and the continuum wave function ψ_f^*, which is normalized to represent one ejected electron per unit time per unit energy range. When account is taken of the fact that the initial and final wave functions must be antisymmetric in their coordinates, one has expressions due to direct ($\chi_i \to \chi_f$; $\psi_i \to \psi_f$) and exchange

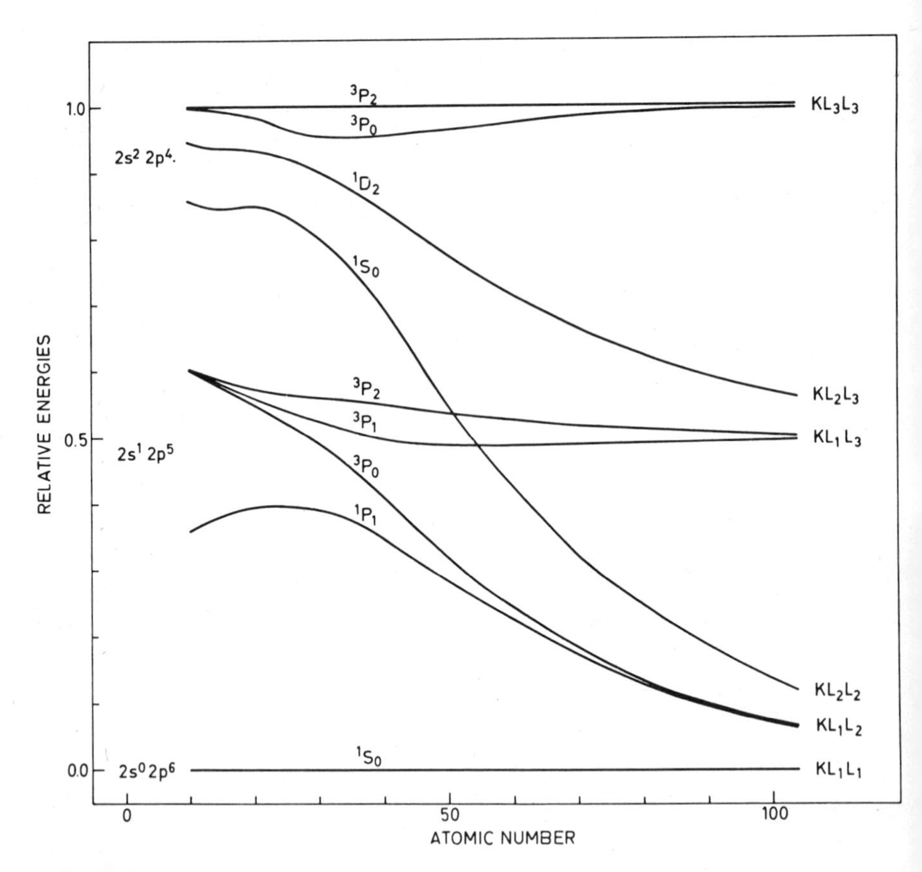

FIGURE 6.1. Relative line positions in *K–LL* Auger transitions as a function of *Z*. At low *Z* one has nearly pure *L–S* coupling and six lines. However, the $2s^2 2p^4\ {}^3P$ is strongly forbidden in the nonrelativistic region. At high *Z* one has nearly pure *jj* coupling and six lines; nine lines are possible in the intermediate coupling region. [Reproduced from Siegbahn *et al.*,[13] Figure 4.1.]

($\chi_i \rightarrow \psi_f$; $\psi_i \rightarrow \chi_f$) transitions. Another theoretical approach to Auger transition has been given by Haymann,[16] in which a dynamical point of view is introduced. A method of wave packets inside a material is presented, and calculations made for the case of surface plasmons interacting with the Auger electrons. Chase *et al.*[17] have carried out a multibody calculation of the Auger process.

Rather comprehensive calculations on Auger transition rates have been made.[18] Comparison with experimental data is generally satisfactory. In some cases, particularly with transitions involving the outer shells, the results are very sensitive to the choice of wave functions, and consideration of configuration interaction and electron correlation appears to be necessary. For example, Chen and Crasemann[19] have carried out calculations on the *KLL* Auger spectra of elements from $Z = 13$ to 47 using configuration interaction, which substantially improves agreement between experiment and theory.

The selection rules governing Auger transition are that transitions are possible only if the initial and final states have the same symmetries, *L*, *S*, *J*, and parity. Thus, $\Delta L = \Delta S = \Delta J = 0$; $\Pi_i = \Pi_f$. Any three subshells (defined by the principal and angular momentum quantum numbers) can be involved in an Auger process, so long as the difference in total energies between the initial and final states indicates the transition is energetically possible. Table 6.1 lists the relative Auger transition rates for Kr as obtained by experiment and theory. Note the importance of Coster–Kronig transitions.*

We have spoken of the Auger process as being an isolated event involving two and only two electrons. But just as electron shakeoff gives rise to multiple electron ejection in photoionization, this same condition may give rise to two or more electrons being ejected in an Auger process (sometimes called a "double Auger process"). Wolfsberg and Perlman[20] were the first to point out that if an Auger electron is ejected from one of the inner shells, an outer shell will experience a sudden change in effective charge and electron shakeoff will occur. The sudden approximation can be invoked to calculate the probability for electron shakeoff (cf. Chapter 3, Section 3.5).

Multiple ionization has also been investigated for Auger processes involving double electron ejection from the valence shell. These studies have been carried out by both charge spectroscopy[21] and electron spectroscopy.[22] The net change in the effective charge under such circumstances is very small (~0.15), yet the probability for double electron ejection runs from 10 to 30%. Just as with photoionization (cf. Chapter 5, Section 5.1.2.), removal or excitation of more than one electron from the valence shell involves electron correlation, and cannot be explained simply by a net change in electron shielding. A more recent study of this problem has been carried out by Mehlhorn *et al.*[23]

Besides the normal Auger process, there also exists what has been called the radiative Auger process or semi-Auger process. In this case there is a simultaneous filling of an inner shell hole and excitation of a second electron with the emission of a single photon. This has been noted[24] for *K*-shell x rays in Al, Si, S, and Ar, involving two-electron processes in the $L_{\text{II, III}}$ shell. Cooper and

* A Coster–Kronig transition is an Auger transition in which the initial vacancy and one of the electrons that fills this vacancy are in a shell with the same principal quantum number. For example, L_{I}–$L_{\text{II}}M$. Because of the large overlap of the wave functions, Coster–Kronig transitions are more than an order of magnitude larger than normal Auger processes. Whether a Coster–Kronig transition will occur depends on whether the differences in the subshell binding energies are sufficient to eject an electron from an orbital in the next higher shell. For example, the above transition will take place only for elements whose *Z* is less than 40.

TABLE 6.1

Comparison of Relative Auger Rates in Kr

Transitions	Theory[18]	Experiment
1*s* 2*s* 2*s*	10.3	7.9[a]
1*s* 2*s* 2*p*	40.0	33.3[a]
1*s* 2*p* 3*p*	100.0	100.0[a]
1*s* 2*s* 3*s*	2.9	3.1[a]
1*s* 2*s* 2*p*	5.3	5.5[a]
1*s* 2*s* 3*d* 1*s* 2*p* 3*s*	5.2	6.3[a]
1*s* 2*p* 3*p*	26.1	24.7[a]
1*s* 2*p* 3*d*	3.1	2.6[a]
1*s* *M* *M*	2.9	—
2*s* 2*p* 3*p*	23.6	—
2*s* 2*p* 3*d*	100.0	—
2*s* 3*s* 3*s*	0.4	—
2*s* 3*s* 3*p*	2.6	—
2*s* 3*s* 3*d*	3.1	—
2*s* 3*p* 3*p*	0.1	—
2*s* 3*p* 3*d*	1.1	—
2*s* 3*d* 3*d*	3.6	—
2*p* 3*s* 3*s*	0.7	—
2*p* 3*s* 3*p*	9.8	—
2*p* 3*s* 3*s*	1.9	—
2*p* 3*p* 3*p*	40.9	22[b]
2*p* 3*p* 3*d*	68.2	46[b]
2*p* 3*d* 3*d*	100.0	100.0[b]
3*s* 3*p* 4*p*	100.0	—
3*s* 3*d* 3*d*	5.8	—
3*s* 3*d* 4*s*	11.3	—
3*s* 3*d* 4*p*	2.9	—
3*s* 3*d* 4*d*	0.7	—
3*p* 3*d* 3*d*	106.5	
3*p* 3*d* 4*s*	21.4	7[40]
3*p* 3*d* 4*p*	100.0	100.0[40]
3*p* 4*s* 4*p*	0.9	—
3*p* 4*p* 4*p*	2.9	—
3*d* 4*s* 4*s*	11.5	
3*d* 4*s* 4*p*	100.0	100.0[42]
3*d* 4*p* 4*p*	59.4	89.2[42]

[a] P. Erman, I. Bergstrom, Y. Y. Chu, and G. T. Emergy, *Nucl. Phys.* **62**, 401 (1965).
[b] M. O. Krause, *Phys. Lett.* **19**, 14 (1965); M. O. Krause and T. A. Carlson, *Phys. Rev.* **158**, 18 (1967).

La Villa[25] have studied a similar process in potassium chloride, involving a $L_{II,III}$ photon emission and excitation in the valence shell. Here the probability as observed in an x-ray spectrum is as high as 10% of the parent line. Since these processes arise from the emission of a photon and not an electron, they will be experimentally observed in x-ray emission spectra, not Auger spectra.

Finally, we have not discussed the fate of the ion which follows the Auger process. If the Auger process has involved only core electrons, there remain inner shell vacancies that can also be filled by additional Auger processes, creating more ionization. Since each Auger process produces a new vacancy, a series of such processes is called a vacancy cascade. An initial vacancy in a heavy atom will result in a large net charge because of the repetitive Auger processes. For example, Figure 6.2 shows a typical vacancy cascade in Xe. Experimentally, the average charge for xenon following a K vacancy is +8, while ions with charge as high as +22 have been measured[27] as the result of a single initial vacancy. Comprehensive studies[28] of the net charge resulting

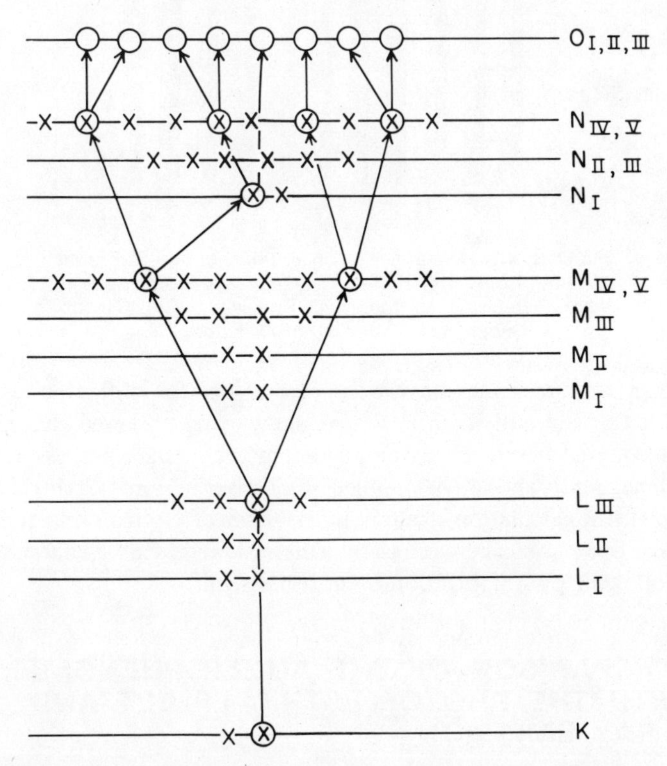

FIGURE 6.2. Schematic representation of a vacancy cascade in Xe. ×, Electrons; ○, vacancies; ⊗, vacancies that were subsequently filled by electrons.

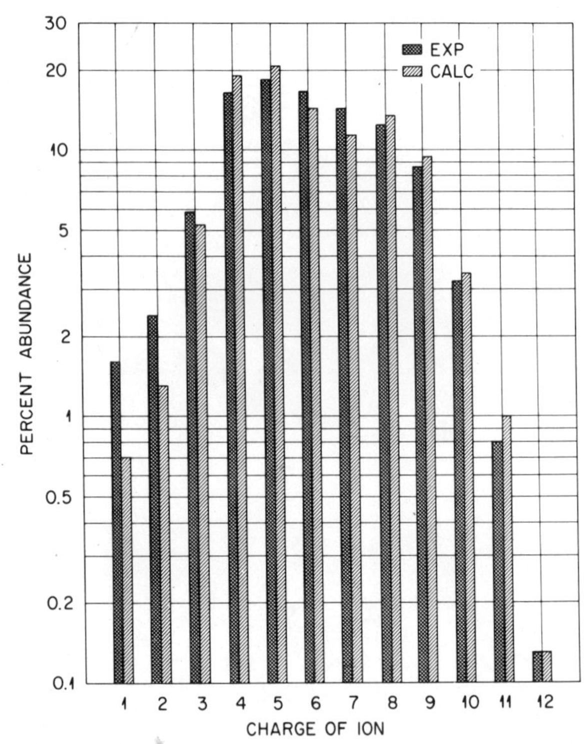

FIGURE 6.3. Experimental charge distribution of Kr ions resulting from irradiation by Mo $K\alpha$ x rays compared with theory. [Reproduced from M. O. Krause and T. A. Carlson, *Phys. Rev.* **158**, 22 (1967), Figure 2.]

from inner shell vacancies have been carried out for both free atoms and molecules. Carlson and Krause[29] have shown that observed charge spectra can be accounted for on the assumption that each Auger process acts essentially independently and can be summed by means of a Monte Carlo calculation. To complete the calculation, account is also taken of electron shakeoff. Figure 6.3 shows a comparison between the calculated and measured charge spectrum for Kr following photoionization in the inner shells.

2. COMPARISON OF THE AUGER PHENOMENON WITH THE PHOTOELECTRIC EFFECT AND X-RAY EMISSION

In this section we shall compare the closely related phenomena of the Auger process, photoionization, and x-ray emission. We shall bring together

Photoionization (one step)

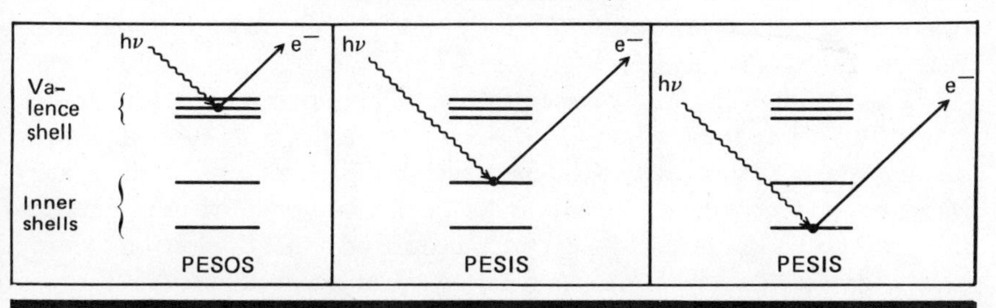

Two-step processes

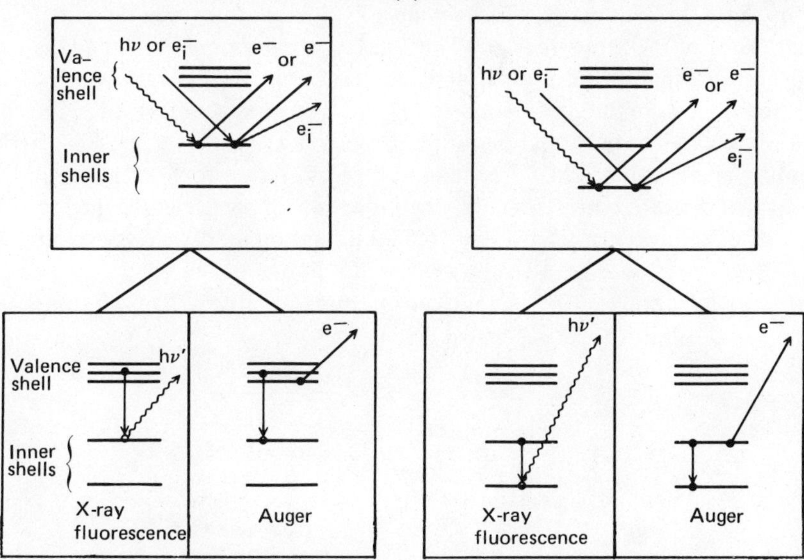

FIGURE 6.4. Basic processes involved in photoelectron, x-ray, and Auger spectroscopy. Comparison of photoelectron, x ray, and Auger spectroscopy distinguishes between one-step (above) and two-step (below) processes. Photoionization is a one-step process; for both outer shell (PESOS) and inner shell (PESIS) emission we have a simple relation between the measured energy of the emitted electron, the photon energy, and the binding energy in the affected shell. From PESOS we determine valence binding energies of the various molecular orbitals. In PESIS we determine core-electron (atomic) binding energies and note how they vary with chemical environment (chemical shift). X-ray fluorescence and Auger spectroscopy are two-step processes. A photon or an electron ejects an electron from an inner shell. (If we use electrons rather than photons, the incident electron e_i is also emitted.) In the second step an electron drops down to fill the hole and either an x ray or a second electron (the Auger electron) is emitted. For x-ray or Auger processes in which an electron drops from a valence shell, the chemical shift tells us about binding energies in both shells. Where the transition is between two inner shells, x-ray spectra are not very useful because the binding energies shift together. Auger spectra should, in principle, be useful for studying inner shell transitions, because the net effect is ejection of an inner shell electron. But complex spectra and short lifetimes limit the usefulness of this type of Auger spectroscopy. [Reproduced from T. A. Carlson, *Physics Today* **25**, 30 (1972), Figure 1.]

material that was touched on before, as well as some new ideas. The task will be completed when we discuss the merits of Auger spectroscopy for studying gases and solids (Sections 3 and 4).

Figure 6.4 illustrates the basic processes. Photoionization is a single-step process involving the ejection of an electron by a photon. Auger and x-ray phenomena are essentially two-step processes. First, an inner shell vacancy is formed, and second, that vacancy is filled either by a radiative or a nonradiative process. The inner shell vacancy can be formed either by electron or x-ray bombardment. If an electron beam is employed, only a portion of the energy of the impact electron is used in the ionization. Both the impact and ejected electrons share the net energy, which equals the difference between the initial and final states of the atom and ion. Thus, unless a coincidence technique is employed, information on atomic and molecular binding energies will not be forthcoming from the continuous-energy electron spectrum formed in the collision process. The cross section for photoejection is usually largest close to the binding energy threshold. The maximum cross section for ionization due to electron impact generally occurs when electron impact energies are several times the binding energy (cf. Figure 6.5 for a typical plot of an electron-impact cross section).

The advantages for using electron impact rather than photons are

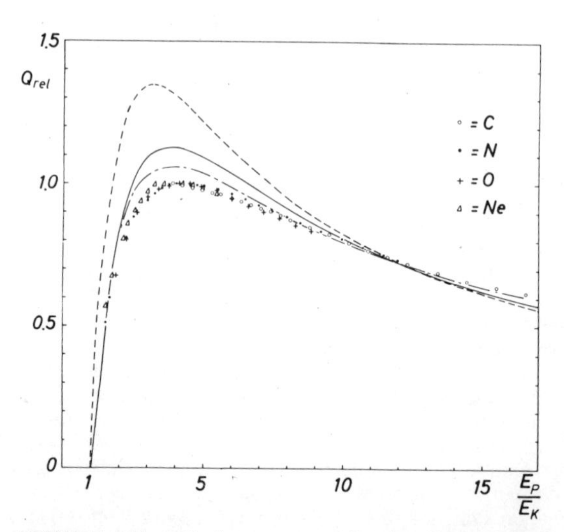

FIGURE 6.5. Relative cross section for K ionization in neon, carbon, nitrogen, and oxygen as function of E_P/E_K, which is the ratio of energy of input electron to K-binding energy. Lines are theoretical predictions. Points are experiment. [Reproduced from Glupe and Mehlhorn,[56] Figure 1.]

(1) an electron beam can be obtained with many order of magnitude greater intensity than is possible for an x-ray source, and (2) a well-defined focal spot can be achieved with an electron beam, of inestimable value in surface topography. Although, in principle, soft x rays can be focused, it is impractical with present technology to produce a beam of sharp focus with reasonable intensity. Soft x rays, however, have their advantages for producing inner shell vacancies. In many applications, the intensity of the x-ray beam is more than sufficient. The production of secondary electrons is less, and the peak-to-background ratio is superior to that for electron impact. Less radiation damage is incurred, and the problem of carbon baking is not present. (If a vacuum system is not clean of hydrocarbon vapor from pump oil and *O*-ring seals, electron bombardment of a surface containing a hydrocarbon film will break down the hydrocarbon and soon a carbon layer will form.) Under less severe radiation the hydrocarbon film will equilibrate on the surface to an amount that is tolerable under many circumstances.

Protons, α-particles, and heavy ions have also been used for promoting inner shell ionization. The chief advantage for using positively charged nuclei rather than electrons occurs in the domain of x-ray emission spectroscopy. Ions do not create the bremsstrahlung background that electrons do. Needham *et al.*[30] have used protons in connection with Auger emission to ensure that only surface atoms are excited.

Each of the three methods (photoelectron spectroscopy, Auger spectroscopy, and fluorescence spectroscopy) yields information on binding energies. Photoelectron spectroscopy gives the simplest and most direct information that relates to the binding energy of a single orbital. X-ray emission depends on the binding energy of two shells, while Auger data are the most complex to interpret, depending on the behavior of three shells. For the last two methods the bulk of the transition energy is characterized by an inner shell (the detailed spectrum reflects the nature of the singly and doubly charged final state); thus with all three methods one can determine the binding energy of an inner shell, which in turn affords elemental analysis. This arises because binding energies for the inner shells differ by large amounts, and in most cases are unique to the different elements. The spectra thus obtained by all three of the methods (AES, PES, and x ray) consist of widely separated peaks that provide an easy method of elemental analysis.

The information contained in electron and x-ray spectra concerns the nature of the chemical environment as well as the identification of the element. In comparing the three methods for the sorts of chemical information that can be extracted, we must first recall the differences between the valence and inner shells. The binding energies of the valence shell orbitals are characteristic of the molecule or solid as a whole. With core electrons one is dealing principally with atomic orbitals, which feel changes in chemistry only as a perturbation in the valence shell potential. Binding energies of the inner shells all change in unison with alterations of the chemical environment. (Cf. Chapter 5, Section 2.)

Let us now turn to each of the three spectroscopic methods to see what chemical information we may gain (cf. Figure 6.4). Photoionization again gives the most direct information by telling us separately either about the valence shell or about the core electron. For Auger or x-ray emissions there are basically two different types of situations. In the first case we have a transition between an inner and an outer shell. The binding energy of the inner shell gives evidence of what element is being investigated. The detailed spectrum depends on the final state of the molecular ion: In the case of x rays it is the singly charged ion, and in the case of Auger processes, the doubly charged ion. The energy spacings in an x-ray spectrum of a given molecule should resemble those found in the direct photoionization spectrum of the outer shell. Which of the states are reached and the intensity of the x-ray peaks corresponding to the states depend on the x-ray transition rates. Until recently, the intensities and resolution of x-ray spectra have not been sufficiently high to make many unambiguous identifications, but technology has advanced to the point[31] that molecular orbitals can now be clearly identified in x-ray spectra. The study of doubly charged molecular ions by Auger spectra will be discussed in Section 3.

The second situation that can occur for x ray and Auger transitions is to have the transitions occur between two inner shells. Since the chemical shifts for all the core electrons are essentially identical, x-ray energies should be essentially independent of the chemical environment if the radiative transitions occur inside the valence shell. This indeed has been found to be the case. Small shifts are sometimes noted, however. These are due to a breakdown in the assumption that changes in chemical bonding will be felt by the core electrons only in terms of alterations in the outer shell potential. Changes in shielding also occur, as well as changes in the relaxation energies. These are usually minor effects, but can be estimated by observing x-ray spectra. (They can also be estimated by observing the *difference* in chemical shifts between two core shells studied by photoelectron spectroscopy.)

When only core electrons are involved in an Auger process, the net result is the ejection of an inner shell electron. Thus, ejected electrons will feel the potential of the valence shell just as in the case of photoionization, and the chemical shifts observed in Auger spectroscopy of inner shells should be similar to those seen with PESIS. This was first demonstrated by Fahlman *et al.*[32] for the K–$L_{\mathrm{II,\,III}}L_{\mathrm{II,\,III}}$ transitions in sulfur, using $Na_2S_2O_3$. The chemical shift observed in the Auger spectrum between the +6 and −2 oxidation states was 4.7 eV (cf. Figure 6.6). Data from PESIS gave 7.0 eV as the chemical shift between the K shells of the two sulfurs of the $Na_2S_2O_3$ and 6.0 eV for the $2p$ shells.

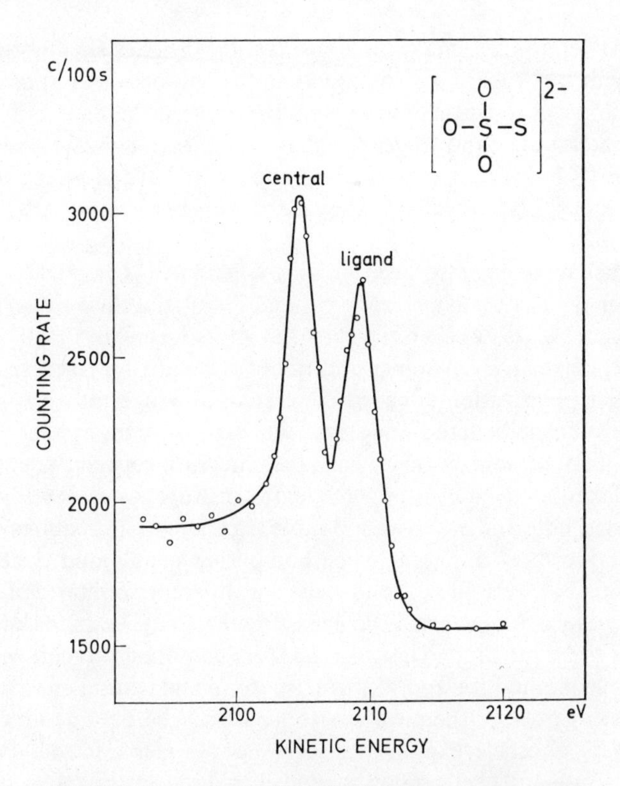

FIGURE 6.6. The *K–LL* [$2s^2 2p^4$ ($^1D_2$)] Auger spectrum of sulfur for $Na_2S_2O_3$, showing two peaks corresponding to differences in chemical environment. [Reproduced from Fahlman *et al.*(32)]

To a good approximation the chemical shift ΔE_A in the Auger transition $K\text{–}L_{II,\,III}L_{II,\,III}$ is

$$\Delta E_A = \Delta E_I - 2\,\Delta E(L_{II,\,III}) \qquad (6.4)$$

Using the data of PESIS, ΔE_A is determined from equation (6.4) to be 5.0 eV, in reasonable agreement with the Auger results. As another example, the *KLL* Auger spectrum of SiH_4 and SiF_4 were measured.[33] The observed shift was 5.5 eV, in reasonable agreement with the chemical shifts observed in the photoelectron spectra of the 2*p* shells of corresponding silicon compounds. The study of chemical shifts of core electrons by Auger spectroscopy is hampered by the fact that one has to produce a vacancy or principal shell deeper than would be the case of photoionization, with a subsequent broadening in the Auger spectrum due to the larger natural widths of the more deeply lying atomic levels. This broadening, coupled with the more complex Auger spectra, makes the observation of inner shell chemical shifts more difficult for Auger spectroscopy than for photoelectron spectroscopy.

The simple correlation described above between chemical shifts in photoelectron spectroscopy and Auger spectroscopy does not always occur. Wagner and Biloen[34] found, when comparing chemical shifts between some metals and their oxides, that the shifts in the case of Auger processes were much larger (cf. Table 6.2).

This behavior has also been noted by Schön.[35] Chemical shifts found from Auger spectra for Na,[36] Mg,[37] and Al[38] and their respective oxides were found to be greater than the chemical shifts seen with PESIS by 5.0, 4.1, and 4.2 eV, respectively. Wagner and Biloen have attributed these differences to the higher polarization afforded the case of an Auger process where there are two core vacancies in the final states, which is to say the extraatomic relaxation effect. In photoionization we have the difference between a neutral species and the relaxation given by a single hole. In the Auger process we begin with a single relaxed hole and end with a double vacancy which requires relaxation. The effects are most apparent when comparing metals and dielectrics, but slight differences were also found between different sodium and zinc compounds.[36] Since Auger and XPS measurements can be made on the same material at the same time, relative shifts from compound to compound might be a valuable method for appraising extraatomic relaxation effects.

Knowledge of the chemical environment can be obtained most directly and easily by photoelectron spectroscopy, but Auger electron spectroscopy and x-ray emission spectroscopy may also be used in the study of chemical bonding. Information from all three methods supplements one another.

For our final comparison of the three spectroscopies, let us turn to the subject of surfaces. The surface depth that is studied depends on penetration

TABLE 6.2

Chemical Shifts between Element and Oxide for Selected Auger and Photoelectron Lines[a]

	Photoline shifts, eV			Auger line shifts, eV	
Element	$2p_{3/2}$	$3d_{5/2}$	$4d$	LMM	MNN
Zn	0.4	0.6	—	4.2	—
Ga	1.7	2.2	—	6.2	—
Ge	3.0	3.3	—	6.7	—
As	2.3	3.6	—	6.4	—
Cd	—	0.4	0.9	—	5.5
In	—	0.8	0.9	—	2.6
Sn	—	1.5	1.2	—	3.9

[a] Taken from Wagner,[34] Table 3.

TABLE 6.3

Comparison of Average Auger and Photoelectron Energies for the First Row Elements

Element	K–LL Auger,[a] eV	E_e,[b] eV
Li	45	1199
Be	100	1143
B	176	1066
C	266	970
N	375	855
O	507	732
F	654	568
Ne	813	387

[a] Postion of most intense peak. C to Ne from Siegbahn *et al.*[42]; Li to B from Palmberg *et al.*[15]
[b] Photoelectron energy using Mg $K\alpha$ x rays.

of the excitation source and the probability for escape of the electron or x ray being measured. From Figure 5.42 it is evident that the mean free path for photons is always several orders of magnitude larger than for electrons of the same energy. We may thus conclude that the escape depth is determined entirely by kinetic energy of the electrons involved. Table 6.3 shows the photoelectron and Auger energies expected for the first row elements. Comprehensive tables[9] and graphs[15] of the approximate photoelectron energies and Auger energies normally encountered in electron spectroscopy are available. For heavier elements the Auger and photoelectron energies are more varied. In both instances the energies will vary from about 100 to 1500 eV. When comparing Auger electrons with photoelectrons formed by irradiating with Mg $K\alpha$ x rays, the Auger electrons are usually found to be lower in energy, but sometimes the reverse is true. In addition, lower photon energies can be employed, such as Y and Zr $M\zeta$ lines (132.3 and 151.4 eV), which will produce correspondingly lower photoelectron energies. Thus, Auger and photoelectron spectroscopy examine essentially the same region of surface. This is certainly true if x rays are employed as the initial excitation source. If electron impact is employed, the depth for forming the initial vacancy is reduced. But in producing the initial vacancy, the energy of the impact electron is generally chosen to be at least several times that of the subsequent Auger electron. Thus, it is still the mean free paths of the escaping Auger electrons that essentially determine the depth of the surface layer measured. When x rays produced by fluorescence are measured, the depth of the surface being studied increases substantially. If photoionization is employed as the initial excitation source, the measurements are made well into the bulk material. If electron impact is used, we must re-

TABLE 6.4
Approximate Surface Depth Studied as a Function of Analysis[a]

Excitation	Emission	Surface thickness, Å
e^-	Auger electron	20
X ray	Photoelectron	20
X ray	Auger electron	20
e^-	X ray	600[b]
X ray	X ray	40,000

[a] Based on mean free path of electrons and x rays in Al, assuming 1-keV ejected electron or x ray, 1.5-keV impact x ray, and 10-keV impact electron.
[b] Mean free path for core ionization taken as five times larger than plasmon loss.

member that one is concerned with the distance an electron can travel while still maintaining sufficient energy for inner shell ionization. This is considerably larger than just the mean free path for electrons undergoing inelastic scattering, which is of importance if we wish to detect a discrete energy electron. It has been estimated,[(39)] in fact, to be approximately five times the mean free path. Table 6.4 summarizes the above discussion by giving a rough idea of the surface depth measured by the various methods. Considerable variations are possible (cf. Section 3 for further discussion of surface problems).

3. USE OF AUGER SPECTROSCOPY FOR GASES

The high resolution possible for measuring electrons ejected from a gas target and the simplifying conditions encountered when dealing with individual molecules or ions have made the study of the Auger spectra of gases a fruitful area for fundamental research. Details of the Auger process itself, the nature of the ground and excited states of the doubly charged ions, and the various phenomena associated with the initial excitation processes are topics that will be discussed in this section. In addition, we shall try to evaluate the potential for gas analysis by Auger spectroscopy.

3.1. Atoms

Mehlhorn[(40, 41)] was the first to demonstrate the advantages of studying the Auger spectra of gaseous atoms with high resolution. These studies gave fine details against which the concepts of the Auger process could be tested. Subsequently, extensive data on the rare gases were obtained by the group at

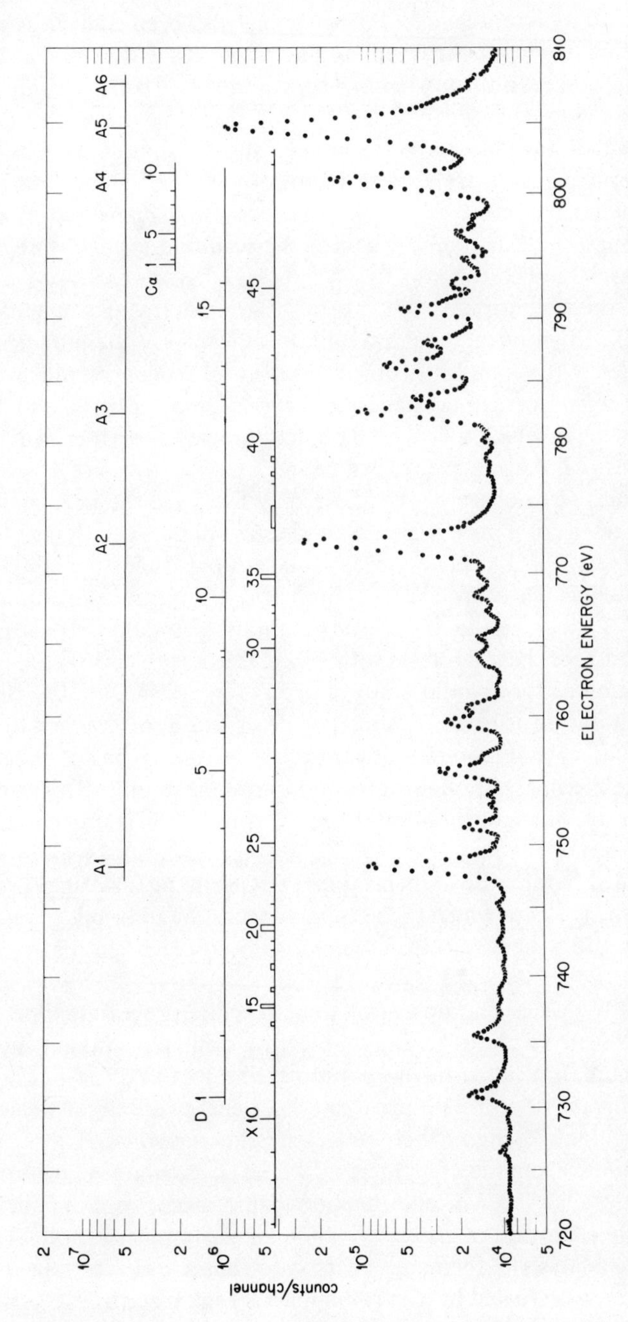

FIGURE 6.7. Neon K Auger spectrum excited by 4.5-keV electrons. The A lines are diagram Auger lines; the $C\alpha$ lines are shakeup processes, and the D lines are from electron shakeoff. [Reproduced from Krause *et al.*,[44] Figure 1.]

Uppsala[42,43] and by Krause *et al.*[44] Sodium has also been studied by Hillig *et al.*[45] Auger spectra arising from filling the M and L shells as well as the K shell were taken, and even the low-energy spectra due to Coster–Kronig transitions have been recorded.[40]

Flügge *et al.*[46] have measured the angular distribution of ejected Auger electrons. They have demonstrated that a nonisotopic distribution relative to a directed unpolarized electron or photon beam is possible if an electron with a quantum number $j > 1/2$ is removed and if the resulting atomic state has a quantum number $J'' > 1/2$.

In addition to the normal Auger lines, whose energy spacings can be accurately checked by optical data, a wealth of other lines (satellite lines) are observed. An idea of the complexity of a well-resolved Auger spectrum is given by Figure 6.7. To give some understanding of where these lines arise and of the type of phenomena on which they shed light, let us review the paper by Krause *et al.*[44] on the K–LL Auger spectrum of neon.

Some 90 lines were observed. Of these only five were the normal K–LL Auger lines expected from pure Russel–Saunders coupling. A sixth line due to the forbidden K–$L_{III}L_{III}$ ^{3}P transition was found with an intensity of 0.05% of the most intense line. The normal K–LL lines arise from transitions in which a single vacancy appears in the K shell for the initial state and two vacancies in the L shell occur for the final states; otherwise, the other electrons are in the same configuration as they would be for a neon atom. A satellite line occurs when the configuration is altered from that described above, either for the initial or the final state. Figure 6.8 illustrates the various types of processes that lead to satellite lines. Though the normal lines make up the majority of the Auger spectra, the contribution from satellite lines is by no means negligible.

As mentioned before, the nomenclature for describing an Auger process is to designate the shells in which the vacancies occur. In addition, we use the letter e to designate a single electron which has been placed into an excited state.

Let us briefly discuss the nature of these satellite processes. First, a K vacancy can be formed without having the electron originally in the K shell go into the continuum, but rather having it promoted into an excited but bound neutral state. This may occur with photons whose energy exactly matches the transition to the highly excited K-hole state, whereupon resonance absorption occurs. When monochromatic x rays are used whose energies are in excess of the K binding energy, such resonance absorption cannot occur; but impact electrons utilize only a portion of its energy and can easily form K-hole excited states. From the analysis of the neon Auger spectrum it was ascertained that when a K vacancy was created by electron impact it took place by excitation in

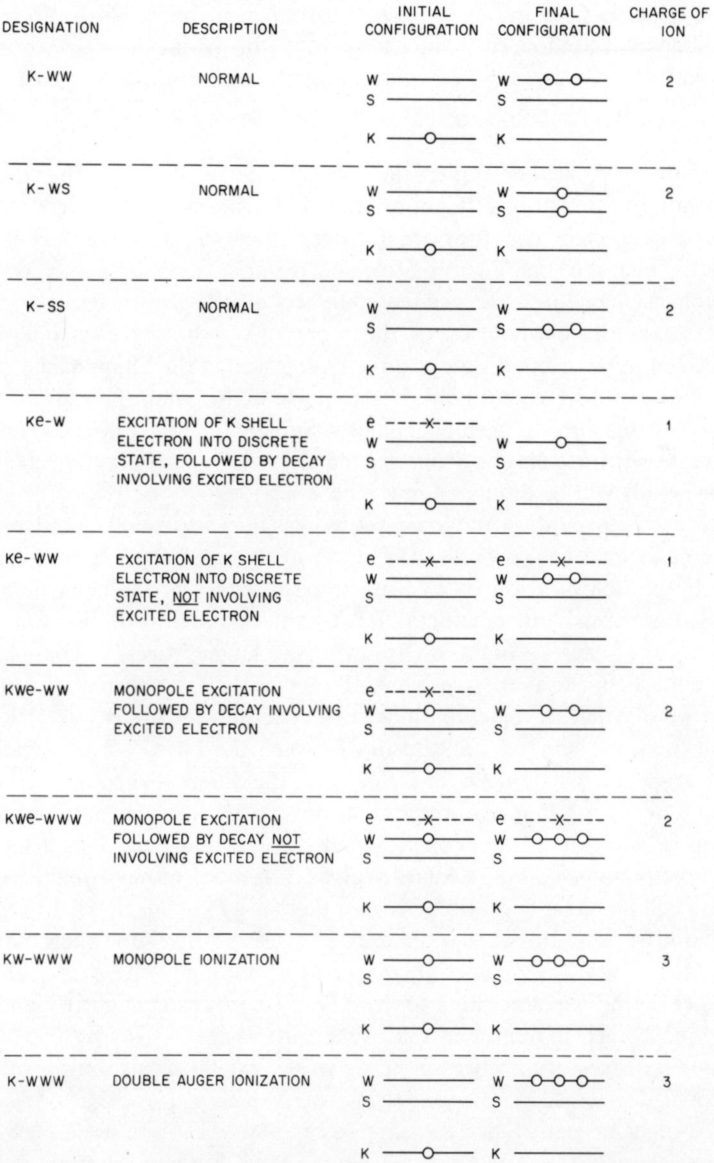

DESIGNATION	DESCRIPTION	INITIAL CONFIGURATION	FINAL CONFIGURATION	CHARGE OF ION
K-WW	NORMAL	W ——— S ——— K —O—	W —O—O— S ——— K ———	2
K-WS	NORMAL	W ——— S ——— K —O—	W —O— S —O— K ———	2
K-SS	NORMAL	W ——— S ——— K —O—	W ——— S —O—O— K ———	2
Ke-W	EXCITATION OF K SHELL ELECTRON INTO DISCRETE STATE, FOLLOWED BY DECAY INVOLVING EXCITED ELECTRON	e ---x--- W ——— S ——— K —O—	W —O— S ——— K ———	1
Ke-WW	EXCITATION OF K SHELL ELECTRON INTO DISCRETE STATE, NOT INVOLVING EXCITED ELECTRON	e ---x--- W ——— S ——— K —O—	e ---x--- W —O—O— S ——— K ———	1
KWe-WW	MONOPOLE EXCITATION FOLLOWED BY DECAY INVOLVING EXCITED ELECTRON	e ---x--- W —O— S ——— K —O—	W —O—O— S ——— K ———	2
KWe-WWW	MONOPOLE EXCITATION FOLLOWED BY DECAY NOT INVOLVING EXCITED ELECTRON	e ---x--- W —O— S ——— K —O—	e ---x--- W —O—O—O— S ——— K ———	2
KW-WWW	MONOPOLE IONIZATION	W —O— S ——— K —O—	W —O—O—O— S ——— K ———	3
K-WWW	DOUBLE AUGER IONIZATION	W ——— S ——— K —O—	W —O—O—O— S ——— K ———	3

FIGURE 6.8. Designation for the various types of Auger processes observed in a *K–LL* Auger spectrum where the *L* shell is the valence shell of the atom. *W* and *S* designate orbitals in the valence shell in which the electrons are, respectively, relatively weakly or strongly bound. For the case of a free atom $W = L_{2,3}$; $S = L_{\mathrm{I}}$. An excited orbital is designated *e*. [Reproduced from Moddeman *et al.*,[48] Figure 1.]

about 2% of the time and by ionization for the remainder. This ratio remains approximately the same over the energy range of the incident electrons from 2 to 6 keV. It was further ascertained that the excitation took place preferentially to the *np* states rather than *ns* states, indicative of a dipole transition.

Following *K*-shell excitation the vacancy may be filled by either radiative or nonradiative transitions. If a nonradiative transition occurs, discrete-energy electrons are ejected, as in the normal Auger processes, although the final state is a singly charged ion. Two possible spectra arise according to whether (1) the electron placed into the excited state acts as a spectator while an Auger process takes place with two of the more tightly bound electrons, or (2) the electron in the excited state is directly involved in the filling of the *K* hole. Both cases result in electron energies that are higher than the normal Auger lines, with case 2 having the higher energy. Substantial data have been taken on satellite Auger lines arising from excited *K* holes for simple molecules, and further details will be discussed in the next section.

Monopole excitation and ionization, or, as they are usually called, electron shakeup and shakeoff, occur as the result of an inner shell vacancy, caused either by photoionization or electron impact. These phenomena have been discussed previously in some detail (cf. Chapters 3 and 5). If electron shakeoff occurs, the final state of the ion following the Auger process will be +3. These satellite lines appear at energies generally lower than the normal Auger lines. (When we say that the satellite lines fall at energies higher or lower than the normal lines, we imply a relationship between the lines of both series. The complete spectrum extends over 169 V, with the different spectra overlapping with one another.) If monopole excitation or electron shakeup occurs, an electron is placed in an excited state, and it either may behave as a spectator or be directly involved in the Auger process. It is much more probable that the electron behaves as a spectator, because there are more electrons in the filled *L* shell and because this shell has a greater overlap with the 1*s* shell than does the excited orbital. The Auger energies corresponding to the *KLe–LLLe* processes are slight lower and those to the *KLe–LLL* processes slightly higher than those for the corresponding normal Auger processes.

Satellite lines occur corresponding to the excitation or ionization of an additional electron accompanying the normal Auger process, that is to say, the double Auger process. Since these processes always require extra energy, the Auger electrons are ejected with energies below those of the corresponding normal Auger processes. In the case of double ionization the two electrons share the energy, and the result is a continuum spectrum. Just as in the case when double electron ejection occurs in photoionization, one of the electrons carries away the bulk of the energy.

3.2. Molecules

The first high-resolution Auger spectra in molecules were taken by Stalherm *et al.*[47] A number of studies have since been carried out at Uppsala[42] and at Oak Ridge.[48] The Auger spectra of molecules are considerably more complex than the corresponding spectra for atoms. This is due in part to the larger number of molecular orbitals and in part to the added complexity of vibrational structure.

As was mentioned earlier, Auger transitions between only inner shells give chemical shifts that are similar to those observed with photoionization. More extensively studied and more interesting because of the unique information that is available are the K–LL Auger transitions in which the L shell is the valence shell, which is to say Auger transitions occurring for the first row elements, including boron, carbon, nitrogen, oxygen, and fluorine. Our attention will be concentrated on Auger spectra involving these elements.

Moddeman *et al.*[48] have provided a generalized scheme for analyzing the various portions of the K–LL Auger spectra of molecules. It consists in (1) determining which portions are due to normal Auger processes and which are satellite peaks; (2) dividing the normal Auger lines into categories as to whether they are derived from molecular orbitals whose electrons are weakly or strongly bound; and (3) identifying specific Auger peaks with the help of theory or optical data. Let us describe in turn each of these different parts of the analysis.

First, one needs to identify the satellite peaks. The basic origin of these extra Auger peaks has been discussed in Section 3.1 on atoms. The most important portion of the spectrum to clarify is the high-energy end. This portion is entirely made up of satellite lines. (In the analysis it is given the letter A, while the normal lines are divided into portions B, C, and D. For example, see Figure 6.9, showing the Auger spectrum for N_2.) The key to analyzing an Auger spectrum is to find the highest energy peak due to a normal Auger process, since this peak, if observable, should correspond to the lowest energy or ground state of the doubly charged molecular ions. Often this identification is aided by the fact that the transition to the ground state is also one of the most intense transitions. For an example of how high-energy peaks can be properly identified as satellite peaks or normal Auger lines, let us examine the case of nitrogen in Figure 6.9. Peaks A-1 and A-2 are believed to be due to K-hole excitation followed by an Auger process in which the excited electrons directly participate (i.e., Ke–LL). This is verified by observing that

$$E_{\mathrm{A}} = E^*(K) - E_B \tag{6.5}$$

where E_{A} is the measured energy for the Auger peaks A-1 and A-2, $E^*(K)$ is the energy required to place the K electron into the first excited state, which has been measured by Nakamura *et al.*[49] for resonance absorption states to be

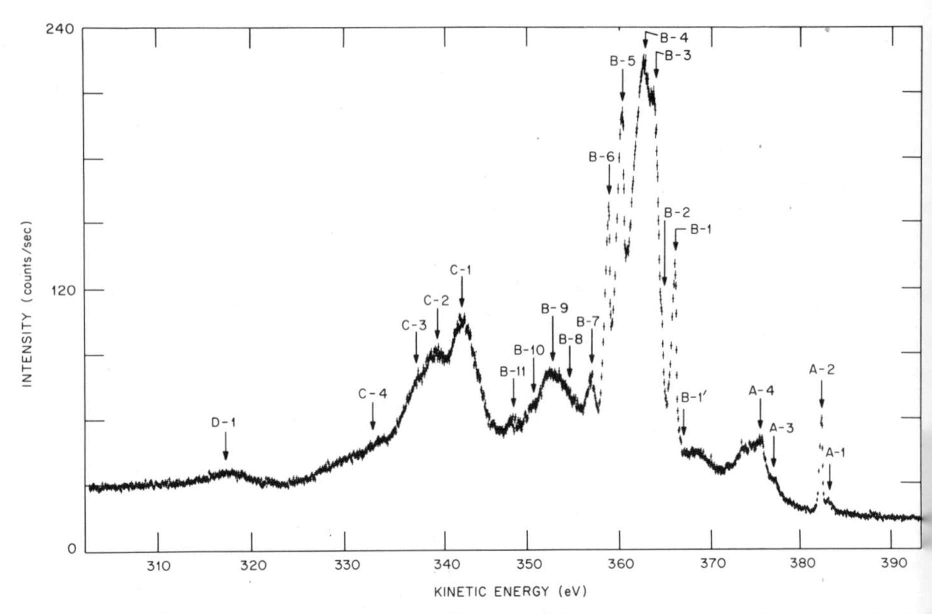

FIGURE 6.9 The *K–LL* Auger spectrum of N_2 excited by electron impact. See Table 6.5 for identification of labeled peaks. [Reproduced from Moddeman *et al.*,[48] Figure 2.]

400.2 eV, and E_B is the binding energy for the two least bound orbitals of N_2. From equation (6.5) one obtains 384.6 and 383.5 eV for E_A, in excellent agreement with the energy of the Auger lines *A*-1 and *A*-2. In addition, it is expected that satellite lines due to *K–LL* excitation will disappear when Al *Kα* x rays are used rather than electron impact to create *K* vacancies. (The 1487-eV photons will only cause ionization and not excitation of the *K*-shell electrons.) Indeed, the *A*-1 and *A*-2 peaks are missing in the Auger spectrum of N_2 generated by Al *Kα* x rays. Other portions of the spectrum are also missing. (The shaded area in Figure 6.10 represents the difference in the Auger spectra of N_2 as produced by electron impact and by Al *Kα* x rays.) These portions are believed due to transitions where the excited electron remains as a spectator (i.e., *Ke–LLe*). In addition, another part of the spectrum in region *A* is believed to be due to electron shakeup (i.e., *KLe–LL*). The energy of these satellite lines is greater than that of the corresponding normal Auger line (*K–LL*) by just the amount involved in electron shakeup. The shakeup energies in turn have been determined by photoelectron spectroscopy.[50] In Figure 6.10 the dashed lines give the normal Auger spectrum shifted by the energy of monopole excitation. The relative intensities are, however, strongly dependent on the relative transition rates and will not necessarily be the same as the normal Auger spectrum.

Assignments for the high-energy satellite Auger lines having been made, the peak *B*-1 is said to be first normal line, which is confirmed by good agreement with the appearance potential for N_2^{2+}, using mass spectrometry. Before giving specific designations to the individual peaks, let us next consider the overall breakdown of the normal Auger spectrum. To accomplish this breakdown, we shall divide the orbitals that make up the *L* valence shell into *W* and *S* orbitals, which are so designated because the electrons in these orbitals are either weakly or strongly bound. Thus, the normal Auger processes *K–LL* will be now subdivided into *K–WW*, *K–WS*, and *K–SS*, and the energy regions to which they correspond will be called *B*, *C*, and *D*, respectively. Our job will next be to calculate at what energy these processes will occur. This can be done by relating the differences in energy to the various ways of removing two electrons from a neutral molecule. Consider an atom or molecule with shells 1, 2, . . ., where 1 represents the least tightly bound orbital, 2 the next least tightly bound, etc. If coupling in the final states of the doubly charged ion is

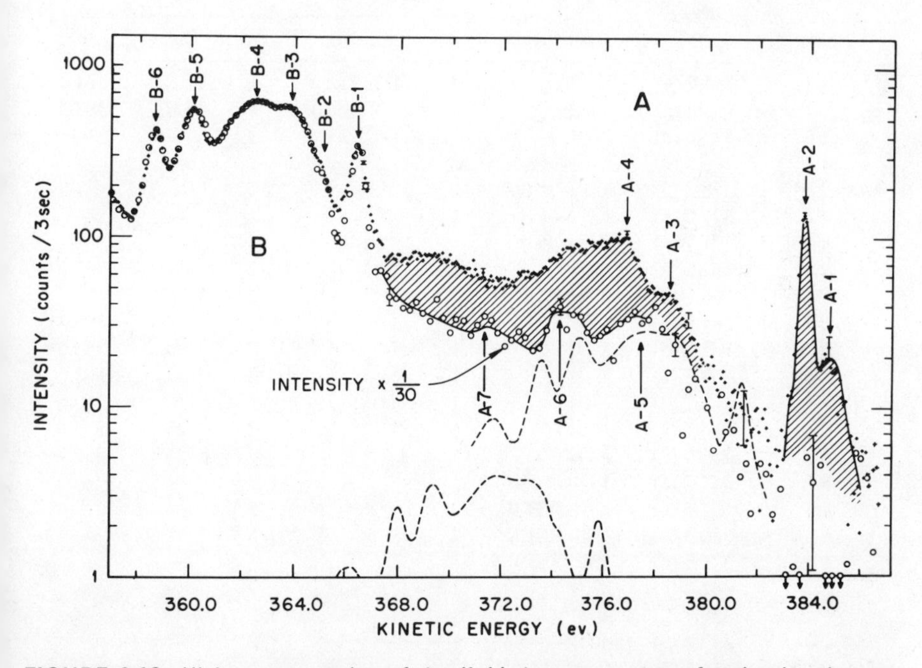

FIGURE 6.10. High energy portion of the *K–LL* Auger spectrum of molecular nitrogen excited by electrons and Al *Kα* x rays. (+) Data from electron impact; (○) data from x-ray bombardment. Shaded area gives the difference between two experiments. Dashed lines give approximate expected shape due to *KWe–WWW* processes. [Reproduced from Moddeman *et al.*,[48] Figure 3.]

ignored, the relative energy of the Auger lines resulting from vacancies in the first two orbitals may be taken as

$$E_A(1) - E_A(2) = \varepsilon_2 - \varepsilon_1 \tag{6.6}$$

$$E_A(1) - E_A(2) = 2(\varepsilon_2 - \varepsilon_1) - C(\varepsilon_1 - \varepsilon_2) \tag{6.7}$$

$E_A(1)$, $E_A(2)$, and $E_A(3)$ are the Auger lines or bands, with $E_A(1)$ representing the highest kinetic energy process, $E_A(2)$ the next, etc.; ε_n is the binding energy of orbital n; and C is a factor which is related to change in the configuration as electrons are successively removed. C may be obtained empirically or from a simple model. It was found that for both atoms and molecules C is approximately 0.1 *K–LL* Auger processes. In any case it is not a sensitive parameter.

TABLE 6.5

Assignment of Peaks Appearing in the *K–LL* Auger Spectrum of N_2[a]

Line	Energy,[b] eV	Transition[c]	Initial state q	Initial state Desig.	Final state q	Final state Desig.	Calc. E, eV
A-1	384.7 (4)	Ke–$2p\sigma_g$	0	$(1s)^3(2p\pi_g)^1$	1	$X\,^2\Sigma_g^+$	384.6
A-2	383.8 (2)	Ke–$2p\pi_u$	0	$(1s)^3(2p\pi_g)^1$	1	$A\,^2\Pi_u$	383.5
A-3	378.6 (5)	Ke–WWe	0	$(1s)^3(2p\pi_g)^1$	1	—	—
A-4	376.7 (6)	Ke–WWe	0	$(1s)^3(2p\pi_g)^1$	1	—	—
A-5	378.3 (5)	KWe–WW	1	—	2	—	—
A-6	375.0 (5)	KWe–WW	1	—	2	—	—
A-7	371.7 (6)	KWe–WW	1	—	2	—	—
B-1′	367.0	K–WW	1	$(1s)^3$	2	—	367.2
B-1	366.5 (2)	K–$2p\sigma_g 2p\sigma_g$	1	$(1s)^3$	2	$X\,^1\Sigma_g^+$	—
B-2	365.0 (4)	K–WW	1	$(1s)^3$	2	$b\,^1\Pi_u$	365.2
B-3	363.5 (5)	K–WW	1	$(1s)^3$	2	$A''\,^3\Sigma_u^+$	364.0
B-4	362.5 (5)	K–WW	1	$(1s)^3$	2	$A\,^3\Pi_g$	363.0
B-5	360.2 (2)	K–WW	1	$(1s)^3$	2	$c\,^1\Pi_g$	360.6
B-6	358.7 (2)	K–WW	1	$(1s)^3$	2	$d\,^1\Sigma_u^+$	358.6
B-7	356.9 (3)	K–WW	1	$(1s)^3$	2	$e\,^1\Sigma_g^+$	356.7
B-8	354.7 (6)	KW–WWW	2	—	3	—	—
B-9	352.5 (6)	KW–WWW	2	—	3	—	—
B-10	350.4 (6)	KW–WWW	2	—	3	—	—
B-11	347.8 (6)	KW–WWW	2	—	3	—	—
C-1	342.4 (4)	K–WS	1	$(1s)^3$	2	—	—
C-2	339.1 (5)	K–WS	1	$(1s)^3$	2	—	—
C-3	337.4 (6)	K–WS	1	$(1s)^3$	2	—	—
C-4	332.7 (7)	—	—	—	—	—	—
D-1	315.0 (9)	K–$2s\sigma_g 2s\sigma_g$	1	$(1s)^3$	2	$^1\Sigma_g^+$	—

[a] Taken from Moddeman *et al.*,[48] Table 2.
[b] Uncertainty in energy given for last significant figure in parentheses.
[c] Compare with Figure 6.8 for description of terms. Specific designations of orbitals are given rather than W or S when they are known with some degree of confidence.

In order to apply equation (6.7) for estimating the relative energies represented by regions B, C, and D, ε_2 is generally taken to be the more tightly bound $2s\sigma$ orbital, while ε_1 is an average of the remaining less tightly bound orbitals. In general, regions C and D were found to be respectively about 20 and 40 V lower than region B (cf. Figure 6.9 for nitrogen).

Returning to the analysis of some of the specific peaks of the Auger spectrum, it should be remembered that the energies of the different lines arising from normal Auger processes are related to one another by the energy separation between the various excited states of the doubly charged molecular ions. The most important peak is designated B-1 and represents an Auger transition to the ground state. The energy differences between this line and the lower energy normal Auger line are the excitation energies for the doubly charged ions. In the case of nitrogen it has been possible to assign states to some of the Auger lines by using the energy differences of the isolated curves of N_2^{2+} as derived by Hurley.[51] The analysis of the K–LL Auger spectrum of nitrogen is given in Table 6.5. Similar analyses have been made for O_2, CO, NO, H_2O, and CO_2.

From the highest-energy normal Auger line designated B-1 one can obtain the minimum energy for double electron removal $E_{II}(\text{min})$ from the relationship

$$E_{II}(\text{min}) = E(K) - E_A(B\text{-}1) \tag{6.8}$$

where $E(K)$ is the binding energy of the K shell and $E_A(B\text{-}1)$ is the measured Auger energy for the highest-energy normal Auger lines. (To be more accurate, the onset of this line (B-1′) is usually used, which ought to correspond more closely to the adiabatic ionization potential) Table 6.6 lists $E_{II}(\text{min})$ for a number of molecules using Auger spectroscopy together with appearance potentials

TABLE 6.6

Minimum Energy Required for Producing Doubly and Triply Charged Molecular Ions

Ion	E(min), eV	
	Auger spectra[48]	Electron impact[a]
$(N_2)^{2+}$	42.9	42.7
$(N_2)^{3+}$	84	—
$(O_2)^{2+}$	37.4	36.5
$(CO)^{2+}$	39.9 (C), 40.2 (O)	41.8
$(NO)^{2+}$	35.7, 40.1 (N)	39.8
	34.7 (O)	
$(H_2O)^{2+}$	39.2	—
$(CO_2)^{2+}$	37.8 (C), 37.4 (O)	36.4

[a] J. L. Franklin *et al.*, *Natl. Std. Ref. Data Ser., Natl. Bur. Std.* 26 (1969).

as obtained from mass spectroscopy. In one case, nitrogen, it was also possible from Auger data to estimate the energy necessary to form the triply charged ion. In general, the agreement between values arrived at by Auger spectroscopy and those obtained by mass spectroscopy is good. Auger spectroscopy has a distinct advantage over mass spectroscopy in determining the minimum energy for forming molecular ions, since in mass spectroscopy the ion must remain intact the 10^{-5} sec required to detect it, while meaningful results can be obtained from Auger spectra without the requirement that the ion remain stable.

Besides the formal methods for analyzing molecular Auger spectra that have been discussed in the preceding paragraph, there are several other observations that can be of assistance. First, it is of value to compare the Auger spectra corresponding to different elements of the same molecule. The energy spacings in both spectra correspond to those that are expected for the various states of the doubly charged molecular ion that is the result of a *K–LL* Auger process. The total Auger energy depends on the element in which the initial *K* vacancy is formed, but the energy spacing depends primarily on the states of the doubly charged molecular ion. The probability for reaching a given state will depend on whether the orbitals involved in the transition are strongly associated with the atom having the *K* vacancy. For example, Figure 6.11, compare the Auger spectrum corresponding to *K* vacancies in carbon and in oxygen of CO. The peaks are labeled identically when it is felt that they both represent the same final state. Neumann and Moskowitz[52] have calculated the atomic population of the molecular orbitals of CO and find the percentage carbon character in each orbital to be as follows: $2p\sigma^b$, 92%; $2p\pi^b$, 33%; $2s\sigma^*$, 20%; and $2s\sigma^b$, 33%. (For example, one would expect that Auger processes filling a vacancy in the *K* shell of carbon would involve the $2p\sigma^b$ orbitals much more readily than would be the case if the *K* vacancy occurred with oxygen.) This information helps us in assigning the origin of the various Auger peaks. Thus, the presence of the strong peaks *B*-1, *B*-2, and *B*-3 found in the carbon spectrum, which are weak or missing in the oxygen spectrum, suggests these peaks involve $2p\sigma^b$ orbitals. The strongest peaks for the oxygen spectrum of CO occur between *B*-5 and *B*-7 and apparently involve the $2p\pi^b$ and $2s\sigma^*$ molecular orbitals. These suggestions are further supported by the fact that the binding energy of the $2p\sigma^b$ orbital is less than that of the $2p\pi^b$ and $2s\sigma^*$.

Another source of information for analyzing the Auger spectrum comes from vibrational structure. Normally such structure is not readily seen because of the complexity of the Auger spectra and the need for very high resolution. However, it has been observed in the carbon spectrum of CO (cf. details of high-energy portion of carbon Auger spectrum of CO, Figure 6.12). These three bands are believed to arise from a *K*-vacancy excitation in which the excited electron participates in the subsequent reorganization, i.e., *Ke–L*. The final

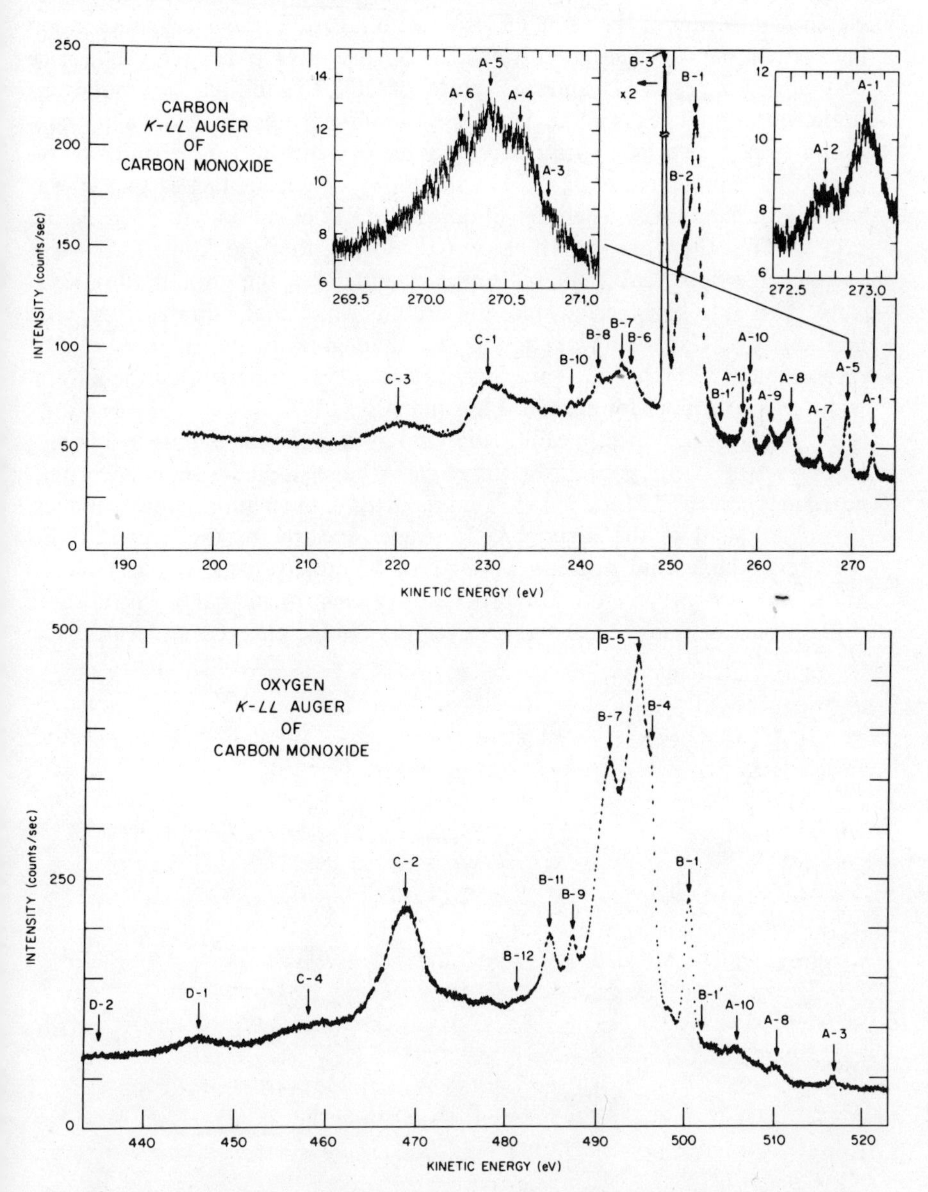

FIGURE 6.11. Comparison of the *K–LL* Auger spectra in CO resulting from *K* vacancies in C and O, respectively. Peaks labeled the same are believed to belong to the same final state of the molecular ion. [Reproduced from Moddeman *et al.*,[48] Figures 6 and 7.]

state is the singly charged carbon monoxide and the bands resemble the photoelectron spectrum of the same molecule with regard to the energy separation between the bands, which is equivalent in both instances to the energy difference between the $X\,^2\Sigma^+$, $A\,^2\Pi$, and $B\,^2\Sigma^+$ states of CO^+. In addition, the separations of the vibrational peaks in the Auger spectra of the bands are identical to those found in photoelectron spectroscopy. Some evidence of a hot band was reported by Moddeman *et al.*,[48] but the more significant fact is that it was small. Hot bands arise when the vibrational states of the initial molecule are excited. Thus the vibrational data on CO indicate that the formation of a K vacancy in carbon leaves the molecules essentially in the ground vibrational state. This is to be expected since the orbitals which make up the K shell are nonbonding. Also significant is the fact that vibrational structures can be resolved, and from the line shape Siegbahn *et al.*[42] estimated that the natural width of Auger lines for carbon is less than 0.18 eV.

Finally, Auger spectra can be analyzed by comparison of data in a homologous series of compounds. Spohr *et al.*[53] have studied a series of brominated methanes and Moddeman[33] has reported data on fluorinated methanes. Figure 6.13 shows the carbon K–LL Auger spectra for the hydrocarbons methane, ethane, and benzene. Spohr *et al.*[53] noted that the sharpest lines are found in benzene while the broadest are seen for methane. The authors explain this in terms of the lifetimes for the doubly charged molecules. The

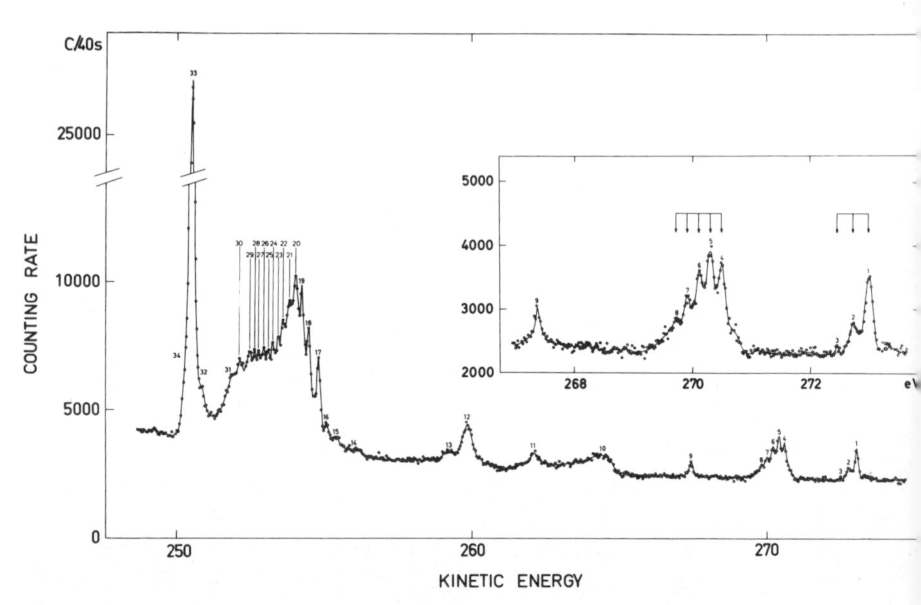

FIGURE 6.12. Details of vibrational structure in the carbon K–LL Auger spectrum of CO. [Rep
duced from Siegbahn *et al.*,[42] Figure 5.28.]

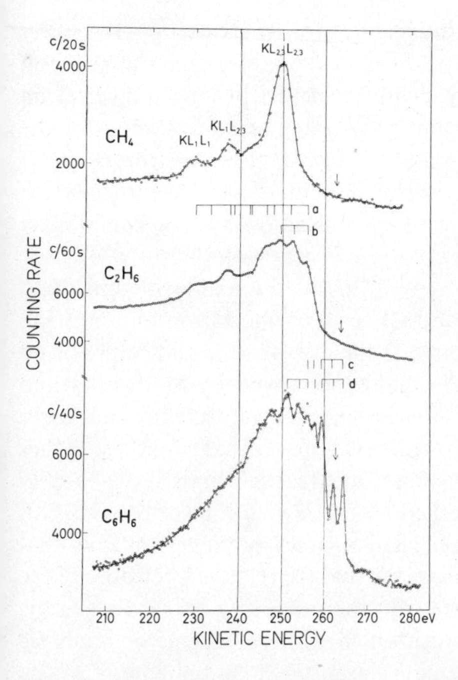

FIGURE 6.13. Comparison of *K–LL* Auger spectra for methane, ethane, and benzene. As one goes from methane to benzene the dissociation energies become less and the sharpness of the Auger line increases. [Reproduced from Spohr *et al.*,[53] Figure 5.]

dissociative energy for methane, ethane, and benzene is estimated to be, respectively, −13, −8, and +1 eV, and the line broadening may also be expected to decrease in that order. The relative sharpness of the lines observed in the bromomethane compounds was also interpreted in terms of whether the Auger transitions involve bonding orbitals leading to dissociative states.

Recently, detailed analyses have been carried out on the high-resolution Auger spectra of H_2O,[37] HF,[54] and C_2O_3.[55]

3.3. Study of Ionization Phenomena by Auger Spectroscopy

In the previous sections on Auger spectroscopy of gases we have emphasized the nature and identification of the various Auger lines. Our aim was to obtain the correct analysis of the observed Auger process. In obtaining this analysis we learned that many of the lines were satellite lines, which arose from some form of initial excitation. These lines can thus be used as a means for following the nature of the excitation process.

We have already discussed in some detail the information gained on the nature of *K*-hole excitation states as the result of electron impact. Glupe and Mehlhorn[56] used the Auger spectra of C, N, O, and Ne to determine the *K*-shell ionization cross section as a function of electron impact energy. The nature of electron shakeoff has been studied by Auger spectroscopy. Krause *et al.*[57] showed that the satellite spectra for neon representing initial shakeoff and shakeup processes were essentially identical whether produced by electron impact or photoionization. By examining the satellite lines relative to the normal Auger lines, Carlson *et al.*[58] studied the extent of electron shakeoff as a function of electron impact energy. From these studies it became clear that electron shakeoff as the result of inner shell ionization by electron impact could be treated by the sudden approximation, just as with photoionization.

Rudd,[59] using Auger spectroscopy, followed the nature of ionization by protons and found, in marked contrast to photoionization and electron impact, much larger contributions from satellite lines than would be predicted by the sudden approximation. Recent studies, however, by Stolterfoht and co-workers[60] show that as the proton energy approaches 500 keV the Auger spectrum approaches that produced by electron impact. That is, at these higher energies the region of the sudden approximation is reached, and the probability for initial multiple ionization is governed by the laws for electron shakeoff in which only the creation of an inner shell vacancy matters, not how that vacancy was formed. From Figure 6.14 we see that the relative intensities of the regions A, B, C, and D for nitrogen remain the same with projectile energy. These are the normal Auger processes, which do not depend on multiple ionization. Region B' increases in relative importance as the energy of the proton decreases. This region must be due to multiple ionization; and, in fact, Moddeman *et al.*[48] had previously suggested that the region B' might be due to KW–WWW transitions. The increase of multiple ionization with lower proton velocity suggests a direct collision mechanism. The use of proton bombard-

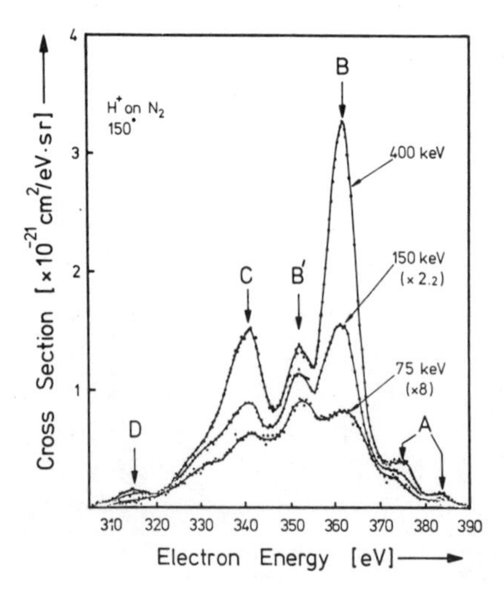

FIGURE 6.14. Comparison of N_2 K–LL Auger spectra as a function of kinetic energy of impact protons. B' arises from multiple ionization and its intensity relative to the other portions of the Auger spectrum changes with impact energy. [Reproduced from Stolterfoht *et al.*,[60] Figure 1.]

ment with varying energy could be of substantial value in helping to distinguish which areas of the Auger spectrum arise from normal Auger processes and which are due to initial multiple ionization.

If an ion is used as a projectile, the ionization it produces will be influenced by the atomic behavior of electrons surrounding the ion unless the ion moves at velocities that are far in excess of electron orbital velocities, which for protons would be far in excess of 20 keV. The present interest in the nature of multiple ionization by high-energy heavy ions could be assisted by Auger spectroscopy. Auger energies in highly charged ions will shift from that observed for an ion with a single inner shell vacancy. From both the Auger energies and relative intensities of Auger peaks one can in principle analyze the extent of ionization and excitation for a given heavy ion. Both the target and projectile could be so analyzed. This type of analysis has been applied extensively to the x rays of highly charged ions,* and could be profitably made using Auger spectra. In fact, Auger energies are more sensitive to the removal of outer shell electrons than is the case with x-ray energies. One drawback is that electrons are more susceptible to recoil energy than photons, and considerable broadening due to Doppler shifts may arise from electron emission from a recoiling heavy ion. Nevertheless, increasing use has been made of Auger spectroscopy in the study of ion–ion collisions.[62]

3.4. Autoionization

When an atom is placed in an excited state which is above the ionization potential, the atom may readjust by a nonradiative process with the ejection of a discrete-energy electron. When excitation occurs only with valence electrons, this process is generally known as autoionization rather than as an Auger process, the latter name being restricted to situations containing a vacancy in the core shells. However, the nature of the Coulombic readjustment is physically identical for the two phenomena.

It is not the purpose of the book to cover the vast amount of material on autoionization, but rather to point out its similarity with Auger spectroscopy. In particular, the studies carried out by Siegbahn *et al.*[42] on the rare gases using 4.0-keV impact electrons to create the excitations and an electron spectrometer to record the discrete-energy electrons are identical in experimental procedures to Auger spectroscopy, except that the energies are lower. Figure 6.15 shows a typical example of the autoionization spectra for Ar. Helium, Ne, and Xe were also studied by these authors. The dips in the spectra are caused by the interference between a discrete autoionization state and the continuum.

* See Ref. 61 for a general review of multiple ionization arising from ion collisions and the use of x-ray spectra for studying the process.

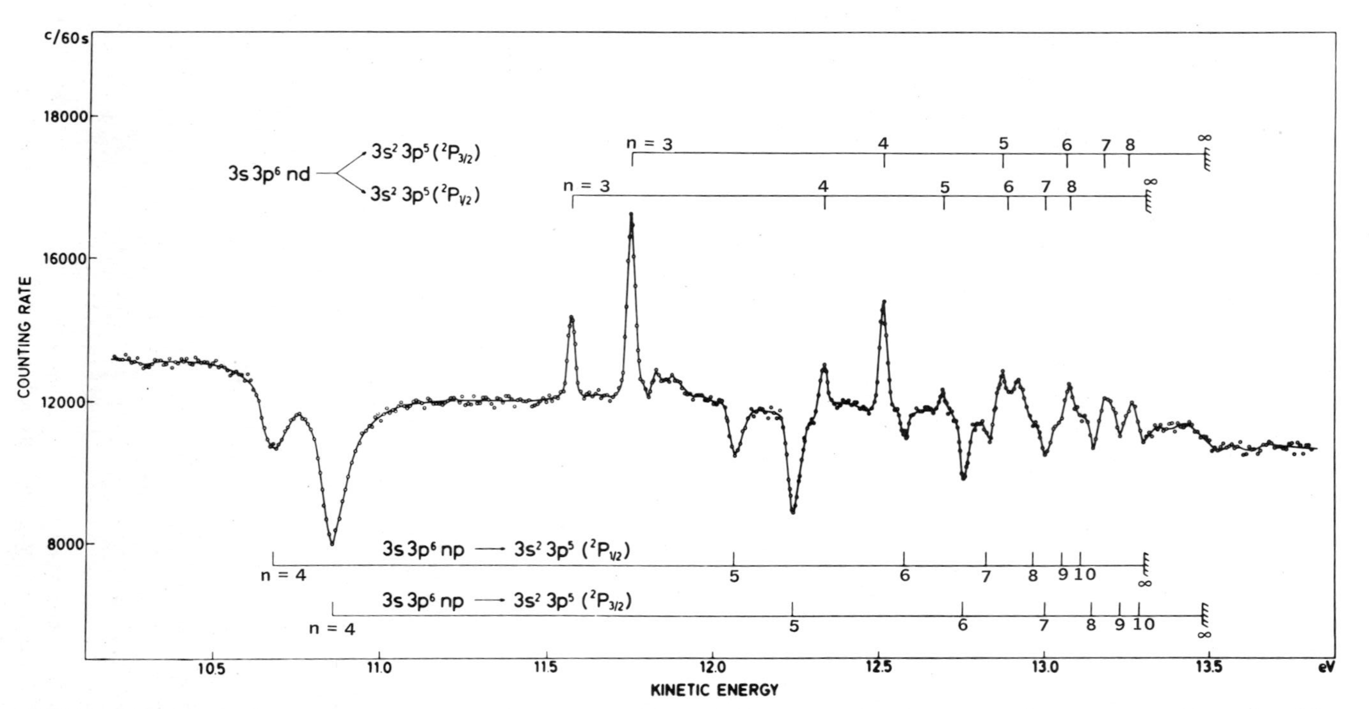

FIGURE 6.15. Autoionization electron spectrum from argon. Initial excitation created by high-energy electron impact. [Reproduced from Siegbahn *et al.*,[42] Figure 4.6.]

Good agreement was obtained with photoabsorption measurements, although some observed lines formed by electron impact are forbidden for photoabsorption.

Another area of recent interest has been the study of discrete-energy electrons ejected from an ion which has passed through a thin foil. If the ion as the result of passing through the foil is excited into a metastable state, the state may decay during its flight by autoionization. Lucas and Harrison[63] have discussed such measurements for an incident 291-keV N^+ beam. Sellin *et al.*[64] have studied a number of highly charged ions for autoionization, using ions with high kinetic energy as projectiles. For example, Cl^{14+} and Ar^{15+}, which are lithiumlike in configuration, have been studied. By examining the autoionization processes during flight, they have also been able to ascertain the half-lives of the excited states.

3.5. Auger Spectroscopy for Use in Gas Analysis

Auger spectroscopy offers a number of advantages for gas analysis. Figure 6.16 shows oxygen *K–LL* Auger spectra from four different molecules. Though the center of the spectrum in each case is approximately 490 eV, indicating that we are dealing with the element oxygen, the details of the spectra are quite dissimilar, yielding essentially "a fingerprint" for a given molecule. Thus, an Auger spectrum offers both an elemental analysis and molecular identification.

Auger spectra of gases often reveal lines with very small natural widths. There is no line broadening from the excitation source, as with photoionization. The principal limitation to resolution comes from the natural widths of the core levels. In the case of carbon this has been shown to be less than 0.2 eV. From calculations on the lifetime for K vacancies the natural width for all elements in the first row ought to be less than 0.2 eV. The state of the doubly charged molecular ions following an Auger process should also be stable for at least 10^{-14} sec if the line broadening is not to be increased, but many molecules should be able to meet this requirement. The intensity of producing Auger lines by electron impact can be made very high. It is easy to produce electron beams of 500 mA. I have found that a beam of only 0.5 mA would produce Auger lines with intensities whose limitation was the saturation of the electron multiplier (~50,000 counts/sec) when the pressure of the target gas was about 10 μm. For higher rates a current integrating device can be used. The peak-to-background ratio is an important consideration if high sensitivity is being sought (parts per million). When x rays were used to produce inner shell vacancies, ratios as high as 30,000 to 1 were obtained. When electron impact was used, the best ratio was 500 to 1, and the importance of scattered electrons in-increased as one went to lower kinetic energies. However, with greater care for

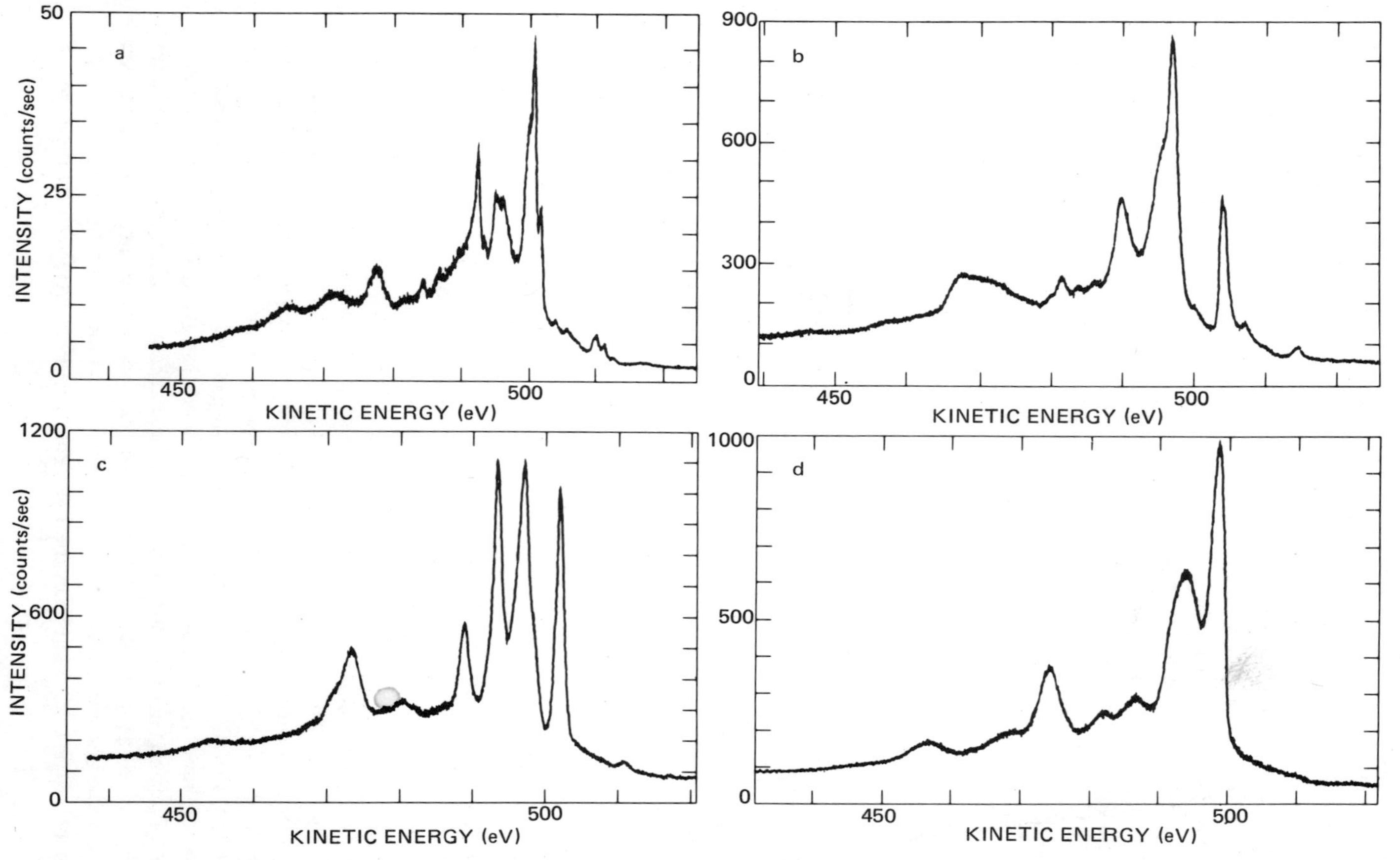

FIGURE 6.16. Comparison of oxygen *K–LL* Auger spectra as a function of molecule. (a) Oxygen molecule; (b) nitric oxide molecule;

reducing secondary electrons, the peak-to-background ratios might be substantially improved.

There are limitations in the use of Auger spectroscopy as an analytical tool for gases. To analyze two or more molecules (say hydrocarbons) simultaneously by studying the Auger spectra of a common element (carbon) would be difficult, since the spectra would overlap. If the gases were present to about the same order of magnitude, and their spectral profiles well known, analysis would be possible, but not for trace amounts. A trace element might be studied, however, without interference, conceivably to a few parts per million. The use of Auger spectroscopy can be greatly enhanced by combining it with gas chromatography, as has been accomplished with mass spectroscopy.

Good resolution, high counting rates, unique identification of molecules, and sensitive elemental identification all make Auger spectroscopy a potentially powerful tool for gas analysis. As yet, no systematic study of its analytical possibilities on gases has been made, but it would appear to be a worthwhile undertaking.

4. USE OF AUGER SPECTROSCOPY IN THE STUDY OF SOLIDS

The practical application of Auger spectroscopy as a tool for elemental analysis on surfaces has been so successful as to almost isolate these endeavors from other applications of Auger spectroscopy. Before undertaking a discussion of surfaces, there are several smaller topics on solids which require our attention. First, there are some problems with AES which are peculiar to solids. These concern, for example, (1) the practical differences between measuring Auger lines in solids when initiating inner shell vacancies by x rays and impact electrons, (2) the presence of high energy Auger lines in solids, (3) characteristic energy losses in solids, and (4) charging of nonconducting materials.

Second, we shall examine those studies that have stressed the nature of the Auger spectrum itself rather than the information it can yield about the surface. These studies have eschewed cleanliness for high resolution. Most surface studies have put the emphasis on cleanliness of the surface, accepting rather modest resolution. Recent improvements in technology have made the decision of cleanliness versus resolution unnecessary, but it is still present in much work. In any case there exists an interest in AES of solids that lies outside of surface science and needs separate treatment. Finally, the use of Auger spectroscopy as an analytical tool for solids, again outside the special domain of surface interest, will be discussed.

When we do turn our attention to surfaces, we shall discuss the types of information about the surface that can be obtained with Auger spectroscopy, giving numerous examples from the recent literature. Comparison will then be made between AES and PES as a surface tool. Also, other competitive methods of surface analysis will be briefly discussed and compared with AES.

4.1. Special Problems Encountered on Using AES with Solids

4.1.1. Variables Concerned with Production of Auger Electrons

One may create an inner shell vacancy through a variety of methods, including irradiation with electrons and x rays. Haas *et al.*[65] have used argon ions to excite the Auger spectrum of Al metal. Electron impact is usually used for producing Auger lines for analytical purposes. It provides an intense beam which can also be brought to a fine focus. X irradiation has its value in providing less radiation damage and (under some conditions) better peak-to-background ratios. Barrie and Brundle[66] have explored the practical problems of measuring Auger spectra from CO adsorbed on Mo induced by both electron impact and soft x rays. Pessa[67] has studied the problem of peak to background using electron impact versus Al $K\alpha$ radiation. Figure 6.17 shows some of his results using 5-keV electron impact. The abscissa is the Auger energy and the ordinate is

$$S = \frac{(I_m - I_b)_x/(I_b)_x}{(I_m - I_b)_e/(I_b)_e} \tag{6.9}$$

where I_m and I_b are the peak and background intensities for x rays (x) and electron impact (e). For Auger processes with energies below 500 eV, electron impact seems superior, but above that figure, Al $K\alpha$ x rays give a better performance. Figure 6.18 shows a signal-to-noise ratio $[(I_m - I_b)/I_b]_e$ as a function of electron impact energy.

Neave *et al.*[68] and Gallon[69] have studied the role played by backscattered electrons in the production of Auger peaks. They concluded that a sizable fraction of the Auger electrons usually observed came not from ionization created by the primary beam of impact electrons, but from backscattered electrons. Neave *et al.* also pointed out the particular importance that backscattered electrons play in the production of Auger electrons as a function of glancing incidence of the initial beam. Figure 6.19 contrasts Auger yields as a function of initial impact energy and angle of incidence.

Staib[70] has compared relative intensities of *K–LL* Auger lines for O, Al, and Si and the *L–MM* lines from potassium under the controlled conditions of looking at only the top layers of a cleaved mica sample where the positions

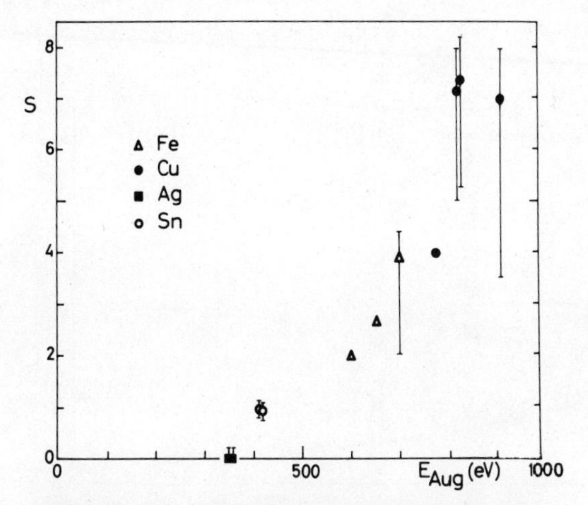

FIGURE 6.17. Peak height ratio S [cf. equation (6.9)] vs Auger energy for Fe, Cu, Ag, and Sn. Sn. Electron impact energy is 5 keV. [Reproduced from Pessa,[67] Figure 3.]

of the different atoms are known from crystallography. Results are compared with the semiempirical formula of Darwin.

The angular dependence of Auger emission from a copper 111 surface has been studied by Holland *et al.*[71] They found marked crystallographic

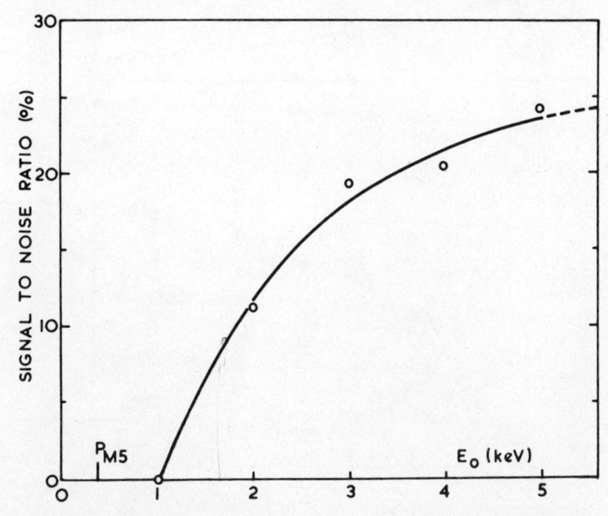

FIGURE 6.18. Signal-to-noise ratio of the 350.5-eV $M_5N_{4,5}N_{4,5}$ Auger line from Ag metal vs kinetic energy of primary electrons. [Reproduced from Pessa,[67] Figure 1.]

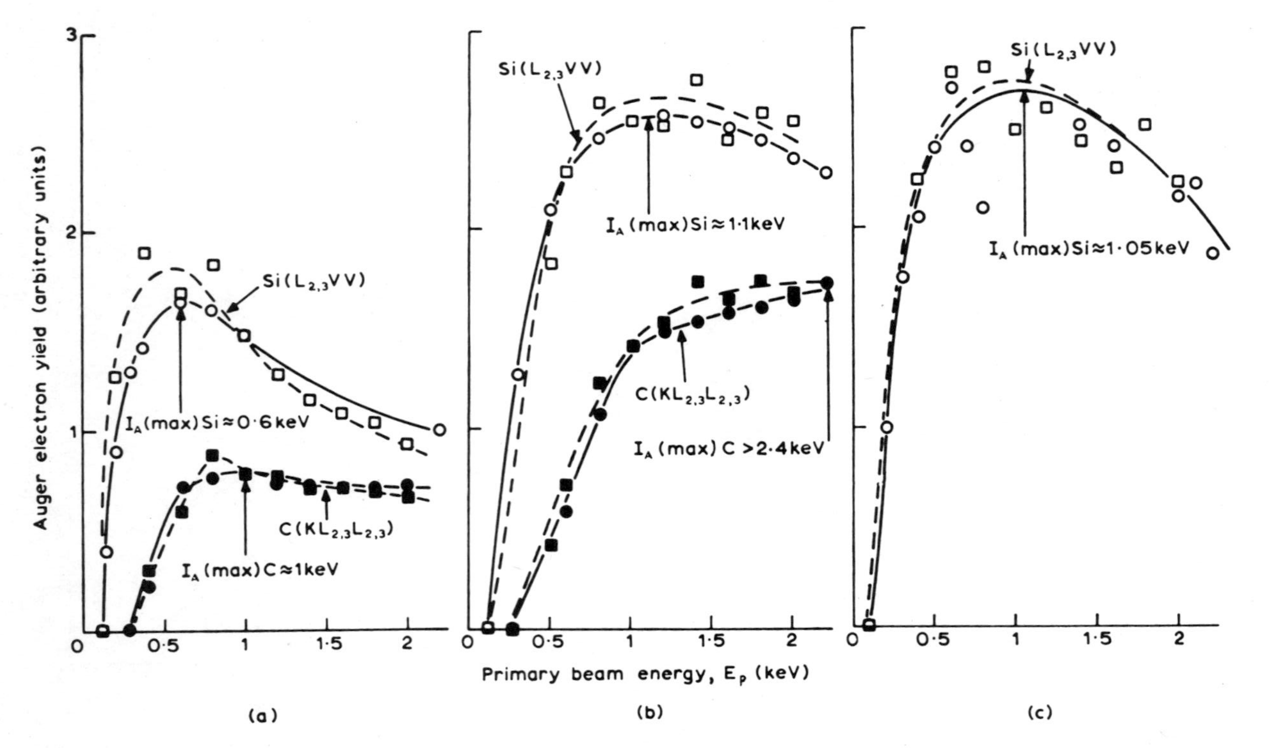

FIGURE 6.19. Measured intensity of Auger electrons as a function of primary bean energy at normal and glancing incidence and with constant beam current or constant target current. (a) Normal incidence, constant beam current (1 μA); (b) glancing incidence (20°), constant beam current (1 μA); (c) glancing incidence (20°), constant target current (0.8 μA). For Si: (○) observed; (□) predicted. For C: (●) observed; (■) predicted. Constant rms voltage 2.5 V. [Reproduced from Neave *et al.*,[68] Figure 1.]

dependence, and feel that such analyses could provide new information on the valence electron wave functions in the surface region of the solid.

4.1.2. High-Energy Satellite Lines

High-energy satellite structures have been observed in the Auger spectra of solids.[72] This has been interpreted by Chung and Jenkins[73] and others[74] as being due to plasmon gains. Pattinson and Harris feel the plasmon gain mechanism is unlikely.[75] It is more likely that the high-energy lines arise from an initial multiple ionization[72] or perhaps resonance absorption as observed in the gas phase. (Cf. Section 3.2.) The question of Auger satellites in solids is still under active consideration.[76]

4.1.3. Characteristic Energy Losses

Electrons ejected from a solid can suffer characteristic energy losses, usually due to plasmon losses. Since Auger spectra are generally rather complex and often not well resolved and are spread over a considerable range of energies, peaks from characteristic energy losses are much more difficult to disentangle from the normal Auger spectrum than is usual in the case of photoelectron spectroscopy. Also, surface contamination will alter the nature of the characteristic loss peaks considerably.

4.1.4. Charging in Nonconducting Samples

Charging as the result of an impinging beam of electrons on a nonconductor is a particularly severe problem in Auger spectroscopy. Often the charging and instability of the charged surface will prevent a meaningful Auger spectrum. However, this problem can be overcome by choosing the proper angle of incidence and bombarding energy. The important factor is the ratio δ (the number of secondary electrons leaving the target to the number impinging on the target). If $\delta = 1$, the charge is stabilized; if <1, the charge is negative, and if >1, positive. Goldstein and Carlson[77] have worked on this problem with glass surfaces. They found the best spectra are taken when δ is initially >1 and stabilizes to a slightly positive charge, which prevents some of the secondary electrons from leaving. To obtain $\delta > 1$, a grazing incidence angle of 10° was used. (The smaller the grazing angle, the greater chance there is for secondary electrons to leave.) The choice of impact energy is also important, δ becoming less than 1 if the energy of the impinging beam of electrons is either too large or too small. In this case the primary beam needed to be between 1.5 and 3.0 keV.

4.2. High-Resolution Auger Spectroscopy with Solids

In this section we shall discuss use of Auger spectroscopy with solids in areas of interest other than surfaces. In addition, attention will be concentrated on data taken with high-resolution analyzers, and on those spectra whose energies are sufficiently low that instrumental broadening and broadening from the natural widths of shells do not exceed a net resolution of a couple of volts. Auger spectra accompanying nuclear decay have received a fair degree of attention in the past from beta spectroscopists (e.g., Ref. 5) and although the data were of great help in providing a basis for understanding the Auger process, they are generally not of practical interest to the chemist.

Fahlman *et al.*(78) studied a number of light elements using either metals (Na, Mg) or simple salts (KCl, K_2SO_4, $Na_2S_2O_3$) and Albridge *et al.*(79) added to these studies work on alkali metals and alkaline earth fluorides. Also, high-resolution studies have been done on the *M–NN* Auger spectra of Ag, Cd, In, Sb, and Te metals and NaI.(80) These spectra, not surprisingly, showed more the character of free elemental ions than molecular orbital structure (cf. Figure 6.20). Some variation in the fluoride spectra due to chemical environment was found. For example, the $K\text{–}L_{\text{II, III}}L_{\text{II, III}}\ {}^1D$ transition was 654.6, 654.4, and

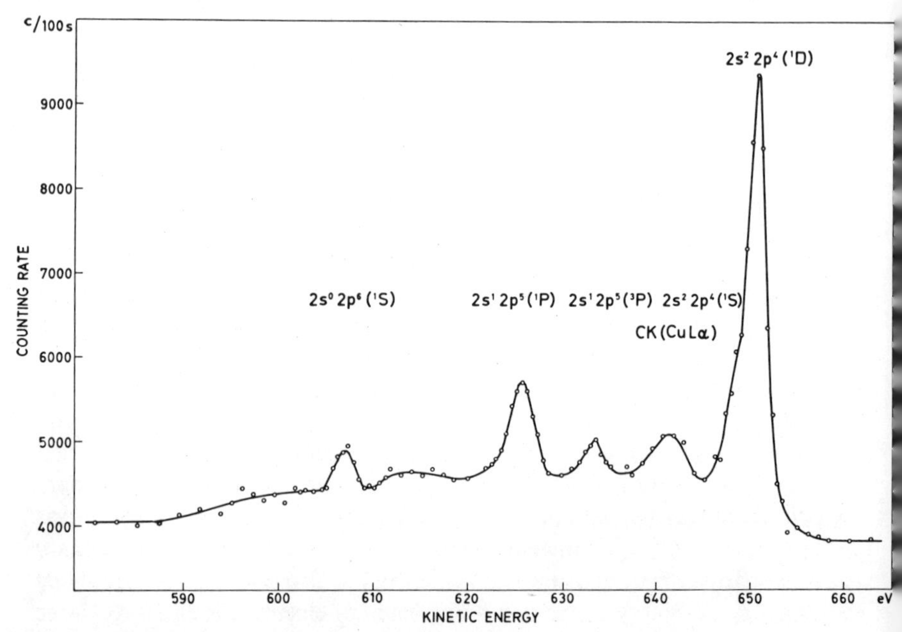

FIGURE 6.20. The *K–LL* Auger spectrum of fluorine in sodium fluoride using Cu(*L*) x rays as excitation source. [Reproduced from Albridge *et al.*,(79) Figure 2.]

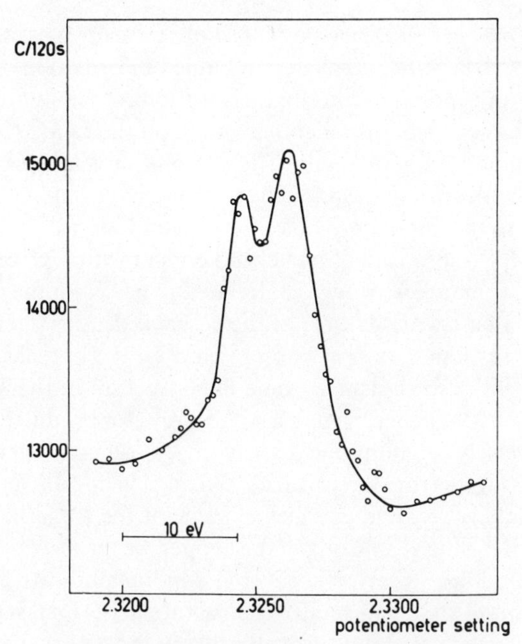

FIGURE 6.21. Spectrum of the $KL_1L_1(^1P_1)$ normal Auger line from KCl and its low energy satellite. [Reproduced from A. Fahlman, K. Hamrin, G. Axelson, C. Nordling, and K. Siegbahn, *Z. Phys.* **192**, 484 (1966), Figure 2.]

657.3 for salts whose cations were Na, Li, Mg. In addition, there was variation in the broadening of the fluoride spectra. Besides the normal Auger spectra, satellite lines were found at energies slightly lower than the normal lines (Figure 6.21). These lines probably arise from some phenomenon similar to that discussed in Section 2.2, such as electron shakeoff. What is surprising here are the relatively large intensities. The *K–LL* spectrum of oxygen from TiO_2 indicates a more complex Auger spectrum than that deduced from only atomic considerations.

The experimental Auger spectra for solids are less satisfactory than the corresponding spectra of gases with regard to resolution and peak-to-background ratio. Line broadening in solids may be due to the more complex band structure or to experimental problems associated with the charging up of nonconductors. The poor peak-to-background ratios with solids are due to the unavoidable secondary electrons from inelastic scattering. The spectra are generally better when the initial inner shell vacancies are created by x radiation than by electron impact. Interference from plasmon loss peaks as produced by the higher energy Auger electrons presents a problem for measuring the lower energy Auger spectra. Still a great deal of added information other than element identification is possible with high-resolution Auger studies on solids.

Kumar[81] has carried out a detailed study on the *L–MM* Auger spectra from solid compounds of germanium. High-resolution *KLL* spectra for Mg[37] and Al[38] and their respective oxides have been taken using Al Kα and Ag $L\alpha$ x rays for creating the initial core holes. Powell and Mandl[82] have observed the L_{III}–$M_{\mathrm{II,\,III}}M_{\mathrm{IV,\,V}}$ spectrum of nickel and copper with high resolution. The data reveal new structure, which was correlated in part with the final states for atomic Cu and in part with features of the 3*d* band density of states as determined by soft x-ray emission spectroscopy and x-ray photoelectron spectroscopy. Yin *et al.*[83] also measured some fine structure in the *L–MM* Auger spectra of Cu and Zn. Wagner,[84] using Al $K\alpha$ x rays for producing Auger lines, studied a wide variety of compounds, including 44 different elements. Instrumental linewidths were from 0.5 to 1.0 eV. Strong chemical effects were noted for the shapes, intensities, and energies of some of the Auger lines. Schön[85] has observed triplet–singlet splitting in the copper Auger spectrum and Bassett *et al.*[86] have found fine structure in silver and indium indicative of spin–orbit splitting in the initial state and multiplet structure in the final state. Jørgensen and Berthou[87] have related multiplet splitting in the Auger spectrum of copper(I) and silver(I) to the effects of linear ligand fields.

4.3. General Analytical Use of Auger Spectroscopy

Although both PESIS and AES are surface measurements, photoelectron spectroscopy has been used extensively to study bulk properties, with lesser emphasis on surface properties, while the reverse is true for Auger spectroscopy. The more complex Auger spectra are harder to measure and interpret for chemical shifts, and studies using electron bombardment on nonconducting solids are difficult because of the extensive charging. A comprehensive analytical tool must obviously be able to handle more than just conductors. The above problems can be overcome in Auger spectroscopy and the advantage of high signal response leading to a rapid analysis makes AES a valuable tool for general analytical purposes.

As will be obvious from Section 4.4.2.5, the interest in chemical shifts in Auger spectroscopy of solids is rapidly growing. This growth will certainly be enhanced by the availability of Auger spectrometers with high resolution. The problem of charging for nonconducting samples may be overcome by the proper choice of grazing incidence radiation and energy of the electron impact beam (cf. Section 4.1.4.).

As an example of the use of AES in general analysis, see the work of Connell *et al.*[88] on some lunar samples. Carter *et al.*[89] have appraised the relative merits of PESIS and AES in analyzing particulate matter. Whereas photoelectron spectroscopy requires from 30 min to a day for a complete detailed analysis, Auger spectra can be taken in seconds. However, because of the greater complexity of Auger spectra and overlapping of Auger lines from different elements, it was not possible to unambiguously detect some of the elements. For example, it is generally not possible to determine lead from its low-energy Auger spectrum, because of interferences, and one has to use impact electrons with energies of 5 keV or greater in order to excite the higher energy Auger lines. Also, since the peak-to-background ratio is generally poorer in Auger spectroscopy, the sensitivity for elements in low concentration is not as good. Charging is a major problem and the requirement for a low grazing incidence means the analysis must confine itself close to the surface. Auger spectroscopy is also not as successful in measuring the chemical state of, for example, the nitrogen and sulfur compounds as is PESIS. However, in spite of limitations, Auger spectroscopy should still be quite valuable as a general analytical tool for studies other than surface work.

4.4. Use of Auger Spectroscopy in the Study of Surfaces

4.4.1. General Considerations

The first use of Auger spectra to aid in the analysis of surfaces was made by Lander[90] in 1953, who studied various surfaces with low-energy electrons (500–1000 eV) and measured the resultant Auger electrons that were emitted. The recent interest, however, was kindled by Scheibner and Tharp[91] and Weber and Peria,[92] who coupled Auger spectra measurements to a conventional low-energy electron diffraction (LEED) system. Since that time hundreds of papers have appeared using Auger spectroscopy as a tool for surface analysis.*

Since the use of Auger spectroscopy for studying surfaces has come primarily from the work of surface scientists, more attention has been placed on surface cleanliness than high resolution. In contrast, photoelectron spectroscopy has been more concerned with better resolution. One field may be said to be clean but imprecise, the other accurate but dirty. Actually, technology is capable of achieving both goals. Physical Electronics now markets an Auger instrument capable of 0.05% instrumental resolution of the initial kinetic energy, which still holds the vacuum in the 10^{-10} Torr region. We can expect to see more and more use of high-resolution AES in surface analysis.

The choice of an analyzer to be used in Auger studies with LEED was dictated by the needs for simplicity and high transmission and the limitations

* See Ref. 93 for a bibliography and Ref. 94 for reviews of this area.

placed by the desire for surface cleanliness. Thus, a retarding grid device was chosen (cf. Chapter 2, Section 2.2.1 for further description). The energy resolution of such a device is usually from 3% to almost 0.3%. One of the biggest handicaps of taking data by the retarding grid method is that the signal is accumulative. Even modulation of the signal still makes analysis of details of some of the lower-energy lines difficult. Since Auger data taken with a retarding grid spectrometer are plotted in terms of the differential intensity, a single Auger line has both a maximum and a minimum. By convention the energies for various lines are usually taken at the minimum.

The detector for the LEED type of retarding grid analyzer must be, by virtue of its geometry, an electrostatic collector. One obvious advantage for a dispersion instrument is that it can focus its signal on an electron multiplier. Thus, each electron event can be individually detected and recorded. In addition, the development of dispersion instruments with high transmission has removed the advantage of the large-solid-angle analysis capability of the retarding grid method.

However, whereas the LEED Auger spectrometers may be open to criticism with regard to the most effective energy analysis, workers in this field have been most studious in their development of clean vacuum systems and in surface preparation. See Chapter 2, Section 1.2.2 for details on obtaining clean surfaces. A pressure of at least 1×10^{-9} Torr is needed if one wishes to prevent complete coverage of the surface in 1 h. Much of the LEED–Auger work is done in the 10^{-10} Torr range or lower.

Modern Auger spectrometers with dispersion analysis have more than ample intensity. Thus, they still collect their signal on an electrostatic accumulator. The signal is processed electronically and is presented in a differentiated form. Grant *et al.*[95] have suggested processing the data electronically by first differentiating and then integrating back the spectrum. The background, which can usually be represented by a polynomial, is removed by such a procedure, while the signal is restored to its original form. The Auger spectra thus processed have a better peak signal-to-noise ratio, and the intensities of the peaks are easier to compare quantitatively in the integrated form.

As an analytical tool, Auger spectroscopy is quite successful for the lighter elements, particularly the first row elements from $Z = 3$ to 10. The fluorescence yield is less than 1%, so that these elements are difficult if not impossible to measure by use of their characteristic x rays. The K vacancies are filled in nearly every case by an Auger process. In addition, the K–LL Auger spectra are relatively simple and quite distinctive as to the element involved. However, any element can be studied (except H and He) by Auger spectroscopy and researchers have now extended the method throughout the periodic table. Comprehensive tables based on atomic binding energies have been prepared[9] for all Auger transitions for energies between 10 and 3000 eV for elements up to $Z = 103$. Comprehensive experimental data for most of the elements have also been reported.[15]

Auger spectroscopy for surface analysis has been used primarily for qualitative analysis. It is quite sensitive; 1–2% of a monolayer can generally be detected, and in some cases as little as 0.2–0.5% has been reported. The sensitivity in fact is limited by surface cleanliness rather than signal strength, since the cleanliness depends on time of exposure. It is also important to note that Auger spectra can be taken in milliseconds. In general, the cross section for inner shell ionization, and thus the sensitivity for detection, decreases with increasing binding energy. Therefore, the lower energy Auger processes that are still characteristic for a given element are the ones usually chosen for study, because of the greater cross section for producing these Auger processes and because such low-energy Auger electrons will emanate from layers closer to the surface.

Quantitative analysis is in principle also possible. The quantitative aspects of Auger spectroscopy have been discussed by Seah.[96] Meyer and Vrakking[97] have studied C, N, O, P, S, and Cl for quantitative analysis, combining AES with ellipsometry. Staib and Kirschner[98] have determined absolute atomic densities using Auger electron spectroscopy. The relative intensities depend on a number of variables: inner shell ionization cross sections, Auger transitions rates, and inelastic scattering of the emitted Auger electrons. The problems, in fact, are similar to those encountered in the use of PESIS for quantitative analysis. However, with careful calibration, one ought to be able to obtain quantitative analysis of a sample, assuming homogeneity, to an accuracy of about 10%. If the sample is not homogeneous at the surface, auxiliary experiments will have to be made before a final interpretation is reached. Use of Auger spectroscopy in quantitative analysis is not a precise measurement, but it can still be of great help in studying the surface.

To emphasize the surface layers, the angle between the electron beam and the surface plane of the target is made as small as possible (usually about 10°). By changing the grazing angle of the electron beam, one can vary the relative contributions of the surface layers, and thus determine if there is any inhomogeneity of the chemical composition of the surface layers. This information could also be correlated with studies of the relative intensities for Auger electrons and photoelectrons having different energies and different mean free paths, as discussed in Chapter 5, Section 5.2. Meyer and Vrakking[99] have shown that it is also possible to gain some in-depth information by variation of the energy of the primary impact electron.

One of the chief advantages of Auger spectroscopy as an analytical tool for solids is that a small but intense beam of electrons can be used to survey the sample. A normal electron beam has a current from 10 to 100 μA with a focal spot of less than 1 mm diameter. Thus, individual small portions of the surface can be studied one at a time. Spatial resolution for Auger analysis has been achieved for areas smaller than 1 μm in diameter[100] (cf. Figure 6.22). Auger spectroscopy has been wedded with other methods of surface analysis, such as LEED (Low Energy Electron Diffraction) and electron microscopy, so as to yield elemental analysis and structural information simultaneously.

4.4.2. Literature Survey of Surface Applications

A complete literature survey of the use of Auger spectroscopy for surface analysis will not be attempted here (see Refs. 93 and 94), but a sufficient number of examples will be cited to illustrate the scope of the field.

4.4.2.1. Surface Impurities. Auger spectroscopy can detect foreign elements down to about 1% of a monomolecular layer. For bulk analysis Thomas and Morabito[101] found that the detectable limits for boron and phosphorus translate into about 8×10^{18} atoms/cm^3 when mixed with silicon. Auger spectroscopy has become one of the criteria for surface cleanliness. For example, in the study of methods for surface cleaning, it has been found that although electrical heating removes oxygen and carbon from nickel[102] and steel,[102, 103] it also promotes diffusion of sulfur to the surface. Jenkins and Chung[104] found both carbon and sulfur diffused to the surface on heating of copper. Taylor[103] and Sickafus[105] demonstrated ion bombardment as a preferable method for cleaning the transition metals. Lambert *et al.*[106] studied the carbon overlay on Pt and also warned about the problem of cleaning Pt by heating.

Lassiter[107] studied contaminants as polycrystalline silver and Dufour *et al.*[108] examined the surface of high-purity magnesium and aluminum crystalline specimens.

4.4.2.2. Surface Coverage. Detailed studies can be made by Auger spectroscopy on the way in which surfaces are covered. For example, Weber and Johnson[109] measured the degree of deposition of K ions into silicon and germanium from 0.1 to 1.0 of the first surface layer, finding a linear relationship. Palmberg[110] discovered that the sticking coefficient of Xe on palladium (100) was independent of surface coverage up to 5.8×10^{14} atoms/cm^2 and then decreased suddenly to zero. A study of the adsorption of oxygen on tungsten was made by Musket and Ferrante.[111] Pollard[112] investigated the growth of thorium on the tungsten (100) plane. The tungsten Auger peak was found not to be covered by thorium even after an equivalent of 25 monolayers was deposited. The conclusion was that the thorium nucleated so as to cover only 5% of the total surface.

Joyce and Neave[113] studied the interaction of oxygen on silicon, obtaining a sticking constant of 8×10^{-4}, which was independent over 10^{-5}–10^{-8} Torr. The adsorption was found to be initially rapid, followed by a substantial change in rate as the monolayer neared completion.

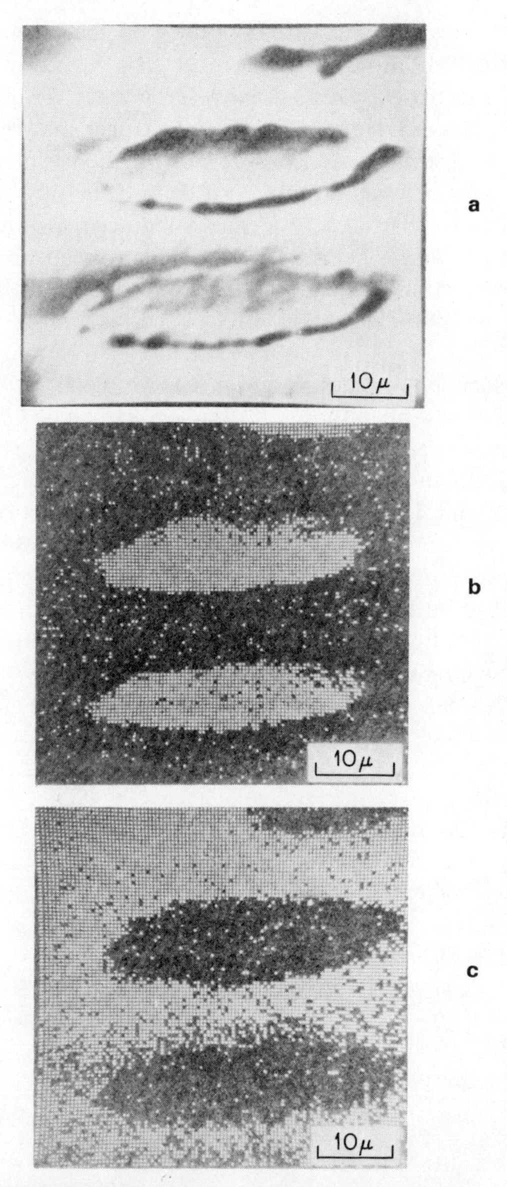

FIGURE 6.22. Study of iron–copper composite sample using scanning electron microscope: (a) secondary electron micrograph, (b) Auger image of iron (oval), (c) Auger image of copper (background). [Reproduced from MacDonald and Waldrop.[100]]

Bonzel[114] carried out careful studies on the reaction of oxygen with sulfur preadsorbed onto a Cu surface. He used this study to demonstrate the use of Auger spectroscopy in studying reaction rates. An example of his results is given in Figure 6.23. He supplemented the Auger analysis with mass spectral analysis of the absorbed gases. From this study Bonzel concluded that (1) fast surface reactions between S and O can be measured by Auger spectroscopy, (2) the kinetics and partial dependence of surface reactions indicated a Langmuir–Hinshelwood mechanism, and (3) as a consequence of high activation energies the reaction could be surface-diffusion controlled. Holloway and Hudson[115] have studied the kinetics of the reaction of oxygen with nickel surfaces.

The adsorption and desorption of CO on metal and alloy surfaces have been studied by several authors.[116] Studies were carried out down to −145°C and Auger measurements were coupled with LEED, work function measurements, and flash desorption.

Thomas and Haas[117] measured the adsorption of alkali metals on Mo(110). Results for cesium are shown in Figure 6.24. Both cesium and rubidium adsorption saturate out at about one monolayer, defined as one adsorbed atom per substrate atom. Potassium, on the other hand, continues to deposit above 1.2×10^{15}, indicating multilayer adsorption. Auger spectroscopy is a sensitive measurement for K, Rb, or Cs. As few as 5×10^{12} ions/cm^2 were detected of Rb. However, AES on sodium was relatively insensitive.

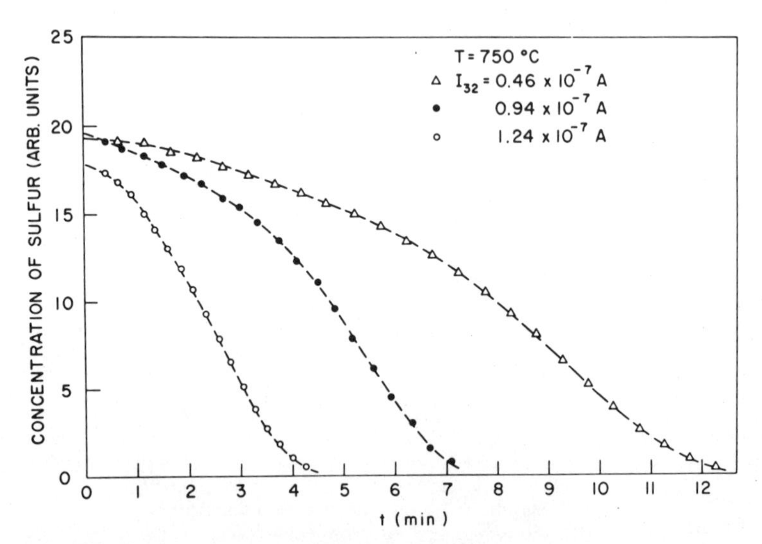

FIGURE 6.23. Concentration of adsorbed sulfur as a function of reaction time for constant temperature and partial pressure of oxygen. I_{32} is a direct measure of the partial pressure of oxygen. [Reproduced from Bonzel.[114]]

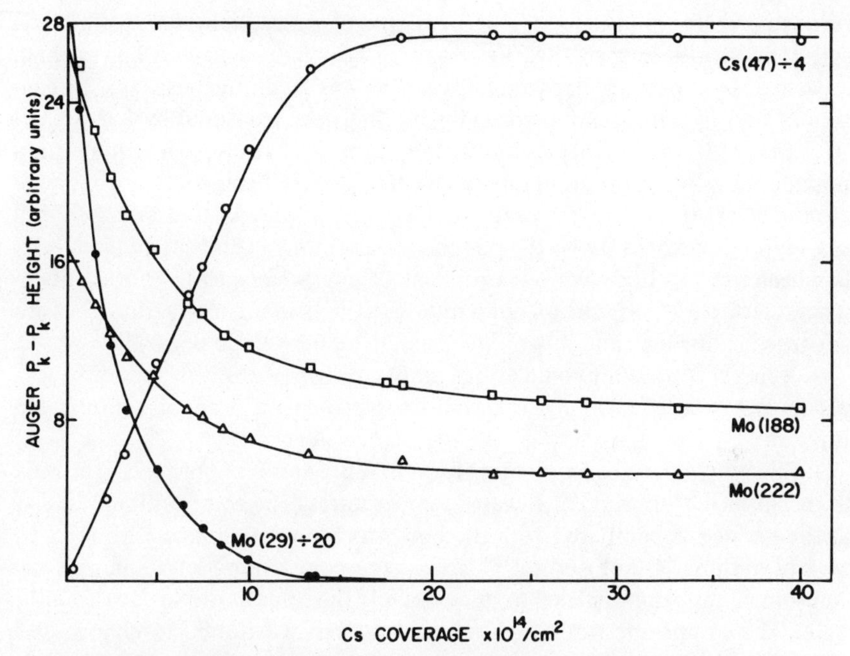

FIGURE 6.24. Variation of Cs and Mo Auger intensity as a function of Cs coverage of Mo(110) surface. [Reproduced from Thomas and Haas,(117) Figure 2.]

Tarng and Wehner(118) studied the deposition and sputter removal of molybdenum from Cu, Au, Al, and W surfaces. The sputtering yields of Mo from foreign surfaces are sometimes an order of magnitude lower than removal of molybdenum from itself; thus, caution must be used in employing ion bombardment for layer removal in determination of depth–composition profiles.

The general area of AES in the study of thin films has been reviewed by Weber,(119, 120) who points out that analysis can be carried out successfully on insulators as well as metals and semiconductors. The silicon–oxygen ratios have been determined in spin-on glass films at various temperatures by Smith *et al.*(121)

4.4.2.3. Inhomogeneity between the Bulk and Surface. The inhomogeneity of a given material can be nicely studied by Auger spectroscopy. For example, rocks of geological interest may be studied in detail over their whole surface without breaking apart the sample. Such an analysis was carried out on lunar samples by Connell and Gupta.(88)

Often the surface analysis of a supposedly homogeneous material is quite different than that of the bulk. Coad and Rivière(122) showed that carbon segregates to the surface of nickel foil and is present as Ni_3C below 673°K but turns to graphite about 573–873°K. This kind of segregation often occurs with sulfur impurities in metals as the result of heat treatment and has been so measured by Auger spectroscopy,[102, 103] as was mentioned earlier. Segregation of boron, sulfur, and nitrogen in iron has been measured by Bishop and Rivière.[123] It was further found that fracturing of steel occurs along grain boundaries containing antimony, these boundaries being only a couple of atomic layers thick. See, for example, the work of Marcus and Palmberg[124] in Figure 6.25, which shows the presence of antimony in embrittled steel. AES has been used to study various problems of interfaces, such as a glass fiber–resin interface,[125] the silicon–gold interface,[126] and the interaction between adsorbed sulfur ions and a two-dimensional copper sulfide phase.[127]

Auger spectroscopy could also be employed for the study of slow diffusion, particularly where knowledge of a very sharp profile, the order of a monolayer, is required.

Ellis[128] has studied the segregation of S, C, and P impurities in the bulk to the surface of thorium at elevated temperatures. The composition of alloys at the surface as compared with the bulk has been determined with AES by several authors. Quinto *et al.*[129] found the composition of a binary solid solution of copper and nickel to be essentially the same at the surface as in the bulk. The composition of Cu–Al alloy, however, was found to change as a function of temperature.[114, 130] Dooley[131] studied the behavior of zircalloys and was able to correlate the surface analysis (both impurities and alloy composition) with previously established mechanical properties.

Narusawa *et al.*[126] studied the nature of the mixed-phase formation of Si–Au (100–300°C) in an oxidizing atmosphere. They concluded that the reaction induces a mixed phase almost identical with the Si–Au alloyed phase obtained by heat treatment in a high vacuum.

4.4.2.4. Radiation Damage. One of the problems in the use of a high-intensity, sharply focused electron beam is radiation damage. This problem can be turned to an advantage in the study of surface effects due to electron bombardment. For example, Palmberg and Rhodin[132] and later Tokutaka *et al.*[133] investigated the surface dissociation of KCl under electron irradiation. The desorbed neutral particles of K, Cl, and Cl_2 were measured by mass spectroscopy, while the surface concentration was analyzed by Auger spectroscopy. For temperatures above 60°C surface stoichiometry was maintained, but below that value there was a net loss of Cl. Desorption of adsorbed gases such as oxygen on tungsten[134] and CO on silicon[135] has been studied by Auger spectroscopy. In the latter case it was determined that the radiation damage occurred in a two-step process: (1) CO is dissociated and (2) the carbon diffuses over the surface while oxygen remains fixed. This was ascertained by monitoring different portions of the surface by Auger spectroscopy, but maintaining constant irradiation only at one point. Kirby and Lichtman[136] have recently

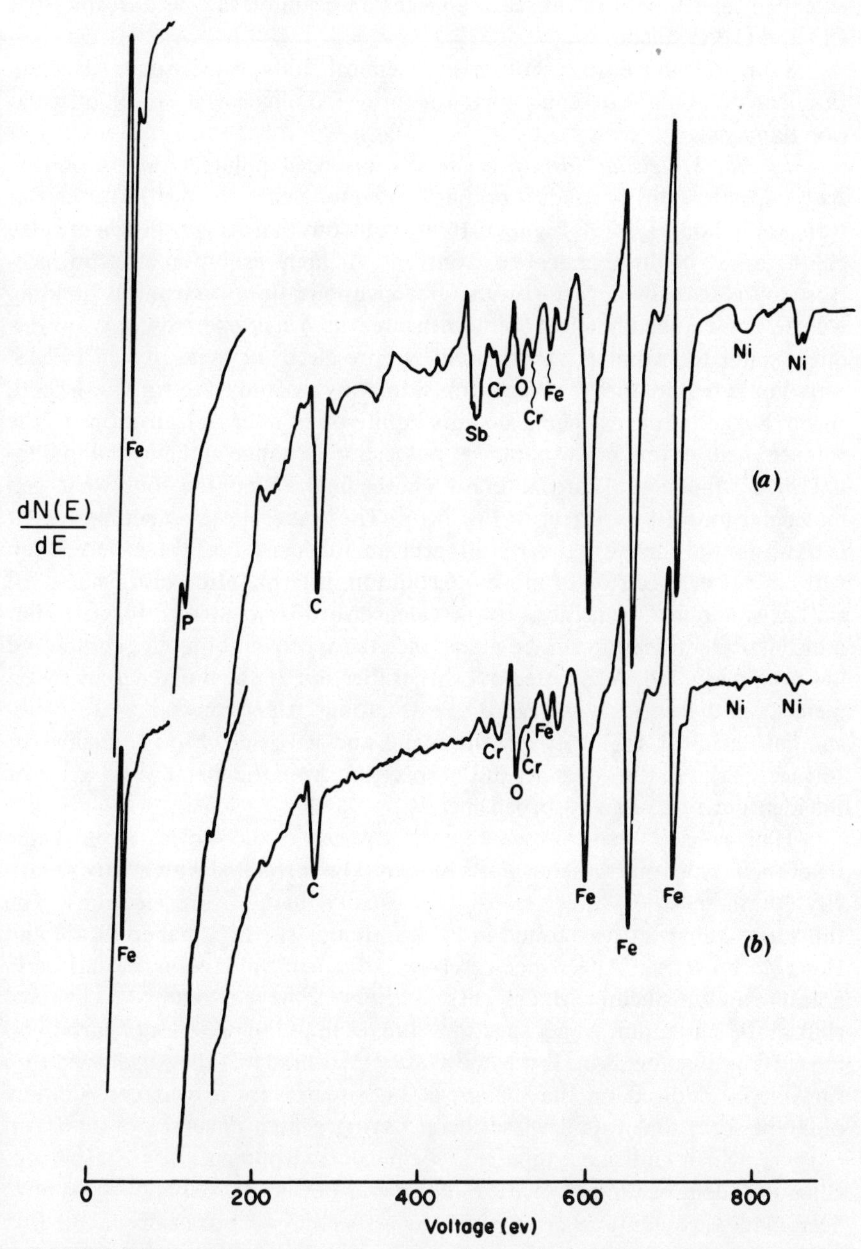

FIGURE 6.25. Auger spectra of fracture surface of AISI 3340 steel (a) embrittled, (b) unembrittled. [Reproduced from Marcus and Palmberg.[124]]

extended the study of the effect of an electron beam on the gas adsorption of CO and O_2 on silicon.

Salmerón and Baró[137] observed chemical shifts in the Auger spectrum of silicon of SiO_2 due to reduction of the surface to silicon as the result of radiation damage.

4.4.2.5. Molecular Identification. As has been pointed out previously, Auger spectroscopy of solids is primarily used for elemental identification. But from studies on gases (cf. Figure 6.16) it is obvious that Auger spectra are also characteristic of the chemical environment. As high-resolution electron spectroscopy is used in surface studies, greater application to chemical bonding will be made. The chemical shifts measured in Auger spectroscopy are frequently not the same as experienced by core electrons measured in PESIS. This would require that the Auger transitions involve only electrons in the core shells. Rather, Auger "chemical shifts" quite often refer to transitions to the valence shell, so that one experiences not a simple change in the overall potential but a rather complicated alteration in the final state of the doubly charged molecular ions. (Again refer to Fig. 6.16.) These alterations sometimes result in a substantial change in the overall spectrum that can be easily measured even with an Auger spectrum of modest resolution. Interpretation of the chemical shift arising out of transitions to the valence shell is not straightforward, but molecular identification can be made once the spectrum of a pure compound has been measured. As pointed out earlier (Section 2), the differences between metals and dielectrics with regard to extraatomic relaxation can create a substantial chemical shift between the metal and its oxide. Most valuable for surface work are thus changes in the spectrum from the metal to its oxide or the identification of an adsorbed species.

Haas *et al.*[138] have presented a short review of the subject. From Table III of their paper one sees that shifts often can be correlated with electronegativity. In these cases Auger transitions usually involve core electrons. The shifts are similar to those found in PESIS, though slightly smaller. Grant and Haas[139] have used Auger spectroscopy to distinguish between the carbon in graphite and in silicon carbide. These authors[140] have also noted a chemical shift in the ruthenium Auger spectra as the result of the presence of carbon on the surface of ruthenium. Joyce and Neave[113] found that the Auger spectrum for silicon oxidized on the surface at high temperatures was considerably different from that found with silicon having a high coverage of adsorbed oxygen, which indicated different chemical environments for the silicon. Chemical identification of mixtures of Be and BeO and of Al and Al_2O_3 have been made through their distinctive Auger spectra.[141] For example, see Fig. 6.26.

LeJeune and Dixon[142] have offered an interpretation of oxidized beryllium in terms of total density of states. Szalkowski and Somorjai[143] have

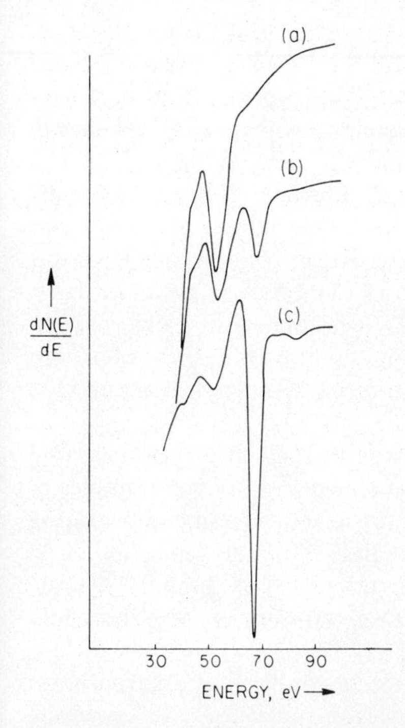

FIGURE 6.26. Auger spectra during progressive cleaning of pure Al sample. Curve (a) is essentially due to Al_2O_3, (b) is due to Al + Al_2O_3, and (c) is due to "clean" Al. [Reproduced from Quinto and Robertson.[141]]

studied the chemical shifts of the L_{II}–$M_{II,\,III}M_{II,\,III}$ Auger transitions as a function of the oxidation state of vanadium, the surface composition of which was in turn ascertained from the Auger spectrum of oxygen. Chemical shifts between uranium oxide and its metal[140] and between Y and its oxide[144] have also been measured. It has been found that though chemisorption onto Pt[146] does not yield chemical shifts in the Auger energies, there is a decided modification of some of the Auger transition probabilities involving platinum valence electrons. Seo *et al.*[147] have studied the Auger chemical shifts of passive films and Dadayan *et al.*[148] have measured the shifts in the Auger spectrum of molybdenum and tungsten during oxidation.

Characteristic energy losses as Auger electrons emerge from the solid have been used to learn about the solid state structure of the material under study. Coad and Rivière[149] have used characteristic energy loss data combined with changes in the Auger spectrum to give information regarding the formation of surface oxide on Cr, V, Fe, and Co. Melles *et al.*[150] have measured characteristic energy losses in the Auger spectrum of phosphorus films.

When Auger transitions involve the valence band, one can in principle learn about the density of states from details of the Auger spectrum. This has been done for Mg_2Sn by Tejeda *et al.*[151] and the results compared with photoemission results using both XPS and UPS.

4.4.2.6. Combination of Auger Spectroscopy with Other Surface Techniques. Auger spectroscopy is a particularly powerful tool when it can be combined with other techniques for studying the surface. Auger spectroscopy in fact received its initial impetus as the result of combining it with LEED. Low-energy electron diffraction gives information on the structure of the surface of a crystal. (It is restricted to measurements of a single crystal.) LEED is generally insensitive to the detection of impurities and gives information on structure, but not the elemental composition of the surface. Auger spectroscopy does this admirably and offers an ideal complement to LEED. For example, on cleaning Pt, a 1×5 pattern occurred in the LEED measurements, which was explained(152) with the help of Auger spectroscopy as due to the raising of oxygen impurities below the surface to close to the surface such as to cause an undulating surface and thus the 1×5 pattern.

Surface electrical properties of ZnO single crystals have been measured and correlated with LEED and Auger measurements.(153) Other examples of the use of Auger spectroscopy as a monitor for surface contamination during LEED studies are Jackson and Hooker's studies(154) on the deposition of Al onto Nb and Holcombe *et al.*'s investigation(155) of lithium hydride surface. For a review of Auger and LEED spectroscopies, see Tracy and Burkstrand.(94c)

As with LEED, scanning electron microscopy utilizes an electron beam so that the opportunity for Auger spectroscopy is also provided simultaneously. As one maps the topographical features of a material by electron microscopy, one can also map the chemical composition of the surface by Auger spectroscopy. MacDonald and Waldrop(100) have given a particularly nice demonstration of the value of combining scanning electron microscopy with Auger spectroscopy, and have discussed its potentialities (cf. Figure 6.22). Griffiths *et al.*(156, 157) have extended this discussion. Physical Electronics is now marketing an Auger microprobe. An x-ray emission microprobe has been for several years an important supplement to scanning electron microscopy. Let us make a few comparisons. X rays measure the bulk concentration better and are less sensitive to surface-cleanliness and high-vacuum requirements. An Auger microprobe sees the surface. Since the depth of material that is being probed is smaller, an Auger microprobe has a better ultimate resolution. More important, an Auger microprobe can study the very light elements, which cannot be handled by the x-ray emission microprobe.

Auger electron spectroscopy has also been combined with high energy electron diffraction, as, for example, in the work of Henderson and Helm,(158) who demonstrated that clean silicon surfaces could be grown in an even homoepitaxial manner by pyrolysis of SiH_4. A study was made of cesium adsorption on tungsten and titanium using the combined measurements of ellipsometry, Auger spectroscopy, and surface potential differences. The deposition of Cs^+

did not occur unless some oxygen was detected on the sample by Auger spectroscopy.

An apparatus has been described by Palmberg in which surface chemical analysis by AES is carried out simultaneously with inert gas sputtering to obtain compositional profiles of the surface normal to the solid–vacuum interface. Robinson and Jarvis[159] have also used sputter etching with Auger spectroscopy to examine the alloying behavior of the Ni/Au–Ge Ohmic contact to *n*-type GaAs.

Nishijima and Murotani[160] have combined Auger electron spectroscopy with electron impact desorption (EID) in a study on silicon surfaces. Coad and Rivière[149] have used characteristic energy loss data combined with changes in the Auger spectrum to give information regarding the formation of surface oxide on Cr, V, Fe, and Co. Likewise, Kawai *et al.*[161] have followed the oxidation of molybdenum surfaces by combining AES with energy loss spectroscopy. Willis *et al.*[162] have combined structure seen in the low-energy portion of the secondary electron energy distribution (3.0–17.5 eV) with Auger analysis to study carbon fiber surfaces.

Auger spectroscopy can be used successfully with a variety of other surface techniques: x-ray photoelectron spectroscopy in which both Auger and photoelectrons are produced in the same spectrum, high-energy electron spectroscopy for finding information regarding the structure of a material in the vicinity of the surface, and mass spectroscopy for analyzing desorbed gases and other particles ejected from the surface. Such an interweaving of these and other physical techniques has given rise to a substantial advance in the field of surface science.

4.5. Other Methods for Surface Analysis

During the recent development of Auger electron spectroscopy as an analytical tool for surfaces, other new methods have also been developed. First, we shall contrast AES with its closest counterpart, PESIS. Then we shall make brief mention of other miscellaneous physical techniques for chemical analysis of the surface.

4.5.1. Comparison of PESIS and Auger Spectroscopy for Surface Studies

The possible use of x-ray photoelectron spectroscopy for studying surfaces has been discussed in Chapter 5, Section 5.2. Comparison between the two techniques, photoelectron spectroscopy and Auger spectroscopy, for surface analysis has been made throughout this chapter. In this section I should like to tie this information together.

Both methods, PESIS and AES, study approximately the same portions of the surface. Although in principle an electron beam can produce a much stronger initial signal, in practice PESIS is almost as sensitive and in some cases more so when there is a large number of elements present. This is due in part to the better peak-to-background ratio found with photoionization as compared with electron bombardment, and with the fact that the limitation of sensitivity for surface detection lies ultimately more with surface cleanliness than with the sensitivity for measurement. Of course, Auger spectra can be obtained much faster, an important consideration if rapid changes on the surface are to be studied.

An important advantage of AES is that a beam with a highly focused spot can be used, so that the surface of a given material can be studied point by point, and AES may be used in conjunction with other surface-measuring devices such as LEED or electron microscopy. Somewhat offsetting these advantages is the care one must take with radiation damage of the sample being caused by an electron beam. In addition, there is the possibility of carbon deposits resulting from radiolysis of deposited hydrocarbon vapor (Chang[94b] warns that the pressure due to hydrocarbon vapor should be held below 5×10^{-8} Torr). Also, there is the difficulty produced by charging of the sample if it is an insulator.

The chief advantage in using PESIS is that the spectra are simpler and easier to analyze, and that chemical shift measurements with their information concerning changes in the chemical environment are much easier to make and interpret. It is possible, however, in some cases to examine the nature of chemical shifts with Auger spectroscopy, and the importance of this should grow with application of high resolution analyzers.

Since both x-ray photoelectron spectroscopy and Auger spectroscopy can be carried out with the same electron spectrometer, it can be expected that in the future the two techniques will be used together frequently in the study of surface phenomena. For example, Coad and Cunningham[163] have combined the two techniques in the study of the oxidation of steel.

4.5.2. Methods of Surface Analysis Other than AES and PESIS

Gerlach *et al.*[164] have proposed the differentiation of the total integrated distribution (a total secondary electron yield) with respect to primary electron energy. What this accomplishes is a measurement of resonance absorption of core electrons into unfilled energy states above the Fermi level. This in turn identifies the atom by means of its binding energy. This method has been called ionization spectroscopy (IS).

Closely related to IS is appearance potential spectroscopy (APS), in which x-ray fluorescence is measured at the ionization threshold of atomic levels by variation of the electron impact energy (see Ref. 165 for reviews). Musket[166] compared APS with Auger electron spectroscopy and found that AES is more sensitive for elemental analysis and that APS should be reserved for the study of chemical bonding of atoms on surfaces. Grant *et al.*[167] have combined AES and APS in their studies on titanium and titanium monoxide.

Secondary ion mass spectroscopy (SIMS) is a technique in which ions are sputtered off a surface and analyzed with a mass spectrometer. Under certain conditions SIMS is extremely sensitive. As few as parts per billion of alkali metals can be determined. Electron spectroscopy has, however, a much greater versatility for general analysis. For a comparison of the two methods, see Niehus and Bauer.[168] Examples[169] also exist of combining AES with SIMS.

Ion scattering spectrometry is (ISS) is a particularly important tool for surface work, since by its very nature it examines essentially only the outermost atomic layer (for a review of ISS, see Ref. 170). It operates simply on the conservation of energy and momentum, i.e.,

$$\frac{E_1}{E_0} = \frac{M_s - M_0}{M_s + M_0}$$

where E_0 and M_0 are the mass and kinetic energy of the impact ions, E_1 is the energy of the ion scattered at 90°, and M_s is the mass of the surface atoms. The mass analysis yields identification of the element. No information on chemical bonding is obtained, but studies on isotopic distribution are possible. The sensitivity is claimed to be approximately 1% of the surface monolayer for most elements. The samples studied must be reasonably flat, but all types of surface can be studied, from metals to insulators, either crystalline or amorphous.

Although new methods for surface analysis proliferate, Auger electron spectroscopy has proven itself highly versatile, reasonably sensitive, and capable of analyzing almost any element in the periodic table. It is unlikely that its usefulness for surface studies will not continue to grow.

Appendix 1

Atomic Binding Energies for Each Subshell for Elements $Z = 1$–106

Values are given in eV for free atoms referenced to the vacuum potential. (For solids one must subtract the work function if referenced to the Fermi level, which is on the average about 4 eV.)

The table is based mainly on Lotz,* with the following exceptions: (1) Values of Kr and Xe have been amended according to Siegbahn *et al.*† (2) Experimental data (+4 eV) of M. O. Krause (private communication) are used for Am, Cm, Bk, Cf, and Es as marked with asterisk [see also Krause and Wuilleumier‡]. (3) Values of subshells K, L, M, and $N_{1,2,3}$ of elements 96–106 taken from Carlson *et al.*§ (4) Values for subshells $N_{4,5,6,7}$, O, P, and Q for $Z = 96$–106 and $P_{1,2,3}$ for $Z = 87$–95 taken from eigenvalues calculated by Lu *et al.*‡‡ (5) Values plus 4 eV marked with dagger taken from Bearden and Burr.**

* W. Lotz, *J. Opt. Soc. Am.* **60**, 206 (1970).
† K. Siegbahn *et al.*, *ESCA Applied to Free Molecules* (North-Holland, Amsterdam, 1969).
‡ M. O. Krause and F. Wuilleumier in *Electron Spectroscopy, Proc. Int. Conf., Asilomar, 1971*, ed. by D. A. Shirley (North-Holland, Amsterdam, 1972), p. 759.
§ T. A. Carlson, C. W. Nestor, Jr., F. B. Malik, and T. C. Tucker, *Nucl. Phys.* **A135**, 57 (1969).
‡‡ C. C. Lu, T. A. Carlson, F. B. Malik, T. C. Tucker, and C. W. Nestor, Jr., *Atomic Data* **3**, 1 (1971).
** J. A. Bearden and A. F. Burr, *Rev. Mod. Phys.* **39**, 125 (1967).

TABLE A1.A

Binding Energies of Electrons in Free Atom (eV) : $Z = 1–53$

Z		K	L_1	L_2	L_3	M_1	M_2	M_3	M_4	M_5	N_1	N_2	N_3	N_4	N_5	O_1	$O_{2,3}$
1	H	13.60															
2	He	24.59															
3	Li	58	5.392														
4	Be	115	9.322														
5	B	192	12.93	8.298													
6	C	288	16.59	11.26													
7	N	403	20.33	14.53													
8	O	538	28.48	13.62													
9	F	694	37.85	17.42													
10	Ne	870.1	48.47	21.66	21.56												
11	Na	1075	66	34	34	5.139											
12	Mg	1308	92	54	54	7.646											
13	Al	1564	121	77	77	10.62	5.986										
14	Si	1844	154	104	104	13.46	8.151										
15	P	2148	191	135	134	16.15	10.49										
16	S	2476	232	170	168	20.20	10.36										
17	Cl	2829	277	208	206	24.54	12.97										
18	Ar	3206.3	326.5	250.6	248.5	29.24	15.94	15.76									
19	K	3610	381	299	296	37	19	18.7			4.341						
20	Ca	4041	441	353	349	46	28	28			6.113						
21	Sc	4494	503	408	403	55	33	33	8		6.540						
22	Ti	4970	567	465	459	64	39	38	8		6.820						
23	V	5470	633	525	518	72	44	43	8		6.740						

24	Cr	5995	702	589	580	80	49	48	8.25		6.765						
25	Mn	6544	755	656	645	89	55	53	9		7.434						
26	Fe	7117	851	726	713	98	61	59	9		7.870						
27	Co	7715	931	800	785	107	68	66	9		7.864						
28	Ni	8338	1015	877	860	117	75	73	10	10	7.635						
29	Cu	8986	1103	958	938	127	82	80	11	10.4	7.726						
30	Zn	9663	1198	1047	1024	141	94	91	12	11.2	9.394						
31	Ga	10371	1302	1146	1119	162	111	107	21	20	11	6.00					
32	Ge	11107	1413	1251	1220	184	130	125	33	32	14.3	7.90					
33	As	11871	1531	1362	1327	208	151	145	46	45	17	9.81					
34	Se	12662	1656	1479	1439	234	173	166	61	60	20.15	9.75					
35	Br	13481	1787	1602	1556	262	197	189	77	76	23.80	11.85					
36	Kr	14327	1924.6	1730.9	1678.4	292.8	222.2	214.4	95.0	93.8	27.51	14.65	14.00				
37	Rb	15203	2068	1867	1807	325	251	242	116	114	32	16	15.3			4.18	
38	Sr	16108	2219	2010	1943	361	283	273	139	137	40	23	22			5.69	
39	Y	17041	2375	2158	2083	397	315	304	163	161	48	30	29	6.38		6.48	
40	Zr	18002	2536	2311	2227	434	348	335	187	185	56	35	33	8.61		6.84	
41	Nb	18990	2702	2469	2375	472	382	367	212	209	62	40	38	7.17		6.88	
42	Mo	20006	2872	2632	2527	511	416	399	237	234	68	45	42	8.56		7.10	
43	Tc	21050	3048	2800	2683	551	451	432	263	259	74	49	45	8.6		7.28	
44	Ru	22123	3230	2973	2844	592	488	466	290	286	81	53	49	8.50		7.37	
45	Rh	23225	3418	3152	3010	634	526	501	318	313	87	58	53	9.56		7.46	
46	Pd	24357	3611	3337	3180	677	565	537	347	342	93	63	57	8.78	8.34		
47	Ag	25520	3812	3530	3357	724	608	577	379	373	101	69	63	11	10	7.58	
48	Cd	26715	4022	3732	3542	775	655	621	415	408	112	78	71	14	13	8.99	
49	In	27944	4242	3943	3735	830	707	669	455	447	126	90	82	21	20	10	5.79
50	Sn	29204	4469	4160	3933	888	761	719	497	489	141	102	93	29	28	12	7.34
51	Sb	30496	4703	4385	4137	949	817	771	542	533	157	114	104	38	37	15	8.64
52	Te	31820	4945	4618	4347	1012	876	825	589	578	174	127	177	48	46	17.84	9.01
53	I	33176	5195	4858	4563	1078	937	881	638	626	193	141	131	58	56	20.61	10.45

TABLE A1.B

Binding Energies of Electrons in Free Atom (eV) : $Z = 54–80$

Z		K	L_1	L_2	L_3	M_1	M_2	M_3	M_4	M_5	N_1	N_2	N_3
54	Xe	34565	5453.2	5107.2	4787.3	1148.7	1002.1	940.6	689.0	676.4	213.2	157	145.5
55	Cs	35987	5717	5362	5014	1220	1068	1000	742	728	233	174	164
56	Ba	37442	5991	5626	5249	1293	1138	1063	797	782	254	193	181
57	La	38928	6269	5894	5486	1365	1207	1124	851	834	273	210	196
58	Ce	40446	6552	6167	5726	1437	1275	1184	903	885	291	225	209
59	Pr	41995	6839	6444	5968	1509	1342	1244	954	934	307	238	220
60	Nd	43575	7132	6727	6213	1580	1408	1303	1005	983	321	250	230
61	Pm	45188	7432	7017	6464	1653	1476	1362	1057	1032	335	261	240
62	Sm	46837	7740	7315	6720	1728	1546	1422	1110	1083	349	273	251
63	Eu	48522	8056	7621	6981	1805	1618	1484	1164	1135	364	286	262
64	Gd	50243	8380	7935	7247	1884	1692	1547	1220	1189	380	300	273
65	Tb	51999	8711	8256	7518	1965	1768	1612	1277	1243	398	315	285
66	Dy	53792	9050	8585	7794	2048	1846	1678	1335	1298	416	331	297
67	Ho	55622	9398	8922	8075	2133	1926	1746	1395	1354	434	348	310
68	Er	57489	9754	9267	8361	2220	2008	1815	1456	1412	452	365	323
69	Tm	59393	10118	9620	8651	2309	2092	1885	1518	1471	471	382	336
70	Yb	61335	10490	9981	8946	2401	2178	1956	1580	1531	490	399	349
71	Lu	63320	10876	10355	9250	2499	2270	2032	1647	1596	514	420	366
72	Hf	65350	11275	10742	9564	2604	2369	2113	1720	1665	542	444	386
73	Ta	67419	11684	11139	9884	2712	2472	2197	1796	1737	570	469	407
74	W	69529	12103	11546	10209	2823	2577	2283	1874	1811	599	495	428
75	Re	71681	12532	11963	10540	2937	2686	2371	1953	1887	629	522	450
76	Os	73876	12972	12390	10876	3054	2797	2461	2035	1964	660	551	473
77	Ir	76115	13422	12828	11219	3175	2912	2554	2119	2044	693	581	497
78	Pt	78399	13883	13277	11567	3300	3030	2649	2206	2126	727	612	522
79	Au	80729	14356	13738	11923	3430	3153	2748	2295	2210	764	645	548
80	Hg	83108	14845	14214	12288	3567	3283	2852	2390	2300	806	683	579

TABLE A1.B (*continued*)

Z		N_4	N_5	N_6	N_7	O_1	O_2	O_3	O_4	O_5	P_1
54	Xe	69.5	67.5			23.39	13.43	12.13			
55	Cs	81	79			25	14	12.3			3.89
56	Ba	94	92			31	18	16			5.21
57	La	105	103			36	22	19	5.75		5.58
58	Ce	114	111	6		39	25	22	6		5.65
59	Pr	121	117	6		41	27	24			5.42
60	Nd	126	122	6		42	28	25			5.49
61	Pm	131	127	6		43	28	25			5.55
62	Sm	137	132	6		44	29	25			5.63
63	Eu	143	137	6		45	30	26			5.68
64	Gd	150	143	6		46	31	27	6		6.16
65	Tb	157	150	6		48	32	28	6		5.85
66	Dy	164	157	6		50	33	28			5.93
67	Ho	172	164	6		52	34	29			6.02
68	Er	181	172	6		54	35	30			6.10
69	Tm	190	181	7		56	36	30			6.18
70	Yb	200	190	8	7	58	37	31			6.25
71	Lu	213	202	13	12	62	39	32	6.6		7.0
72	Hf	229	217	21	20	68	43	35	7.0		7.5
73	Ta	245	232	30	28	74	47	38	8.3		7.9
74	W	261	248	38	36	80	51	41	9.0		8.0
75	Re	278	264	47	45	86	56	45	9.6		7.9
76	Os	295	280	56	54	92	61	49	9.6		8.5
77	Ir	314	298	67	64	99	66	53	9.6		9.1
78	Pt	335	318	78	75	106	71	57	9.6		9.0
79	Au	357	339	91	87	114	76	61	12.5	11.1	9.23
80	Hg	382	363	107	103	125	85	68	14	12	10.4

TABLE A1.C

Binding Energies of Electrons in Free Atom (eV) : $Z = 81\text{–}106$

Z	K	L_1	L_2	L_3	M_1	M_2	M_3	M_4	M_5	N_1	N_2	N_3
81 Tl	85536	15350	14704	12662	3710	3420	2961	2490	2394	852	726	615
82 Pb	88011	15867	15206	13041	3857	3560	3072	2592	2490	899	769	651
83 Bi	90534	16396	15719	13426	4007	3704	3185	2696	2588	946	813	687
84 Po	93106	16937	16244	13816	4161	3852	3301	2802	2687	994	858	724
85 At	95729	17490	16782	14212	4320	4005	3420	2910	2788	1044	904	761
86 Rn	98404	18055	17334	14615	4483	4162	3452	3019	2890	1096	951	798
87 Fr	101134	18637	17903	15028	4652	4324	3666	3134	2998	1153	1003	839
88 Ra	103919	19237	18488	15449	4827	4491	3793	3254	3111	1214	1060	884
89 Ac	106759	19850	19086	15874	5005	4661	3921	3374	3223	1274	1116	928
90 Th	109654	20475	19696	16303	5185	4833	4049	3494	3335	1333	1171	970
91 Pa	112604	21112	20318	16735	5368	5008	4178	3613	3446	1390	1225	1011
92 U	115611	21762	20953	17171	5553	5187	4308	3733	3557	1446	1278	1050
93 Np	118676	22427	21602	17612	5742	5370	4440	3854	3669	1504	1331	1089
94 Pu	121800	23109	22267	18059	5936	5557	4574	3977	3783	1563	1384	1128
95 Am	124984	23803	22949	18512	6135	5748	4710	4102	3898	1623	1439	1167
96 Cm	128261	24523	23654	18974	6313	5946	4828	4236	4014	1664	1493	1194
97 Bk	131586	25260	24372	19440	6523	6149	4968	4366	4133	1729	1554	1236
98 Cf	134967	26008	25103	19907	6733	6352	5103	4492	4247	1789	1610	1273
99 Es	138440	26781	25859	20383	6954	6566	5247	4225	4369	1857	1674	1316
100 Fm	141962	27581	26642	20872	7187	6791	5399	4767	4498	1933	1746	1366
101 Md	145571	28390	27434	21354	7413	7010	5540	4898	4616	1998	1807	1404
102 No	149273	29228	28253	21849	7651	7242	5689	5037	4743	2071	1876	1449
103 Lw	153040	30091	29101	22356	7903	7487	5848	5185	4877	2153	1954	1501
104	156889	30974	29973	22870	8161	7738	6009	5336	5014	2237	2034	1554
105	160820	31884	30869	23389	8427	7997	6172	5489	5152	2324	2117	1609
106	164820	32811	31787	23915	8698	8263	6338	5644	5292	2413	2202	1664

TABLE A1.C (*continued*)

Z	N_4	N_5	N_6	N_7	O_1	O_2	O_3	O_4	O_5	O_6	O_7	P_1	P_2	P_3	$P_{4,5}$	Q_1
81 Tl	411	391	127	123	139	98	79	21	19			8	6.11			
82 Pb	441	419	148	144	153	111	90	27	25			10	7.42			
83 Bi	472	448	170	165	167	125	101	34	32			12	7.29			
84 Po	503	478	193	187	181	139	112	41	38			15	8.43			
85 At	535	508	217	211	196	153	123	48	44			19	11	9.3		
86 Rn	567	538	242	235	212	167	134	55	51			24	14	10.7		
87 Fr	603	572	268	260	231	183	147	65	61			33	19	14		4.0
88 Ra	642	609	296	287	253	201	161	77	73			40	25	19		5.28
89 Ac	680	645	322	313	274	218	174	88	83			45	29	22	5.7	6.3
90 Th	717	679	347	338	293	233	185	97	91			50	33	25	6	6
91 Pa	752	712	372	362	312	248	195	104	97	6		50	32	24	6	6
92 U	785	743	396	386	329	261	203	110	101	6		52	34	24	6.1	6
93 Np	819	774	421	410	346	274	211	116	106	6		54	35	25	6	6
94 Pu	853	805	446	434	356†	287	219	122	111	6		53	34	23		6
95 Am	887*	836*	467*	452*	355*	301	220*	123*	112*	6		54*	44*	36*		6.0
96 Cm	919	864	494*	479*	384	314	239	126*	119*	11		60	39	27	5	6
97 Bk	955	898	520*	504*	401	329	248	142	124*	12		63	41	27	4	6
98 Cf	987	925	546*	529*	412	338	251	142	129	9		61	39	25		6
99 Es	1024	959	573*	554*	429	353	260	148	135*	9		63	40	25		6
100 Fm	1068	1000	606	587	453	375	275	160	145	15		69	45	29	4	7
101 Md	1100	1029	627	607	464	384	278	160	144	11		67	43	26		6
102 No	1139	1064	654	633	483	400	287	166	149	14	11	69	44	26		6
103 Lw	1186	1106	689	666	509	424	303	178	160	20	17	76	49	30	4	7
104	1233	1149	724	700	535	448	318	190	171	26	23	82	55	33	5	8
105	1281	1193	760	735	562	473	335	203	183	32	29	89	60	36	6	8
106	1330	1238	797	770	590	499	350	216	194	39	35	96	65	39	7	9

Appendix 2

Energy Separation between j Subshells in Atoms

The energy separation in eV between the filled j subshells $(l+\frac{1}{2}, l-\frac{1}{2})$ is tabulated for all the elements from $Z=10$ to 105. $\Delta L_{2,3}$, $\Delta M_{2,3}$, $\Delta M_{4,5}$, $\Delta N_{2,3}$, $\Delta N_{4,5}$, $\Delta N_{6,7}$, $\Delta O_{2,3}$, $\Delta O_{4,5}$, $\Delta O_{6,7}$, $\Delta P_{2,3}$ correspond, respectively, to differences in the subshell binding energies: $(2p_{1/2}-2p_{3/2})$, $(3p_{1/2}-3p_{3/2})$, $(3d_{3/2}-3d_{5/2})$, $(4p_{1/2}-4p_{3/2})$, $(4d_{3/2}-4d_{5/2})$, $(4f_{5/2}-4f_{7/2})$, $(5p_{1/2}-5p_{3/2})$, $(5d_{3/2}-5d_{5/2})$, $(5f_{5/2}-5f_{7/2})$, and $(6p_{1/2}-6p_{3/2})$. The values are taken from differences in eigenvalues listed by Lu *et al.** as obtained from relativistic Hartree–Fock–Slater wave functions for the free atoms. In many instances, particularly for the less tightly bound core shells, these values are more reliable than those available from experiment. The values should be of help to the electron spectroscopist in identifying the pairs of lines associated with a given subshell, and in deconvoluting that pair when not clearly separated in energy. Generally, the pair of lines has the relative intensities: $[2(l+\frac{1}{2})+1]/[2(l-\frac{1}{2})+1]$, or $p_{3/2}/p_{1/2}=2:1$, $d_{5/2}/d_{3/2}=3:2$, and $f_{7/2}/f_{5/2}=4:3$. These generalizations can break down when the photon energy is close to the ionization threshold or when Coster–Kronig transitions broaden the peak associated with the lower j subshell.

* C. C. Lu, T. A. Carlson, F. B. Malik, T. C. Tucker, and C. W. Nestor, Jr., *Atomic Data* 3, 1 (1971).

Z	$\Delta L_{2,3}$	$\Delta M_{2,3}$	$\Delta M_{4,5}$	$\Delta N_{2,3}$	$\Delta N_{4,5}$	$\Delta N_{6,7}$	$\Delta O_{2,3}$	$\Delta O_{4,5}$	$\Delta O_{6,7}$	$\Delta P_{2,3}$
10 Ne	0.12									
11 Na	0.20									
12 Mg	0.32									
13 Al	0.48									
14 Si	0.70									
15 P	0.99									
16 S	1.35									
17 Cl	1.81									
18 Ar	2.23	0.17								
19 K	3.05	0.30								
20 Ca	3.88	0.43								
21 Sc	4.87	0.56								
22 Ti	6.04	0.71								
23 V	7.41	0.89								
24 Cr	9.00	1.08								
25 Mn	10.8	1.35								
26 Fe	12.9	1.63								
27 Co	15.3	1.95								
28 Ni	18.1	2.32								
29 Cu	21.2	2.69	0.31							
30 Zn	24.6	3.20	0.40							
31 Ga	28.5	3.79	0.51							
32 Ge	32.8	4.47	0.65							
33 As	37.5	5.25	0.80							
34 Se	42.8	6.13	0.97							
35 Br	48.7	7.13	1.17							
36 Kr	55.1	8.25	1.40	0.72						
37 Rb	62.2	9.50	1.66	0.99						
38 Sr	70.0	10.9	1.95	1.29						
39 Y	78.4	12.5	2.28	1.59						
40 Zr	87.7	14.2	2.65	1.93						
41 Nb	97.8	16.1	3.06	2.25						

42 Mo	108.7	18.2	3.51	2.65			
43 Tc	120.6	20.5	4.01	3.14			
44 Ru	133.5	22.9	4.55	3.58			
45 Rh	147.3	25.7	5.15	4.11			
46 Pd	162.3	28.6	5.81	4.65	0.50		
47 Ag	178.5	31.9	6.52	5.33	0.63		
48 Cd	195.8	35.4	7.30	6.10	0.78		
49 In	214.5	39.2	8.14	6.97	0.95		
50 Sn	234.5	43.3	9.06	7.92	1.14		
51 Sb	255.9	47.7	10.1	8.98	1.35		
52 Te	278.9	52.5	11.1	10.1	1.59		
53 I	303.4	57.7	12.3	11.4	1.84		
54 Xe	329.6	63.2	13.5	12.8	2.13		1.41
55 Cs	358	69.2	14.9	14.3	2.44		1.86
56 Ba	387	75.8	16.3	16.0	2.78		2.32
57 La	419	82.4	17.8	17.8	3.16		2.77
58 Ce	453	89.8	19.5	19.5	3.49		3.01
59 Pr	489	97.7	21.3	21.2	3.79		3.02
60 Nd	527	106.1	23.2	23.2	4.16		3.27
61 Pm	568	115.1	25.2	25.3	4.57		3.53
62 Sm	611	124.6	27.3	27.5	4.99		3.80
63 Eu	657	134.8	29.6	29.9	5.44		4.08
64 Gd	705	145.6	32.0	32.6	5.98		4.68
65 Tb	756	157.2	34.6	35.1	6.42		4.69
66 Dy	810	169.4	37.3	38.0	6.97		5.01
67 Ho	868	182.4	40.2	41.1	7.54		5.35
68 Er	928	196.1	43.2	44.3	8.14		5.71
69 Tm	992	210.7	46.4	47.7	8.78		6.09
70 Yb	1060	226.2	49.8	51.4	9.45	1.45	6.48
71 Lu	1131	242.5	53.4	55.4	10.2	1.67	7.34
72 Hf	1206	259.8	57.1	59.7	11.1	1.90	8.28
73 Ta	1285	278.0	61.0	64.4	12.0	2.14	9.30
74 W	1368	297.3	65.1	69.3	12.9	2.41	10.4

Z	$\Delta L_{2,3}$	$\Delta M_{2,3}$	$\Delta M_{4,5}$	$\Delta N_{2,3}$	$\Delta N_{4,5}$	$\Delta N_{6,7}$	$\Delta O_{2,3}$	$\Delta O_{4,5}$	$\Delta O_{6,7}$	$\Delta P_{2,3}$
75 Re	1456	317.7	69.4	74.5	13.9	2.69	11.6			
76 Os	1548	339.2	74.0	80.2	15.0	3.00	12.9			
77 Ir	1646	361.9	78.7	86.1	16.2	3.33	14.3			
78 Pt	1748	385.9	83.7	92.5	17.4	3.67	15.7			
79 Au	1855	411	89.0	99.3	18.7	4.04	17.3	1.73		
80 Hg	1968	438	94.4	106.4	20.1	4.44	19.2	2.06		
81 Tl	2087	466	100.1	114.0	21.5	4.86	21.2	2.45		
82 Pb	2212	496	106.1	122.1	23.0	5.31	23.4	2.83		
83 Bi	2343	527	112.4	130.7	24.6	5.78	25.8	3.29		
84 Po	2481	560	118.9	139.8	26.3	6.29	28.4	3.72		
85 At	2626	594	125.7	149.4	28.1	6.82	31.1	4.22		
86 Rn	2778	631	132.8	159.6	30.0	7.39	34.1	4.74		4.07
87 Fr	2938	669	140.2	170.4	32.0	7.98	37.3	5.31		5.17
88 Ra	3105	709	147.9	181.8	34.1	8.61	40.7	5.91		6.30
89 Ac	3281	752	156.0	193.9	36.3	9.28	44.3	6.55		7.39
90 Th	3466	797	164.3	206.6	38.5	9.98	48.2	7.24		8.53
91 Pa	3660	844	173.1	220.1	41.0	10.7	52.0	7.80		8.81
92 U	3863	893	182.2	234.3	43.5	11.5	56.2	8.46		9.53
93 Np	4077	945	191.6	249.3	46.1	12.3	60.6	9.14		10.3
94 Pu	4301	1000	201.4	265.1	48.8	13.1	65.1	9.79		10.5
95 Am	4536	1057	211.6	281.8	51.7	14.0	70.0	10.5		11.3
96 Cm	4783	1118	222.2	299.5	54.7	14.9	75.3	11.4		12.6
97 Bk	5042	1182	233.2	318	57.8	15.9	80.9	12.2		13.5
98 Cf	5314	1248	244.6	338	61.0	16.9	86.5	13.0		13.7
99 Es	5600	1319	256.5	358	64.4	18.0	92.7	13.9		14.6
100 Fm	5899	1393	268.8	380	67.9	19.1	99.3	14.9		16.3
101 Md	6214	1471	281.5	403	71.6	20.2	106.0	15.8		16.5
102 No	6545	1552	294.7	427	75.4	21.4	113.3	16.8	2.56	17.6
103 Lw	6892	1639	308	453	79.4	22.7	121.2	17.9	2.89	19.5
104	7256	1729	322	480	83.6	24.0	129.5	19.1	3.23	21.7
105	7639	1824	337	509	87.9	25.3	138.4	20.3	3.59	24.0
106	8041	1924	352	539	92.4	26.7	147.8	21.7	3.97	26.4

Appendix 3

Compilation of Data on Shifts in Core Binding Energies

This appendix is a compilation of data relating to core binding energies as a function of their chemical environment. For each element references are given to papers where inner shell binding energies have been measured by means of x-ray photoelectron spectroscopy for a specific chemical state. For common elements, e.g., carbon or sulfur, I have been more selective in listing references. For elements where fewer studies have been carried out, I have tried to list each instance in which that element was measured. The references underlined, e.g., 10, are given on pp. 408–410. The other references, which constitute the majority, correspond to those given for Chapter 5.

For each element a selection of binding energies in eV for various compounds is given. These are chosen arbitrarily to give some qualitative indication of the scope of the chemical shifts. (The most commonly studied subshells are given in parentheses.) The energies of the solids are given relative to the Fermi level, while those of gases are given relative to the vacuum level. Unless indicated the results are for solids. Different investigators use different standards, but I have not tried to renormalize the results to some common standard, except in the case of Jørgensen (Refs. 32, 94, 104, 140, 143). Here, I have subtracted 5.0 eV, since this author relates his data to C(1*s*) of hydrocarbon equals 290 eV, whereas 285 eV would be more in line for that of the Fermi level. Although most authors quote their binding energies to a couple of tenths of an eV, the reproducibility between different laboratories is often much poorer. These tables are thus of value for obtaining a qualitative picture rather than for giving definitive standards. For some compounds, values relating to the calculated charge or the oxidation state are given. *q*(P), *q*(CNDO), *q*(EH) stand for calculated charge using, respectively, Pauling's electronegativity, CNDO calculations, and extended Hückel calculations.

As with the main body of this book, Appendix 3 is based on a literature research up through 1974. To supplement this table, the reader is strongly advised to see K. Siegbahn, D. A. Allison, and J. H. Allison, *Handbook of Spectroscopy*, Vol. 1 (ed. by J. W. Robinson; CRC Press, Cleveland, Ohio, 1974), Table III (pp. 518–572). The literature appears to be covered through 1972, and 181 references are cited. About 4000 binding energies are listed.

Li (1s) Z = 3

Compound	E	ΔE	q(P)	Ref.
Li (metal)	48.7	0	0	79
Li acetate	54.6	5.9	0.36	79
LiCl	55.4	6.7	0.64	79
Li_2SO_4	55.6	6.9	0.63	79
LiBr	56.1	7.4	0.65	79
LiF	56.3	7.6	0.98	79

Refs.: 39, 90, 92; 79.

Be (1s) Z = 4

Compound	E	ΔE	Ref.
Be	111.4	0	24
BeO	114.2	2.8	24
BeF_2	116.3	4.9	24

Refs.: 39; 24, 71.

B (1s) Z = 5

Compound	E	ΔE	q(CNDO)	Ref.
Solid				
B_4C	186.7	−0.8	—	72*a*
B(amorph)	187.5	0.0	0.0	72*a*
$B_{10}H_{14}$	188.0	0.5	—	72*a*
BN	190.2	2.7	—	72*a*
$Na_2B_3O_6$	192.2	4.7	0.32	72*a*
H_3BO_3	193.2	5.7	0.65	72*a*
$NaBF_4$	195.1	7.3	0.75	72*a*
Gas				
$BH_3N(CH_3)_3$	193.7	0	−0.08	73
BH_3CO	195.2	1.5	−0.43	73
$B(CH_3)_3$	196.4	2.7	0.13	73
B_2H_6	196.5	2.8	−0.03	73
$B(OCH_3)_3$	198.4	4.7	0.49	73
BCl_3	200.5	6.8	0.25	73
BF_3	202.8	9.1	0.70	73

Refs.: 29, 39, 72, 73; 6, 39.

C (1*s*) $Z = 6$

Compound	E_B	ΔE_B	q(CNDO)	q(P)	Ref.
Solid					
HfC	281.0	−4.0	—	—	46*a*
TiC	281.7	−3.3	—	—	46*a*
WC	282.9	−2.1	—	—	46*a*
C (graphite)	284.3	−0.7	0	0	26
$C^*H_3CH_2NH_2$	285.0	0.0	−0.06	−0.12	24
C^*H_3COOH	285.0	0.0	−0.16	−0.12	24
C_6H_6	285.0	0.0	0.01	−0.04	24
KCN	285.0	0.0	—	−0.31	24
$CH_3C^*H_2NH_2$	285.7	0.7	0.14	−0.01	24
CH_3OH	287.2	2.2	0.18	0.10	24
CH_2Br_2	—	2.3	—	0.27	24
HCOH	287.9	2.9	0.27	0.36	24
$CH_3C^*OCH_3$	288.1	3.1	0.30	0.44	24
C_6F_6	288.9	3.9	0.15	0.43	24
CH_3C^*OONa	289.0	4.0	0.43	0.56	24
$CHCl_3$	289.8	4.8	0.29	0.68	24
$(CH_3)C^{*+}$	290.1	5.1	—	—	184
$C^*O(OCH_3)_2$	291.4	6.4	0.55	0.88	24
COF_2	294.1	9.1	0.71	1.30	24
CHF_3	294.9	9.9	0.68	1.25	24
CF_4	296.9	11.9	0.85	1.72	24
Gas					
CH_4	290.7	0	−0.08	−0.16	48
$C^*H_3CH_2OH$	290.9	0.2	−0.16	−0.12	48
$C^*H_3(COCH_3)$	291.2	0.5	−0.13	−0.12	48
CH_3OH	292.3	1.6	0.17	0.10	48
$CH_3C^*H_2OH$	292.3	1.6	0.14	0.22	48
CS_2	293.1	2.4	0.00	0.25	48
$CH_3C^*OCH_3$	293.8	3.1	0.30	0.44	48
$^*CO(C_2H_5)(OC_2H_5)$	294.5	3.8	—	0.66	48
$CH_3{}^*COOH$	295.4	4.7	0.44	0.66	48
CO	295.9	5.2	0.02	0.55	48
CO_2	297.5	6.8	0.61	0.88	48
CHF_3	298.8	8.1	0.68	1.26	48
CF_4	301.8	11.1	0.85	1.73	48

Refs.: 1, 10, 22, 24, 30, 34c, 41, 42, 43, 44, 46, 48, 71, 81, 99, 100, 101, 102, 103, 104, 106, 180, 184, 185, 186, 198, 321, 322; 3, 4, 37, 48, 49, 62, 66.

N (1*s*) $Z = 7$

Compound	E_B	ΔE_B	q(CNDO)	q(P)	Ref.
Solid					
CrN	396.6	0.0	—	—	47
VN	397.2	0.6	—	—	47
P_3N_5	397.8	1.2	—	—	47
C_5H_5N	398.0	1.4	−0.166	—	1
BN	398.2	1.6	—	—	47
C_6H_5CN	398.4	1.8	−0.226	—	1
KSCN	398.5	1.9	—	—	47
NaN*NN*	399.3	2.7	−0.548	—	1
$(CONH_2)_2$	400.0	3.4	−0.133	—	47
$Na_2(ON^*NO_2)$	400.9	4.3	−0.195	—	1
C_4H_5N	401.1	4.5	—	—	44
$(C_2H_5)_4NCl$	401.1	4.5	—	—	50
NaN_2O_7	401.3	4.7	−0.256	—	47
S_4N_4	402.1	5.5	—	—	47
NH_3OHCl	402.1	5.5	0.219	—	47
$(CH_3)_4NB_3H_8$	402.2	5.6	0.185	—	47
$(CH_3)_3NO$	402.2	5.6	0.079	—	47
Na(NN*N)	403.7	7.1	0.096	—	47
$Na_2(ONN^*O_2)$	403.9	7.3	0.140	—	47
$NaNO_2$	404.1	7.5	0.100	—	47
$C_6H_5NO_2$	405.1	8.5	0.347	—	1
p-$^*NO_2C_6H_4CONH_2$	405.9	9.3	—	—	47
trans$[Co(en)_2(NO_2)_2]N^*O_3$	406.8	10.2	—	—	47
NaN^*O_3	407.4	10.8	0.429	—	47
Gas					
$(CH_3)_3N$	404.7	−5.2	−0.14	−0.15	17
$(CH_3)_2NH$	404.9	−5.0	−0.17	−0.23	17
CH_2NH_2	405.1	−4.8	−0.21	−0.32	17
NH_3	405.6	−4.3	−0.26	−0.39	17
N_2H_4	406.1	−3.8	−0.14	−0.26	17
HCN	406.8	−3.1	−0.10	−0.06	17
N*NO	408.6	−1.3	−0.24	−0.34	17
N_2	409.9	0.0	0.0	0.0	17
NO	410.7	0.8	0.05	−0.37	17
ONCl	411.4	1.5	0.17	0.05	17
N_2F_4	—	2.4	0.23	0.32	17
NN*O	412.3	2.6	0.52	0.51	17
NO_2	412.9	3.0	0.41	0.53	17
NF_3	414.2	4.3	0.30	0.45	17
ONF_3	417.0	7.1	0.70	1.24	17

Refs.: 1, 10, 28, 29, 34*c*, 39, 44, 47, 48, 50, 52, 54, 57*a*, 59, 64, 89, 90, 120, 131, 145, 164, 183, 197, 297, 322, 324; 3, 4, 5, 8, 12, 17, 18, 19, 29, 54, 57, 59, 60, 66, 68.

O (1s) $Z = 8$

Compound	E_B	ΔE_B	q(P)	q(CNDO)	Ref.
Solid					
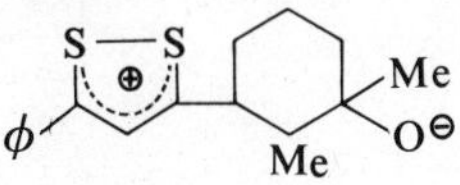 S—S ⊕ ϕ Me Me O⊖	530.3	0.0	—	—	38
$(C_6H_5)_3PO$	531.3	1.0	—	—	55
Na_2SO_3	531.9	1.6	—	—	61
$FeSO_4$	532.5	2.2	—	—	61
$C_3H_6O_2SO_2$*	532.5	2.2	—	—	61
SOF_2	533.0	2.7	—	—	61
ϕNO_2	533.0	2.7	—	—	64
$SOCl_2$	533.6	3.3	—	—	61
$C_3H_6O_2$*SO_2	534.1	3.7	—	—	61
Gas					
CH_3CHO	537.6	−5.5	−0.44	−0.25	48
C_2H_5OH	538.6	−4.5	−0.61	−0.29	48
CH_3OH	538.9	−4.2	−0.61	−0.27	48
CH_3COCH_3	539.0	−4.1	−0.44	−0.28	48
SO_2	539.6	−3.5	−0.44	−0.44	48
H_2O	539.7	−3.4	−0.66	−0.33	48
CH_3COOH	540.8	−3.1	−0.44	−0.32	48
CO_2	540.8	−2.3	−0.44	−0.30	48
N_2O	541.2	−1.9	—	−0.34	48
NO_2	541.3	−1.8	—	—	48
CO	542.1	−1.0	−0.55	−0.02	48
O_2	543.1	0	0.00	0.00	48
NO	543.3	0.2	—	—	48

Refs.: 1, 10, 30, 48, 55, 61, 64, 70, 71, 125, 164; 8, 12, 37, 38, 55, 66.

F (1s) $Z = 9$

Compound	E	ΔE	q(CNDO)[180]	Ref.
Solid				
NaF	685.4	0.0	—	107
LiF	686.3	0.9	—	107
KBF_4	688.0	3.4	—	85
C_6H_5F	689.6	4.2	—	106
C_6F_6	690.9	5.5	—	106
Gas				
CH_3F	692.4	0.0	0.0	49
C_6H_5F	692.4	0.0	0.1	12
C_6F_6	693.7	1.3	1.8	12
CHF_3	694.1	1.7	2.1	48
SF_6	694.6	2.2	—	48
CF_4	695.0	2.6	3.1	12

Refs.: 39, 41, 48, 85, 98, 100, 101, 104, 106, 107, 108, 176, 180; 12, 49, 55, 75.

Na (1s, 2p) $Z = 11$

Compound	E_B (2p)	ΔE_B	Ref.
NaF	63.5	0	85
NaCl	64.1	0.6	85
NaBr	64.4	0.9	85
NaI	64.5	1.0	85

Refs.: 1, 39, 76, 85, 112, 113.

Mg (1s, 2p) $Z = 12$

Compound	E_B (2p)	ΔE_B	Ref.
MgO	52.0	0.0	111
MgF_2	55.7	2.9	111

Refs.: 39, 76, 111; 20.

Al (2s, 2p) Z = 13

Compound	E	ΔE	q(P)	Ref.
Al	72.4	0.0	0.0	85
$Al(C_5H_7O_2)_3$	73.1	0.7	—	85
$Na(AlSiO_4)$	—	—	—	76
Al_2O_3	74.5	2.1	1.90	85
AlN	74.6	2.2	1.29	85
Al_2S_3	74.8	2.4	0.66	85
AlI_3	74.8	2.4	0.66	85
$AlCl_3$	74.9	2.5	1.29	85
AlP	75.2	2.8	—	92
$AlBr_3$	75.4	3.0	1.02	85
K_3AlF_6	75.6	3.2	2.37	85
$LiAlH_4$	75.8	3.4	0.25	85
AlF_3	76.5	4.1	2.37	85

Refs.: 74, 75, 85, 92, 111; 56.

Si (2p) Z = 14

Compound	E_B	ΔE_B	Ref.
Solid			
Si	99.0	0	77
$(C_6H_5)_2SiSi(C_6H_5)_3$	101.3	2.3	77
$(SiN_4)_\infty$	101.8	2.8	77
$(C_6H_5)_2Si(OH)_2$	101.8	2.8	77
$(Na_2SiO_3)_n$	102.0	3.0	77
SiC	102.2	3.2	77
SiB_4	102.4	3.4	77
$[(CH_3)_2SiO]_n$	102.8	3.8	77
$(SiO_2)_\infty$ quartz	103.0	4.0	77
SiI_4	103.8	4.8	77
$(HSiO_{3/2})_\infty$	104.2	5.2	77
Na_2SiF_6	107.4	8.4	77
Gas			
$Si(CH_3)_4$	105.8	0	78
SiH_4	107.1	1.3	78
$(SiH_3)_2O$	107.7	1.9	78
SiH_3Br	108.0	2.2	78
SiH_3Cl	108.0	2.2	78
$SiCl_4$	110.3	4.5	78
SiF_4	111.5	5.7	78

Refs.: 39, 76, 77, 78, 81; 85.

P (2p) $Z = 15$

Compound	E_B	ΔE_B	q(EH)	q(P)	Ref.
CrP	128.8	−1.3	—	—	53
BP	129.5	−0.6	—	—	53
P (red)	130.1	0	0	0	53
$(C_6H_5)_3P$	131.3	1.2	—	0.12	27
$[(C_6H_5)_3P]_2HgI_2$	131.9	0.8	—	—	90
Na_3PO_4	132.1	2.0	3.62	—	53
$(C_6H_5)_3PO$	132.8	2.7	—	0.89	55
$Na_4P_2O_7$	133.3	3.2	3.35	—	53
$(CH_3PS_2)_2$	133.4	3.3	1.59	—	53
KH_2PO_4	133.9	3.8	3.67	—	53
$POBr_3$	134.4	4.3	—	1.68	27
KPF_2O_2	134.8	4.7	3.55	—	53
NH_4PF_6	137.3	7.2	3.81	3.54	53

Refs.: 27, 39, 53, 54, 55, 56, 57*a*, 58*b*, 59, 60, 90, 113; 80.

S (2p) $Z = 16$

Compound	E_e	ΔE_B	q(P)	q(CNDO)	Ref.
Solid					
Na_2S	162.0	−2.2	−0.94	—	61
(cyclohexyl)–SH	163.3	−1.9	−0.04	—	61
S_2Cl_2	163.7	−0.5	0.06	—	61
S_8	164.2	0.0	0.00	—	61
(thiophene ring, S)	164.5	0.3	0.00	—	61
(cyclohexyl)–S(=O)–(cyclohexyl)	166.2	2.2	0.44	—	61
Na_2SO_3	166.7	2.5	0.68	—	61
$SOCl_2$	168.3	4.1	0.56	—	61
Na_2SO_4	168.9	4.7	1.12	—	61
$Fe_2(SO_4)_3$	169.3	5.1	1.12	—	61
SOF_2	170.2	6.0	1.30	—	61
S_2F_{10}	174.6	10.4	2.15	—	61
SF_6	177.4	13.2	2.58	—	61
Gas ($2p_{3/2}$)					
CS_2	169.8	0	0.00	0.12	48
H_2S	170.2	0.4	0.08	0.11	48
SO_2	174.8	5.0	0.88	0.89	48
SOF_2	176.2	6.4	1.30	1.15	48
SF_6	180.4	10.6	2.58	2.36	48

Refs.: 1, 39, 44, 46, 48, 55, 57*a*, 61, 64, 66, 67, 68, 69, 95, 154, 198; 2, 15, 23, 25, 30, 38, 57, 70.

Cl (2s, 2p) $Z = 17$

Compound	E_B (2p)	ΔE_B	Ref.
Solid			
NaCl	199.1	0	85
K_2PtCl_6	199.1	0	141
$AlCl_3$	199.7	0.7	85
C_6H_5Cl	201.4	2.3	104
CCl_4	202.7	3.6	42
$NaClO_2$	203.1	4.0	1
$NaClO_3$	205.7	6.6	1
$NaClO_4$	208.7	9.6	1
Gas			
	E_B (2s)		
CH_3Cl	277.2	0.0	49
CH_2Cl_2	277.6	0.4	49
$CHCl_3$	277.7	0.5	49
CCl_4	278.0	0.8	49

Refs.: 1, 35, 39, 42, 56, 85, 101, 104, 110, 112, 113, 120, 135, 136, 145, 147, 179; 11, 14, 49.

K (2s, 2p) $Z = 19$

Compound	E_B (2$p_{3/2}$)	ΔE_B	Ref.
K_2SiF_6	296.2	3.3	108
KI	296.0	3.1	108
KF	295.4	2.5	108
KNO_3	293.8	0.9	108
$K_3Fe(CN)_6$	292.9	0.0	108

Refs.: 1, 39, 85, 108, 112, 113, 120, 135, 145, 147; 10, 55.

Ca(2$p_{3/2}$) $Z = 20$

Compound	E_B	ΔE_B	Ref.
$CaCrO_4$	347.3	0	39
$Ca(IO_3)_2$	349.7	2.4	39
CaF_2	351.0	3.7	39

Ref.: 39.

Sc (2$p_{3/2}$) $Z = 21$

Compound	E_B	ΔE_B	Ref.
Sc_2O_3	403.8	0	39
$(NH_4)_3ScP_6$	405.0	1.2	39
$Sc_2(SO_4)_3$	405.1	1.3	39

Ref.: 39.

Ti (2s, $2p_{3/2}$) $Z = 22$

Compound	E_B	ΔE_B	Ref.
Ti	—	0	46*a*
TiB_2	—	0.3	46a
TiC	—	1.3	46*a*
TiN	—	1.5	46*a*
TiO_2	—	4.9	46*a*
TiO_2	460.3	(4.9)	39
K_2TiF_6	464.0	8.6	39

Refs.: 39, 46*a*, 115.

V ($2p_{3/2}$) $Z = 23$

Compound	E_B ($2p_{3/2}$)	ΔE_B	Ref.
V	512.4	0	36
$V(acac)_2$	514.0	1.6	36
$(NH_4)_3VS_4$	514.1	1.7	39
WV_2O_6	515.6	3.2	116
	516.6	4.2	116
	517.6	5.2	116
V_2O_5	516.6	4.2	36
V_2O_5	517.2	4.8	39
V_2O_5	517.6	5.2	116
YVO_4	518.4	6.0	39

Refs.: 39, 46*b*, 46*c*, 87, 115, 116; 36.

Cr ($2p_{3/2}$, 3p) $Z = 24$

Compound	*E* 3p	*E* $2p_{3/2}$	ΔE	*q*(P)	Ref.
Cr	43.2	—	0	0	71
CrN	43.2	—	0	—	71
Cr_2S_3	—	575.0	0.4	0.54	117
$CrBr_3$	—	576.4	1.8	0.91	117
Cr_2O_3	45.6	577.0	2.4	1.78	87, 117
$CrCl_3$	—	577.6	3.0	1.16	117
$K_3Cr(SCN)_6$	—	577.7	3.1	1.74	117
$Cr(CO)_6$	45.8	578.5	3.9	—	213
CrF_3	—	580.5	5.1	2.29	117
$K_2Cr_2O_7$	48.7	—	5.5	3.44	71

Refs.: 39, 71, 87, 115, 117, 118, 119, 213; 45, 54.

Mn ($2p_{3/2}$, $3p$) $Z = 25$

Compound	E_B ($2p_{3/2}$)	ΔE_B	q(P)	Ref.
MnS	640.5	0	0.44	117
Mn_2O_3	640.9	0.4	1.89	117
$K_4Mn(CN)_6$	641.4	0.9	1.13	117
MnN	641.5	1.0	1.29	117
MnO	641.7	1.2	1.26	117
$MnCl_2$	642.1	1.6	0.86	117
MnO_2	642.4	1.9	2.53	117
MnF_3	642.8	2.3	2.37	117

Refs.: 39, 117, 201; 59, 66.

Fe ($2p_{3/2}$, $3p$) $Z = 26$

Compound	E_B		ΔE_B		q(P)	Ref.
	$3p$	$2p_{3/2}$	$2p$	$2p_{3/2}$		
Fe	52.0	—	0	0	0	68
FeS_2	53.0	—	1.0	—	0.45	68
$FeBr_3$	—	710.3	—	1.2	0.66	117
$Fe(C_5H_5)_2$	53.7	710.8	1.7	1.7	1.00	68, 125
Fe_2O_3	—	711.5	—	—	1.54	117
$Fe(CO)_5$	54.0	713.0	2.0	4.5	1.02	68, 125
$K_4Fe(CN)_6$	54.0	—	2.0	—	1.03	68
$Fe_2(CO)_9$	54.6	—	2.6	—	1.30	68
$K_3Fe(CN)_6$	55.0	—	3.0	—	1.24	68
K_3FeF_6	57.7	714.1	5.7	5.0	1.81	68, 117
FeF_3	—	714.4	—	5.3	2.11	117

Refs.: 39, 67, 68, 69, 117, 126, 127, 128, 188, 190, 294; 32.

Co ($2p_{3/2}$, $3p$) $Z = 27$

Compound	E_B ($2p_{3/2}$)	ΔE_B	q(P)	Ref.
$K_3Co(CN)_6$	778.8	0	0.72	117
$K_3Co(C_2O_4)_3$	781.1	2.3	1.45	117
$Co(NH_3)_6Cl_3$	781.3	2.5	2.06	117
$Co(C_5H_5)_2$	782.2	3.4		125
CoF_2	782.6	3.8	1.40	117
CoF_3	783.2	4.4	2.11	117

Refs.: 1, 39, 117, 125, 219, 295; 31, 61, 66.

Ni ($2p_{3/2}$, $3p$) $Z = 28$

Compound	E_B ($2p_{3/2}$)	ΔE_B	Ref.
Ni	852.8	0.0	120
$K_2Ni(CN)_3$	854.7	1.9	123
Ni_2O_3	855.0	2.2	120
NiO	856.2	3.4	123
$Ni(\pi C_5H_5)_2$	857.7	4.9	125
$Ni(CO)_4$	859.2	6.4	125
$KNiF_3$	859.0	6.2	120

Refs.: 39, 120, 121, 122, 123, 125, 244, 245, 306; 31, 57, 65, 66, 70, 78.

Cu ($2p_{3/2}$, $3p$) $Z = 29$

Compound	E_B ($2p_{3/2}$)	ΔE_B	Ref.
Cu	932.8	0	28
$CuCl \cdot \phi_3P$	934.1	1.3	150
CuCl	934.6	1.8	150
Cu_2S	934.6	1.8	150
Cu_2O	934.7	1.9	150
$CuBr_2$	935.8	3.0	150
$K_2Cu(C_2O_4)_2$	935.8	3.0	150
CuO	936.2	3.4	150
$CuCl_2$	936.4	3.6	150
$CuSO_4$	937.2	4.4	150
$Cu(ClO_4)_2 \cdot 6H_2O$	938.5	5.7	150
CuF_2	939.7	6.9	150

Refs.: 39, 150, 314; 28, 44, 47, 65.

Zn ($2p_{1/2}$, $3p_{3/2}$) $Z = 30$

Compound	E_B ($3p_{3/2}$)	Ref.
ZnO	88.8	154
ZnS	89.0	154
ZnSe	89.0	154
ZnTe	88.7	154

Refs.: 39, 154; 35, 55.

Ga ($2p_{3/2}$, $3p_{3/2}$) $Z = 31$

Compound	E_B	ΔE_B	q(P)	Ref.
Ga_2O_3	105.7	0	1.78	84
Ga_2S_2	106.2	0.5	0.36	84
$Ga(acac)_3$	106.3	0.6	—	84
Ga_2S_3	106.8	1.1	0.54	84
GaI_3	106.8	1.1	0.54	84
$GaBr_3$	107.0	1.3	0.91	84
GaF_3	108.2	2.5	2.29	84

Refs.: 39, 74, 85; 33, 34.

Ge ($3p_{3/2}$, $3d$) $Z = 32$

Compound	E_B ($3p_{3/2}$)	ΔE_B	Ref.
Solid			
GeO_2	127.9		39
K_2GeF_6	129.2		39
Gas			
$Ge(CH_3)_4$	120.6	0	78
GeH_3CH_3	122.6	2.0	78
GeH_3Br	123.4	2.8	78
GeH_4	123.7	3.1	78
GeH_3Cl	123.8	3.2	78
$GeCl_4$	124.2	3.6	78
GeF_4	128.9	8.1	78

Refs.: 39, 78, 79, 80, 81, 82, 86; 86.

As ($3d$, $3p_{3/2}$) $Z = 33$

Compound	E_B ($3d$)	ΔE_B	q(P)[94]	Ref.
InAs	41.3	−0.7	−0.1	94
As	42.0	0	0.0	90
As_2S_3	43.7	1.7	0.2	94
As_2S_4	44.3	2.3	—	90
$(C_6H_5)_3AsO$	44.6	2.6	—	90
K_3AsO_4	45.2	3.2	—	90
$Na_4As_2O_7$	45.6	3.6	—	90
As_2O_3	46.0	4.0	1.4	90
As_2O_5	46.4	4.4	2.2	90
$KAsF_6$	47.9	5.9	—	90
$LiAsF_6$	49.3	6.3	—	90

Refs.: 28, 39, 79, 90, 94.

Se ($3p_{3/2}$, $3d$) $Z = 34$

Compound	E_B ($3p_{3/2}$)	ΔE_B	q(P)	Ref.
KSeCN	159.3	–0.6	—	95
Se	159.9	0	0.0	95
$CH_3(CH_2)_{14}H_2Se\text{-}CN$	163.2	3.3	0.01	96
$C_6H_{11}\text{–}H_2CSe(=O)\text{–}CH_2\text{–}C_6H_{11}$	164.0	4.1	0.53	96
Na_2SeO_3	164.1	4.2	—	96
Na_2SeO_4	164.6	4.7	—	95
$CH_3C_6H_4SeOOH$	164.9	5.0	0.73	96
$BrC_6H_4SeO_2OH$	165.9	6.0	1.31	96

Refs.: 39, 95, 96, 154.

Br ($3p_{3/2}$, $3d$) $Z = 35$

Compound	E ($3d$)	ΔE	Oxidation state	Ref.
KBr	70.0	0	–1	94
$InBr_3$	71.5	1.5	–1	85
$KBrO_3$	75.9	5.9	+5	94
$KBrO_4$	77.6	7.6	+7	94

Refs.: 39, 42, 85, 94, 134, 135.

Rb ($3p_{3/2}$, $3d_{5/2}$) $Z = 37$

Compound	E_B ($3d_{5/2}$)	q(P)	Ref.
RbF	110.8	0.92	85
RbCl	110.8	0.70	85
RbBr	110.8	0.63	85
RbI	110.8	0.51	85

Refs.: 39, 85, 113.

Sr ($3p_{3/2}$, $3d_{5/2}$) $Z = 38$

Compound	E_B ($3d_{5/2}$)	ΔE_B	Ref.
$Sr(NO_3)_2$	135.6	0	39
$SrSO_4$	136.4	0.8	39
SrF_2	138.2	2.2	39

Refs.: 39, 120.

Y ($3p_{3/2}$, $3d_{5/2}$) $Z = 39$

Compound	E_B ($3d_{5/2}$)	ΔE_B	Ref.
Y_2O_3	158.6	0	39
$Y_2(SO_4)_3$	160.3	1.7	39
$Y(IO_3)_3$	161.3	2.7	39
YF_3	163.5	4.9	39

Ref.: 39.

Zr ($3p_{3/2}$, $3d_{5/2}$) $Z = 40$

Compound	E ($3d_{5/2}$)	Ref.
$Zr(IO_3)_4$	185.5	39
ZrF_4	188.2	39

Refs.: 39, 129.

Nb ($3p_{3/2}$, $3d_{5/2}$) $Z = 41$

Compound	E_B ($3d_{5/2}$)	ΔE_B	q(P)	Ref.
Nb	202.3	0.0	0.0	85
$KNbO_3$	206.7	4.4	2.96	85
$NbBr_5$	207.3	5.0	1.51	85
$NbCl_5$	208.2	5.9	1.94	85
K_2NbF_7	209.3	7.0	3.80	85
NbF_5	209.6	7.3	3.82	85

Refs.: 39, 46*b*, 46*c*, 74, 85, 130.

Mo ($3p_{3/2}$, $3d_{5/2}$) $Z = 42$

Compound	E_B ($3d_{5/2}$)	ΔE_B	Oxidation state	Ref.
Mo	226.1	0	0	87
$Mo(CO)_6$	226.6	0.5	0	87
$MoO_2(C_5H_7O_2)_2$	230.8	4.7	+6	87
MoO_2	230.9	4.8	+4	87
$MoCl_5$	231.0	4.9	+5	87
$H_4(SiMo_{12}O_{40})$	231.6	5.5	+6	87
$(NH_4)_4(NiMo_6O_{24}H_6)$	231.6	5.5	+6	87
Na_2MoO_4	232.1	6.0	+6	87
MoO_3	232.5	6.4	+6	87

Refs.: 39, 87, 110, 316; 60.

Tc ($3p_{3/2}$, $3d_{5/3}$) $Z = 43$

Ref.: 13

Ru ($3p_{3/2}$, $3d_{5/2}$) $Z = 44$

Compound	E_B ($3d_{5/2}$)	ΔE_B	Ref.
Ru (metal)	278.8	0	132
$Ru(NH_3)_6Cl_2$	279.5	0.7	132
RuI_3	279.9	1.1	132
$RuBr_3$	280.4	1.7	132
$RuCl_3$	281.4	2.6	132
$Ru(acac)_3$	281.6	2.8	132
$[Ru(NH_3)_5Br]Br_2$	281.8	3.0	132
$K_2[RuCl_6]$	282.4	3.6	132
RuO_2	282.8	4.0	132
$[RuCl_3(NO)(P\phi_3)_2]$	283.6	4.8	132
$BaRuO_4$	285.4	6.6	132

Refs.: 39, 131, 132; 63.

Rh ($3d_{5/2}$) $Z = 45$

Compound	E_B	ΔE_B	Ref.
Rh	307.1	0	307
Rh_2O_3	208.4	1.3	307
$RhCl_3$	310.5	3.4	39

Refs.: 39, 142, 307; 27, 28, 42, 77.

Pd ($3d_{5/2}$) $Z = 46$

Compound	E_B	ΔE_B	q(P)	Ref.
Pd	335.7	0.0	0.0	134
PdI_2	336.6	0.9	0.09	134
$PdBr_2$	337.3	1.6	0.36	134
$Pd(P(C_6H_5)_3)_2Cl_2$	337.9	2.2	0.30	134
$K_2[PdCl_4]$	338.4	2.7	0.42	134
$K_2[Pd(NO_2)_4]$	339.2	3.5	0.60	134
$Pd(CN)_2$	339.6	3.9	0.34	134
$K_2[PdCl_6]$	340.3	4.6	0.72	134

Refs.: 39, 134, 135, 136; 28, 40, 42, 70, 89.

Ag ($3d_{5/2}$) $Z = 47$

Compound	E	ΔE	Oxidation state	Ref.
Ag	366.8	0.0	0	152
AgBr	368.2	1.4	+1	152
$Ag(bipyr)_2(NO_3)_2 \cdot H_2O$	368.5	1.7	+2	152
Ag_2O	368.6	1.8	+1	152
Ag_2SO_4	368.6	1.8	+1	152
$Na_5H_2Ag(IO_6)_2 \cdot 16H_2O$	371.0	4.2	+3	152

Refs.: 39, 151, 152; 28.

Cd ($3d_{5/2}$) $Z = 48$

Compound	E_B	ΔE_B	Ref.
Cd	403.7	0	154
CdTe	404.7	1.0	154
CdS	405.0	1.3	154
CdSe	405.1	1.4	154
CdO	406.3	2.6	154

Refs.: 39, 79, 93, 120, 154; 46.

In ($3d_{5/2}$) $Z = 49$

Compound	E_B	ΔE_B	q(P)	Ref.
In	444.8	0	0	85
InCl	445.1	0.3	0.34	85
In_2S_3	445.1	0.3	0.44	85
In_2O_3	445.2	0.4	1.67	85
$In(acac)_3$	445.6	0.8	—	85
InI_3	446.3	1.5	0.44	85
InF_3	446.7	1.9	2.20	85
$InBr_3$	446.8	2.0	0.78	85
$InCl_3$	447.1	2.3	1.03	85

Refs.: 39, 74, 85; 33, 88.

Sn ($3d_{5/2}$, $4d$) $Z = 50$

Compound	E_B	ΔE_B	q(P)	Ref.
Sn	486.4	0	0.0	84
$(Et_4N)_2SnI_6$	487.2	0.8	0.16	84
$(Et_4N)_2SnI_4Cl_2$	487.4	1.0	0.15	84
$(Et_4N)_2SnCl_4F_2$	487.8	1.4	0.23	84
K_2SnBr_6	488.1	1.7	—	84
K_2SnF_6	488.8	2.4	0.34	84

Refs.: 39, 83, 84, 86, 147; 67, 87.

Sb ($3d_{5/2}$, $4d$) $Z = 51$

Compound	E_B ($3d_{5/2}$)	ΔE_B	Ref.
Sb	528.5	0	92
AlSb	528.8	0.3	92
Sb_2S_3	529.7	1.2	92
Sb_2O_3	530.1	1.6	92
SbI_3	530.6	2.1	92
Sb_2O_5	531.0	2.5	92
SbF_3	531.9	3.4	92
$KSbF_6$	532.5	4.0	92

Refs.: 39, 92, 93, 120; 33, 52.

Te ($3d_{5/2}$, $4d$) $Z = 52$

Compound	E_B ($3d_{5/2}$)	ΔE_B	Oxidation state	Ref.
Na_2Te	572.4	−0.5	−2	95
Te	572.9	0	0	95
ZnTe	573.3	0.4	−2	95
K_2TeO_3	574.7	1.8	+4	95
K_2TeO_4	576.4	3.5	+6	95
TeO_3	576.8	3.9	+6	95
$(NH_4)_2TeO_4$	576.7	3.8	+6	95

Refs.: 39, 95, 97, 154.

I ($3d_{5/2}$, $4d$) $Z = 53$

Compound	E_B	ΔE_B	Oxidation state	Ref.
TlI	619.0	−0.3	−1	85
KI	619.3	0.0	−1	85
AlI_3	620.6	1.3	−1	85
KIO_3	623.9	4.6	+5	155
KIO_4	625.4	6.1	+7	155

Refs.: 39, 85, 108, 109, 155.

Xe ($3d_{5/2}$, $4d_{5/2}$) $Z = 54$

Compound	E_B ($3d_{5/2}$)	ΔE_B (gas)	Oxidation state	Ref.
Xe	676.4	0	0	48
XeF_2	679.4	3.0	+2	48
XeF_4	681.9	5.5	+4	48
$XeOF_4$	683.4	7.0	+6	48
XeF_6	684.3	7.9	+6	48

Refs.: 48, 98.

Cs ($3d_{5/2}$, $4d_{5/2}$) $Z = 55$

Compound	E_B ($4d_{5/2}$)	ΔE_B	q(P)	Ref.
CsF	75.6	0.0	0.93	85
CsCl	76.1	0.5	0.73	85
CsBr	75.7	0.1	0.67	85
CsI	76.2	0.6	0.56	85

Refs.: 39, 85, 113, 120; 10, 55.

Ba ($3d_{5/2}$, $4d_{5/2}$) $Z = 56$

Compound	E_B ($4d_{5/2}$)	ΔE_B	Ref.
$Ba(NO_3)_2$	92.1	0	39
$BaSO_4$	93.8	1.7	39
BaF_2	94.1	2.0	39

Ref.: 39.

La ($3d_{5/2}$, $4d_{5/2}$) $Z = 57$

Compound	E_B ($3d_{5/2}$)	ΔE_B	Ref.
La_2O_3	835.9	0.0	159
LaOCl	837.4	1.5	159
$LaAsO_4$	838.1	2.2	159
$La_2(CO_3)_3$	838.2	2.3	159
$La_2(SO_4)_3$	838.4	2.5	159
LaF_3	842.0	6.1	159

Refs.: 39, 159, 160; 73.

Ce ($3d_{5/2}$, $4d$) $Z = 58$

Compound	E_B ($3d_{5/2}$)	ΔE_B	Oxidation state	Ref.
$Ce_2(CO_3)_3$	885.0	0	+3	39
CeF_3	888.5	3.5	+3	39
$Ce(CrO_4)_2$	900.4	15.4	+4	39
$CeF_4 \cdot H_2O$	902.9	17.9	+4	39

Refs.: 39, 160; 73.

Pr ($3d_{5/2}$, $4d$) $Z = 59$

Compound	E_B ($3d_{5/2}$)	ΔE	Ref.
$Pr_2C_2O_4)_3$	933.8	0	39
$Pr_2(SO_4)_3$	935.2	2.4	39
PrF_3	938.7	4.9	39

Refs.: 39, 160.

Nd ($3d_{5/2}$, $4d$) $Z = 60$

Compound	E_B (4d)	ΔE	Ref.
Nd_2O_3	130.3	0.0	39
$Nd_2(SO_4)_3$	130.3	0.0	39
NdF_3	130.8	0.5	39

Ref.: 39.

Sm ($3d_{5/2}$, $4d$) $Z = 62$

Compound	E_B ($3d_{5/2}$)	ΔE_B	Ref.
Sm_2O_3	1085.3	0.0	39
$Sm(IO_3)_3$	1086.4	1.1	39
SmF_3	1086.6	1.3	39

Refs.: 39; 73.

Eu ($3d_{5/2}$, $4d_{5/2}$) $Z = 63$

Compound	E_B ($3d_{5/2}$)	ΔE_B	Oxidation state	Ref.
Eu	1126.7[a]	0.0	0	155
$EuAl_2$	1126.7[a]	0.0	+2	155
Eu_2O_3	1136.3[a]	9.6	+3	155
EuF_3	1138.0	12.3	+3	39

[a] Energies normalized to 1136.3 eV as taken from Ref. 39.
Refs.: 39, 155, 160; 73.

Gd ($3d_{5/2}$, $4d_{5/2}$) $Z = 64$

Compound	E_B ($4d_{5/2}$)	ΔE_B	Ref.
Gd_2O_3	144.3	0	39
$Gd_2(SO_4)_3$	145.9	1.6	39
$Gd(IO_3)_3$	146.6	2.3	39
GdF_3	147.5	3.2	39

Refs.: 39, 160; 73.

Tb ($3d_{5/2}$, $4d_{5/2}$) $Z = 65$

Compound	E_B ($3d_{5/2}$)	ΔE_B	Oxidation state	Ref.
$TbCl_3$	1243.1	0.0	+3	39
$Tb(OH)_3$	1243.9	0.8	+3	39
Tb_4O_7	1244.1	1.0	+3	39
TbO_2	1255.0	11.9	+4	39

Ref.: 39.

Dy ($3d_{5/2}$, $4d_{5/2}$) $Z = 66$

Compound	E_B ($3d_{5/2}$)	ΔE	Ref.
$DyVO_4$	1296.7	0.0	39
Dy_2O_3	1298.6	1.9	39

Ref.: 39.

Ho ($4d_{5/2}$) $Z = 67$

Compound	E_B ($4d_{5/2}$)	ΔE_B	Ref.
Ho_2O_3	164.4	0.0	39
$Ho(IO_3)_3$	165.4	1.0	39

Ref.: 39.

Er ($4d_{5/2}$) $Z = 68$

Compound	E_B	ΔE_B	Ref.
Er_2O_3	172	0.0	39
$Er(IO_3)_3$	172.1	0.0	39
ErF_3	173.8	1.8	39

Ref.: 39.

Tm ($4d_{5/2}$) $Z = 69$

Compound	E_B	ΔE_B	Ref.
Tm_2O_3	179.5	0.0	39
$Tm(IO_3)_3$	180.1	0.6	39

Ref.: 39.

Yb ($4d_{5/2}$) $Z = 70$

Compound	E_B	Ref.
Yb_2O_3	184.2	39
$Yb_2Ti_2O_7$	186.4	39
$Yb_2(SO_4)_3$	187.4	39
$Yb_2(WO_4)_3$	188.0	39
$Yb(IO_3)_3$	188.9	39
YbF_3	189.7	39

Refs.: 39, 160; 73, 74.

Lu ($4d_{5/2}$) $Z = 71$

Compound	E_B	ΔE_B	Ref.
Lu_2O_3	198.9	0.0	39
$Lu(IO_3)_3$	199.8	0.9	39
LuF_3	203.7	4.8	39

Refs.: 39; 73, 74.

Hf ($4d_{5/2}$, $4f_{7/2}$) $Z = 72$

Compound	E_B ($4d_{5/2}$)	ΔE	Ref.
$Hf(C_6H_5CHOHCO_2)_4$	214.5	0.0	39
HfO_2	215.8	1.3	39
$Hf(IO_3)_4$	216.9	2.4	39
K_2HfF_6	217.5	2.9	39

Refs.: 39, 129, 163.

Ta ($4d_{5/2}$, $4f_{7/2}$) $Z = 73$

Compound	E ($4d_{5/2}$)	ΔE	q(P)	Ref.
Ta	226.9	0.0	0.00	85
$KTaO_3$	230.0	3.1	3.16	85
Ta_2O_5	230.8	3.9	3.16	85
TaS_2	231.0	4.1	0.88	85
TaS	231.1	4.2	0.44	85
Ta_5Si_3	231.1	4.2	0.03	85
$TaSi_2$	231.3	4.4	0.08	85
$TaBr_5$	231.4	4.5	1.72	85
TaF_5	231.8	4.9	3.95	85
K_2TaF_7	235.1	8.2	3.95	85

Refs.: 39, 46*b*, 46*c*, 74, 163.

W ($4d_{5/2}$, $4f_{7/2}$) $Z = 74$

Compound	E_B ($4f_{7/2}$)	ΔE_B	q(P)	Ref.
W	31.8	0	0	85
WC	32.4	0.6	0.60	85
Cs_2WS_4	35.9	4.1	—	163
WO_3	36.5	4.7	3.33	85
$WOCl_4$	37.1	5.3	2.06	85
Na_2WO_4	37.9	6.1	3.33	163

Refs.: 39, 74, 85, 116, 137, 163; 60.

Re ($4d_{5/2}$, $4f_{7/2}$) $Z = 75$

Compound	$E\ (4f_{7/2})$	ΔE	Ref.
$N(CH_3)_4[ReS_4]$	43.2	0.0	163
K_2ReCl_6	44.2	1.0	163
$As(C_6H_5)_4[ReO_4]$	45.6	2.4	163
$KReO_4$	47.9	4.7	163
$AgReO_4$	48.4	5.2	163
$TlReO_4$	48.9	5.7	163
$CsReO_4$	49.3	6.1	163

Refs.: 39, 110, 138, 139, 141, 163, 179; 50, 51, 83, 84.

Os ($4d_{5/2}$, $4f_{7/2}$) $Z = 76$

Compound	$E\ (4f_{7/2})$	ΔE	Oxidation state	Ref.
$Fe_4[Os(CN)_6]_3$	50.8	0.0	+2	39
Tl_2OsBr_6	52.9	2.1	+4	163
K_2OsCl_6	53.9	3.1	+4	163
Cs_2OsBr_6	54.1	3.3	+4	163
Cs_2OsCl_6	54.8	4.0	+4	163
$K_2OsO_2(OH)_4$	55.7	4.9	+6	163
$K[OsO_3N]$	58.7	7.9	+8	39

Refs.: 39, 163, 179.

Ir ($4d_{5/2}$, $4f_{7/2}$) $Z = 77$

Compound	$E_B\ (4f_{7/2})$	ΔE_B	Ref.
$Ir[S_2P(OC_3H_7)_2]_3$	61.9	0.0	163
$[Ir(NH_3)_5Cl]Cl_2$	63.9	2.0	163
Cs_2IrCl_6	65.7	3.8	163

Refs.: 39, 51, 140, 142, 163, 179, 191.

Pt ($4d_{5/2}$, $4f_{7/2}$) $Z = 78$

Compound	E_B ($4f_{7/2}$)	ΔE_B	Ref.
Pt	72.0	0.0	135
	71.1	0.0	136
	70.7	0.0	146
$(Ph)_3P)Pt(Me)_2$	72.2	1.1	136
K_2PtBr_4	73.7	1.7	135
$KPt(CN)_4$	74.0	2.0	135
$PtCl_2$	74.3	2.3	135
K_2PtCl_4	74.4	2.4	135
PtO	73.3	2.6	146
$K_2Pt(NO_3)_4$	74.8	2.8	135
PtO_2	74.1	3.4	146
K_2PtBr_6	75.8	3.8	135
$K_2Pt(CN)_6$	76.2	4.2	135
$K_2Pt(CN)_4Cl_2$	76.8	4.8	135
K_2PtCl_6	76.8	4.8	135

Refs.: 1, 29, 39, 108, 135, 136, 142, 144, 145, 146, 148, 306; 9, 42, 58, 70, 81.

Au ($4f_{7/2}$) $Z = 79$

Compound	E_B	ΔE_B	Ref.
Au	83.2	0.0	53
$AuAl_2$	84.5	1.3	53
$AuSCH_2CO_2H$	85.8	2.6	163
$KAu(CN)_2$	86.2	3.0	163
$KAuCl_4$	88.9	5.7	163
Au_2Cl_6	89.6	6.4	39

Refs.: 39, 143, 163; 28, 53.

Hg ($4f_{7/2}$) $Z = 80$

Compound	E_B	ΔE_B	Ref.
Hg	98.5	0.0	154
HgS	100.6	2.1	154
$K_2Hg(SCN)_4$	101.8	3.3	163
K_2HgI_4	102.8	4.3	163
HgO	103.6	5.1	163

Refs.: 39, 154, 163; 46.

Tl $(4f_{7/2})$ $Z = 81$

Compound	E_B	ΔE	Ref.
Tl_2O_3	118.0	0.0	163
Tl_2OsBr_6	119.4	1.4	163
Tl_2WS_4	119.9	1.9	163
TlI	120.1	2.1	163
TlBr	120.5	2.5	163
Tl_2CO_3	120.9	2.9	163
Tl_2WO_4	121.4	3.4	163
Tl_2SO_4	121.8	3.8	163
$TlIO_3$	122.0	4.0	163
TlF	122.2	4.2	163

Refs.: 39, 74, 75, 85, 120, 163; 72.

Pb $(4f_{7/2})$ $Z = 82$

Compound	E_B	ΔE	Ref.
PbO_2	137.4	0.0	163
$Pb[S_2P(OC_2H_5)_2]_2$	138.3	0.9	163
$PbCrO_4$	138.8	1.4	163
PbO	139.2	1.8	163
$PbCO_3$	140.0	2.6	163
PbF_2	140.9	3.5	163
$PbSO_4$	141.1	3.7	163
PbI_2	141.8	4.4	163
$Pb(NO_3)_2$	142.2	4.8	163

Refs.: 39, 86, 87, 120, 613, 309; 6.

Bi $(4f_{7/2})$ $Z = 83$

Compound	E_B	ΔE_B	Ref.
Bi	157.1	0.0	92
Bi_2S_3	159.1	2.0	92
$NaBiO_3$	159.3	2.1	92
BiI_3	159.5	2.4	92
Bi_2O_3	159.5	2.4	92
$Bi_2Ti_2O_7$	159.9	2.8	92
BiOCl	160.1	3.0	92
BiF_3	161.0	3.9	92
$Bi(SO_4)_3 \cdot H_2O$	161.4	4.3	92

Refs.: 39, 91, 92, 93, 120, 163; 46.

Th ($4f_{7/2}$) $Z = 90$

Compound	E_B	ΔE_B	Ref.
$Th(C_6H_5CHOHCO_2)_4$	336.5	0.0	163
$Th_2Eu_2O_7$	336.8	0.3	163
ThO_2	337.7	1.2	163
$Th(C_2O_4)_2$	339.1	2.6	39
ThF_4	340.4	3.9	163

Refs.: 39, 163; 69.

U ($4f_{7/2}$, $5d_{5/2}$) $Z = 92$

Compound	E_B ($4f_{7/2}$)	ΔE_B	q(P)	Ref.
US	379.4	−0.8	0.30	85
U	380.2	0.0	0.0	161
UO_2	380.8	0.6	2.22	85
$[P(C_6H_5)_3]_2UCl_6$	381.0	0.8	—	163
U_3O_8	381.5	1.3	2.94	85
UO_3	382.3	2.1	3.33	85
$UO_2(C_9H_6NO)_2$	383.2	3.0	—	164
$(NH_4)_2U_2O_7$	384.1	3.9	—	163
K_2UF_6	384.5	4.3	—	163
$RbUO_2(NO_3)_3$	384.7	4.5	—	163
$UO_2(IO_3)_2$	385.8	5.6	—	39

Refs.: 39, 85, 161, 162, 163, 164; 69.

Pu ($4f_{7/2}$) $Z = 94$

Ref.: 19.

Am ($4f_{7/2}$) $Z = 95$

Compound	E_B	Ref.
$Am(OH)_3$	449.3	2

Ref.: 2.

Appendix 4

Acronyms and Definitions of Special Interest in Electron Spectroscopy

AES	Auger electron spectroscopy
APS	Appearance potential spectroscopy
Autoionization	Nonradiative transition akin to Auger process but arising from excitation of outer atomic orbitals rather than from vacancy in core shell (Chapter 6, Section 3.4)
Characteristic energy losses	Inelastic losses suffered by electrons passing through solid or gas at high pressure; such losses are characteristic of the material they pass through; in photoelectron spectra they are found as satellite structure on the low-energy side of the main peak
CNDO	Complete neglect of differential overlap, a semi-empirical method for calculating molecular orbitals; other modifications are INDO, MINDO, SPINDO (Chapter 6, Section 5.1.2.2)
Configuration interaction	Mixing of excited configurations of an atom or molecule with ground state wave function; an important mechanism for creating excited states (Chapter 5, Section 5.1.3)
Coster–Kronig transition	An Auger process involving electrons from the same principal shell as the initial vacancy; e.g., L_{II}–$L_{III}M$

Crystal potential, molecular potential	Potential felt by core electrons in atom i for all other atoms and ions in molecule or solid: $V = \sum_j (q_i/r_{ij})$ (Chapter 5, Section 2.2)
EIS	Electron impact spectroscopy; primarily involves investigation of excited states of neutral gas molecules (cf. Chapter 1, Section 4.1)
Electron shakeup and electron shakeoff	Excitation of a bound electron into an excited state (shakeup) or continuum (shakeoff) as the result of a sudden change in central potential (such as photoejection of a shielding electron) (Chapter 3, Section 3.4; Chapter V, Section 5.1.2)
ESCA	Electron spectroscopy for chemical analysis; acronym sometimes used to include all fields of chemistry served by electron spectroscopy; often restricted to only x-ray photoelectron spectroscopy
FWHM	Full width at half maximum of a spectral peak (usually Gaussian in shape), usual definition of resolution
Hartree–Fock	*Ab initio* wave function calculations based on self-consistent procedure of having each electron move in potential of nuclear charge plus other electrons (Chapter 3)
Helmholtz coils	A set of parallel coils through which electric current is passed in order to create a magnetic-field-free volume (cf. Chapter 2, Section 2.1.1)
Hückel MO	A semiempirical molecular orbital calculation (Chapter 3, Section 5.1.2.1)
IEE	Induced electron emission; acronym covers the general field of electron spectroscopy, including Auger and photoelectron spectroscopy
Ion neutralization spectroscopy	The study of autoionization by ion impinging on a solid surface (cf. Chapter 1, Section 4.4)
Jahn–Teller splitting	When a molecule possesses a high degree of symmetry, ejection of a photoelectron will destroy that symmetry, leading to a breakdown in degeneracy and more than one final electronic state
Koopmans' theorem	Eigenvalues can be used as binding energies on the assumption that all other orbitals do not relax or are frozen upon the removal of an electron from a given orbital

LEED — Low energy electron diffraction; an important method for characterizing surfaces of crystals; often used in conjunction with Auger spectroscopy

Lone-pair orbitals — A lone-pair orbital is a pair of molecular orbitals made up of two electrons that are not strongly involved in molecular bonding but are highly localized on a given atom

MO — Molecular orbital (Chapter 3, Section 5)

Mössbauer isomer shift — Chemical shifts in the Mössbauer energy due to changes in the electron density at the nucleus (Chapter 5, Section 4.4.1)

Multiplet or exchange splitting — Splitting of photoelectron peaks due to coupling of unpaired spins created by photoelectron ejection with incompletely filled valence orbital (Chapter 5, Section 5.1.1)

NMR — Nuclear magnetic resonance (Chapter V, Section 4.4.1)

Penning ionization — An excited atom is used to deliver a known quantity of excitation to a molecule, giving rise to ionization (cf. Chapter 1, Section 4.3)

Perfluoro effect — Changes in the photoelectron spectrum of hydrocarbons as the result of substituting all the hydrogens by fluorine (Chapter IV, Section 3.1.4)

PES — Photoelectron spectroscopy; this term can be used for all ranges of photon energy, but in general applies primarily to low-energy photons (uv and vacuum uv)

PESIS — Photoelectron spectroscopy of the inner shells; this field involves the study of the binding energies of core electrons in atoms and molecules by x-ray photoelectron spectroscopy, which may be perturbed by the chemical environment

PESOS — Photoelectron spectroscopy of the outer shells; this field of photoelectron spectroscopy is primarily involved in directly determining the binding energies of electrons in the valence shell; it deals with both solids and gases, and use is made of both uv and x-ray sources

Photoemission spectroscopy — Usually restricted to photoelectron measurements on solids using uv or x-ray source for the purpose

	of studying the structure of valence and conduction bands (cf. Chapter 1, Section 4.2)
Plasmon loss	One of the principal sources of inelastic scattering in solids by electrons in the range 0.1–10 keV
Relaxation energy	The difference in energy between the eigenvalue and adiabatic binding energy (cf. Chapter 3, Section 2)
Rydberg series	A series of excited atomic or molecular levels that are hydrogenic in nature; they converge to an ionization limit
SIMS	Secondary ion mass spectrometry
Spin–orbit splitting	Splitting in photoelectron spectra due to coupling of the spin and orbital angular momentum; splitting becomes most pronounced when orbitals of heavier elements are involved (Chapter 3, Section 3.1; Chapter 4, Section 2.2.1)
Synchrotron radiation	Continuous radiation source created by the acceleration of high-energy electrons in a vacuum such as in a synchrotron; a most valuable source for photoelectron spectroscopy (cf. Chapter 2, Section 1.1.3)
UPS	Ultraviolet photoelectron spectroscopy; this term has commonly been restricted to pertain to photoelectron studies made with the He I (21.2 eV) and He II (40.8) resonance lines, although it may include all uv and vacuum-uv sources
X_α scattering	Method used for *ab initio* molecular orbital calculation, usually using muffin-tin potential (Chapter 3, Section 5.1.1)
XPS	X-ray photoelectron spectroscopy; usually involves studies with Al and Mg Kα x rays

References

CHAPTER 1

1. A. Einstein, *Ann. Physik* **17**, 132 (1905).
2. P. Auger, *Compt. Rend.* **177**, 169 (1923); *Compt. Rend.* **180**, 65 (1925); *Ann. Phys. (Paris)* **6**, 183 (1926).
3. K. Siegbahn, ed., *Alpha-, Beta-, and Gamma-Ray Spectroscopy* (North-Holland, Amsterdam) 1965.
4. K. D. Sevier, *Low Energy Electron Spectrometry* (Wiley–Interscience, New York, 1972).
5. R. G. Steinhardt, Jr. and E. J. Serfass, *Ann. Chem.* **23**, 1585 (1951); **25**, 697 (1953).
6. R. G. Steinhardt, Jr., F. A. D. Granados, and G. I. Post, *Anal. Chem.* **27**, 1046 (1955).
7. H. Robinson and W. F. Rawlinson, *Phil. Mag.* **28**, 277 (1914).
8. K. Siegbahn, C. Nordling, and E. Sokolowski, "Chemical Shifts of Photo and Auger Electron Lines," in *Proc. Rehovoth Conf. on Nuclear Structure, 1957*, ed. by H. J. Lipkin (North-Holland, Amsterdam, 1958), p. 291; E. Sokolowski, *Ark. Fys.* **15**, 1 (1959).
9. D. W. Turner and M. L. Al-Joboury, *J. Chem. Phys.* **37**, 3007 (1962).
10. J. J. Lander, *Phys. Rev.* **91**, 1382 (1953).
11. L. N. Tharp and E. J. Scheibner, *J. Appl. Phys.* **38**, 3320 (1967).
12. R. E. Weber and W. T. Peria, *J. Appl. Phys.* **38**, 4355 (1967).
13. K. Siegbahn, C. Nordling, A. Fahlman, R. Nordberg, K. Hamerin, J. Hedman, G. Johansson, T. Bergmark, S.-E. Karlsson, I. Lindgren, and B. Lindberg, *Electron Spectroscopy for Chemical Analysis—Atomic, Molecular, and Solid State Structure Studies by Means of Electron Spectroscopy* (Almqvist and Wiksells, Boktryekeri AB, Stockholm, Sweden, 1967); *Nova Acta Regiae Soc. Sci. Upsaliensis, Ser. IV* **20** (1967).
14. K. Siegbahn, C. Nordling, G. Johansson, J. Hedman, P. F. Heden, K. Hamrin, U. Gelius, T. Bergmark, L. O. Werme, R. Manne, and Y. Baer, *ESCA Applied to Free Molecules* (North-Holland, Amsterdam–London, 1969).
15. D. W. Turner, A. D. Baker, C. Baker, and C. R. Brundle, *Molecular Photoelectron Spectroscopy; a Handbook of He 584 Å Spectra* (Interscience, London–New York, 1970).
16. A. D. Baker and D. Betteridge, *Photoelectron Spectroscopy: Chemical and Analytical Aspects*, Vol. 53 of International Series of Monographs in Analytical Chemistry (Pergamon Press, Oxford, 1972).
17. J. H. D. Eland, *Photoelectron Spectroscopy* (Wiley–Halsted, New York, 1974).

18. V. I. Nevedov, *Application of X-ray Photoelectron Spectroscopy in Chemistry*, in series *Structure of Molecules and the Chemical Bond*, Vol. I, ed. by V. V. Bondar (All Union Institute of Scientific and Technical Information, Moscow, 1973) (in Russian).
19. W. Dekeyser, L. Fiermans, G. Vanderkelen, and J. Vennik, eds. *Electron Emission Spectroscopy* (Reidel, Holland, 1973).
20. K. Siegbahn, D. A. Allison, and J. H. Allison, in *Handbook of Spectroscopy*, Vol. 1, ed. by J. W. Robinson (CRC Press, Cleveland, Ohio, 1974), pp. 257–752.
21. K. Siegbahn, *Phil. Trans. R. Soc. London* **A268**, 33 (1970); K. Siegbahn, D. Hammond, H. Fellner-Feldegg, and E. F. Barnett, *Science* **176**, 245 (1972); K. Siegbahn, *J. Electron Spectros.* **5**, 3 (1974).
22. D. M. Hercules, *Anal. Chem.* **42**, 20A (1970); **44**, 106R (1972); D. M. Hercules and J. C. Carver, *Anal. Chem.* **46**, 133R (1974).
23. D. A. Shirley, *Advan. Chem. Phys.* **23**, 85 (1973).
24. R. G. Albridge, "Photoelectron Spectroscopy," in *Techniques of Chemistry*, Vol. I, ed. by A. Weissberger and B. W. Rossiter (Wiley–Interscience, New York, 1972), p. 307.
25. W. L. Jolly, *Coord. Chem. Rev.* **13**, 47 (1974).
26. U. Gelius, *Physica Scripta* **9**, 133 (1974); *J. Electron Spectros.* **5**, 985 (1974).
27. D. Betteridge and A. D. Baker, *Anal. Chem.* **42**, 43A (1970); D. Betteridge, *Anal. Chem.* **44**, 100R (1972); D. Betteridge and M. A. Williams, *Anal. Chem.* **46**, 125R (1974).
28. R. S. Berry, *Ann. Rev. Phys. Chem.* **20**, 357 (1969).
29. D. W. Turner, "Molecular Photoelectron Spectroscopy," in *Physical Methods in Advanced Inorganic Chemistry*, ed. by H. A. O. Hill and P. Day (Interscience, London–New York, 1968), p. 74; *Ann. Rev. Phys. Chem.*, **21**, 107 (1970); *Phil. Trans. R. Soc. London* **A268**, 7 (1970).
30. S. D. Worley, *Chem. Rev.* **71**, 295 (1971).
31. C. R. Brundle, *Appl. Spect.* **25**, 8 (1971); C. R. Brundle and M. B. Robin, in *Determination of Organic Structures by Physical Methods*, Vol. III, ed. by F. Nachod and G. Zuckerman (Academic Press, New York, 1971).
32. E. Heilbronner, in *The World of Quantum Chemistry*, ed. by R. Daudel and B. Pullman (Reidel, Holland, 1974), p. 211.
33. E. Heilbronner, J. P. Maier, and E. Haselbach, in *Physical Methods in Heterocyclic Chemistry*, Vol. 6, ed. by A. R. Katritzky (Academic Press, New York, 1974), p. 1.
34. W. C. Price, *Advan. Atomic Molecular Phys.* **10**, 131 (1974).
35. C. C. Chang, *Surface Sci.* **25**, 53 (1971).
36. C. C. Chang, in *Characterization of Solid Surfaces*, ed. by P. F. Kane (Plenum, New York, 1974), p. 509.
37. A. Pentenero, *Catalysis Rev.* **5**, 199 (1971).
38. C. R. Brundle, *J. Vac. Sci. Technol.* **11**, 212 (1974); *J. Electron Spectros.* **5**, 291 (1974).
39. D. A. Shirley, ed., *Electron Spectroscopy—Proc. Int. Conf. at Asilomar, Calif., 1971* (North-Holland, Amsterdam, 1972).
40. "The Photoelectron Spectroscopy of Molecules" (contributions from meeting held in Brighton, 1972), *Faraday Disc. Chem. Soc.* No. 54 (1972).
41. R. Cavdano and J. Verbist, eds., *Electron Spectroscopy: Progress in Research and Applications, Proc. Int. Conf. Namur, April 16–19, 1974* (Elsevier Press, Holland, 1974); also *J. Electron Spectros.* **5**, 1–1136 (1974).
42. J. Frank and G. Hertz, *Verhandl. deut. physik. ges.* **15**, 613 (1913); **16**, 457 (1914); *Physik. Z.* **17**, 409 (1916).
43. J. Geiger and M. Topschowsky, *Z. Naturforsch.* **21a**, 626 (1966).
44. E. N. Lassettre, A. Skerbele, M. A. Dillon, and K. J. Ross, *J. Chem. Phys.* **48**, 5066 (1968).

45. S. Trajmar, J. K. Rice, and A. Kuppermann, *Adv. Chem. Phys.* **18**, 15 (1970).
46. G. R. Wight and C. E. Brion, *J. Electron Spectros.* **4**, 25, 313, 327, 335, 347 (1974).
47. H. Ehrhardt, M. Shulz, T. Tekaat, and K. Willman, *Phys. Rev. Lett.* **22**, 89 (1969).
48. M. J. Van der Wiel and C. E. Brion, *J. Electron Spectros.* **1**, 309, 443 (1972/73).
49. C. E. Kuyatt and J. A. Simpson, *Rev. Sci. Instr.* **38**, 103 (1967).
50. D. E. Eastman, "Photoemission Spectroscopy of Metals," in *Techniques of Metals Research VI*, ed. by E. Passaglia (Wiley–Interscience, New York, 1971).
51. W. E. Spicer, *J. Res. Natl. Bur. Std.* **74A**, 397 (1970); D. T. Pierce and W. E. Spicer, *Phys. Rev.* B **5**, 2125 (1971); R. S. Bauer and W. E. Spicer, in *Electron Spectroscopy*, ed. by D. A. Shirley (North–Holland, Amsterdam, 1972), p. 569; W. E. Spicer, in *Band Structure Spectroscopy of Metals and Alloys*, ed. by D. J. Fabian and L. M. Watson (Academic Press, New York, 1973), p. 7.
52. D. E. Eastman, in *Electron Spectroscopy*, ed. by D. A. Shirley (North-Holland, Amsterdam, 1972), p. 487.
53. C. E. Brion, C. A. McDowell, and W. B. Stewart, *J. Electron Spectros.* **1**, 113 (1972/73).
54. V. Čermák, "Penning Ionization Electron Spectroscopy," in *Advances in Mass Spectrometry*, Vol. 4, ed. by E. Kendrick (The Institute of Petroleum, London, 1968), p. 697; V. Čermák and J. B. Ozenne, *Int. J. Mass Spectrom. Ion Phys.* **7**, 399 (1971); V. Čermák, in *Electron Spectroscopy*, ed. by D. A. Shirley (North-Holland, Amsterdam, 1972), p. 385.
55. H. D. Hagstrum, *Phys. Rev.* **150**, 495 (1966); H. D. Hagstrum and G. E. Becker, *Phys. Rev. Lett.* **22**, 1054 (1969); H. D. Hagstrum, *J. Res. Natl. Bur. Std.* **74A**, 433 (1970); H. D. Hagstrum and G. E. Becker, *Phys. Rev.* B **4**, 4187 (1971); *Proc. R. Soc. London* A **331**, 395 (1972).
56. G. Innorta, S. Torroni, and S. Pignataro, *Organic Mass Spectrometry* **6**, 113 (1972).
57. J. A. Bearden, *Rev. Mod. Phys.* **39**, 78 (1967); J. A. Bearden and A. F. Burr, *Rev. Mod. Phys.* **39**, 125 (1967).
58. B. K. Agarwal and L. P. Verma, *J. Phys. C; Solid State Phys.* **3**, 535 (1970); L. O. Werme, B. Grennberg, J. Nordgren, C. Nordling, and K. Siegbahn, *Phys. Rev. Lett.* **30**, 523 (1973).
59. G. D. Stucky, D. A. Matthews, J. Hedman, M. Klasson, and C. Nordling, *J. Am. Chem. Soc.* **94**, 8009 (1972).

CHAPTER 2

1. J. R. Pierce, *Theory and Design of Electron Beams* (D. Van Nostrand, New York, 1949), p. 186.
2. V. Hughes and H. L. Schultz, ed., *Methods of Experimental Physics*, IVA, *Atomic Sources and Detectors* (Academic Press, New York, 1967).
3. J. E. Holliday, in *Soft X-Ray Spectrometry and Band Structures of Metals and Alloys*, ed. by D. J. Farbian (Academic Press, London, New York, 1968), p. 101.
4. M. O. Krause, *Chem. Phys. Lett.* **10**, 65 (1971).
5. M. O. Krause, private communication.
6. H. Fellner-Feldegg, U. Gelius, B. Wannberg, A. G. Nilsson, E. Basilier, and K. Siegbahn, *J. Electron Spectros.* **5**, 643 (1974).
7. K. Siegbahn, in *Electron Spectroscopy—Proc. Int. Conf., Asilomar, September 1971*, ed. by D. A. Shirley (North-Holland, Amsterdam, 1972), p. 15.
8. B. L. Henke and M. A. Tester, *Advances in X-Ray Analysis*, Vol. 18 (1975).

9. K. Siegbahn, C. Nordling, A. Fahlman, R. Nordberg, K. Hamerin, J. Hedman, G. Johansson, T. Bergmark, S.-E. Karlsson, I. Lindgren, and B. Lindberg, *Electron Spectroscopy for Chemical Analysis—Atomic, Molecular, and Solid State Structure Studies by Means of Electron Spectroscopy* (Almqvist and Wiksells Boktryekeri AB, Stockholm, Sweden, 1967); *Nova Acta Regiae Soc. Sci. Upsaliensis, Ser. IV* **20** (1967).
10. K. Siegbahn, D. Hammond, H. Fellner-Feldegg, and E. F. Barnett, *Science* **176**, 245 (1972).
11. R. P. Godwin, "Synchrotron Radiation as a Light Source," in *Springer Tracts in Modern Physics* 51 (1970).
12. K. Codling, *Rep. Prog. Phys.* **36**, 544 (1973).
13. D. L. Ederer, NBS, private communication.
14. K. Thimm, *J. Electron Spectros.* **5**, 755 (1974).
15. E. M. Purcell (private communication), p. 31 in Godwin.[11]
16. D. E. Eastman and W. D. Grobman, *Phys. Rev. Lett.* **21**, 1378 (1972); P. Mitchell and K. Codling, *Phys. Lett.* **38A**, 31 (1972).
17. M. J. Lynch, K. Codling, and A. B. Gardner, *Phys. Lett.* **43A**, 213 (1973); D. E. Eastman, W. D. Grobman, J. J. Freeouf, and M. Erbudak, *Phys. Rev. B* **9**, 3473 (1974); T. Sagawa, R. Kato, S. Sato, M. Watanabe, T. Ishii, I. Nagakura, S. Kono, and S. Suzuki, *J. Electron Spectros.* **5**, 551 (1974).
18. I. Lindau, P. Pianetta, S. Doniach, and W. E. Spicer (to be published).
19. J. W. Taylor, in *Chemical Spectroscopy and Photochemistry in the Vacuum-Ultraviolet*, ed. by S. C. Sandorfy, P. J. Ausloos, and M. B. Robin (Reidel, Boston, 1974), p. 543; K. Codling, *Physica Scripta* **9**, 247 (1974).
20. J. A. R. Samson, *Techniques of Vacuum Ultraviolet Spectroscopy* (Wiley, New York, 1967).
21. L. Åsbrink and J. W. Rabalais, *Chem. Phys. Lett.* **12**, 182 (1971).
22. J. A. R. Samson, *Rev. Sci. Instrum.* **40**, 1174 (1969).
23. R. T. Poole, J. Liesegang, R. C. G. Leckey, and J. G. Jenkin, *J. Electron Spectros.* **5**, 773 (1974).
24. F. Burger and J. P. Maier, *J. Electron Spectros.* **5**, 783 (1974).
25. D. W. Turner, private communication.
26. J. A. Kinsinger, W. L. Stebbings, R. A. Valenzi, and J. W. Taylor, *Anal. Chem.* **44**, 773 (1972).
27. R. D. Deslattes, Jr., private communication.
28. K. Siegbahn, C. Nordling, G. Johansson, J. Hedman, P. F. Heden, K. Hamrin, U. Gelius, T. Bergmark, L. O. Werme, R. Manne, and Y. Baer, *ESCA Applied to Free Molecules* (North-Holland, Amsterdam–London, 1969).
29. C. R. Brundle, M. W. Roberts, D. Latham, and K. Yates, *J. Electron Spectros.* **3**, 241 (1974).
30. R. M. Friedman, J. Gobel, J. Hudis, and M. L. Perlman, *J. Electron Spectros.* **1**, 300 (1972/73).
31. D. J. Hnatowich, J. Hudis, M. L. Perlman, and R. C. Ragaini, *J. Appl. Phys.* **42**, 4883 (1971).
32. D. Betteridge, J. C. Carver, and D. M. Hercules, *J. Electron Spectros.* **2**, 327 (1973); D. S. Urch and M. Webber, *J. Electron Spectros.* **5**, 792 (1974).
33. C. K. Jørgensen, *Chimia* **25**, 213 (1971).
34. W. E. Swartz, Jr., P. H. Watts, Jr., J. C. Watts, J. M. Brasch, and E. R. Lippincott, *Anal. Chem.* **44**, 2001 (1972).
35. J. L. Ogilvie and A. Wolberg, *Appl. Spect.* **26**, 401 (1972).
36. S. Evans, *Chem. Phys. Lett.* **23**, 134 (1973).

37. P. H. Citrin and T. D. Thomas, *J. Chem. Phys.* **57**, 4446 (1972).
38. G. Grimvall, *Physica Scripta* **9**, 43 (1974).
39. D. A. Huchital and R. T. McKeon, *Appl. Phys. Lett.* **20**, 158 (1972).
40. D. T. Clark, D. Kilcast, and W. K. R. Musgrave, *J. Chem. Soc. D: Chem. Comm.* **1971**, 516.
41. C. S. Fadley and D. A. Shirley, *J. Res. NBS, Phys. Chem.* **74A**, 543 (1970).
42. H. Siegbahn, L. Asplund, P. Kelfve, K. Hamrin, L. Karlsson, and K. Siegbahn, *J. Electron Spectros.* **5**, 1059 (1974); H. Siegbahn and K. Siegbahn, *J. Electron Spectros.* **2**, 319 (1973).
43. J. Berkowitz and H. Ehrhardt, *Phys. Lett.* **21**, 531 (1966); J. Berkowitz, H. Ehrhardt, and T. Tekaat, *Z. Physik* **200**, 69 (1967).
44. J. A. R. Samson, *Phil. Trans. R. Soc. Lond.* **A268**, 141 (1970).
45. D. A. Vroom, A. R. Comeaux, and J. W. McGowan, *Chem. Phys. Lett.* **3**, 476 (1969).
46. T. A. Carlson and A. E. Jonas, *J. Chem. Phys.* **55**, 4913 (1971).
47. D. C. Mason, A. Kuppermann, and D. M. Mintz, in *Electron Spectroscopy*, ed. by D. A. Shirley (North-Holland, Amsterdam, 1972), p. 269.
48. R. Morgenstern, A. Niehaus, and M. W. Ruf, *Chem. Phys. Lett.* **4**, 635 (1970).
49. M. O. Krause, *Phys. Rev.* **177**, 151 (1969).
50. D. L. Ames, J. P. Maier, F. Watt, and D. W. Turner, *Faraday Disc. Chem. Soc.* **54**, 277 (1972).
51. B. P. Pullen, T. A. Carlson, W. E. Moddeman, G. K. Schweitzer, W. E. Bull, and F. A. Grimm, *J. Chem. Phys.* **53**, 768 (1970).
52. J. C. Helmer and N. H. Weichert, *Appl. Phys. Lett.* **13**, 266 (1968).
53. M. J. Weiss, *J. Electron Spectros.* **1**, 179 (1972/73).
54. B. Wannberg, U. Gelius, and K. Siegbahn, *J. Phys. E* **7**, 149 (1974); M. E. Gellender and A. D. Baker, *J. Electron Spectros.* **4**, 249 (1974); F. H. Read, J. Comer, R. E. Imhof, J. N. H. Brunt, and E. Harting, *J. Electron Spectros.* **4**, 293 (1974).
55. H. Z. Sar-el, *Rev. Sci. Instr.* **38**, 1210 (1967).
56. H. Hafner, J. A. Simpson, and C. E. Kuyatt, *Rev. Sci. Instr.* **39**, 33 (1968).
57. D. E. Golden and A. Zecca, *Rev. Sci. Instr.* **42**, 210 (1971).
58. J. D. Lee, *Rev. Sci. Instr.* **43**, 2191 (1972); W. M. Riggs and R. P. Fedchenko, *Am. Lab.* **4**, 65 (1972).
59. K. D. Sevier, *Low Energy Electron Spectrometry* (Wiley–Interscience, New York, 1972).
60. K. Siegbahn, ed., *Alpha-, Beta-, and Gamma-Ray Spectroscopy* (North-Holland, Amsterdam, 1965).
61. C. S. Fadley, R. N. Healey, J. M. Hollander, and C. E. Miner, *Electron Spectroscopy*, ed. by D. A. Shirley (North-Holland, Amsterdam, 1972), p. 121.
62. D. Betteridge, A. D. Baker, P. Bye, S. K. Hasannudin, N. R. Kemp, and M. Thompson, *J. Electron Spectros.* **4**, 163 (1974).
63. P. H. Citrin, R. W. Shaw, Jr., and T. D. Thomas, in *Electron Spectroscopy*, ed. by D. A. Shirley (North-Holland, Amsterdam, 1972), p. 105; S. Aksela, *Rev. Sci. Instr.* **42**, 810 (1971); J. L. Gardner and J. A. R. Samson, *J. Electron Spectros.* **2**, 267 (1973).
64. J. D. Allen, Jr., J. P. Wolfe, and G. K. Schweitzer, *Int. J. Mass. Spectrom. Ion Phys.* **8**, 81 (1972).
65. H. Boersch, J. Geiger, and W. Stickel, *Z. Physik* **180**, 415 (1964).
66. B. Tsai, T. Baer, and M. L. Horovitz, *Rev. Sci. Instrum.* **45**, 494 (1974).
67. J. Berkowitz, *J. Chem. Phys.* **56**, 2766 (1972); S. Evans, A. F. Orchard, and D. W. Turner, *Int. J. Mass Spectrom. Ion Phys.* **7**, 261 (1971); Y. S. Khodeyev, H. Siegbahn, K. Hamrin, and K. Siegbahn, University of Uppsala report UUIP-802 (1972).

68. C. J. Danby and J. H. D. Eland, *Int. J. Mass Spectrom. Ion Phys.* **8**, 153 (1972); J. Daintith, J. P. Maier, D. A. Sweigart, and D. W. Turner, in *Electron Spectroscopy*, ed. by D. A. Shirley (North-Holland, Amsterdam, 1972), p. 289.
69. R. E. Ballard, and G. A. Griffiths, in *Electron Spectroscopy*, ed. by D. A. Shirley (North-Holland, Amsterdam, 1972), p. 151.
70. C. D. Moak, S. Datz, F. Garcia Santibáñez, and T. A. Carlson, *J. Electron Spectros.* **6**, 151 (1975).
71. J. A. Decker and M. O. Harwit, *Appl. Optics* **7**, 2205 (1968); **8**, 2552 (1969).
72. J. D. Allen, Jr. and G. K. Schweitzer, *J. Electron Spectros.* **1**, 507 (1972/73).
73. L. D. Hulett and M. T. Kelley, unpublished calculations.
74. T. D. Thomas, *J. Electron Spectros.* **6**, 81 (1975).
75. M. O. Krause and J. Tarrant, private communication.
76. H. Ebel and N. Gurker, *J. Electron Spectros.* **5**, 799 (1974); G. K. Wertheim, *J. Electron Spectros.* **6**, 239 (1975).

CHAPTER 3

1. H. Hall, *Rev. Mod. Phys.* **8**, 358 (1936).
2. A. J. Bearden, *J. Appl. Phys.* **37**, 1681 (1966).
3. S. T. Manson and J. W. Cooper, *Phys. Rev.* **165**, 126 (1968).
4. D. J. Kennedy and S. T. Manson, *Phys. Rev. A* **5**, 227 (1972).
5. J. C. Tully, R. S. Berry, and B. J. Dalton, *Phys. Rev.* **176**, 95 (1968).
6. J. Cooper and R. N. Zare, *J. Chem. Phys.* **48**, 942 (1968).
7. G. Purvis, private communication.
8. E. J. McGuire, *J. Phys. Chem. Solids* **33**, 577 (1972).
9. J. H. Scofield, "Theoretical Cross Sections from 1–1500 keV," Lawrence Livermore Laboratory Report, UCRL-51326 (1973).
10. C. S. Fadley, *Chem. Phys. Lett.* **25**, 225 (1974).
11. T. Koopmans, *Physica* **1**, 104 (1934).
12. S. Lundqvist and G. Wendin, *J. Electron Spectros.* **5**, 513 (1974).
13. H. Basch, *J. Electron Spectros.* **5**, 463 (1974).
14. D. A. Shirley, *J. Electron Spectros.* **5**, 135 (1974).
15. L. Hedin and A. Johansson, *J. Phys. B* **2**, 1336 (1969).
16. A. J. Freeman, P. S. Bagus, and J. V. Mellow, *Int. J. Magn.* **4**, 35 (1973); P. S. Bagus, A. J. Freeman, and F. Sasaki, *Int. J. Quantum Chem.* **7**, 83 (1973).
17. A. Migdal, *J. Phys. (USSR)* **4**, 449 (1941).
18. E. L. Feinberg, *J. Phys. (USSR)* **4**, 423 (1941).
19. M. Wolfsberg and M. L. Perlman, *Phys. Rev.* **99**, 1833 (1955).
20. M. O. Krause, M. L. Vestal, W. H. Johnston, and T. A. Carlson, *Phys. Rev.* **133**, A385 (1964); T. A. Carlson and M. O. Krause, *Phys. Rev.* **137**, A1655 (1965).
21. T. A. Carlson, C. W. Nestor, Jr., T. C. Tucker, and F. B. Malik, *Phys. Rev.* **169**, 27 (1968).
22. T. Åberg, *Phys. Rev.* **156**, 35 (1967).
23. E. L. Feinberg, *J. Nucl. Phys. (USSR)* **1**, 438 (1965).
24. T. A. Carlson and M. O. Krause, *Phys. Rev.* **140**, A1057 (1965).
25. T. A. Carlson and C. W. Nestor, Jr., *Phys. Rev. A* **8**, 2887 (1973).
26. T. A. Carlson, W. E. Moddeman, and M. O. Krause, *Phys. Rev. A* **1**, 1406 (1970); M. O. Krause, F. A. Stevie, L. J. Lewis, T. A. Carlson, and W. E. Moddeman, *Phys. Lett.* **31A**, 81 (1970).

27. N. Stolterfoht, F. J. de Heer, and J. van Eck, *Phys. Rev. Lett.* **30**, 1159 (1973); N. Stolterfoht, "Cross Sections for Inner Shell Ionization of Gaseous Molecules by 50–500 keV Protons," in *Proc. Int. Conf. Inner Shell Ionization Phenomena and Future Applications*, ed. by R. W. Fink *et al.* (USAEC Tech. Information Center, Oak Ridge, 1973), CONF-720404, p. 1043.
28. T. A. Carlson, *Phys. Rev.* **156**, 142 (1967); T. A. Carlson, M. O. Krause, and W. E. Moddeman, *J. Phys. (Paris)* **32**, CA-76 (1971); M. O. Krause and F. Wuilleumier, *J. Phys. B. Atom. Molec. Phys.* **5**, 2143 (1972).
29. T. Åberg, *Phys. Rev. A* **1**, 341 (1970); F. W. Byron and C. J. Joachain, *Phys. Rev.* **164**, 1 (1967).
30. R. Manne and T. Åberg, *Chem. Phys. Lett.* **7**, 282 (1970).
31. Y. Itikawa, *J. Electron Spectros.* **2**, 125 (1973).
32. P. Citrin, P. Eisenberger, and D. R. Hamann, *Phys. Rev. Lett.* **33**, 965 (1974).
33. J. A. D. Matthew and M. G. Devey, *J. Phys. C* **7**, L335 (1974).
34. D. R. Hartree, *The Calculation of Atomic Structure* (Wiley, New York, 1955).
35. J. C. Slater, *Quantum Theory of Atomic Structure*, Vol. II (McGraw-Hill, New York, 1960).
36. C. W. Nestor, T. C. Tucker, T. A. Carlson, L. D. Roberts, F. B. Malik, and C. Froese, Oak Ridge National Laboratory Report, ORNL-4027.
37. F. Herman and S. Skillman, *Atomic Structure Calculations* (Prentice-Hall Englewood Cliffs, New Jersey, 1963).
38. J. B. Mann, Los Alamos Scientific Laboratory Report No. LA-3690 (1967); LA-3691 (1968).
39. C. Froese, University of British Columbia Report, Vancouver, B.C. (1968).
40. E. Clementi, *IBM J. Res. Development* **9**, 2 (1965); see also supplement.
41. C. C. Lu, T. A. Carlson, F. B. Malik, T. C. Tucker, and C. W. Nestor, Jr., *Atomic Data* **3** (1971).
42. J. B. Mann, private communication; see also J. B. Mann and J. T. Waber, *J. Chem. Phys.* **53**, 2397 (1970).
43*a*. C. F. Fischer, *Comp. Phys. Comm.* **1**, 151 (1970).
43*b*. C. F. Fischer, *Comp. Phys. Comm.* **5**, 147 (1973).
44. C. W. Nestor, Jr., Mathematics Division, Oak Ridge National Laboratory, Oak Ridge, Tennessee; also see J. P. Desclaux, D. F. Mayers, and F. O'Brien, *J. Phys. B* **4**, 631 (1971).
45. K. H. Johnson and F. C. Smith, *Phys. Rev. B* **5**, 831 (1972).
46. J. W. D. Connolly and K. H. Johnson, *Chem. Phys. Lett.* **10**, 616 (1971); J. W. D. Connolly, H. Siegbahn, U. Gelius, and C. Nordling, *J. Chem. Phys.* **58**, 4265 (1973); J. C. Slater and K. H. Johnson, *Phys. Rev. B* **5**, 844 (1972).
47. L. S. Cederbaum, G. Hohlneicher, and S. Peyerimhoff, *Chem. Phys. Lett.* **11**, 421 (1971); F. Ecker and G. Hohlneicher, *Theoret Chim. Acta* **25**, 289 (1972); L. S. Cederbaum, G. Hohlneicher, and W. Von Niessen, *Mol. Phys.* **26**, 1405 (1973); D. P. Chong, F. G. Herring, and D. McWilliams, *J. Chem. Phys.* **61**, 78 (1974).
48. L. S. Cederbaum and W. Domcke, *J. Chem. Phys.* **60**, 2878 (1974).
49. G. D. Purvis and Y. Öhrn, *J. Chem. Phys.* **60**, 4063 (1974).
50. R. Hoffmann, *J. Chem. Phys.* **39**, 1397 (1963).
51. F. Brogli and E. Heilbronner, *Theoret. Chim. Acta* **26**, 289 (1972).
52. P. A. Cox, S. Evans, A. F.Orchard, N. V. Richardson, and P. J. Roberts, *Faraday Disc. Chem. Soc.* **54**, 26 (1972).
53. J. A. Pople, D. P. Santry, and G. A. Segal, *J. Chem. Phys.* **43**, S129 (1965); J. A. Pople and G. A. Segal, *J. Chem. Phys.* **43**, S136 (1965), **44**, 3289 (1966); D. P. Santry and G. A. Segal, *J. Chem. Phys.* **47**, 158 (1967).

54. R. M. Friedman, *J. Electron Spectros.* **5**, 501 (1974).
55. M. J. S. Dewar and E. Haselbach, *J. Am. Chem. Soc.* **92**, 590 (1970); M. J. S. Dewar, D. H. Lo, D. B. Patterson, N. Trinajstic, and G. E. Peterson, *J. Chem. Soc. D: Chem. Comm.* **1971**, 238; M. J. S. Dewar and S. Kirschner, *J. Am. Chem. Soc.* **93**, 4290 (1971); **93**, 4291 (1971).
56. G. Höjer and S. Meza, *Acta Chim. Scand.* **26**, 3723 (1972).
57. E. Lindholm, C. Fridh, and L. Åsbrink, *Faraday Disc. Chem. Soc.* **54**, 127 (1972).
58. W. E. Bull, B. P. Pullen, F. A. Grimm, W. E. Moddeman, G. K. Schweitzer, and T. A. Carlson, *Inorg. Chem.* **9**, 2474 (1970).
59. P. J. Bassett and D. R. Lloyd, *J. Chem. Soc. (London) A* **1971**, 641.
60. I. H. Hillier and V. R. Saunders, *Mol. Phys.* **22**, 193 (1971).
61. M. C. Green, M. F. Lappert, J. B. Pedley, W. Schmidt, and B. T. Wilkins, *J. Organometal. Chem.* **31**, C55 (1971).
62. C. R. Brundle, N. A. Kuebler, M. B. Robin, and H. Basch, *Inorg. Chem.* **11**, 20 (1972).
63. C. G. Pitt and H. Bock, *J. Chem. Soc. Chem. Comm.* **1972**. 28.
64. F. A. Cotton, *Chemical Applications of Group Theory* (Interscience, New York, 1963, 1971).

CHAPTER 4

1. F. A. Grimm, *J. Electron Spectros.* **2**, 475 (1973).
2. E. Ishiguro and M. Kobori, *J. Phys. Soc. (Japan)* **22**, 263 (1967).
3. J. L. Berkosky, F. O. Ellison, T. H. Lee, and J. W. Rabalais, *J. Chem. Phys.* **59**, 5342 (1973).
4. T. H. Lee and J. W. Rabalais, *J. Chem. Phys.* **60**, 1172 (1974).
5. K. Wittel, *Chem. Phys. Lett.* **15**, 555 (1972).
6. M. Jungen, *Theoret. Chim. Acta* **27**, 33 (1972).
7. K. Wittel, B. S. Mohanty, and R. Manne, *J. Electron Spectros.* **5**, 1115 (1974).
8. F. Brogli and E. Heilbronner, *Helv. Chim. Acta* **54**, 1423 (1971).
9. O. Edqvist, E. Lindholm, L. E. Selin, and L. Åsbrink, *Physica Scripta* **1**, 25 (1970).
10. O. Edqvist, L. Åsbrink, and E. Lindholm, *Z. Naturforsch.* **A26**, 1407 (1971).
11. J. W. Rabalais, L. O. Werme, T. Bergmark, L. Karlsson, M. Hussain, and K. Siegbahn, *J. Chem. Phys.* **57**, 1185 (1972).
12. P. A. Cox and A. F. Orchard, *Chem. Phys. Lett.* **7**, 273 (1970); P. A. Cox, S. Evans, and A. F. Orchard. *Chem. Phys. Lett.* **13**, 386 (1972).
13. R. N. Dixon, *Mol. Phys.* **20**, 113 (1971).
14. F. A. Grimm and J. Godoy, *Chem. Phys. Lett.* **6**, 336 (1970).
15. A. W. Potts and W. C. Price, *Proc. R. Soc. Lond.* **A236**, 165 (1972).
16. A. E. Jonas, G. K. Schweitzer, F. A. Grimm, and T. A. Carlson, *J. Electron Spectros.* **1**, 29 (1972/73).
17. S. Evans, J. C. Green, P. J. Joachim, A. F. Orchard, D. W. Turner, and J. P. Maier, *J. Chem. Soc., Faraday Trans. II* **68**, 905 (1972).
18. J. H. D. Eland, *Photoelectron Spectroscopy* (Halsted/Wiley, New York, 1974).
19. M. O. Krause and F. Wuilleumier, *J. Phys. B: Atom. Molec. Phys.* **5**, L143 (1972).
20. T. A. Carlson, M. O. Krause, and W. E. Moddeman, *J. Phys. (Paris) CA*-76 (1971).
21. D. P. Spears, H. J. Fischbeck, and T. A. Carlson, *Phys. Rev.* **9**, 1603 (1974).
22. K. Siegbahn, C. Nordling, G. Johansson, J. Hedman, P. F. Heden, K. Hamrin, U. Gelius, T. Bergmark, L. O. Werme, R. Manne, and Y. Baer, *ESCA Applied to Free Molecules* (North-Holland, Amsterdam, 1969).

23. S. Cradock and W. Duncan, *Mol. Phys.* **27**, 837 (1974).
24. A. W. Potts and T. A. Williams, *J. Electron Spectros.* **3**, 3 (1974).
25. J. C. Lorquet and C. Cadet, *Int. J. Mass Spectrom. Ion Phys.* **7**, 245 (1971).
26. J. E. Collin, J. Delwiche, and P. Natalis, *Int. J. Mass Spectrom. Ion Phys.* **7**, 19 (1971).
27. J. E. Collin, J. Delwiche, and P. Natalis, in *Electron Spectroscopy, Int. Conf., Asilomar, September 1971*, ed. by D. A. Shirley (North-Holland, Amsterdam, 1972), p. 401.
28. P. Natalis, J. Delwiche, and J. E. Collin, *Faraday Disc. Chem. Soc.* **54**, 98 (1972).
29. V. Cermák, M. Smutek, and J. Srámek, *J. Electron Spectros.* **2**, 1 (1973).
30. C. Sluse-Goffart and P. Natalis, *J. Electron Spectros.* **2**, 215 (1973).
31. T. Bear and B. P. Tsai, *J. Electron Spectros.* **2**, 25 (1973).
32. J. L. Gardner and J. A. R. Samson, *J. Electron Spectros.* **2**, 153 (1973).
33. D. G. Streets, A. W. Potts, and W. C. Price, *Int. J. Mass Spectrom. Ion Phys.* **10**, 123 (1972/73).
34. D. Betteridge and M. Thompson, *J. Mol. Struc.* **21**, 341 (1974).
35. D. A. Sweigart and J. Daintith, *Sci. Prog. Oxf.* **59**, 325 (1971); D. A. Sweigart and D. W. Turner, *J. Am. Chem. Soc.* **94**, 5599 (1972).
36. A. D. Baker and D. Betteridge, *Photoelectron Spectroscopy: Chemical and Analytical Aspects*, Vol. 53 of International Series of Monographs in Analytical Chemistry (Pergamon Press, Oxford, 1972).
37. D. W. Turner, A. D. Baker, C. Baker, and C. R. Brundle, *Molecular Photoelectron Spectroscopy: a Handbook of He 584 Å Spectra* (Interscience, London–New York, 1970), p. 386.
38. C. R. Brundle, M. B. Robin, and H. Basch, *J. Chem. Phys.* **53**, 2196 (1970).
39. A. W. Potts, H. J. Lempka, D. G. Streets, and W. C. Price, *Phil. Trans. R. Soc. Lond.* **A268**, 59 (1970).
40. F. Brogli, J. K. Crandall, E. Heilbronner, E. Kloster-Jensen, and S. A. Sojka, *J. Electron Spectros.* **2**, 455 (1973).
41. G. W. Mines and R. K. Thomas, *Proc. R. Soc. Lond.* **A336**, 355 (1974).
42. T. P. Debies and J. W. Rabalais, *J. Electron Spectros.* **1**, 355 (1972/73).
43. F. T. Chau and C. A. McDowell, *J. Electron Spectros.* **6**, 357 (1975).
44. K. Yoshikawa, M. Hashimoto, and I. Morishima, *J. Am. Chem. Soc.* **96**, 288 (1974).
45. J. C. Bunzli, D. C. Frost, and L. Weiler, *J. Am. Chem. Soc.* **96**, 1952 (1974); W. Schaefer, A. Schweig, G. Maier, and T. Sayrac, *J. Am. Chem. Soc.* **96**, 279 (1974).
46. D. G. Streets and T. A. Williams, *J. Electron Spectros.* **3**, 71 (1974).
47. R. Hoffmann, *Acc. Chem. Res.* **4**, 1 (1971); R. Hoffmann, A. Imamura, and W. J. Hehre, *J. Am. Chem. Soc.* **90**, 1499 (1969).
48. E. Heilbronner, J. P. Maier, and E. Haselbach, in *Physical Methods in Heterocyclic Chemistry*, Vol. 6, ed. by A. R. Katritzky (Academic Press, New York, 1974), p. 1; D. M. W. Van Den Ham and D. Van Der Meer, *J. Electron Spectros.* **2**, 247 (1973).
49. K. Kimura, S. Katsumata, Y. Achiba, H. Matsumoto, and S. Nagakura, *Bull. Chem. Soc. Japan* **46**, 373 (1973); S. Katsumata and K. Kimura, *Bull. Chem. Soc. Japan* **46**, 1342 (1973); T. Yamazaki, S. Katsumata, and K. Kimura, *J. Electron Spectros.* **2**, 335 (1973); K. Kimura, S. Katsumata, T. Yamazaki, and W. Wakabayashi, *J. Electron Spectros.* **6**, 41 (1975); K. Osafune and K. Kimura, *Chem. Phys. Lett.* **25**, 47 (1974); S. Katsumata, T. Iwai, and K. Kimura, *Bull. Chem. Soc. Japan* **46**, 3391 (1973); K. Osafune, S. Katsumata, and K. Kimura, *Chem. Phys. Lett.* **19**, 369 (1974).
50. C. R. Brundle, M. B. Robin, N. A. Kuebler, and H. Basch, *J. Am. Chem. Soc.* **94**, 1451 (1972); C. R. Brundle, M. B. Robin, and N. A. Kuebler, *J. Am. Chem. Soc.* **94**, 1466 (1972).
51. J. P. Maier and D. W. Turner, *Faraday Disc. Chem. Soc.* **54**, 149 (1972).

52. J. P. Maier and D. W. Turner, *J. Chem. Soc. Faraday Trans. II* **69**, 196 (1973); 521 (1973).
53. S. A. Cowling and R. A. W. Johnstone, *J. Electron Spectros.* **2**, 161 (1973).
54. C. Batich, O. Ermer, E. Heilbronner, and J. R. Wiseman, *Angew. Chem. Int. Ed.* **12**, 312 (1973).
55. T. Kobayashi, K. Yokota, and S. Nagakura, *J. Electron Spectros.* **2**, 449 (1973).
56. A. D. Baker, M. Brisk, and M. Gellender, *J. Electron Spectros.* **3**, 227 (1974).
57. S. F. Nelsen and J. M. Buschek, *J. Am. Chem. Soc.* **95**, 2011 (1973).
58. A. H. Cowley, M. J. S. Dewar, D. W. Goodman, and M. C. Padolina, *J. Am. Chem. Soc.* **96**, 2648 (1974).
59. R. Boshi, E. Clar, and W. Schmidt, *J. Chem. Phys.* **60**, 4406 (1974).
60. T. G. Edwards, *Theoret. Chim. Acta* **27**, 1 (1972).
61. W. E. Bull, B. P. Pullen, F. A. Grimm, W. E. Moddeman, G. K. Schweitzer, and T. A. Carlson, *Inorg. Chem.* **9**, 2474 (1970).
62. M. B. Hall, M. F. Guest, I. H. Hillier, O. R. Lloyd, A. F. Orchard, and A. W. Potts, *J. Electron Spectros.* **1**, 497 (1972/73).
63. J. M. Hollas and T. A. Sutherley, *Mol. Phys.* **24**, 1123 (1972).
64. J. Berkowitz, *Chem. Phys. Lett.* **11**, 21 (1971).
65. W. C. Price, A. W. Potts, and D. G. Streets, in *Electron Spectroscopy*, ed. by D. A. Shirley (North-Holland, Amsterdam, 1972), p. 187.
66. U. Gelius, in *Electron Spectroscopy*, ed. by D. A. Shirley (North-Holland, Amsterdam, 1972), p. 311.
67. L. L. Lohr, Jr., in *Electron Spectroscopy*, ed. by D. A. Shirley (North-Holland, Amsterdam, 1972), p. 245; L. L. Lohr and M. B. Robin, *J. Am. Chem. Soc.* **92**, 7241 (1970).
68. W. B. Perry and W. L. Jolly, *J. Electron Spectros.* **4**, 219 (1974).
69. J.-T. J. Huang and F. O. Ellison, *J. Electron Spectros.* **4**, 233 (1974).
70. H. P. Kelly, *Chem. Phys. Lett.* **20**, 547 (1973).
71. A. Schweig and W. Thiel, *J. Chem. Phys.* **60**, 951 (1974); *Mol. Phys.* **27**, 265 (1974); *J. Electron Spectros.* **3**, 27 (1974).
72. F. O. Ellison, *J. Chem. Phys.* **61**, 507 (1974).
73. J. J. Huang, F. O. Ellison, and J. W. Rabalais, *J. Electron Spectros.* **3**, 339 (1974).
74. R. J. Colton and J. W. Rabalais, *J. Electron Spectros.* **3**, 345 (1974).
75. J. W. Rabalais, T. P. Debies, J. L. Berkosky, J.-T. J. Huang and F. O. Ellison, *J. Chem. Phys.* **61**, 516 (1974).
76. M. B. Robin, N. A. Kuebler, and C. R. Brundle, in *Electron Spectroscopy*, ed. by D. A. Shirley (North-Holland, Amsterdam, 1972), p. 351; A. Katrib, T. P. Debies, R. J. Colton, T. H. Lee, and J. W. Rabalais, *Chem. Phys. Lett.* **22**, 196 (1973).
77. J. N. Murrell, *Chem. Phys. Lett.* **15**, 296 (1972).
78. T. H. Lee and J. W. Rabalais, *J. Chem. Phys.* **61**, 2747 (1974).
79. J. Berkowitz and H. Ehrhardt, *Phys. Lett.* **21**, 531 (1966).
80. J. Berkowitz, H. Ehrhardt, and T. Tekaat, *Z. Physik* **200**, 69 (1967).
81. T. A. Carlson, *Chem. Phys. Lett.* **9**, 23 (1971).
82. T. A. Carlson and C. P. Anderson, *Chem. Phys. Lett.* **10**, 561 (1971).
83. T. A. Carlson, G. E. McGuire, A. E. Jonas, K. L. Cheng, C. P. Anderson, C. C. Lu, and B. P. Pullen, in *Electron Spectroscopy*, ed. by D. A. Shirley (North-Holland, Amsterdam, 1972), p. 207.
84. T. A. Carlson and A. E. Jonas, *J. Chem. Phys.* **55**, 4913 (1971).
85. T. A. Carlson and R. M. White, *Faraday Disc. Chem. Soc.* **54**, 285 (1972).
86. T. A. Carlson and G. E. McGuire, *J. Electron Spectros.* **1**, 209 (1972/73).
87. R. Morgenstern, A. Niehaus, and M. W. Ruf, *Chem. Phys. Lett.* **4**, 635 (1970).

88. A. Niehaus and M. W. Ruf, *Chem. Phys. Lett.* **11**, 55 (1971).
89. J. A. R. Samson, *J. Opt. Soc. Am.* **59**, 356 (1969).
90. D. C. Mason, A. Kuppermann, and D. M. Mintz, in *Electron Spectroscopy*, ed. by D. A. Shirley (North-Holland, Amsterdam, 1972), p. 269.
91. J. W. McGowan, D. A. Vroom, and A. R. Comeaux, *J. Chem. Phys.* **51**, 5626 (1969).
92. J. C. Tully, R. S. Berry, and B. J. Dalton, *Phys. Rev.* **176**, 95 (1968); B. Schneider and R. S. Berry, *Phys. Rev.* **182**, 141 (1969).
93. S. Iwata and S. Nagakura, *Mol. Phys.* **27**, 425 (1974).
94. J. W. Rabalais, T. P. Debies, J. L. Berkosky, J. L. Huang, and F. O. Ellison, *J. Chem. Phys.* **61**, 529 (1974).
95. B. Richie, *J. Chem. Phys.* **60**, 898 (1974).
96. S. T. Manson, *J. Electron Spectros.* **2**, 413 (1972/73).
97. J. A. Kensinger and J. W. Taylor, *Int. J. Mass Spectrom. Ion Phys.* **10**, 445 (1972/3).
98. R. M. White and T. A. Carlson, unpublished results.
99. G. Herzberg, *Molecular Spectra and Molecular Structure*, Vol. III, *Electronic Spectra and Electronic Structure of Polyatomic Molecules* (D. Van Nostrand, Princeton, New Jersey, 1966).
100. J. L. Gardner and J. A. R. Samson, *J. Chem. Phys.* **60**, 3711 (1974).
101. A. W. Potts, private communication.
102. J. M. Sichel, *Mol. Phys.* **18**, 95 (1970).
103. D. Dill, *Phys. Rev.* **6**, 160 (1972).
104. L. Åsbrink, C. Fridh, B. Ö. Jonsson, and E. Lindholm, *Int. J. Mass Spectrom. Ion. Phys.* **8**, 229 (1972); **8**, 215 (1972); C. Fridh, L. Åsbrink, B. Ö Jonsson, and E. Lindholm, *Int. J. Mass Spectrom. Ion Phys.* **8**, 85 (1972); **8**, 101 (1972); **9**, 485 (1972).
105. G. Herzberg, *Molecular Spectra and Molecular Structure*, II. *Infrared and Raman Spectra of Polyatomic Molecules* (D. Van Nostrand, Princeton, New Jersey, 1945).
106. G. Herzberg, *Molecular Spectra and Molecular Structure*, I. *Spectra of Diatomic Molecules*, 2nd ed. (D. Van Nostrand, Princeton, New Jersey, 1950).
107. E. Haselbach and A. Schmelzer, *Helv. Chim. Acta* **55**, 1745 (1972).
108. J. Daintith, J. P. Maier, D. A. Sweigart, and D. W. Turner, in *Electron Spectroscopy*, ed. by D. A. Shirley (North-Holland, Amsterdam, 1972), p. 289.
109. L. O. Werme, B. Grennberg, J. Nordgren, C. Nordling, and K. Siegbahn, *Phys. Rev. Lett.* **30**, 523 (1973).
110. K. Siegbahn, *J. Electron Spectros.* **5**, 3 (1974); L. O. Werme, J. Nordgren, H. Ågren, C. Nordling, and K. Siegbahn, submitted to *Z. Physik*.
111. J. W. Rabalais, T. Bergmark, L. O. Werme, L. Karlsson, and K. Siegbahn, *Physica Scripta* **3**, 13 (1971).
112. G. Innorta, S. Torroni, and S. Pignataro, *Organic Mass Spectrometry* **6**, 113 (1972).
113. C. J. Danby and J. H. D. Eland, *Int. J. Mass Spectrom. Ion Phys.* **8**, 153 (1972).
114. K. Siegbahn, D. A. Allison, and J. H. Allison, *Handbook of Spectroscopy*, Vol. I, ed. by J. W. Robinson (CRL Press, Cleveland, 1974), pp. 258–749.
115. S. D. Worley, *Chem. Reviews* **71**, 295 (1971).
116. D. Betteridge, *Anal. Chem.* **44**, 100R (1972); D. Betteridge and M. A. Williams, *Anal. Chem.* **46**, 125R (1974).
117. A. Hammett and A. F. Orchard, *Electronic Structure and Magnetism of Inorganic Compounds*, Vol. 1, ed. by P. Day (Chem. Soc., London, 1972), pp. 1–62.
118. S. Evans and A. F. Orchard, *Electronic Structure and Magnetism of Inorganic Compounds*, Vol. II, ed. by P. Day (Chem. Soc., London, 1973), pp. 1–96.
119. H. Boch and B. G. Ramsey, *Angew. Chem. Int. Ed.* **12**, 734 (1973).
120. R. L. DeKock and D. R. Lloyd, *Advan. Inorg. Chem. Radiochem.* **16**, 65 (1974).

121. E. Heilbronner, in *The World of Quantum Chemistry*, ed. by R. Daudel and B. Pullman (Reidel, Holland, 1974), p. 211.
122. F. Wuilleumier, in *Advances in X-Ray Analysis*, Vol. 16, ed. by L. S. Birks (Plenum, New York, 1973), p. 63.
123. J. H. Scofield, Lawrence Livermore Laboratory Report, UCRL-51326 (1973).
124. S. T. Manson, *J. Electron Spectros.* **1**, 413 (1972/73).
125. A. Niehaus and M. W. Ruf, *Z. Physik* **252**, 84 (1972); D. C. Frost, C. A. McDowell, and D. A. Vroom, *Chem. Phys. Lett.* **1**, 93 (1967); J. L. Dehmer and J. Berkowitz, *Phys. Rev. A* **10**, 484 (1974).
126. J. A. R. Samson and V. E. Petrosky, *Phys. Rev. A* **9**, 2449 (1974).
127. Y. Itikawa, *J. Electron Spectros.* **2**, 125 (1973).
128. H. J. Lempka, T. R. Passmore, and W. C. Price, *Proc. R. Soc. Lond.* **A304**, 53 (1968).
129. J. Delwiche, P. Natalis, J. Momigny, and J. E. Collin, *J. Electron Spectros.* **1**, 219 (1972/73).
130. C. P. Anderson, G. Mamatov, W. E. Bull, F. A. Grimm, J. C. Carver, and T. A. Carlson, *Chem. Phys. Lett.* **12**, 137 (1971); R. L. DeKock, B. R. Higginson, D. R. Lloyd, A. Breeze, D. W. J. Cruickshank, and D. R. Armstrong, *Mol. Phys.* **24**, 1059 (1972); S. Evans and A. F. Orchard, *Inorg. Chim. Acta* **5**, 81 (1971); A. W. Potts and W. C. Price, *Trans. Faraday Soc.* **67**, 1242 (1971).
131. C. R. Brundle and D. W. Turner, *Int. J. Mass Spectrom. Ion Phys.* **2**, 195 (1969).
132. L. Åsbrink and J. W. Rabalais, *Chem. Phys. Lett.* **12**, 182 (1971).
133. R. Botter and H. M. Rosenstock, *Advan. Mass Spec.* **49**, 579 (1968).
134. C. R. Brundle and D. W. Turner, *Proc. R. Soc. Lond.* **A307**, 27 (1968).
135. T. Bergmark, L. Karlsson, R. Jadrny, L. Mattsson, R. G. Albridge, and K. Siegbahn, *J. Electron Spectros.* **4**, 85 (1974).
136. B. P. Pullen, T. A. Carlson, W. E. Moddeman, G. K. Schweitzer, W. E. Bull, and F. A. Grimm, *J. Chem. Phys.* **53**, 768 (1970).
137. J. W. Rabalais and A. Katrib, *Mol. Phys.* **27**, 923 (1974).
138. J. N. Murrell and W. Schmidt, *J. Chem. Soc., Faraday Trans. II* **1972**, 1709.
139. R. M. White, T. A. Carlson, and D. P. Spears, *J. Electron Spectros.* **3**, 59 (1974).
140. L. Åsbrink, C. Fridh, and E. Lindholm, *J. Am. Chem. Soc.* **94**, 5502 (1972).
141. L. Åsbrink, O. Edqvist, E. Lindholm, and L. E. Selin, *Chem. Phys. Lett.* **5**, 192 (1970).
142. A. W. Potts, W. C. Price, D. G. Streets, and T. W. Williams, *Faraday Disc. Chem. Soc.* **54**, 168 (1972).
143. T. P. Debies and J. W. Rabalais, *J. Electron Spectros.* **1**, 355 (1972/73).
144. J. W. Rabalais and R. J. Colton, *J. Electron Spectros.* **1**, 83 (1972/73).
145. D. G. Streets, W. E. Hall, and G. P. Ceasar, *Chem. Phys. Lett.* **17**, 90 (1972).
146. R. Glutes, E. Heilbronner, and V. Hornung, *Helv. Chem. Acta* **55**, 255 (1972).
147. A. D. Baker, D. Betteridge, N. R. Kemp, and R. E. Kirby, *Anal. Chem.* **42**, 1064 (1970).
148. J. W. Rabalais, L. O. Werme, T. Bergmark, L. Karlsson, and K. Siegbahn, *Int. J. Mass Spectrom. Ion Phys.* **9**, 185 (1972).
149. T. Kobayashi and S. Nagakura, *J. Electron Spectros.* **4**, 207 (1974).
150. C. Batich, P. Bischof, and E. Heilbronner, *J. Electron Spectros.* **1**, 333 (1972/73).
151. D. M. W. Van den Ham and D. Van der Meer, *Chem. Phys. Lett.* **12**, 447 (1972).
152. D. Chadwick. D. C. Frost, and L. Weiler, *J. Am. Chem. Soc.* **93,** 4962 (1971).
152*a*. R. Boschi, J. N. Murrell, and W. Schmidt, *Faraday Disc., Chem. Soc.* **54**, 116 (1972).
153. V. Boekelheide, J. N. Murrell, and W. Schmidt, *Tetrahedron Lett.* **7**, 575 (1972).
154. E. Heilbronner and H.-D. Martin, *Helv. Chim. Acta* **55**, 1490 (1972); F. Brogli, E. Heilbronner, and J. Ipaktschi, *Helv. Chim. Acta* **55**, 2447 (1972).
155. V. Boekelheide and W. Schmidt, *Chem. Phys. Lett.* **17**, 410 (1972).
156. M. Allan, E. Heilbronner, and E. Kloster-Jensen, *J. Electron Spectros.* **6**, 181 (1975).

157. S. Evans, A. F. Orchard, and D. W. Turner, *Int. J. Mass Spectrom. Ion Phys.* **7**, 261 (1971).
158. M. J. S. Dewar and D. W. Goodman, *J. Chem. Soc., Faraday Trans. II* **68**, 1784 (1972).
159. E. Heilbronner, V. Hornung, J. P. Maier, and E. Kloster-Jensen, *J. Am. Chem. Soc.* **96**, 4262 (1974).
160. G. H. King, J. N. Murrell, and R. J. Suffolk, *J. Chem. Soc., Dalton Trans.* **1972**, 564.
161. J. C. Green, M. L. H. Green, P. J. Joachim, A. F. Orchard, and D. W. Turner, *Phil. Trans. R. Soc. Lond.* **A268**, 111 (1970).
162. M. B. Robin and N. A. Kuebler, *J. Electron Spectros.* **1**, 13 (1972/73).
163. D. Chadwick, D. C. Frost, and L. Weiler, *J. Am. Chem. Soc.* **93**, 4320 (1971).
163*a*. D. A. Sweigart and D. W. Turner, *J. Am. Chem. Soc.* **94,** 5592 (1972).
164. W. Schäfer, and A. Schweig, *Angew. Chem. Int. Ed.* **11**, 836 (1972).
165. A. Schweig, W. Schäfer, and K. Dimroth, *Angew. Chem. Int. Ed.* **11**, 631 (1972).
166. C. Batich, E. Heilbronner, V. Hornung, A. J. Ashe, III, D. T. Clark, U. T. Cobley, D. Kilcast, and I. Scanlan, *J. Am. Chem. Soc.* **95**, 928 (1973).
167. D. Betteridge, M. Thompson, A. D. Baker, and N. R. Kemp, *Anal. Chem.* **44**, 2005 (1972).
168. J. W. Rabalais, L. Karlsson, L. O. Werme, T. Bergmark, and K. Siegbahn, *J. Chem. Phys.* **58**, 3370 (1973).
169. C. R. Brundle, N. A. Kuebler, M. B. Robins, and H. Basch, *Inorg. Chem.* **11**, 20 (1972).
170. J. W. Rabalais, T. Bergmark, L. O. Werme, L. Karlsson, M. Hussain, and K. Siegbahn, in *Electron Spectroscopy*, ed. by D. A. Shirley (North-Holland, Amsterdam, 1972), p. 425.
171. S. Evans, A. Hamnett, A. F. Orchard, and D. R. Lloyd, *Faraday Disc. Chem. Soc.* **54**, 227 (1972).
172. S. Evans, J. C. Green, and S. E. Jackson, *J. Chem. Soc., Faraday Trans. II* **68**, 249 (1972).
173. S. Evans, J. C. Green, S. E. Jackson, and B. Higginson, *J. Chem. Soc., Dalton Trans.* **1974**, 304.
174. S. Elbel, H. Bergmann, and W. Ensslin, *J. Chem. Soc., Faraday Trans. II* **70**, 555 (1974).
175. R. L. DeKock, B. R. Higginson, and D. R. Lloyd, *Faraday Disc. Chem. Soc.* **54**, 84 (1972).
176. P. J. Bassett and D. R. Lloyd, *J. Chem. Soc., Dalton Trans.* **1972**, 248.
177. D. C. Frost, F. G. Herring, K. A. R. Mitchell, and I. A. Stenhouse, *J. Am. Chem. Soc.* **93**, 1596 (1971); A. B. Cornford, D. C. Frost, F. G. Herring, and C. A. McDowell, *Chem. Phys. Lett.* **10**, 345 (1971); *J. Chem. Phys.* **54**, 1872 (1971); **55**, 2820 (1971).
178. P. J. Bassett and D. R. Lloyd, *J. Chem. Soc. Lond. A* **1971**, 641; G. R. Branton, D. C. Frost, C. A. McDowell, and I. A. Stenhouse, *Chem. Phys. Lett.* **5**, 1 (1970); S. Cradock, E. A. V. Ebsworth, W. J. Savage, and R. A. Whiteford, *J. Chem. Soc., Faraday Trans. II* **1972**, 934; S. Cradock, E. A. V. Ebsworth, and J. D. Murdoch, *J. Chem. Soc., Faraday Trans. II* **1972**, 86; A. Schweig, U. Weidner, and G. Manuel, *Angew. Chem. Int. Ed.* **11**, 837 (1972).
179. P. J. Bassett and D. R. Lloyd, *J. Chem. Soc. Lond. A* **1971**, 1551; H. Bock and W. Fuss, *Chem. Ber.* **104**, 1687 (1971); R. J. Boyd and D. C. Frost, *Chem. Phys. Lett.* **1**, 649 (1968); B. Cetinkaya, G. H. King, S. S. Krishnamurthy, M. F. Lappert, and J. B. Pedley, *J. Chem. Soc. D:* Chem. Comm. **1971**, 1370; A. K. Holliday, W. Reade, R. A. W. Johnstone, and A. F. Neville, *J. Chem. Soc. D: Chem. Comm.* **1971**, 51; G. H. King, S. S. Krishnamurthy, M. F. Lappert, and J. B. Pedley, *Faraday Disc. Chem Soc.* **54**, 70 (1972); R. F. Lake, *Spectrochim. Acta* **27A**, 1220 (1971); D. R. Lloyd and N. Lynaugh, *J. Chem. Soc. D: Chem. Comm.* **1970**, 1545; **1971**, 125; **1971**, 627; *J. Chem. Soc., Faraday Trans. II* **1972**, 947; N. Lynaugh, D. R. Lloyd, M. F. Guest, M. B. Hall, and I. H. Hillier, *J. Chem. Soc., Faraday Trans. II* **1972**, 2192; T. Rose, R. Frey, and B.

Brehm, *J. Chem. Soc. D: Chem. Comm.* **1969**, 1518; T. E. H. Walker and J. A. Horsley, *Mol. Phys.* **21**, 939 (1971).
180. J. Kroner, H. Noeth, and K. Niedenzu, *J. Organometal. Chem.* **71**, 165 (1974).
181. M. C. Green, M. F. Lappert, J. B. Pedley, W. Schmidt, and B. T. Wilkins, *J. Organometal. Chem.* **31**, C55 (1971).
182. C. G. Pitt and H. Bock, *J. Chem. Soc., Chem. Comm.* **1972**, 28.
183. G. Bieri, F. Brogli, E. Heilbronner, and E. Kloster-Jensen, *J. Electron Spectros.* **1**, 67 (1972/73); H. Bock and W. Ensslin, *Angew. Chem. Int. Ed.* **10**, 404 (1971); S. Cradock, *J. Chem. Phys.* **55**, 980 (1971); D. C. Frost, F. G. Herring, A. Kantrib, R. A. N. McLean, J. E. Drake, and N. P. C. Westwood, *Chem. Phys. Lett.* **10**, 347 (1971); U. Weidner and A. Schweig, *Angew. Chem. Int. Ed.* **11**, 537 (1972); *J. Organometal Chem.* **39**, 261 (1972).
184. U. Weidner and A. Schweig, *Angew. Chem. Int. Ed.* **11**, 536 (1972); *J. Organometal. Chem.* **37**, C29 (1972).
185. D. C. Frost, F. G. Herring, A. Kantrib, R. A. N. McLean, J. E. Drake, and N. P. C. Westwood, *Can. J. Chem.* **49**, 4033 (1971).
186. J. H. D. Eland, *Int. J. Mass Spectrom. Ion Phys.* **4**, 37 (1970).
187. C. R. Brundle, M. B. Robin, and G. R. Jones, *J. Chem. Phys.* **52**, 3383 (1970); C. R. Brundle and G. R. Jones, *J. Chem. Soc. D: Chem. Comm.* **1971**, 1198; C. R. Brundle, G. R. Jones, and H. Basch, *J. Chem. Phys.* **55**, 1098 (1971); C. R. Brundle and G. R. Jones, *Faraday Disc. II* **68**, 959 (1972); *J. Electron Spectros.* **1**, 403 (1972/73).
188. R. B. Cairns, H. Harrison, and R. I. Schoen, *Adv. Atoms Mol. Phys.* **8**, 131 (1972).
189. N. Jonathan, D. J. Smith, and K. J. Ross, *Chem. Phys. Lett.* **9**, 217 (1971).
190. N. Jonathan, A. Morris, D. J. Smith, and K. J. Ross, *Chem. Phys. Lett.* **7**, 497 (1970).
191. N. Jonathan, A. Morris, M. Okuda, D. J. Smith, and K. J. Ross, *Chem. Phys. Lett.* **13**, 334 (1972); N. Jonathan, A. Morris, M. Okuda, K. J. Ross, and D. J. Smith, *Faraday Soc. Disc. Chem Soc.* **54**, 48 (1972); D. C. Frost, S. T. Lee, and C. A. McDowell, *Chem. Phys. Lett.* **17**, 153 (1972).
192. H. W. Kroto and R. J. Suffolk, *Chem. Phys. Lett.* **15**, 545 (1972); *Chem. Phys. Lett.* **17**, 213 (1972); G. H. King, H. W. Kroto, and R. J. Suffolk, *Chem. Phys. Lett.* **13**, 457 (1972).
193. A. B. Cornford, D. C. Frost, F. G. Herring, and C. A. McDowell, *Faraday Disc. Chem. Soc.* **54**, 56 (1972).
194. I. Morishima, K. Yoshikawa, T. Yonezawa, and H. Matsumoto, *Chem. Phys. Lett.* **16**, 336 (1972).
195. C. R. Brundle, *Chem. Phys. Lett.* **26**, 25 (1974); D. C. Frost, S. T. Lee, and C. A. McDowell, *Chem. Phys. Lett.* **24**, 149 (1974).
196. D. L. Ames, J. P. Maier, F. Watt, and D. W. Turner, *Faraday Disc. Chem. Soc.* **54**, 277 (1972).
197. J. Berkowitz and W. A. Chupka, *J. Chem. Phys.* **45**, 1287 (1966); J. Berkowitz, *J. Chem. Phys.* **56**, 2766 (1972); J. Berkowitz and J. L. Dehmer, *J. Chem. Phys.* **57**, 3194 (1972); J. Berkowitz, in *Advances in High Temperature Chemistry*, Vol. 3, ed. by L. Eyring (Academic Press, New York, 1971), p. 123; J. L. Dehmer, J. Berkowitz, L. C. Cusachs, and H. S. Aldrich, *J. Chem. Phys.* **61**, 594 (1974); J. Berkowitz, *J. Chem. Phys.* **61**, 407 (1974).
198. J. D. Allen, Jr., G. W. Boggess, T. D. Goodman, A. S. Wachtel, Jr., and G. K. Schweitzer, *J. Electron Spectros.* **3**, 289 (1974).
199. A. Schweig, H. Vermeer, and U. Weidner, *Chem. Phys. Lett.* **26**, 229 (1974).
200. Y. Baer, P. F. Heden, J. Hedman, M. Klasson, C. Nordling, and K. Siegbahn, *Physica Scripta* **1**, 55 (1970); C. S. Fadley and D. A. Shirley, *J. Res. NBS Phys. Chem.* **74A**, 543 (1970).

201. S. Hüfner, G. K. Wertheim, R. L. Cohen, and J. H. Wernick, *Phys. Rev. Lett.* **28**, 488 (1972); S. Hüfner, G. K. Wertheim, N. V. Smith, and M. M. Traum, *Solid State Comm.* **11**, 323 (1972); R. A. Pollak, S. Kowalczyk, L. Ley, and D. A Shirley, *Phys. Rev. Lett.* **29**, 274 (1972); R. A. Pollak, L. Ley, S. Kowalczyk, D. A. Shirley, J. D. Joannopoulos, D. J. Chadi, and M. L. Cohen, *Phys. Rev. Lett.* **29**, 1103 (1972); D. A. Shirley, *Phys. Rev. B* **5**, 4709 (1972); L. Ley, S. Kowalczyk, R. Pollak, and D. A. Shirley, *Phys. Rev. Lett.* **29**, 1088 (1972); L. Ley, R. A. Pollak, F. R. McFeely, S. P. Kowalczyk, and D. A. Shirley, *Phys. Rev. B* **9**, 600 (1974); S. Hüfner and G. K. Wertheim, *Phys. Lett. A* **47**, 349 (1974).
202. D. E. Eastman, *Phys. Rev. Lett.* **26**, 846 (1971).
203. C. R. Brundle, M. W. Roberts, D. Latham, and K. Yates, *J. Electron Spectros.* **3**, 241 (1974).
204. D. E. Eastman, in *Electron Spectroscopy*, ed. by D. A. Shirley (North-Holland, Amsterdam, 1972), p. 487; W. E. Spicer, in *Band Structure Spectroscopy of Metals and Alloys*, ed. by D. J. Fabian and L. M. Watson (Academic Press, New York, 1973), p. 7; S. B. M. Hagström, in *Band Structure Spectroscopy of Metals and Alloys*, ed. by D. J. Fabian and L. M. Watson (Academic Press, New York, 1973), p. 73.
205. K. Sehi, Y. Harada, K. Ohno, and H. Inokuchi, *Bull. Chem. Soc. Japan* **47**, 1608 (1974).
206. J. E. Demuth and D. E. Eastman, *Phys. Rev. Lett.* **32**, 1123 (1974).
207. C. R. Brundle and M. W. Roberts, *Surface Sci.* **38**, 234 (1973).
208. S. J. Atkinson, C. R. Brundle, and C. R. Roberts, *Chem. Phys. Lett.* **24**, 175 (1974).
209. W. F. Egelhoff, Jr., D. L. Perry, and J. W. Linnett, *J. Electron Spectros.* **5**, 339 (1974).
210. R. Prins, *J. Chem. Phys.* **61**, 2580 (1974).
211. A. Calabrese and R. G. Hayes, *J. Electron Spectros.* **6**, 1 (1975).
212. V. I. Nefedov, Y. A. Buslaev, N. P. Sergushin, L. Bayer, Y. V. Kokunov, and V. V. Kovalev, *J. Electron Spectros.* **6**, 221 (1975).
213. E. Diemann and A. Müller, *Chem. Phys. Lett.* **27**, 351 (1974).
214. D. Betteridge and A. D. Baker, *Anal. Chem.* **42**, 43A (1970).

CHAPTER 5

1. K. Siegbahn, C. Nordling, A. Fahlman, R. Nordberg, K. Hamrin, J. Hedman, G. Johansson, T. Bergmark, S.-E. Karlsson, I. Lindgren, and B. Lindberg, *Electron Spectroscopy for Chemical Analysis—Atomic, Molecular, and Solid State Structure Studies by Means of Electron Spectroscopy* (Almqvist and Wiksells Boktryekeri AB, Stockholm, Sweden, 1967); *Nova Acta Regiae Soc. Sci., Upsaliensis, Ser. IV* **20** (1967).
2. M. O. Krause and F. Wuilleumier, in *Electron Spectroscopy: Proc. Int. Conf., Asilomar, September 1971*, ed. by D. A. Shirley (North-Holland, Amsterdam, 1972), p. 759.
3. J. A. Bearden and A. F. Burr, *Rev. Mod. Phys.* **39**, 125 (1967).
4. M. O. Krause, Oak Ridge National Laboratory report ORNL-TN-2943 (1970).
5. W. Lotz, *J. Opt. Soc. Am.* **60**, 206 (1970).
6. Y. S. Khodeyev, H. Siegbahn, K. Hamrin, and K. Siegbahn, University of Uppsala report UUIP-802 (1972).
7. M. E. Schwartz, *Chem. Phys. Lett.* **5**, 50 (1970).
8. M. E. Schwartz, J. D. Switalski, and R. E. Stronski, in *Electron Spectroscopy*, ed. by D. A. Shirley (North-Holland, Amsterdam, 1972), p. 605.
9. M. E. Schwartz, *J. Am. Chem. Soc.* **94**, 6899 (1972); M. E. Schwartz and J. D. Switalski, *J. Am. Chem. Soc.* **94**, 6298 (1972).

10. F. O. Ellison and L. L. Larcom, *Chem. Phys. Lett.* **10**, 580 (1971); **13**, 399 (1972).
11. D. A. Shirley, *Chem. Phys. Lett.* **16**, 220 (1972).
12. L. Ley, S. P. Kowalczyk, F. R. McFeely, R. A. Pollak, and D. A. Shirley, *Phys. Rev. B* **8**, 2392 (1973).
13. D. W. Davis and D. A. Shirley, *Chem. Phys. Lett.* **15**, 185 (1972).
14. D. W. Davis and D. A. Shirley, *J. Electron Spectros.* **3**, 137 (1974).
15. L. Hedin and G. Johansson, *J. Phys. B* **2**, 1336 (1969).
16. D. B. Adams and D. T. Clark, *J. Electron Spectros.* **2**, 201 (1973); D. B. Adams, *J. Electron Spectros.* **4**, 72 (1974).
17. W. Van Gool and A. G. Piken, *J. Materials Sci.* **4**, 95 (1969); W. R. Busing, Oak Ridge National Laboratory report ORNL-4976, p. 154 (1974).
18. R. R. Slater, *Surface Sci.* **23**, 403 (1970).
19. M. A. Butler, G. K. Wertheim, D. L. Rousseau, and S. Hüfner, *Chem. Phys. Lett.* **13**, 473 (1972).
20. L. D. Hulett and T. A. Carlson, unpublished.
21. M. Barber, J. A. Connor, I. H. Hillier, and V. R. Saunders, in *Electron Spectroscopy*, ed. by D. A. Shirley (North-Holland, Amsterdam, 1972), p. 379.
22. J. Bus and S. de Jong, *Rec. Trav. Chim.* **91**, 251 (1972).
23. W. L. Jolly, *J. Am. Chem. Soc.* **92**, 3260 (1970); D. A. Johnson, *Some Thermodynamic Aspects of Inorganic Chemistry* (Cambridge University Press, London, 1968).
24. U. Gelius, P. F. Hedén, J. Hedman, B. J. Lindberg, R. Manne, R. Nordberg, C. Nordling, and K. Siegbahn, *Physica Scripta* **2**, 70 (1970).
25. W. L. Jolly, *Faraday Disc. Chem. Soc.* **54**, 13 (1972).
26. W. L. Jolly, W. Perry, and G. Anderson, Lawrence Berkeley Laboratory report LBL-1417 (1972); W. L. Jolly and W. B. Perry, *J. Am. Chem. Soc.* **95**, 5442 (1973).
27. J. Hedman, M. Klasson, C. Nordling, and B. J. Lindberg, in *Electron Spectroscopy*, ed. by D. A. Shirley (North-Holland, Amsterdam, 1972), p. 681.
28. B. J. Lindberg and J. Hedman, Univ. Uppsala Institute of Phys. report UUIP-764 (1972).
29. W. M. Riggs, *Anal. Chem.* **44**, 830 (1972).
30. J. C. Carver, R. C. Gray, and D. M. Hercules, *J. Am. Chem. Soc.* **96**, 6851 (1974); R. C. Gray and D. M. Hercules (to be published).
31. R. T. Sanderson, *Chemical Bonds and Bond Energy* (Academic Press, New York) 1971.
32. R. S. Mulliken, *J. Chem. Phys.* **23**, 1833 (1955).
33. D. W. Davis and D. A. Shirley, *J. Electron Spectros.* **3**, 137 (1974).
34. (*a*) W. L. Jolly and D. N. Hendrickson, *J. Am. Chem. Soc.* **92**, 1863 (1970); (*b*) W. L. Jolly, *J. Am. Chem. Soc.* **92**, 3260 (1970); (*c*) W. L. Jolly, in *Electron Spectroscopy*, ed. by D. A. Shirley (North-Holland, Amsterdam, 1972), p. 769.
35. D. T. Clark and D. B. Adams, *Nature Phys. Sci.* **234**, 95 (1971).
36. D. C. Frost, F. G. Herring, C. A. McDowell, and I. S. Woolsey, *Chem. Phys. Lett.* **13**, 391 (1972).
37. R. L. Martin and D. A. Shirley, *J. Am. Chem. Soc.* **96**, 5299 (1974); D. W. Davis and J. W. Rabalais, *J. Am. Chem. Soc.* **96**, 5305 (1974).
38. K. Siegbahn, D. A. Allison, and J. H. Allison, in *Handbook of Spectroscopy*, Vol. I (CRC Press, Cleveland, Ohio, 1974), pp. 512–752.
39. C. K. Jørgensen and H. Berthou, *Det. K. Dan. Vidensk. Selsk. Mat.-Fys. Medd.* **38**(15), 1 (1972).
40. E. Clementi, *IBM J. Res. Devel.* **9**, 2 (1965); see also supplement.
41. D. W. Davis, M. S. Banna, and D. A. Shirley, *J. Chem. Phys.* **60**, 237 (1974).
42. D. T. Clark and D. Kilcast, *J. Chem. Soc. A* **21**, 3286 (1971).
43. D. B. Adams and D. T. Clark, *Theoret. Chim. Acta* **31**, 171 (1973).

44. D. T. Clark and D. M. J. Lilley, *Chem. Phys. Lett.* **9**, 234 (1971).
45. J. L. Nelson and A. A. Frost, *Chem. Phys. Lett.* **13**, 610 (1972).
46. (*a*) L. Ramqvist, K. Hamrin, G. Johansson, A. Fahlman, and C. Nordling, *J. Phys. Chem. Solids* **30**, 1835 (1969); (*b*) L. Ramqvist, K. Hamrin, G. Johansson, U. Gelius, and C. Nordling, *J. Phys. Chem. Solids* **31**, 2669 (1970); (*c*) L. Ramqvist, *J. Appl. Phys.* **42**, 2113 (1971).
47. D. N. Hendrickson, J. M. Hollander, and W. L. Jolly, *Inorg. Chem.* **8**, 2642 (1969).
48. K. Siegbahn, C. Nordling, G. Johansson, J. Hedman, P. F. Heden, K. Hamrin, U. Gelius, T. Bergmark, L. O. Werme, R. Manne, and Y. Baer, *ESCA Applied to Free Molecules* (North-Holland, Amsterdam–London, 1969).
49. J. F. Wyatt, I. H. Hillier, V. R. Saunders, J. A. Connor, and M. Barber, *J. Chem. Phys.* **54**, 5311 (1971).
50. J. J. Jack and D. M. Hercules, *Anal. Chem.* **43**, 729 (1971).
51. L. E. Cox, J. J. Jack, and D. M. Hercules, *J. Am. Chem. Soc.* **94**, 6575 (1972).
52. J. Sharma, T. Gora, J. D. Rimstidt, and R. Staley, *Chem. Phys. Lett.* **15**, 232 (1972).
53. M. Pelavin, D. N. Hendrickson, J. M. Hollander, and W. L. Jolly, *J. Phys. Chem.* **74**, 1116 (1970).
54. W. E. Swartz, Jr. and D. M. Hercules, *Anal. Chem.* **43**, 1066 (1971).
55. W. E. Morgan, W. J. Stec, R. G. Albridge, and J. R. Van Wazer, *Inorg. Chem.* **10**, 926 (1971).
56. J. R. Blackburn, R. Nordberg, F. Stevie, R. G. Albridge, and M. M. Jones, *Inorg. Chem.* **9**, 2374 (1970).
57. (*a*) W. J. Stec, W. E. Moddeman, R. G. Albridge, and J. R. Van Wazer, *J. Phys. Chem.* **75**, 3975 (1971); (*b*) W. J. Stec, W. E. Morgan, J. R. Van Wazer, and W. G. Proctor, *J. Inorg. Nucl. Chem.* **34**, 1100 (1972).
58. M. Barber, J. A. Connor, M. F. Guest, I. H. Hillier, and V. R. Saunders, *J. Chem. Soc. D: Chem. Comm.* **1971**, 943.
59. W. E. Swartz, Jr., J. K. Ruff, and D. M. Hercules, *J. Am. Chem. Soc.* **94**, 5227 (1972).
60. W. E. Swartz, Jr., R. C. Gray, J. C. Carver, R. C. Taylor, and D. M. Hercules, *Spectrochim. Acta* **30A**, 1561 (1974).
61. B. J. Lindberg, K. Hamrin, G. Johansson, U. Gelius, A. Fahlman, C. Nordling, and K. Siegbahn, *Physica Scripta* **1**, 286 (1970).
62. U. Gelius, B. Ross, and P. Siegbahn, *Chem. Phys. Lett.* **4**, 471 (1970).
63. U. Gelius, B. Roos, and P. Siegbahn, *Theoret. Chim. Acta* (*Berl.*) **23**, 59 (1971).
64. B. J. Lindberg and K. Hamrin, Acta Chemica Scandinavica **24**, 3661 (1970).
65. D. T. Clark, D. Kilcast, and D. H. Reid, *J. Chem. Soc. D: Chem. Comm.* **1971**, 638.
66. R. Gleiter, V. Hornung, B. Lindberg, S. Högberg, and N. Lozac'h, *Chem. Phys. Lett.* **11**, 401 (1971).
67. L. N. Kramer and M. P. Klein, *J. Chem. Phys.* **51**, 3618 (1969).
68. L. N. Kramer and M. P. Klein, *Chem. Phys. Lett.* **8**, 183 (1971).
69. L. N. Kramer and M. P. Klein, in *Electron Spectroscopy*, ed. by D. A. Shirley (North-Holland, Amsterdam, 1972), p. 733.
70. L. I. Yin, S. Ghose, and I. Adler, *Science*, **173**, 633 (1971).
71. J. Bus, *Rec. Trav. Chim.* **91**, 552 (1972).
72. (*a*) D. N. Hendrickson, J. M. Hollander, and W. L. Jolly, *Inorg. Chem.* **9**, 612 (1970); (*b*) D. A. Allison, G. Johansson, C. J. Allan, U. Gelius, H. Siegbahn, J. Allison, and K. Siegbahn, *J. Electron Spectros.* **1**, 269 (1972/3).
73. P. Finn and W. L. Jolly, *J. Am. Chem. Soc.* **94**, 1540 (1972).
74. G. E. McGuire, G. K. Schweitzer, and T. A. Carlson, *Inorg. Chem.* **12**, 2450 (1973).
75. A. Müller, C. K. Jørgensen, and E. Diemann, *Z. Anorg. Allg. Chem.* **391**, 38 (1972).

76. V. I. Nefedov, V. S. Urusov, and M. M. Kakhana, *Geokhimiya* **1**, 11 (1972).
77. R. Nordberg, H. Brecht, R. G. Albridge, A. Fahlman, and J. R. Van Wazer, *Inorg. Chem.* **9**, 2469 (1970).
78. W. B. Perry and W. L. Jolly, *Chem. Phys. Lett.* **17**, 611 (1972).
79. J. Sharma, R. H. Staley, J. D. Rimstidt, H. D. Fair, and T. F. Gora, *Chem. Phys. Lett.* **9**, 564 (1971).
80. G. Hollinger, P. Kumurdjian, J. M. Mackowski, P. Pertosa, L. Porte, and T. M. Duc, *J. Electron Spectros.* **5**, 237 (1974).
81. W. B. Perry and W. L. Jolly, *Inorg. Chem.* **13**, 1211 (1974).
82. W. B. Perry and W. L. Jolly, *Chem. Phys. Lett.* **23**, 529 (1973).
83. M. Barber, P. Swift, D. Cunningham, and M. J. Frazer, *J. Chem. Soc. D: Chem. Comm.* **1970**, 1338.
84. W. E. Swartz, Jr., P. H. Watts, Jr., E. R. Lippincott, J. C. Watts, and J. E. Huheey, *Inorg. Chem.* **11**, 2632 (1972).
85. G. E. McGuire, Thesis, University of Tennessee, Oak Ridge National Laboratory report ORNL-TM-3820 (1972).
86. W. E. Morgan and J. R. Van Wazer, *J. Phys. Chem.* **77**, 964 (1973).
87. W. E. Swartz, Jr. and D. M. Hercules, *Anal. Chem.* **43**, 1774 (1971).
88. Y. E. Araktingi, N. S. Bhacca, W. G. Proctor, and J. W. Robinson, *Spectr. Lett.* **4**, 365 (1971).
89. T. Novakov, P. K. Mueller, A. E. Alcocer, and J. W. Otvos, *J. Coll. Interface Sci.* **39**, 225 (1972).
90. W. J. Stec, W. E. Morgan, R. G. Albridge, and J. R. Van Wazer, *Inorg. Chem.* **11**, 219 (1972).
91. W. E. Morgan, W. J. Stec, and J. R. Van Wazer, *Inorg. Chem.* **12**, 953 (1973).
92. W. E. Morgan, Thesis, Vanderbilt University, Nashville, Tennessee (1972).
93. V. I. Nefedov, Y. V. Salyn, L. Baier, R. L. Davidovich, B. V. Levin, and L. A. Zemnukhova, *Izv. Akad. Nauk SSSR, Ser. Fiz.* **38**, 582 (1974).
94. L. D. Hulett and T. A. Carlson, *Appl. Spect.* **25**, 33 (1971).
95. W. E. Swartz, Jr., K. J. Wynne, and D. M. Hercules, *Anal. Chem.* **43**, 1884 (1971).
96. G. Malmsten, I. Thoren, S. Hogberg, J.-E. Bergmark, S.-E. Karlsson, and E. Rebane, *Physica Scripta* **3**, 96 (1971).
97. A. H. Norbury, M. Thompson, and J. Songstad, *Inorg. Nucl. Chem. Lett.* **9**, 347 (1973).
98. T. X. Carroll, R. W. Shaw, Jr., T. D. Thomas, C. Kindle, and N. Bartlett, *J. Am. Chem. Soc.* **96**, 1989 (1974).
99. D. B. Adams, D. T. Clark, W. J. Feast, D. Kilcast, W. K. R. Musgrave, and W. E. Preston, *Nature Phys. Sci.* **239**, 47 (1972).
100. D. T. Clark and D. Kilcast, *J. Chem. Soc. B* **11**, 2243 (1971).
101. D. T. Clark, D. Kilcast, and W. K. R. Musgrave, *J. Chem. Soc. D: Chem. Comm.* **1971**, 516.
102. D. T. Clark and D. Kilcast, *Nature Phys. Sci.* **233**, 77 (1971).
103. D. T. Clark, D. B. Adams, and D. Kilcast, *Chem Phys. Lett.* **13**, 439 (1972).
104. D. T. Clark, R. D. Chambers, D. Kilcast, and W. K. R. Musgrave, *J. Chem. Soc., Faraday Trans. II* **1972**, 309.
105. D. T. Clark, D. Kilcast, D. B. Adams, and I. Scanlan, *J. Electron Spectros.* **1**, 153 (1972/73).
106. D. T. Clark, D. Kilcast, D. B. Adams, and W. K. R. Musgrave, *J. Electron Spectros.* **1**, 227 (1972/73).
107. R. G. Hayes and N. Edelstein, in *Electron Spectroscopy*, ed. by D. A. Shirley (North-Holland, Amsterdam, 1972), p. 771.

108. C. Jørgensen, H. Berthou, and L. Balsenc, *J. Fluorine Chem.* **1**, 327 (1972).
109. J. A. Hashmall, B. E. Mills, D. A. Shirley, and A. Streitwieser, Jr., *J. Am. Chem. Soc.* **94**, 4445 (1972).
110. A. D. Hamer and R. A. Walton, *Inorg. Chem.* **13**, 1446 (1974).
111. C. J. Nicholls, D. S. Urch, and A. N. L. Kay, *J. Chem. Soc., Chem. Comm.* **1972**, 1198.
112. P. H. Citrin and T. D. Thomas, *J. Chem. Phys.* **57**, 4446 (1972).
113. W. E. Morgan, J. R. Van Wazer, and W. J. Stec., *J. Am. Chem. Soc.* **95**, 751 (1973).
114. G. E. McGuire, T. A. Carlson, and G. K. Schweitzer (unpublished); D. E. Parry and M. J. Tricker, *Chem. Phys. Letts.* **20**, 124 (1973).
115. C. J. Groenenboom, G. Sawatzky, H. J. De Liefde Meijer, and F. Jellinek, *J. Organometal. Chem.* **76**, C4 (1974).
116. K. Hamrin, C. Nordling, and L. Kihlborg, *Ann. Acad. Reg. Sci. Upsaliensis* **14**, (1970).
117. J. C. Carver, G. K. Schweitzer, and T. A. Carlson, *J. Chem. Phys.* **57**, 973 (1972).
118. J. C. Helmer, *J. Electron Spectros.* **1**, 259 (1972/73).
119. M. Barber, J. A. Connor, J. H. Hillier, and W. N. E. Meredith, *J. Electron Spectros.* **1**, 110 (1972/73).
120. C. K. Jørgensen, *Chimia* **25**, 213 (1971).
121. C. A. Tolman, W. M. Riggs, W. J. Linn, C. M. King, and P. C. Wendt, *Inorg. Chem.* **12**, 2770 (1973).
122. L. J. Matienzo, L. I. Yin, S. O. Grim, and W. E. Swartz, *Inorg. Chem.* **12**, 2762 (1973).
123. S. O. Grim, L. J. Matienzo, and W. E. Swartz, Jr., *J. Am Chem Soc.* **94**, 5116 (1972).
124. M. Barber, J. A. Connor, I. H. Hillier, and V. R. Saunders, *J. Chem. Soc. D: Chem. Comm.* **1971**, 682.
125. D. T. Clark and D. B. Adams, *J. Chem. Soc. D: Chem. Comm.* **1971**, 740.
126. D. O. Cowan, J. Park, M. Barber, and P. Swift, *J. Chem. Soc. D: Chem. Comm.* **1971**, 1444.
127. D. N. E. Buchanan, M. Robbins, H. J. Guggenheim, G. K. Wertheim, and V. G. Lambrecht, Jr., *Solid State Comm.* **9**, 583 (1971).
128. G. K. Wertheim and A. Rosencwaig, *J. Chem. Phys.* **54**, 3255 (1971).
129. G. S. Kharitonova, V. I. Nefedov, L. N. Pankratova, and V. L. Pershin, *Zh. Neorgan. Khim.* **19**, 860 (1974).
130. T. Novakov and T. H. Geballe, *Solid State Comm.* **10**, 225 (1972).
131. B. C. Lane, J. E. Lester, and F. Basolo, *J. Chem. Soc. D: Chem. Comm.* **1971**, 1618.
132. J. Prather, II and D. A. Zatko (to be published).
133. J. Hedman, M. Klasson, R. Nilsson, C. Nordling, M. F. Sorokina, O. I. Kljushnikov, S. A. Nemnonov, V. A. Trapeznikov, and V. G. Zyryanov, *Physica Scripta* **4**, 195 (1971).
134. G. Kumar, J. R. Blackburn, R. G. Albridge, W. E. Moddeman, and M. M. Jones, *Inorg. Chem.* **11**, 296 (1971).
135. W. E. Moddeman, J. R. Blackburn, G. Kumar, K. A. Morgan, M. M. Jones and R. G. Albridge, in *Electron Spectroscopy*, ed. by D. A. Shirley (North-Holland, Amsterdam, 1972), p. 725.
136. D. T. Clark, D. B. Adams, and D. Briggs, *J. Chem. Soc. D: Chem. Comm.* **1971**, 602.
137. T. A. Carlson and G. E. McGuire, *J. Electron Spectros.* **1**, 161 (1972/73).
138. L. C. Cain, Thesis, University of Tennessee, Knoxville (1971).
139. V. I. Nefedov, I. A. Zakharova, M. A. Porai-Koshits, and M. E. Dyatkina, *Izv. Akad. Nauk SSSR. Ser. Khim.* **8**, 1846 (1971).
140. V. I. Nefedov and I. B. Baranovskii, *Zh. Neorgan. Khim.* **17**, 466 (1972).
141. V. I. Nefedov, M. A. Porai-Koshits, I. A. Zakharova, and M. E. Dyatkina, *Doklady Akad. Nauk SSSR* **202**, 605 (1972).

142. R. Mason, D. M. P. Mingos, G. Rucci, and J. A. Connor, *J. Chem. Soc. Dalton* **1972**, 1729.
143. F. Holsboer and W. Beck, *Z. Naturforsch.* **27b**, 884 (1972).
144. C. D. Cook, K. Y. Wan, U. Gelius, K. Hamrin, G. Johansson, E. Olsson, H. Siegbahn, C. Nordling, and K. Siegbahn, *J. Am. Chem. Soc.* **93**, 1904 (1971).
145. V. I. Nefedov, M. A. Poraikos, I. A. Zakharov, I. S. Kolomnik, and N. N. Kuzmina, *Izv. Akad. Nauk SSR, Ser. Fiz.* **36**, 381 (1972).
146. K. S. Kim, N. Winograd, and R. E. Davis, *J. Am. Chem. Soc.* **93**, 6296 (1971).
147. W. M. Riggs, in *Electron Spectroscopy*, ed. by D. A. Shirley (North-Holland, Amsterdam, 1972), p. 713.
148. M. A. Butler, D. L. Rousseau, and D. N. E. Buchanan, *Phys. Rev. B* **7**, 61 (1973).
149. T. Novakov and R. Prins, in *Electron Spectroscopy*, ed. by D. A. Shirley (North-Holland, Amsterdam, 1972). p. 821.
150. D. C. Frost, A. Ishitani, and C. A. McDowell, *Mol. Phys.* **24**, 861 (1972).
151. D. P. Murtha and R. A. Walton, *Inorg. Chem.* **12**, 368 (1973).
152. D. A. Zatko and J. W. Prather, *J. Electron Spectros.* **2**, 191 (1973).
153. D. Betteridge, J. C. Carver, and D. M. Hercules, *J. Electron Spectros.* **2**, 327 (1973).
154. C. J. Vesely and D. W. Langer, *Phys. Rev. B* **4**, 451 (1971).
155. C. S. Fadley, S. B. M. Hagstrom, M. P. Klein, and D. A. Shirley, *J. Chem. Phys.* **48**, 3779 (1968).
156. C. Bonnelle, R. C. Karnatak, and C. K. Jørgensen, *Chem. Phys. Lett.* **14**, 145 (1972).
157. G. K. Wertheim, A. Rosencwaig, R. L. Cohen, and H. J. Guggenheim, *Phys. Rev. Lett.* **27**, 505 (1971).
158. R. L. Cohen, G. K. Wertheim, A. Rosencwaig, and H. J. Guggenheim, *Phys. Rev. B* **5**, 1037 (1972).
159. C. K. Jørgensen and H. Berthou, *Chem. Phys. Lett.* **13**, 186 (1972).
160. A. J. Signorelli and R. G. Hayes, *Phys. Rev. B* **8**, 81 (1973).
161. D. Chadwick and J. Graham, *Nature Phys. Sci.* **237**, 127 (1972); G. C. Allen, J. A. Crofts, M. T. Curtis, P. M. Tucker, D. Chadwick, and P. J. Hampson, *J. Chem. Soc. Dalton Trans.* **1974**, 1296.
162. J. Vervist, J. Riga, J. J. Pireaux, and R. Caudano, *J. Electron Spectros.* **5**, 193 (1974).
163. C. K. Jørgensen, *Theoret. Chim. Acta* (*Berl.*) **24**, 241 (1972).
164. D. B. Adams, D. T. Clark, A. D. Baker, and M. Thompson, *J. Chem. Soc. D: Chem. Comm.* **1971**, 1600.
165. G. Johansson, J. Hedman, A. Berndtsson, M. Klasson, and R. Nilsson, *J. Electron Spectros.* **2**, 295 (1973).
166. T. D. Thomas and R. W. Shaw, Jr., *J. Electron Spectros.* **5**, 1081 (1974).
167. H. W. Harrington, private communication.
168. J. Hedman, Y. Baer, A. Berndtsson, M. Klasson, G. Leonhardt, R. Nilsson, and C. Nordling, *J. Electron Spectros.* **1**, 101 (1972/73).
169. R. M. Friedman, J. Hudis, and M. L. Perlman, *Phys. Rev. Lett.* **29**, 692 (1972).
170. R. W. Shaw, Jr., and T. D. Thomas, *Phys. Rev. Lett.* **29**, 689 (1972).
171. U. Gelius, *J. Electron Spectros.* **5**, 985 (1974).
172. P. H. Citrin, P. Eisenberger, and D. R. Hamann, *Phys. Rev. Lett.* **33**, 965 (1974).
173. J. A. D. Matthew and M. G. Devey, *J. Phys. C: Solid State Phys.* **7**, L335 (1974).
174. S. Hüfner, G. K. Wertheim, D. N. E. Buchanan, and K. W. West, *Phys. Lett.* **46A**, 420 (1974).
175. J. R. Lindsay, H. J. Rose, Jr., W. E. Swartz, Jr., P. H. Watts, Jr., and K. A. Rayburn, *Appl. Spectros.* **27**, 1 (1973).
176. R. W. Shaw, Jr., T. X. Carroll, and T. D. Thomas, *J. Am. Chem. Soc.* **95**, 2033 (1973).

177. F. A. Cotton and G. Wilkinson, *Advanced Inorganic Chemistry*, 1st ed. (Interscience, New York, 1962), p. 616.
178. K. B. Harvey and G. B. Porter, Introduction to Physical Inorganic Chemistry (Addison-Wesley, Reading, Massachusetts, 1963), p. 229.
179. L. E. Cox and D. M. Hercules, *J. Electron Spectros.* **1**, 193 (1972/73).
180. D. W. Davis, D. A. Shirley, and T. D. Thomas, *J. Chem. Phys.* **56**, 671 (1972); *J. Am. Chem. Soc.* **94**, 6565 (1972).
181. J. Hedman, P.-F. Heden, R. Nordberg, C. Nordling, and B. J. Lindberg, *Spectrochim. Acta* **26A**, 761 (1970).
182. D. T. Clark, W. J. Feast, M. Foster, and D. Kilcast, *Nature Phys. Sci.* **236**, 107 (1972).
183. M. Patsch and P. Thieme, *Angew. Chem. Int. Ed.* **10**, 569 (1971); P. Thieme, M. Patsch, and H. König, *Liebigs Ann. Chem.* **764**, 94 (1972).
184. G. D. Mateescu and J. L. Riemenschneider, in *Electron Spectroscopy*, ed. by D. A. Shirley (North-Holland, Amsterdam, 1972), p. 661.
185. G. D. Mateescu, J. L. Reimenschneider, J. J. Svoboda, and G. A. Olah, *J. Am. Chem. Soc.* **94**, 7191 (1972).
186. G. A. Olah and G. D. Mateescu, *J. Am. Chem. Soc.* **92**, 7231 (1970); G. A. Olah, G. D. Mateescu, and J. L. Reimenschneider, *J. Am. Chem. Soc.* **94**, 2529 (1972).
187. M. H. Palmer and R. H. Findlay, *Chem. Phys. Lett.* **15**, 416 (1972).
188. I. Adams, J. M. Thomas, G. M. Bancroft, K. D. Butler, and M. Barber, *J. Chem. Soc. Chem. Comm.* **1972**, 751.
189. D. N. E. Buchanan, M. Robbins, H. J. Guggenheim, G. K. Wertheim, and V. G. Lambrecht, Jr., *Solid State Comm.* **9**, 583 (1971).
190. L. Y. Johansson, L. Larsson, J. Blomquist, C. Cederström, S. Grapengiesser, U. Helgeson, L. C. Moberg, and M. Sundbom, *Chem. Phys. Lett.* **24**, 508 (1974); M. J. Tricker, J. M. Thomas, M. H. Omar, A. Osman, and A. Bishay, *J. Materials Sci.* **9**, 1115 (1974).
191. F. Holsboer, W. Beck, and H. D. Bartunik, *Chem. Phys. Lett.* **18**, 217 (1973).
192. R. E. Block, *J. Magnetic Resonance* **5**, 155 (1971).
193. H. Basch, *Chem. Phys. Lett.* **5**, 337 (1970).
194. U. Gelius, G. Johansson, H. Siegbahn, C. J. Allan, D. A. Allison, J. Allison, and K. Siegbahn, *J. Electron Spectros.* **1**, 285 (1972/73).
195. D. Zeroka, *Chem. Phys. Lett.* **14**, 471 (1972).
196. B. J. Lindberg, *J. Electron Spectros.* **5**, 149 (1974).
197. G. D. Stucky, D. A. Matthews, J. Hedman, M. Klasson, and G. Nordling, *J. Am. Chem. Soc.* **94**, 8009 (1972).
198. B. J. Lindberg and B. Schröder, *Acta Chim. Scand.* **24**, 3089 (1970).
199. B. Folkesson, *Acta Chim. Scand.* **A28**, 491 (1974).
200. J. Hedman, P.-F. Heden, C. Nordling, and K. Siegbahn, *Phys. Lett.* **29A**, 178 (1969).
201. C. S. Fadley, D. A. Shirley, A. J. Freeman, P. S. Bagus, and J. V. Mallow, *Phys. Rev. Lett.* **23**, 1397 (1969); C. S. Fadley and D. A. Shirley, *Phys. Rev. A* **2**, 1109 (1970).
202. D. W. Davis and D. A. Shirley, *J. Chem. Phys.* **56**, 669 (1972).
203. P. S. Bagus, M. Schrenk, D. W. Davis, and D. A. Shirley, *Phys. Rev. A* **9**, 1090 (1974).
204. D. W. Davis, R. L. Martin, M. S. Banna, and D. A. Shirley, *J. Chem. Phys.* **59**, 4235 (1973).
205. I. W. Drummond and H. Harker, *Nature Phys. Sci.* **232**, 71 (1971).
206. P. S. Bagus, A. J. Freeman, and F. Sasaki, *Phys. Rev. Lett.* **30**, 850 (1973).
207. S. P. Kowalczyk, L. Ley, R. A. Pollak, F. R. McFeely, and D. A. Shirley, *Phys. Rev B* **7**, 4009 (1973).
208. S. Hüfner and G. K. Wertheim, *Phys. Rev. B* **7**, 2333 (1973).

209. D. E. Ellis and A. J. Freeman (unpublished), reported in Ref. 201.
210. G. K. Wertheim, S. Hüfner, and H. J. Guggenheim, *Phys. Rev. B* **7**, 556 (1973).
211. F. Garcia Santibáñez and T. A. Carlson (to be published).
212. M. J. Tricker, *J. Inorg. Nucl. Chem.* **36**, 1543 (1974).
213. D. T. Clark and D. B. Adams, *Chem. Phys. Lett.* **10**, 121 (1971).
214. J. C. Carver, T. A. Carlson, L. C. Cain, and G. K. Schweitzer, in *Electron Spectroscopy*, ed. by D. A. Shirley (North-Holland, Amsterdam, 1972), p. 803.
215. C. W. Nestor, Jr., and T. A. Carlson, unpublished results.
216. B. Ekstig, E. Källne, E. Noreland, and R. Manne, *Physica Scripta* **2**, 38 (1970)
217. R. P. Gupta and S. K. Sen, *Phys. Rev. B* **10**, 71 (1974).
218. S. P. Kowalczyk, L. Ley, F. R. McFeely, and D. A. Shirley (to be published).
219. D. C. Frost, C. A. McDowell, and I. S. Woolsey, *Chem. Phys. Lett.* **17**, 320 (1972); *Mol. Phys.* **27**, 1473 (1974).
220. G. K. Wertheim and R. L. Cohen, in *Electron Spectroscopy*, ed. by D. A. Shirley (North-Holland, Amsterdam, 1972), p. 813.
221. T. A. Carlson and M. O. Krause, *Phys. Rev.* **140**, A1057 (1965).
222. T. A. Carlson, *Phys. Rev.* **156**, 142 (1967).
223. M. O. Krause, T. A. Carlson, and R. D. Dismukes, *Phys. Rev.* **170**, 37 (1968).
224. J. S. Levinger, *Phys. Rev.* **90**, 11 (1953).
225. J. A. R. Samson and G. N. Haddad, *Phys. Rev. Lett.* **33**, 875 (1974).
226. F. W. Byron, Jr. and C. J. Joachain, *Phys. Rev.* **164**, 1 (1967).
227. T. Åberg, *Phys. Rev. A* **2**, 1726 (1970).
228. T. A. Carlson, M. O. Krause, and W. E. Moddeman, *J. Phys.* (*Paris*) **32**, C4-76 (1971).
229. D. P. Spears, H. J. Fischbeck, and T. A. Carlson, *Phys. Rev. A* **9**, 1603 (1974).
230. M. O. Krause and F. Wuilleumier, *J. Phys. B: Atom. Molec. Phys.* **5**, L143 (1972).
231. V. L. Jacobs and P. G. Burke, *J. Phys. B: Atom. Molec. Phys.* **4**, L67 (1972).
232. F. Wuilleumier and M. O. Krause, *Phys. Rev. A* **10**, 242 (1974).
233. U. Gelius, C. J. Allan, D. A. Allison, H. Siegbahn, and K. Siegbahn, *Chem. Phys. Lett.* **11**, 224 (1971).
234. D. P. Spears, H. J. Fischbeck, and T. A. Carlson, *J. Electron Spectros.* **6**, 411 (1975).
235. F. R. Gilmore, *J. Quant. Spectros. Radiat. Transfer* **5**, 369 (1965).
236. C. J. Allan, U. Gelius, D. A. Allison, G. Johansson, H. Siegbahn, and K. Siegbahn, *J. Electron Spectros.* **1**, 131 (1972/73).
237. L. J. Aarons, M. F. Guest, and I. H. Hillier, *J. Chem. Soc., Faraday Trans. II* **68**, 1866 (1972).
238. L. J. Aarons, M. Barber, M. F. Guest, I. H. Hillier, and J. H. McCartney, *Mol. Phys.* **26**, 1247 (1973).
239. H. Basch, *J. Electron Spectros.* **5**, 463 (1974).
240. T. X. Carroll and T. D. Thomas, *J. Electron Spectros.* **4**, 270 (1974).
241. S. Pignataro, *Z. Naturforsch.* **27a**, 816 (1972).
242. R. W. Shaw, Jr. and T. D. Thomas, *Chem. Phys. Lett.* **14**, 121 (1972).
243. A. Rosencwaig, G. K. Wertheim, and H. J. Guggenheim, *Phys. Rev. Lett.* **27**, 479 (1971).
244. L. J. Matienzo, W. E. Swartz, Jr., and S. O. Grim, *Inorg. Nucl. Chem. Lett.* **8**, 1085 (1972).
245. L. J. Matienzo, Lo I. Yin, S. O. Grim, and W. E. Swartz, Jr., *Inorg. Chem.* **12**, 2762 (1973).
246. J. E. Castle, *Nature Phys. Sci.* **234**, 93 (1971).
247. S. Pignataro, R. DiMarino, G. Distefano, and A. Mangini, *Chem. Phys. Lett.* **22**, 352 (1973).

248. D. T. Clark, D. B. Adams, I. W. Scanlan, and I. S. Woolsey, *Chem. Phys. Lett.* **25**, 263 (1974).
249. I. Ikemoto, J. M. Thomas, H. Kuroda, M. Barber, and I. H. Hillier, *Mol. Cryst. Liq. Cryst.* **18**, 87 (1972); I. Ikemoto, J. M. Thomas, and H. Kuroda, *Faraday Disc. Chem. Soc.* **54**, 208 (1972); L. J. Aarons, M. Barber, J. A. Connor, M. F. Guest, I. H. Hillier, I. Ikemoto, J. M. Thomas, and H. Kuroda, *J. Chem. Soc. Trans. Faraday Soc. II* **1973**, 270.
250. C. R. Ginnard, R. S. Swingle, II, and B. M. Monroe, *J. Electron Spectros.* **6**, 77 (1975).
251. M. A. Butler, J. P. Ferraris, A. N. Block, and D. O. Cowan, *Chem. Phys. Lett.* **24**, 600 (1974).
252. S. Pignataro and G. Distefano, *J. Electron Spectros.* **2**, 171 (1973).
253. B. Wallbank, C. E. Johnson, and I. G. Mair, *J. Phys. C: Solid State Phys.* **6**, L340 (1973).
254. B. Wallbank, C. E. Johnson, and I. G. Mair, *J. Phys. C: Solid State Phys.* **6**, L493 (1973).
255. T. A. Carlson, J. C. Carver, L. J. Saethre, F. Garcia Santibáñez, and G. A. Vernon, *J. Electron Spectros.* **5**, 247 (1974).
256. B. Wallbank, I. G. Main, and C. E. Johnson, *J. Electron Spectros.* **5**, 259 (1974).
257. K. S. Kim, *J. Electron Spectros.* **3**, 217 (1974).
258. C. K. Jørgensen, *Progr. Inorg. Chem.* **12**, 101 (1970).
259. G. A. Vernon, G. Stucky, and T. A. Carlson (to be published); also see G. A. Vernon, Thesis, Univ. of Illinois (1974).
260. G. K. Wertheim and A. Rosencwaig, *Phys. Rev. Lett.* **26**, 1179 (1971).
261. J. J. Quinn, *Phys. Rev.* **126**, 1453 (1962); N. V. Smith and W. E. Spicer, *Phys. Rev.* **188**, 593 (1969).
262. R. A. Pollak, L. Ley, F. R. McFeely, S. P. Kowalczyk, and D. A. Shirley, *J. Electron Spectros.* **3**, 381 (1974).
263. T. A. Carlson, J. C. Carver, and G. A. Vernon, *J. Chem. Phys.* **62**, 932 (1975).
264. S. Asada, C. Satoko, and S. Sugano, Tech. Report of ISSP, Univ. of Tokyo, A, No. 671 (1974).
265. S. Glasstone and M. C. Edlund, *The Elements of Nuclear Reaction Theory* (Van Nostrand, New York, 1952), p. 106.
266. C. J. Powell, *Surface Sci.* **44**, 29 (1974); I. Lindau and W. E. Spicer, *J. Electron Spectros.* **3**, 409 (1974).
267. W. M. Riggs, private communication.
268. C. J. Tung and R. H. Richie (to be published).
269. M. Klasson, A. Berndtsson, J. Hedman, R. Nilsson, R. Nyholm, and C. Nordling, *J. Electron Spectros.* **3**, 427 (1974).
270. H. Thomas, *Z. Physik* **147**, 395 (1957).
271. C. J. Todd and R. Heckingbottom, *Phys. Lett.* **42A**, 455 (1973).
272. D. T. Clark, W. J. Feast, W. K. R. Musgrave, and I. Ritchie (to be published).
273. W. A. Fraser, J. V. Florio, W. N. Delgass, and W. D. Robertson, *Surface Sci.* **36**, 661 (1973).
274. J. Brunner and H. Zogg, *J. Electron Spectros.* **5**, 911 (1974).
275. C. S. Fadley, R. J. Baird, W. Siekhaus, T. Novakov, S. Å. L. Bergström, *J. Electron Spectros.* **4**, 93 (1974).
276. C. S. Fadley and S. Å. L. Bergström, *Phys. Lett.* **35A**, 375 (1971).
277. R. T. Poole, R. C. G. Leckey, J. G. Jenkin, and J. Liesegang, *J. Electron Spectros.* **1**, 371 (1972/73).
278. C. S. Fadley and S. Å. L. Bergström, in *Electron Spectroscopy*, ed. by D. A. Shirley (North-Holland, Amsterdam, 1972), p. 233.
279. S. T. Manson and D. J. Kennedy, *Chem. Phys. Lett.* **7**, 387 (1970); S. T. Manson, *J. Electron Spectros.* **1**, 413 (1973).

280. M. O. Krause, *Phys. Rev.* **177**, 151 (1969).
281. J. S. Brinen and J. E. McClure, *Anal. Lett.* **5**, 737 (1972); *J. Electron Spectros.* **4**, 243 (1974).
282. D. M. Hercules, L. E. Cox, S. Onisick, G. D. Nichols, and J. C. Carver, *Anal. Chem.* **45**, 1973 (1973).
283. W. J. Carter, G. K. Schweitzer, and T. A. Carlson (to be published).
284. C. D. Wagner, *Anal. Chem.* **44**, 1050 (1972).
285. C. K. Jørgensen and R. Berthou, *Faraday Disc. Chem. Soc.* **54**, 269 (1972).
286. V. I. Nefedov, N. P. Sergushin, I. M. Band, and M. B. Trzhaskovskaya, *J. Electron Spectros.* **2**, 383 (1973).
287. J. H. Scofield, Lawrence Livermore Laboratory Report UCRL-51326 (1973).
288. W. J. Carter, G. K. Schweitzer, and T. A. Carlson, *J. Electron Spectros.* **5**, 827 (1974).
289. R. S. Swingle, II, *Anal. Chem.* **47**, 21 (1975).
290. D. M. Hercules, *J. Electron Spectros.* **5**, 811 (1974).
291. R. W. Phillips, *J. Colloid Interface Sci.* **47**, 687 (1974).
292. M. P. Klein and L. N. Kramer, in *Proc. Conf. on Improving Plant Proteins by Nuclear Techniques, Vienna, 1970* (IAEA, Vienna, 1970), p. 243.
293. M. M. Millard and M. S. Masri, *Anal. Chem.* **46**, 1820 (1974).
294. D. Liebfritz, *Angew. Chem. Int. Ed.* **11**, 232 (1972).
295. J. W. Lauher and J. E. Lester, *Inorg. Chem.* **12**, 244 (1973).
296. M. V. Zeller and R. G. Hayes, *J. Am. Chem. Soc.* **95**, 3855 (1973).
297. L. D. Hulett and T. A. Carlson, *Clinical Chem.* **16**, 677 (1970).
298. L. J. Saethre, T. A. Carlson, J. J. Kaufman, and W. S. Koski (to be published).
299. W. T. Huntress, Jr. and L. Wilson, *Earth Planetary Sci. Lett.* **15**, 59 (1972).
300. A. P. Vinogradov, V. I. Nefedov, V. S. Urusov, and N. M. Zhavoronkov, *Dokl. Akad. Nauk SSSR* **201**, 957 (1971).
301. L. D. Hulett, T. A. Carlson, B. R. Fish, and J. L. Durham, in *Determination of Air Quality*, ed. by G. Mamantov and W. D. Shultz (Plenum, New York, 1971), p. 179.
302. Y. E. Araktingi, N. S. Bhacca, W. G. Proctor, and J. W. Robinson, *Spect. Lett.* **4**, 365 (1971).
303. W. J. Carter, III, G. K. Schweitzer, T. A. Carlson, L. D. Hulett, and B. Fish (to be published).
304. T. Novakov, S. G. Chang, and A. B. Harker, *Science* **186**, 259 (1974).
305. H. D. Schultz and W. G. Proctor, *Appl. Spectros.* **27**, 347 (1973); D. C. Frost, W. R. Leeder, and R. L. Tapping, *Fuel* **53**, 206 (1974).
305*a*. C. R. Brundle, *J. Electron Spectros.* **5**, 291 (1974); *J. Vac. Sci. Tech.* **11**, 212 (1974).
306. W. N. Delgass, T. R. Hughes, and C. S. Fadley, *Catalysis Rev.* **4**, 179 (1970).
307. J. S. Brinen and A. Melera, *J. Phys. Chem.* **76**, 2525 (1972).
308. N. S. McIntyre, N. H. Sagert, R. M. L. Pouteau, and W. G. Proctor, *Can. J. Chem.* **51**, 1670 (1973); T. A. Clarke, I. D. Gay, and R. Mason, *J. Chem. Soc.: Chem. Comm.* **1974**, 331; J. Escard, B. Pontvianne, and J. P. Contour, *J. Electron Spectros.* **6**, 17 (1975); R. M. Friedman, R. I. Declerck-Grimee, and J. J. Fripiat, *J. Electron Spectros.* **5**, 437 (1974).
309. K. S. Kim and N. Winograd, *Chem. Phys. Lett.* **19**, 209 (1973).
310. K. S. Kim and N. Winograd, *Surface Sci.* **43**, 625 (1974).
311. G. Schön and S. T. Lundin, *J. Electron Spectros.* **1**, 105 (1972/73).
312. W. Dianis and J. E. Lester, *Surface Sci.* **43**, 602 (1974).
313. H. Fischmeister and I. Olefjord, *Monatsh. Chem.* **102,** 1486 (1971).
314. L. D. Hulett, A. L. Bacarella, L. LiDonnici, and J. C. Griess, *J. Electron Spectros.* **1**, 169 (1972/73).
315. L. I. Yin, S. Ghose, and I. Adler, *Appl. Spect.* **26**, 355 (1972).

316. D. A. Whan, M. Barber, and P. Swift, *J. Chem. Soc. Chem. Comm.* **1972**, 198.
317. J. M. Thomas, E. L. Evans, M. Barber, and P. Swift, in *Conf. Ind. Carbons and Graphite*, ed. by J. G. Gregory (Soc. Chem. Ind., London, England, 1971), p. 411.
318. J. T. Yates, T. E. Madey, and N. E. Erickson, *Surface Sci.* **43**, 257 (1974); T. E. Madey, J. T. Yates, and N. E. Erickson, *Surface Sci.* **43**, 526 (1974); J. T. Yates, Jr., and N. E. Erickson, *Surface Sci.* (to be published).
319. C. R. Brundle, *J. Electron Spectros.* **5**, 291 (1974).
320. P. E. Larson, *Anal. Chem.* **44**, 1678 (1972).
321. C. R. Ginnard and W. M. Riggs, *Anal. Chem.* **44**, 1310 (1972).
322. D. T. Clark, D. Kilcast, W. J. Feast, and W. K. R. Musgrave, *J. Poly. Sci. A-1* **10**, 1637 (1972).
323. D. T. Clark, W. J. Feast, I. Ritchie, W. K. R. Musgrave, M. Modena, and M. Ragazzini, *J. Polym. Sci., Polym. Chem. Ed.* **12**, 1049 (1974).
324. K. L. Cheng, J. C. Carver, and T. A. Carlson, *Inorg. Chem.* **12**, 1702 (1973).
325. M. P. Klein, private communication.
326. J. S. Brinen and L. A. Wilson, *J. Chem. Phys.* **56**, 6256 (1972).
327. O. Aita, M. Yokota, and Y. Siota, *Phys. Lett.* **46A**, 425 (1974).
328. K. S. Kim, W. E. Baitinger, J. W. Amy, and N. Winograd, *J. Electron Spectros.* **5**, 351 (1974).
329. J. S. Brinen, *J. Electron Spectros.* **5**, 377 (1974).
330. M. Millard, in *Electron Spectroscopy*, ed. by D. A. Shirley (North-Holland, Amsterdam, 1972), p. 765; *Anal. Chem.* **44**, 828 (1972).

CHAPTER 6

1. M. P. Auger, *Compt. Rend.* **180**, 65 (1925); *J. de Phys. Radium* **6**, 205 (1925); *Compt. Rend.* **182**, 773, 1215 (1926).
2. W. Bambynek, B. Crasemann, R. W. Fink, H.-U. Freund, H. Mark, C. D. Swift, R. E Price, and P. V. Rao, *Rev. Mod. Phys.* **44**, 716 (1972).
3. E. H. S. Burhop, *The Auger Effect and Other Radiationless Transitions*, (Cambridge Univ. Press, 1952).
4. W. Mehlhorn, "The Auger Effect," Report from the Behlen Lab. of Phys., Univ. of Nebraska (1970).
5. K. D. Sevier, *Low Energy Electron Spectroscopy* (Wiley–Interscience, New York, 1972).
6. E. C. Parilis, *The Auger Effect* (Acad. Sci. Uzbek SSR, 1969) (in Russian).
7. E. H. S. Burhop and W. M. Asaad, *Adv. Atom. Mol. Phys.* **8**, 163 (1972).
8. I. Bergström and R. D. Hill, *Ark. Fys.* **8**, 21 (1954).
9. W. A. Coghlan and R. E. Clausing, Oak Ridge National Lab. Report ORNL-TM-3576 (1971); *Surf. Sci.* **33**, 411 (1972).
10. M. E. Packer and J. M. Wilson, *Auger Transitions* (Institute of Phys., London, 1971).
11. D. A. Shirley, *Chem. Phys. Lett.* **17**, 312 (1972); *Phys. Rev. A* **7**, 1520 (1973).
12. C. A. Nicolaides and D. R. Beck, *Chem. Phys. Lett.* **27**, 269 (1974).
13. K. Siegbahn, C. Nordling, A. Fahlman, R. Nordberg, K. Hamrin, J. Hedman, G. Johansson, T. Bergmark, S.-E. Karlsson, I. Lindgren, and B. Lindberg, *Electron Spectroscopy for Chemical Analysis—Atomic, Molecular and Solid State Structure Studied by Means of Electron Spectroscopy* (Almqvist and Wiksells Boktryekeri AB, Stockholm, Sweden, 1967); *Nova Acta Regiae Soc. Sci. Upsaliensis, Ser. IV* **20**, 1967.
14. Y. E. Strausser and J. J. Uebbing, *Varian Chart of Auger Electron Energies* (Varian Vacuum Division, Palo Alto, California), 1971.

15. P. W. Palmberg, G. E. Riach, R. E. Weber, N. C. MacDonald, *Handbook of Auger Electron Spectroscopy* (Phys. Elec. Ind. Inc., Edina, Minnesota) 1972.
16. P. Haymann, *Surf. Sci.* **30**, 536 (1972).
17. R. L. Chase, H. P. Kelly, and H. S. Köhler, *Phys. Rev. A* **3**, 1550 (1971).
18. E. J. McGuire, *Phys. Rev.* **185**, 1 (1969); *Nucl. Phys.* **A172**, 127 (1971); *Phys. Rev. A* **3**, 587 (1971); *Phys. Rev. A* **3**, 1801 (1971); *Phys. Rev. A* **5**, 1043 (1972); *Phys. Rev. A* **5**, 1052 (1972); C. P. Bhalla and D. J. Ramsdale, *Z. Physik* **239**, 95 (1970); E. J. McGuire, *Phys. Rev. A* **9**, 1840 (1974).
19. M. H. Chen and B. Crasemann, *Phys. Rev. A* **8**, 7 (1973).
20. M. Wolfsberg and M. L. Perlman, *Phys. Rev.* **99**, 1833 (1955).
21. T. A. Carlson and M. O. Krause, *Phys. Rev. Lett.* **14**, 390 (1965).
22. T. A. Carlson and M. O. Krause, *Phys. Rev. Lett.* **17**, 1079 (1966).
23. W. Mehlhorn, W. Schmitz, and D. Stalherm, *Z. Physik* **252**, 399 (1972).
24. T. Åberg and J. Utriainen, *Phys. Rev. Lett.* **22**, 1346 (1969); O. Keshi-Rahkonen and J. Utriainen, *J. Phys. B* **7**, 55 (1974).
25. J. W. Cooper and R. E. La Villa, *Phys. Rev. Lett.* **25**, 1745 (1970).
26. J. Utriainen, M. Linkoaho and T. Åberg, to be published in *Proc. Int. Symp. X-Ray Spectra and Electronic Structure of Matter*, Munich, 1972.
27. F. Pleasonton and A. H. Snell, *Proc. R. Soc. Lond.* **241A**, 141 (1957).
28. T. A. Carlson, W. E. Hunt, and M. O. Krause, *Phys. Rev.* **151**, 41 (1966).
29. T. A. Carlson and M. O. Krause, *Phys. Rev.* **137**, A1655 (1965); M. O. Krause and T. A. Carlson, *Phys. Rev.* **158**, 18 (1967).
30. P. B. Needham, Jr., T. J. Driscoll, and N. G. Rao, *Appl. Phys. Lett.* **21**, 502 (1972).
31. K. Siegbahn, L. Werme, B. Grennberg, J. Nordgren, and C. Nordling, *Phys. Lett.* **41A** 111 (1972).
32. A. Fahlman, K. Hamrin, R. Nordberg, C. Nordling, and K. Siegbahn, *Phys. Lett.* **20**, 159 (1966).
33. W. E. Moddeman, thesis, University of Tennessee (1970); also see Oak Ridge National Laboratory report No. ORNL-TM-3013.
34. C. D. Wagner, in *Electron Spectroscopy, Proc. Int. Conf., Asilomar, September 1971*, ed. by D. A. Shirley (North-Holland, Amsterdam, 1972), p. 861; C. D. Wagner and P. Biloen, *Surf. Sci.* **35**, 82 (1973).
35. G. Schön, *J. Electron Spectros.* **2**, 75 (1973).
36. S. P. Kowalczyk, L. Ley, F. R. McFeely, R. A. Pollak, and D. A. Shirley, *Phys. Rev. B* **9**, 381 (1974).
37. K. Siegbahn, *J. Electron Spectros.* **5**, 3 (1974).
38. T. A. Carlson (unpublished).
39. H. Seiber, *Z. Angew Phys.* **22**, 249 (1967).
40. W. Mehlhorn, *Z. Physik* **187**, 21 (1965); **208**, 1 (1968); W. Mehlhorn and D. Stalherm, *Z. Physik* **217**, 294 (1968); W. Mehlhorn, D. Stalherm, and H. Verbeck, *Z. Naturforsch.* **23a**, 287 (1968); S. Hagmann, G. Hermann, and W. Mehlhorn, *Z. Phys.* **266**, 189 (1974).
41. H. Körber and W. Mehlhorn, *Z. Physik* 191, 217 (1966).
42. K. Siegbahn, C. Nordling, G. Johansson, J. Hedman, P. F. Heden, K. Hamrin, U. Gelius, T. Bergmark, L. O. Werme, R. Manne, and Y. Baer, *ESCA Applied to Free Molecules* (North-Holland, Amsterdam–London, 1969).
43. L. O. Werme, T. Bergmark, and K. Siegbahn, *Physica Scripta* **8**, 149 (1973).
44. M. O. Krause, T. A. Carlson, and W. E. Moddeman, *J. Phys.* (*Paris*) **32**, C4-139 (1971).
45. H. Hillig, B. Cleff, W. Mehlhorn, and W. Schmitz, *Z. Phys.* **268**, 225 (1974).
46. S. Flügge, W. Mehlhorn, and V. Schmidt, *Phys. Rev. Lett.* **29**, 7 (1972); C. B. Mehlhorn, *J. Phys. B* **7**, 605 (1974).

47. D. Stalherm, B. Cleff, H. Hillig, and W. Mehlhorn, *Z. Naturforsch.* **24a**, 1728 (1969).
48. W. E. Moddeman, T. A. Carlson, M. O. Krause, B. P. Pullen, W. E. Bull, and G. K. Schweitzer, *J. Chem. Phys.* **55**, 2317 (1971).
49. M. Nakamura *et al.*, *Phys. Rev.* **178**, 80 (1969).
50. T. A. Carlson, M. O. Krause, and W. E. Moddeman, *J. Phys.* (*Paris*) **32**, C4-76 (1971).
51. A. C. Hurley, *J. Mol. Spec.* **9**, 18 (1962).
52. D. B. Neumann and J. W. Moskowitz, *J. Chem. Phys.* **50**, 2216 (1969).
53. R. Spohr, T. Bergmark, N. Magnusson, L. O. Werme, C. Nordling, and K. Siegbahn, *Physica Scripta* **2**, 31 (1970).
54. R. W. Shaw and T. D. Thomas (to be published).
55. L. Karlsson, L. O. Werme, T. Bergmark, and K. Siegbahn, *J. Electron Spectros.* **3**, 181 (1974).
56. G. Glupe and W. Mehlhorn, *Phys. Lett.* **25A**, 274 (1967).
57. M. O. Krause, F. A. Stevie, L. J. Lewis, T. A. Carlson, and W. E. Moddeman, *Phys. Lett* **31A**, 81 (1970).
58. T. A. Carlson, W. E. Moddeman, and M. O. Krause, *Phys. Rev.* **A 1**, 1406 (1970).
59. A. K. Edwards and M. E. Rudd, *Phys. Rev.* **170**, 140 (1968); D. J. Volz and M. E. Rudd, *Phys. Rev. A* **2**, 1395 (1970).
60. N. Stolterfoht, D. Schneider, and K. G. Harrison, *Phys. Rev. A* **8**, 2363 (1973); N. Stolterfoht, D. Schneider, and P. Ziem, *Phys. Rev. A* **10**, 81 (1974).
61. R. W. Fink, S. T. Manson, J. M. Palms, and P. V. Rao, eds. *Proc. Int. Conf. Inner Shell Ionization Phenomena and Future Applications, Atlanta, April 17–22, 1972* (USAEC Technical Information Center, Oak Ridge, Tennessee, CONF-720404, 1973).
62. N. Stolterfoht, in *Proc. VIII Intern. Conf. on the Physics of Electron and Atomic Collisions, Invited Lectures and Progress Reports*, ed. by B. C. Čobić and M. U. Kurepa (Institute of Phys., Beograd, Yugoslavia, 1973), p. 117, and reference therein; N. Stolterfoht, D. Schneider, and H. Gabler, *Phys. Lett. A* **47**, 271 (1974); D. L. Matthews, B. M. Johnson, J. J. Mackey, and C. F. Moore, *Phys. Rev. Lett.* **31**, 1331 (1973).
63. M. W. Lucas and K. G. Harrison, *J. Phys. B: Atom. Molec. Phys.* **5**, L20 (1972).
64. I. A. Sellin, D. J. Pegg, P. M. Griffin, and W. W. Smith, *Phys. Rev. Lett.* **28**, 1229 (1972); D. J. Pegg, H. H. Haselton, P. M. Griffin, R. Laubert, J. R. Mowat, R. Peterson, and I. A. Sellin, *Phys. Rev. A* **9**, 1112 (1974).
65. T. W. Haas, R. W. Springer, M. P. Hooker, and J. T. Grant, *Phys. Lett. A* **47**, 317 (1974).
66. A. Barrie and C. R. Brundle, *J. Electron Spectros.* **5**, 321 (1974).
67. M. Pessa, *J. Appl. Phys.* **42**, 5831 (1971).
68. J. H. Neave, C. T. Foxon, and B. A. Joyce, *Surf. Sci.* **29**, 411 (1972).
69. T. E. Gallon, *J. Phys. D: Appl. Phys.* **5**, 822 (1972).
70. P. Staib, *Phys. Lett.* **41A**, 3 (1972).
71. B. W. Holland, L. McDonnell, and D. P. Woodruff, *Solid State Comm.* **11**, 991 (1972); L. McDonnell and D. P. Woodruff, *Vacuum* **22**, 477 (1972).
72. C. J. Powell, *Appl. Phys. Lett.* **20**, 335 (1972), and references therein.
73. M. F. Chung and L. H. Jenkins, *Surf. Sci.* **28**, 649 (1971); L. H. Jenkins and M. F. Chung, *Surf. Sci.* **26**, 151 (1971); **28** 409 (1971); **33** 159 (1972).
74. J. A. D. Matthew and C. M. K. Watts, *Phys. Lett.* **37A**, 239 (1971).
75. E. B. Pattinson and P. R. Harris, *J. Electron Spectros.* **1**, 500 (1972/73).
76. M. Salmeron, *Surf. Sci.* **41**, 584 (1974); J. J. Melles, L. E. Davis, and L. L. Levenson, *Phys. Rev. B* **9**, 4618 (1974).
77. B. Goldstein and D. E. Carlson, *J. Am. Ceram. Soc.* **55**, 51 (1972); also see B. Carriere, J.-P. Deville, and S. Goldsztaub, *C. R. Acad. Sci. Paris* **274B**, 415 (1972); *Vacuum* **22**, 485 (1972).

78. A. Fahlman, R. Nordberg, C. Nordling, and K. Siegbahn, *Z. Physik* **192**, 476 (1966).
79. R. G. Albridge, K. Hamrin, G. Johansson, and A. Fahlman, *Z. Physik* **209**, 419 (1968).
80. S. Aksela, *Z. Physik* **244**, 268 (1971).
81. G. Kumar, Thesis, Univ. of Vanderbilt, 1973.
82. C. J. Powell and A. Mandl, *Phys. Rev. Lett.* **29**, 1153 (1972).
83. L. Yin, T. Tsang, I. Adler, and E. Yellin, *J. Appl. Phys.* **43**, 3464 (1972).
84. C. D. Wagner, *Anal. Chem.* **44**, 967 (1972).
85. G. Schön, *Phys. Lett.* **42A**, 381 (1973).
86. P. J. Bassett, T. E. Gallon, J. A. D. Matthew, and M. Prutton, *Surf. Sci.* **35**, 63 (1973).
87. C. K. Jørgensen and H. Berthou, *Chem. Phys. Lett.* **25**, 21 (1974).
88. G. L. Connell, R. F. Schneidmiller, P. Kraatz, and Y. P. Gupta, in *Proc. 2nd Lunar Sci. Conf.* (MIT Press, 1971), p. 2083.
89. W. J. Carter, III, G. K. Schweitzer, T. A. Carlson, L. D. Hulett, and B. Fish (to be published).
90. J. J. Lander, *Phys. Rev.* **91**, 1382 (1953).
91. E. J. Scheibner and L. N. Tharp, *Surf. Sci.* **8**, 247 (1967).
92. R. E. Weber and W. T. Peria, *J. Appl. Phys.* **38**, 4355 (1967).
93. T. W. Haas, G. J. Dooley, J. T. Grant, A. G. Jackson, and M. P. Hooker, in *Progress in Surface Science*, Vol. 1, ed. by S. G. Davison (Pergamon Press, London, 1971), p. 155.
94. (*a*) A. Pentenero, *Catalysis Rev.* **5**, 199 (1971); (*b*) C. C. Chang, *Surf. Sci.* **25**, 53 (1971); (*c*) J. C. Tracy and J. M. Burkstrand, *Crit. Rev. Solid State Sci.* **4**, 381 (1974); (*d*) C. C. Chang, in *Characterization of Solid Surfaces*, ed. by P. F. Kane (Plenum, New York, 1974), pp. 509–75.
95. J. T. Grant, T. W. Haas, and J. E. Houston, *Phys. Lett. A* **45A**, 309 (1973); J. T. Grant, T. W. Haas, and J. E. Houston, *Surf. Sci.* **42**, 1 (1974); J. T. Grant and T. W. Haas, *Surf. Sci.* **44**, 617 (1974).
96. M. P. Seah, *Surf. Sci.* **32**, 703 (1972).
97. F. Meyer and J. J. Vrakking, *Surf. Sci.* **33**, 271 (1972).
98. P. Staib and J. Kirschner, *Appl. Phys.* **3**, 421 (1974).
99. F. Meyer and J. J. Vrakking, *Surf. Sci.* **45**, 409 (1974).
100. N. C. MacDonald and J. R. Waldrop, *Appl. Phys. Lett.* **19**, 315 (1971).
101. J. H. Thomas, III and J. M. Morabito, *Surf. Sci.* **41**, 629 (1974).
102. L. A. Harris, *J. Appl. Phys.* **39**, 1428 (1968).
103. N. J. Taylor, *J. Vac. Sci. Technol.* **6**, 241 (1969).
104. L. H. Jenkins and M. F. Chung, *Surf. Sci.* **24**, 125 (1971).
105. E. N. Sickafus, *Surf. Sci.* **19**, 181 (1970).
106. R. M. Lambert, W. H. Weinberg, C. M. Comrie, and J. W. Linnett, *Surf. Sci.* **27**, 653 (1971).
107. W. S. Lassiter, *J. Vac. Sci. Technol.* **9**, 310 (1972).
108. G. Dufour, H. Guennou, and C. Bonnelle, *Surf. Sci.* **32**, 731 (1972).
109. R. E. Weber and A. L. Johnson, *J. Appl. Phys.* **40**, 314 (1969).
110. P. W. Palmberg, in *Editions du Centre National de la Recherche Scientifique* No. 187 (Paris, 1970), p. 31.
111. R. G. Musket and J. Ferrante, *J. Vac. Sci. Technol.* **7**, 14 (1970).
112. J. H. Pollard, *Surf. Sci.* **20**, 269 (1970).
113. B. A. Joyce and J. H. Neave, *Surf. Sci.* **27**, 499 (1971).
114. H. P. Bonzel, *Surf. Sci.* **27**, 387 (1971).
115. P. H. Holloway and J. B. Hudson, *Surf. Sci.* **43**, 123, 141 (1974).

116. K. Christmann and G. Ertl, *Surf. Sci.* **33**, 254 (1972); J. M. Martinez and J. B. Hudson, *J. Vac. Sci. Technol.* **10**, 35 (1973); T. N. Taylor and P. J. Estrup, *J. Vac. Sci. Technol.* **10**, 26 (1973).
117. S. Thomas and T. W. Haas, *Surf. Sci.* **28**, 632 (1971); *J. Vac. Sci. Technol.* **9**, 840 (1972).
118. M. L. Tarng and G. K. Wehner, *J. Appl. Phys.* **42**, 2449 (1971).
119. R. E. Weber, *J. Crystal Growth* **17**, 342 (1972).
120. R. E. Weber, *Res. Devel.* **1972**, 22.
121. J. N. Smith, S. Thomas, and K. Ritchie, *J. Electrochem. Soc.* **121**, 827 (1974).
122. J. P. Coad and J. C. Rivière, *Surf. Sci.* **25**, 609 (1971).
123. H. E. Bishop and J. C. Rivière, *Acta Met.* **18**, 813 (1970).
124. H. L. Marcus and P. W. Palmberg, *Trans. Met. Soc. AIME* **245**, 1664 (1969).
125. R. Wong, *J. Adhesion* **4**, 171 (1972).
126. T. Narusawa, S. Komiya, and A. Hiraki, *Appl. Phys. Lett.* **20**, 272 (1972).
127. J. L. Domange, *J. Vac. Sci. Technol.* **9**, 682 (1972).
128. W. P. Ellis, *J. Vac. Sci. Technol.* **9**, 1207 (1972).
129. D. T. Quinto, V. S. Sundaram, and W. D. Robertson, *Surf. Sci.* **28**, 504 (1971).
130. H. P. Bonzel and H. B. Aaron, *Scripta Met.* **5**, 1057 (1971); J. Ferrante, *Acta Met.* **19**, 743 (1971); *Scripta Met.* **5**, 1129 (1971).
131. G. J. Dooley, III, *J. Vac. Sci. Technol.* **9**, 145 (1972).
132. P. W. Palmberg and T. N. Rhodin, *J. Phys. Chem. Solids* **29**, 1917 (1968).
133. H. Tokutaka, M. Prutton, I. G. Higginbotham, and T. E. Gallon, *Surf. Sci.* **21**, 233 (1970).
134. R. G. Musket, *Surf. Sci.* **21**, 440 (1970).
135. J. P. Coad, H. E. Bishop, and J. C. Rivière, *Surf. Sci.* **21**, 253 (1970).
136. R. E. Kirby and D. Lichtman, *Surf. Sci.* **41**, 447 (1974).
137. M. Salmerón and A. M. Baró, *Surf. Sci.* **29**, 300 (1972).
138. T. W. Haas, J. T. Grant, and G. J. Dooley, III, *J. Appl. Phys.* **43**, 1853 (1972).
139. J. T. Grant and T. W. Haas, *Phys. Lett.* **33A**, 386 (1970).
140. J. T. Grant and T. W. Haas, *Surf. Sci.* **21**, 76 (1970).
141. D. T. Quinto and W. D. Robertson, *Surf. Sci.* **27**, 645 (1971); R. J. Fortner and R. G. Musket, *Surf. Sci.* **28**, 339 (1971); M. Suleman and E. B. Pattinson, *J. Phys. F: Metal Phys.* **1**, L21 (1971).
142. E. J. LeJeune, Jr., and R. D. Dixon, *J. Appl. Phys.* **43**, 1998 (1972).
143. F. J. Szalkowski and G. A. Somorjai, *J. Chem. Phys.* **56**, 6097 (1972).
144. G. C. Allen and R. K. Wild, *Chem. Phys. Lett.* **15**, 279 (1972).
145. J. M. Baker and J. L. McNatt, *J. Vac. Sci. Technol.* **9**, 792 (1972).
146. T. A. Clarke, R. Mason and M. Tescari, *Proc. Roy. Soc. Lond. A* **331**, 321 (1972).
147. M. Seo, J. B. Lumsden, and R. W. Staehle, *Sur. Sci.* **42**, 337 (1974).
148. K. A. Dadayan, Y. G. Kriger, V. M. Tapilin, and V. I. Savchenko, *Izv. Akad. Nauk SSR, Ser. Fiz.* **38**, 273 (1974).
149. J. P. Coad and J. C. Rivière, *Phys. Stat. Sol. A* **7**, 571 (1971).
150. J. J. Melles, L. E. Davis, and L. L. Levenson, *J. Vac. Sci. Technol.* **10**, 140 (1973).
151. J. Tejeda, N. J. Shevchik, D. W. Langer, and M. Cardona, *Phys. Rev. Lett.* **30**, 370 (1973).
152. J. T. Grant and T. W. Haas, *Surf. Sci.* **18**, 457 (1969).
153. J. D. Levine, A. Willis, W. R. Bottoms, and P. Mark, *Surf. Sci.* **29**, 144 (1972).
154. A. G. Jackson and M. P. Hooker, *J. Vac. Sci. Technol.* **9**, 784 (1972).
155. C. E. Holcombe, Jr., G. L. Powell, and R. E. Clausing, *Surf. Sci.* **30**, 561 (1972).
156. B. W. Griffiths, A. V. Jones, and I. R. M. Wardell, in *Scanning Electron Microsc: Supt. Appl., Proc. Conf. 1973*, ed. by W. C. Nixon (Inst. Phys. London, 1973), p. 42.

157. B. W. Griffiths, K. Henrich, B. D. Powell, and D. P. Woodruff, *Messtechnik* **82**, 135 (1974).
158. R. C. Henderson and R. F. Helm, *Surf. Sci.* **30**, 310 (1972); *J. Vac. Sci. Tech.* **9**, 164 (1972).
159. G. Y. Robinson and N. L. Jarvis, *Appl. Phys. Lett.* **21**, 507 (1972).
160. M. Nishijima and T. Murotani, *Surf. Sci.* **32**, 459 (1972).
161. T. Kawai, K. Kunimori, T. Kondow, T. Onishi, and K. Tamaru, *J. Chem. Soc. Faraday Trans. I* **70**, 137 (1974).
162. R. F. Willis, B. Fitton, and D. K. Skinner, *J. Appl. Phys.* **43**, 4412 (1972).
163. J. P. Coad and J. G. Cunningham, *J. Electron Spectros.* **3**, 435 (1974).
164. R. L. Gerlach, J. E. Houston, and R. L. Park, *Appl. Phys. Lett.* **16**, 179 (1970); R. L. Gerlach, *J. Vac. Sci. Technol.* **8**, 599 (1971).
165. R. L. Park and J. E. Houston, *J. Vac. Sci. Technol.* **11**, 1 (1974); A. M. Bradshaw, in *Surface and Defect Properties of Solids* (Chem. Soc., London, 1974), Vol. 4.
166. R. G. Musket, *J. Vac. Sci. Technol.* **9**, 603 (1972).
167. J. T. Grant, T. W. Haas, and J. E. Houston, *J. Vac. Sci. Technol.* **11**, 227 (1974).
168. H. Niehus and E. G. Bauer, *Electron Fis. Apl.* **17**, 53 (1974).
169. T. Narusawa and S. Komiya, *J. Vac. Sci. Technol.* **11**, 312 (1974); J. Morabito and M. J. Rand, *Thin Solid Films*, **22**, 293 (1974).
170. D. J. Ball, T. M. Buck, D. Macnair and G. H. Wheatley, *Surf. Sci.* **30**, 69 (1972); F. W. Karasek, *Res. Devel.* **24**, 25 (1973).

APPENDIX 3

1. G. C. Allan and P. M. Tucker, *J. Chem. Soc., Dalton Trans.* **1973**, 470.
2. G. Axelson, K. Hamrin, A. Fahlman, C. Nordling, and B. J. Lindberg, *Spectrochim. Acta* **23A**, 2015 (1967).
3. M. Barber and D. T. Clark, *J. Chem. Soc. D: Chem. Comm.* **1970**, 23.
4. M. Barber and D. T. Clark, *J. Chem. Soc. D: Chem. Comm.* **1970**, 24.
5. M. Barber, S. J. Broadbent, J. A. Connor, M. F. Guest, I. H. Hillier, and H. J. Puxley, *J. Chem. Soc. Perkin Trans.* **2**, 1517 (1972).
6. W. Bremser and F. Linnemann, *Chemiker-Zeitung* **96**, 36 (1972).
7. J. H. Burness, J. G. Dillard, and L. T. Taylor, *Inorg. Nucl. Chem. Lett.* **9**, 825 (1973).
8. J. Bus, *Rec. Trav. Chim.* **91**, 552 (1972).
9. D. Cahen and J. E. Lester, *Chem. Phys. Lett.* **18**, 108 (1973).
10. P. H. Citrin, R. W. Shaw, Jr., A. Packer, and T. D. Thomas, in *Electron Spectroscopy, Int. Conf., Asilomar, September 1971*, ed. by D. A. Shirley (North-Holland, Amsterdam, 1972), p. 691.
11. D. T. Clark, D. Briggs, and D. B. Adams, *J. Chem. Soc., Dalton Trans.* **1973**, 169.
12. D. W. Davis, J. M. Hollander, D. A. Shirley, and T. D. Thomas, *J. Chem. Phys.* **52**, 3295 (1970).
13. A. Fahlman, O. Hörnfeldt, and C. Nordling, *Ark. Fys.* **23**, 75 (1963).
14. A. Fahlman, R. Carlsson, and K. Siegbahn, *Ark. Kemi* **25**, 301 (1966).
15. A. Fahlman, K. Hamrin, J. Hedman, R. Nordberg, C. Nordling, and K. Siegbahn, *Nature* **210**, 4 (1966).
16. A. Fahlman, K. Hamrin, R. Nordberg, C. Nordling, K. Siegbahn, and L. W. Holm, *Phys. Lett.* **19**, 643 (1966).
17. P. Finn, R. K. Pearson, J. M. Hollander, and W. L. Jolly, *Inorg. Chem.* **10**, 378 (1971).

18. P. Finn and W. L. Jolly, *Inorg. Chem.* **11**, 893 (1972).
19. P. Finn and W. L. Jolly, *Inorg. Chem.* **11**, 1434 (1972).
20. F. Freund, M. Hamich, *Fortschr. Mineral.* **48(2)**, 243 (1971).
21. T. Gora, R. Staley, J. D. Rimstidt, and J. Sharma, *Phys. Rev. B* **5**, 2309 (1972).
22. P. A. Grutsch, M. V. Zeller, and T. P. Fehlner, *Inorg. Chem.* **12**, 1431 (1973).
23. S. Hagström, C. Nordling, and K. Siegbahn, *Z. Physik* **178**, 439 (1964).
24. K. Hamrin, G. Johansson, A. Fahlman, C. Nordling, and K. Siegbahn, Inst. of Phys., Uppsala Univ. report UUIP-548 (1967).
25. K. Hamrin, G. Johansson, A. Fahlman, C. Nordling, K. Siegbahn, and B. Lindberg, *Chem. Phys. Lett.* **1**, 557 (1968).
26. K. Hamrin, G. Johansson, U. Gelius, C. Nordling, and K. Siegbahn, *Physica Scripta* **1**, 277 (1970).
27. A. D. Hamer, D. G. Tisley, and R. A. Walton, *J. Chem. Soc., Dalton Trans.* **1973**, 116.
28. J. Hedman, M. Klasson, R. Nilsson, C. Nordling, M. F. Sorokina, O. I. Kljushnikov, S. A. Nemnonov, V. A. Trapeznikov, and V. G. Zyryanov, *Physica Scripta* **4**, 1 (1971).
29. J. M. Hollander, D. N. Hendrickson, and W. L. Jolly, *J. Chem. Phys.* **49**, 3315 (1968).
30. H. Iwamura and M. Fukunaga, *J. Chem. Soc. Chem. Comm.* **1972**, 450.
31. K. Kishi and S. Ikeda, *Chem. Lett.* **3**, 245 (1972).
32. L. N. Kramer and M. P. Klein, *J. Chem. Phys.* **51**, 3620 (1969).
33. T. Lane, C. J. Vesely, and D. W. Langer, *Phys. Rev. B* **6**, 3770 (1972).
34. D. W. Langer, *Z. Naturforsch.* **24a**, 1555 (1969).
35. D. W. Langer and C. J. Vesely, *Phys. Rev. B* **2**, 4885 (1970).
36. R. Larsson, B. Folkesson, and G. Schön, *Chemica Scripta* **3**, 88 (1973).
37. B. J. Lindberg, *Acta Chem. Scand.* **24**, 2242 (1970).
38. B. J. Lindberg, S. Högberg, G. Malmsten, J.-E. Bergmark, Ö. Nilsson, S.-E. Karlsson, A. Fahlman, and U. Gelius, *Chemica Scripta* **1**, 183 (1971).
39. G. Mavel, J. Escard, P. Costa, and J. Castaing, *Surf. Sci.* **35**, 109 (1973).
40. W. E. Moddeman, J. R. Blackburn, G. Kumar, K. A. Morgan, R. G. Albridge, and M. M. Jones, *Inorg. Chem.* **11**, 1715 (1972).
41. V. I. Nefedov, I. B. Baranovskii, A. K. Molodkin, and U. O. Omuralieva, *Zh. Neorg. Khim.* **18**, 1295 (1973).
42. V. I. Nefedov and M. A. Porai-Koshits, *Mat. Res. Bull.* **7**, 1543 (1972).
43. Ö. Nilsson, C.-H. Norberg, J.-E. Bergmark, A. Fahlman, C. Nordling, and K. Siegbahn, *Helv. Physica Acta* **41**, 1064 (1968).
44. T. Novakov, *Phys. Rev. B* **3**, 2693 (1971).
45. S. Pignataro, A. Foffani, and G. Distefano, *Chem. Phys. Lett.* **20**, 350 (1973).
46. R. D. Seals, R. Alexander, L. T. Taylor, and J. G. Dillard, *Inorg. Chem.* **12**, 2485 (1973).
47. E. Sokolowski, C. Nordling, and K. Siegbahn, *Phys. Rev.* **110**, 776 (1958).
48. T. D. Thomas, *J. Chem. Phys.* **52**, 1373 (1970).
49. T. D. Thomas, *J. Am. Chem. Soc.* **92**, 4184 (1970).
50. D. G. Tisley and R. A. Walton, *Inorg. Chem.* **12**, 373 (1973).
51. D. G. Tisley and R. A. Walton, *J. Chem. Soc., Dalton Trans.* **10**, 1039 (1973).
52. M. J. Tricker, I. Adams, and J. M. Thomas, *Inorg. Nucl. Chem. Lett.* **8**, 633 (1972).
53. R. E. Watson, J. Hudis, and M. L. Perlman, *Phys. Rev. B* **4**, 4139 (1971).
54. M. V. Zeller and R. G. Hayes, *J. Am. Chem. Soc.* **95**, 3855 (1973).
55. N. M. Zhavoron, V. I. Nefedov, Y. A. Buslaev, Y. V. Kokunov, E. G. Ilin, and Y. N. Mikhailo, *Akad. Nauk SSR, Izv. Fiz.* **36**, 376 (1972).
56. P. R. Anderson and W. E. Swartz, Jr., *Inorg. Chem.* **13**, 2293 (1974).
57. C. Battistoni, M. Bossa, C. Furlani, and G. Mattogno, *J. Electron Spectros.* **2**, 355 (1973).

58. W. Beck and F. Holsboer, *Z. Naturforsch.* **286**, 511 (1973).
59. H. Binder and D. Sellmann, *Angew. Chem.* **85**, 1120 (1973).
60. D. Briggs, D. T. Clark, H. R. Keable, and M. Kilner, *J. Chem. Soc., Dalton Trans.* **1973**, 2143.
61. D. Briggs and V. A. Gibson, *Chem. Phys. Lett.* **25**, 493 (1974).
62. R. G. Cavell, *J. Electron Spectros.* **6**, 281 (1975).
63. P. Citrin, *J. Am. Chem. Soc.* **95**, 6472 (1973).
64. J. A. Connor, L. M. R. Derrick, and I. H. Hillier, *J. Chem. Soc. Faraday Trans. II* **70**, 941 (1974).
65. J. G. Dillard and L. T. Taylor, *J. Electron Spectros.* **3**, 455 (1974).
66. J. Escard, G. Mavel, J. E. Guerchais, and R. Kergoat, *Inorg. Chem.* **13**, 695 (1974).
67. R. M. Friedman, R. E. Watson, J. Hudis, and M. L. Perlman, *Phys. Rev. B* **8**, 3569 (1973).
68. H. P. Fritzer, D. T. Clark, and I. S. Woolsey, *Inorg. Nucl. Chem. Lett.* **10**, 247 (1974).
69. J. C. Fuggle, A. F. Burr, L. M. Watson, D. J. Fabian, and W. Lang, *J. Phys. F* **4**, 335 (1974).
70. S. O. Grim, L. J. Matienzo, and W. E. Swartz, Jr., *Inorg. Chem.* **13**, 447 (1974).
71. A. I. Grigoriev and V. I. Nefedov, *Dokl. Akad. Nauk SSSR* **210**, 343 (1973).
72. H. Kolind-Andersen, S. O. Lawesson, and R. Larsson, *Rec. Trav. Chim. Pays/Bas* **92**, 609 (1973).
73. S. P. Kowalczyk, N. Edelstein, F. R. McFeely, L. Ley, and D. A. Shirley, *Chem. Phys. Lett.* (in press).
74. W. C. Lang, B. D. Padalia, D. J. Fabian, and L. M. Watson, *J. Electron Spectros.* **5**, 207 (1974).
75. V. I. Nefedov, Y. A. Buslaev, and Y. V. Kokunov, *Zh. Neorg. Khim.* **19**, 1166 (1974).
76. V. I. Nefedov, Y. A. Buslaev, A. A. Kuznetsova, and L. F. Yankina, *Zh. Neorg. Khim.* **19**, 1416 (1974).
77. V. I. Nefedov, E. F. Shubochkina, I. S. Kolomnikov, I. B. Baranovskii, V. P. Kukolev, and M. A. Golubnichaya, *Zh. Neorg. Khim.* **18**, 845 (1973).
78. L. O. Pont, A. R. Siedle, M. S. Lazarus, and W. L. Jolly, *Inorg. Chem.* **13**, 483 (1974).
79. A. F. Povey and P. M. A. Sherwood *J. Chem. Soc. Faraday Trans. II* **70**, 1240 (1974).
80. M. Seno, S. Tsuchiya, and T. Asahara, *Chem. Lett.* **1974**, 405.
81. A. I. Stetsenko, V. I. Nefedov, T. G. Abzaeva, and Y. V. Salyn, *Izv. Akad. Nauk SSSR, Ser. Khim.* **1974**, 530.
82. W. E. Swartz, Jr. and R. A. Alfonso, *J. Electron Spectros.* **4**, 351 (1974).
83. D. G. Tisley and R. A. Walton, *J. Inorg. Nucl. Chem.* **35**, 1905 (1973).
84. D. G. Tisley and R. A. Walton, *J. Mol. Struct.* **17**, 401 (1973).
85. J. A. Tossell, *J. Phys. Chem. Solids* **34**, 307 (1973).
86. R. M. Waghorne and J. R. Bosnell, *J. Non Cryst. Solids* **15**, 107 (1974).
87. P. H. Watts and J. E. Huheey, *Inorg. Nucl. Chem. Lett.* **10**, 287 (1974).
88. K. Yoshihara, T. Skiokawa, K. Maeda, Y. Nakai, and K. Seto, *Chem. Lett.* **1973**, 879.
89. I. A. Zakharova, L. A. Leites, and V. T. Aleksanyan, *J. Organometal. Chem.* **72**, 283 (1974).

Index

777029